Principles of Geographical
Information Systems

Principles of Geographical Information Systems

Third edition

Peter A. Burrough
Rachael A. McDonnell
AND
Christopher D. Lloyd

OXFORD
UNIVERSITY PRESS

Great Clarendon Street, Oxford, OX2 6DP,
United Kingdom

Oxford University Press is a department of the University of Oxford.
It furthers the University's objective of excellence in research, scholarship,
and education by publishing worldwide. Oxford is a registered trade mark of
Oxford University Press in the UK and in certain other countries

First edition 1986
Second edition 1998

Impression: 1

Published in the United States of America by Oxford University Press
198 Madison Avenue, New York, NY 10016, United States of America

British Library Cataloguing in Publication Data
Data available

Library of Congress Control Number: 2014959022

ISBN 978–0–19–874284–5

Printed in Great Britain by Bell & Bain Ltd., Glasgow

For Joy, Alexander, Nicholas, Charles, Oliver, and Gemma
This edition was written in memory of Peter Burrough (1944–2009)

Preface

The availability of digital data has exploded exponentially over the last decade. Thousands of terabits are wittingly or unwittingly collected each day, contributing to the phenomenon of Big Data. Very often, these data have locations recorded with them, offering unprecedented possibilities of gaining new understanding of what is happening at places on the earth and why. The applications and uses of these data defy any sort of categorization, with governments, businesses, and society using (and even abusing) them in so many more ways than could have been envisaged even a decade ago.

The need to understand the theories, methods, practices, and limitations to handling, analysing, and modelling these data in a computing environment is therefore crucial if real insight is to be gained—and most importantly trusted. Decisions affecting people, environments, and finance rest on sound knowledge being generated and used. Thus a text that guides the student, researcher, or practitioner through the theoretical and practical principles for handling spatial data in geographical information systems (GIS) is as essential today as when the first edition of this book was published in 1986.

The planning and early writing for this third edition began in 2001, three years after the publication of *Principles II*. The digital world of geographic data collection and provision was already rapidly changing, fuelled by advances in Internet proliferation and network speeds, and the new information technologies were changing what we could do and the questions we could ask using GIS. The advances in methods and techniques for exploring the rapidly increasing data sets were bringing possibilities of developing new insight through inductive as well as deductive reasoning. In the third edition we also wanted to ensure that time as well as space dynamics were included, and we developed new ideas on modelling to be included. Tragically, Peter died in Leiden in the Netherlands in January 2009 before the third edition could be completed. Christopher Lloyd accepted the invitation to contribute and to help bring the book to publication.

Peter was a pioneer in so many ways. In his early days after completing his PhD at Oxford, this involved hazardous jungle treks in Sabah, Borneo with a team of porters and assistants, to undertake soil surveys. On returning to Europe, his research at Wageningen and then Utrecht Universities in the Netherlands focused on harnessing emerging technologies to bring new insight to managing soil and other land resources. It was during this time that he wrote his seminal work, *Principles of GIS for Land Resources Management*, a pioneering publication that introduced to thousands of students, researchers, and practitioners the theoretical and computational structures and frameworks behind GIS. In the subsequent years, Peter's research concentrated on dynamic modelling, particularly with the PCRaster software package, as well as on developing ideas on using fuzzy representations of real-world objects with non-crisp outlines. The numerous research papers that resulted are testament to his rigorous experimental and theoretical thinking. Throughout his career, Peter worked tirelessly with many young researchers at Utrecht University and beyond, inspiring and developing their thinking in geocomputational science. It is through their continued work that his ideas and scientific approach lives on.

We trust that this third edition is also part of the continuation of his pioneering work.

Rachael McDonnell, Dubai
Christopher Lloyd, Liverpool

July 2014

Acknowledgements

The material contained in this book is the sole responsibility of the authors. But this book would not have been possible without the support of many people.

As acknowledged in previous editions, we thank the following for base data used in examples and analyses: we thank Andrew Skidmore (International Institute for Aerospace Survey and Earth Sciences (ITC), Enschede), Brenda Howard (Institute of Agricultural Radiology, Kiev), Robert MacMillan (formerly Alberta Research Council), Alejandro Mateos (University of Maracai, Venezuela), R. P. L. Legis (formerly Wageningen Agricultural University), Arnold Bregt (Winand Staring Centre, Wageningen, the Netherlands), and Ad de Roo, Steven de Jong, Victor Jetten, Ruud van Rijn, and Marten Rikken (all formerly or currently University of Utrecht). We thank the Ordnance Survey for provision of data through the OS OpenData scheme. We are also grateful to the British Atmospheric Data Centre (BADC) for making available United Kingdom Meteorological Office Land Surface Observations Data (the UKMO is the originator of these data). Data were also provided by the British Geological Survey and the US Geological Survey. The Northern Ireland Statistics and Research Agency (NISRA) provided access to data from the 2001 Census of Population of Northern Ireland. The source of census boundary data is NISRA 2001 Census: Digitized Boundary Data (Northern Ireland) [computer file]; ESRC/JISC Census Programme, Census Geography Data Unit (UKBORDERS), EDINA (University of Edinburgh)/Census Dissemination Unit, Mimas (University of Manchester). Census output is Crown copyright and is reproduced with the permission of the Controller of Her Majesty's Stationery Office and the Queen's Printer for Scotland.

For help with artwork: Rasha Hammad (Space Imaging Dubai), Elizabeth Fredericks (Geoscience Australia), Archie Donovan (Geological Survey of Ireland), Lan Nguyen (PSMA Australia), Tim Weisselberg (Centre for Sustainable Energy, Bristol), Paul White (Paul White Photography), and the Centre for Advanced Spatial Analysis at University College London (data supplied by the Ordnance Survey of Great Britain and Infoterra). For case study examples: Richard Grenyer (University of Oxford), Goranij Nonejuie (Ministry of Commerce, Thailand), Gary Eilerts (USAID-FEWS-net), Willem van Deursen (Carthago, Utrecht), Mike Batty (University College London), and Suzana Dragicevic (Simon Fraser University, Burnaby, British Columbia).

For the many discussions that have brought insight and shaped thinking of the various editions of this book: Helen Couclelis, Waldo Tobler, and Michael Goodchild (University of California, Santa Barbara); Andrew Frank (Technical University, Vienna); Vanessa Lawrence (United Nations Committee of Experts on Global Geospatial Information Management) and Steven Ramage (what3words); Carla Mellor, Karim Bergaoui, Fiona Chandler, Adla Khalaf, Rashyd Zaaboul, Makram BelHaj Fraj, and Richard Sulit (ICBA); Jim Verdin (USGS); Mark Peters, Kalim Hanna, and John Wilson (USAID); Shahid Habib, Matt Rodell, and Christa Peters-Ligard (NASA).

At Oxford University Press we would like to thank Jonathan Crowe for supporting us over the long term as well as Dewi Jackson and Claire Mullen. Finally, special thanks to Joy Burrough, Charlie and Oliver McDonnell, and Gemma Catney.

Contents

Preface vii

Acknowledgements ix

1. Geographical Information Systems and Society 1
2. Spatial Data and their Models: Formal Abstractions of Reality 21
3. Geographical Data in the Computer 45
4. Data Input and Verification 69
5. Visualization 91
6. Exploring Geographical Data 111
7. Analysis of Discrete Entities in Space 127
8. Interpolation 1: Deterministic and Spline-based Approaches 147
9. Interpolation 2: Geostatistical Approaches 171
10. Analysis of Continuous Fields 201
11. Digital Elevation Models 231
12. Space–Time Modelling and Error Propagation 251
13. Fuzzy Sets and Fuzzy Geographical Objects 267
14. GIS, Transformations, and Future Developments 287

Appendix 1 Glossary of Commonly Used GIS Terms 297

References 315

Index 327

Detailed Contents

Preface vii

Acknowledgements ix

Chapter 1 **Geographical Information Systems and Society**

1.1 The changing world of possibilities from spatial data 1
1.2 Definitions of GIS 3
1.3 Components of a geographical information system 3
1.4 Geographical information science and systems: through history 9
1.5 Geographical information systems today 12
1.6 Choosing to use GIS 16
1.7 The structure of this book 17
1.8 Summary 18
Questions 18
Further reading 19

Chapter 2 **Spatial Data and their Models: Formal Abstractions of Reality**

2.1 Fundamentals of geographic phenomena 22
2.2 Exploring absolute georeferencing systems 24
2.3 Structuring the geographical world 28
2.4 The human view of real-world geographical phenomena 29
2.5 Conceptual models of space: entities or fields 30
2.6 Geographical data models and geographical data primitives 32
2.7 Overlap between the two geographic data models 36
2.8 Representation changes with scale—granularity, generalization, and hierarchies 37
2.9 Representing changes in time with geographic data models 38
2.10 Data modelling and spatial analysis 39
2.11 Examples of the use of data models 39
2.12 Summary 42
Questions 43
Further reading 44

Chapter 3 **Geographical Data in the Computer**

3.1 Geographical data and computers 45
3.2 Overview of data in computers 46
3.3 Database structures: data organization in the computer 47
3.4 Coding the basic data models for input to the computer 52
3.5 Points, lines, and areas: vector data structures 53
3.6 Grid cells: raster data structures 60
3.7 GIS and time 66
3.8 Summary 67
Questions 67
Further reading 67

Chapter 4 **Data Input and Verification**

4.1 Creating a digital database 69
4.2 Sources of geographical data 71
4.3 Geographical data collectors 76
4.4 Geographical data providers, metadata, and data exchange standards 79
4.5 Creating digital data sets by manual input 82
4.6 Data transformation and structuring 84
4.7 Data quality 86
4.8 Data updating 87
4.9 Considering local tacit knowledge 87
4.10 Summary 88
Questions 88
Further reading 89

Chapter 5 **Visualization**

5.1 Mapping points 92
5.2 Continuous or discrete categories 94
5.3 Cartographic mapping principles 96
5.4 Distorting space: cartograms 101
5.5 Displaying multiple characteristics 101
5.6 Visualization 101
5.7 Non-cartographic output 106
5.8 Dynamic visualization 106
5.9 Multimedia and GIS 106
5.10 Spatial interaction data: mapping movement 107
5.11 Visualization and opening up access to data 108
5.12 Summary 108
Questions 109
Further reading 109

Chapter 6 **Exploring Geographical Data**
6.1 Summarizing and analysing spatial data 112
6.2 Statistical methods 112
6.3 Geographical data: problems and properties 117
6.4 Spatial autocorrelation 117
6.5 Statistics and GIS 120
6.6 Exploring spatial relations: geographically weighted regression 120
6.7 Point pattern analysis 121
6.8 Summary 125
Questions 125
Further reading 125

Chapter 7 **Analysis of Discrete Entities in Space**
7.1 Spatial analysis is more than asking questions 127
7.2 The basic classes of operations for spatial analysis 128
7.3 Operations on the attributes of geographic entities 128
7.4 Examples of deriving new attributes for spatial entities 134
7.5 Operations that depend on a simple distance between A and B: buffering 137
7.6 Operations that depend on connectivity 137
7.7 Operations on attributes of multiple entities that overlap in space 139
7.8 General aspects of data retrieval and modelling using entities 143
7.9 Summary 144
Questions 144
Further reading 145

Chapter 8 **Interpolation 1: Deterministic and Spline-based Approaches**
8.1 Interpolation: what it is and why it is necessary 148
8.2 The rationale behind interpolation 148
8.3 Data sources for interpolation 149
8.4 Methods for interpolation 150
8.5 The example data sets 151
8.6 Global interpolation 151
8.7 Global prediction using classification models 152
8.8 Global interpolation using trend surfaces 155
8.9 Spatial prediction using global regression on cheap-to-measure attributes 158
8.10 Local, deterministic methods for interpolation 160
8.11 Nearest neighbours: Thiessen (Dirichlet/Voronoi) polygons 160
8.12 Linear interpolators: inverse distance interpolation 163
8.13 Splines 163
8.14 A comparison of simple global and local methods 166
8.15 A comparison of IDW and TPS using cross-validation and grids 166
8.16 Summary 168
Questions 168
Further reading 169

Chapter 9 **Interpolation 2: Geostatistical Approaches**
9.1 A brief introduction to regionalized variable theory and kriging 172
9.2 Fitting variogram models 174
9.3 Using the variogram for spatial analysis 175
9.4 Isotropic and anisotropic variation 176
9.5 Variograms showing spatial variation at several scales 176
9.6 Local variograms 176
9.7 Using the variogram for interpolation: ordinary kriging 177
9.8 Using kriging to validate the variogram model 179
9.9 Block kriging 179
9.10 Other forms of kriging 181
9.11 Kriging using extra information 182
9.12 Probabilistic kriging 187
9.13 Simulation 188
9.14 The relative merits of different interpolation methods 191
9.15 Using variograms to optimize sampling 197
9.16 Summary 198
Questions 199
Further reading 199

Chapter 10 **Analysis of Continuous Fields**
10.1 Basic operations for spatial analysis with discretized continuous fields 202
10.2 Interpolation 203
10.3 Spatial analysis using square windows 204
10.4 Filtering case studies 207
10.5 Other grid operators 217
10.6 Other cell-based analysis operations 218
10.7 First and higher order derivatives of a continuous surface 219
10.8 Deriving surface topology and drainage networks 223
10.9 Using the local drain direction network for spatial analysis 226
10.10 Dilation/spreading with or without friction 228
10.11 Summary 229
Questions 230
Further reading 230

Chapter 11 **Digital Elevation Models**

11.1 Methods of representing DEMs 233
11.2 DEM data sources 235
11.3 Quality of DEMs 242
11.4 Viewsheds, shaded relief, and irradiance 243
11.5 Applications of DEMs 247
11.6 Future developments 247
11.7 Summary 248
Questions 249
Further reading 249

Chapter 12 **Space–Time Modelling and Error Propagation**

12.1 Introducing computational modelling 252
12.2 Capturing spatio-temporal dynamics in computation modelling 253
12.3 GIS-based computational modelling 255
12.4 Accounting for errors in modelling 261
12.5 Summary 264
Questions 265
Further reading 266

Chapter 13 **Fuzzy Sets and Fuzzy Geographical Objects**

13.1 Imprecision as a way of thought 267
13.2 Fuzzy sets and fuzzy objects 268
13.3 Choosing the membership function 1: the semantic import approach 270
13.4 Operations on several fuzzy sets 272
13.5 Error analysis of selections made using Boolean and fuzzy logic 276
13.6 Applying the SI approach to polygon boundaries 277
13.7 Combining fuzzy boundaries and fuzzy attributes 280
13.8 Choosing the membership function 2: fuzzy ***k***-means 280
13.9 Class overlap, confusion, and geographical boundaries 282
13.10 Discussion: the advantages, disadvantages, and applications of fuzzy classification 284
13.11 Summary 285
Questions 286
Further reading 286

Chapter 14 **GIS, Transformations, and Future Developments**

14.1 The fundamental axioms and procedures of GIS use 288
14.2 Policies and legal frameworks of geographical data 289
14.3 Future GIS transformations 293
14.4 Summary 295
Further reading 296

Appendix 1 Glossary of Commonly Used GIS Terms 297

References 315

Index 327

Geographical Information Systems and Society

We have more data than ever before, informing of us where we are, where the people we know are, and more generally where things in the world are happening. We no longer need to carry paper maps on our journeys as the information is readily available from our GPS-enabled mobile phones. Becoming 'lost' is less likely, and knowing where things are is more straightforward, as location-based services guide us to the nearest coffee shop, speed up pizza delivery, or, more importantly, direct emergency services to our homes by the quickest route. As long as there is a relatively open sky to satellites or access to a WiFi signal we can navigate our way or find out about where we are. We have also become, knowingly or unknowingly, **geographical data collectors** as we add our **location** to a social media entry or to the photograph we just took. We are in the era of **Big Data**, with all the opportunities, complexities, and challenges that brings. In a recent report, it was estimated that in the five largest European economies 50 per cent of Internet users access maps online and 35 per cent of smartphone users do so on handsets (Oxera 2013). In the same report, it was estimated that the **geoservices** industry has global revenues of \$150–\$270 billion per year and growing at a rate of 30 per cent per annum globally (Oxera 2013).

Learning objectives

By the end of this chapter, you will:

- **understand some of the current and historical context of GIS**
- **have considered the different definitions of GIS**
- **be able to explain the key components of a GIS and the types of questions these systems can address**
- **have gained a familiarity with the structure of this book.**

1.1 The changing world of possibilities from spatial data

Spatial data (see Box 1.1)—put simply as information about something, somewhere—is part of the Big Data explosion, with vast volumes of digital information being generated, influencing decision-making and understanding from local **scale** through to global. It also changes the types of questions we are able to ask. **Geographical data**

Box 1.1 Spatial data

Spatial data represent phenomena from the real world in terms of:

a) their position with respect to a known locational **coordinate system**;

b) their **attributes**, which are not related to position (such as cost, pH, incidence of disease, air pressure, river flow); and

c) their spatial interrelations with each other, which describe how they are linked together or what they are close to (this is known as **topology** and is described in more detail in Chapter 2).

and **geographical information systems** (GIS), the specialist information technology (IT) systems used in handling this mapped data, are employed increasingly across all levels of government, by many businesses, civil society, and other organizations trying to understand how, why, and what is happening, and what can be done.

In all sciences—natural, economic, social, political—the same technology is bringing new insight and understanding of our past, present, and future, through an ability to interrogate and explore our environments. For the natural scientists and engineers—the geologist, the hydrologist, the ecologist, the land use specialist—information is now available across the earth's surface which supports investigations into processes, drivers of change, resources availability, and environmental damage, to name but a few of the many application areas (Skidmore et al. 2012). Experimentations and scientific exploration are enhanced and even revolutionized by high-capacity computers, large spatial data sets, and dynamic models that represent the processes and environments of interest (Wright and Wang 2011).

In urban landscapes, planners, civil engineers, emergency services, and social and economic scientists are able to manage and understand our cities and towns more effectively. Agencies responsible for **cadastral maps** are today able to supply detailed information on land resources and ownership in settled areas, a fundamental starting point for any planning and management. Civil engineers are supported in planning new routes of roads and railways, and better able to estimate construction costs, including those of cutting away hillsides and filling in valleys. Police departments are able to determine the **spatial distribution** of various kinds of crime, while medical organizations and epidemiologists are able to explore the occurrence and spread of sickness and diseases. Commercial business is able to improve profitability by optimizing the distribution of sales outlets and the identification of potential market areas. In the areas of service provision, the enormous infrastructures behind 'utilities'—that is, water, gas, electricity, telephone lines, sewerage systems—are in many countries now digitally recorded and managed through spatial databases.

This spatial data explosion is influencing the lives of citizens across the globe by informing political and economic processes. In many ways, some of the most remarkable breakthroughs in the last decade centre on the rapidly available data that are now available to understand human, as opposed to the natural, environment. Never before have so many organizations held and analysed so much data to understand their people. It is no longer about where we live and what economic activities are taking place, but about how we respond as individuals, giving enhanced insight on the complexities and simplicities of us as **agents** interacting in our societies and communities.

In a recent study aimed at quantifying the impact of geoservices on our lives (Oxera 2013), three broad categories were identified which highlight the ever-increasing use and influence of geographical data:

- Direct effects—the benefits to organizations, including companies, that generate revenue directly through developing and providing geoservices.
- Consumer effects—the benefits that accrue to consumers, businesses, and government from using geoservices over and above the value that may be paid for any services, i.e. in addition to that from direct effects (for example retail locations, health benefits).
- Wider economic effects—the benefits that come from geoservices improving efficiency elsewhere in the economy, by creating new products and services or through cost savings (for example drivers saving fuel on journeys because of in-car navigation devices).

The major developments in areas of application have been supported by the technological transformation of GIS in the last decade, with these once niche systems becoming more mainstream. Some applications today do not require expert GIS skill, just an understanding of basic spatial analytical concepts and the area being studied, as user interfaces of both commercial and open-source systems have become more intuitive and more similar to popular office software systems. Improvements in both ease of use and ease of access to spatial data have led to a diffusion of systems across many user areas and geographies, ensuring GIS have become part of everyday life across the world.

The major developments of the last decade require us to consider ever more how we think and theorize about spatial data, how the representations we have created in the digital world reflect the same concepts we have as humans, and how we should view and visualize these data. New data sources, including those at the individual person or mobile object level, challenge us to consider in more detail ideas of privacy and access to information, and how to represent movement and changes in these features over time. As ever, it is an exciting time to be a GIS researcher or user.

1.2 Definitions of GIS

It is useful at this point to consider in more detail just what a GIS is. The definition used in this book is 'a powerful set of tools for collecting, storing, retrieving at will, transforming, and displaying spatial data from the real world for a particular set of purposes' (Burrough 1986; Burrough and McDonnell 1998).

Others have provided alternative definitions of GIS (Box 1.2), focusing on either the spatial database, their scientific or technological basis, their value to society, or their impact on an organization (Goodchild 2001; Reitsma 2012). The database definition emphasizes the differences in data organization needed to handle spatial data, with their location, attributes, and topology, in comparison to most other kinds of information that are only defined as entities and attributes. The organizational definition emphasizes the role of institutes and people in handling spatial information rather than the tools they need. In this sense GIS have been serving society for a very long time.

1.3 Components of a geographical information system

We can conceptualize a GIS as being made of up three important components—computer **hardware**, software **application programs** and **modules**, and users linked together through **databases** and **network** infrastructure (see Figure 1.1).

Computer hardware

The main hardware components of GIS serve to provide the computing power and data access needed for all the actions of the software system. Today the hardware platforms used range from mobile or handheld devices, through stand-alone computers to complex server-client networks involving many linked units (as shown in Figure 1.2). The physical size as well as the local computing power and memory of these systems vary greatly, influencing how and where the data processing and analysis takes place.

Box 1.2 Definitions of GIS, GISc, and GISoc

Definitions of GIS

a) Toolbox-based definitions

a powerful set of tools for collecting, storing, retrieving at will, transforming, and displaying spatial data from the real world (Burrough 1986; Burrough and McDonnell 1998).

b) User communities as well as tools definitions

a computer system capable of assembling, storing, manipulating, and displaying geographically referenced information, i.e. data identified according to their locations. Practitioners also regard the total GIS as including operating personnel and the data that go into the system (United States Geological Survey 2013).

c) Larger context in which GIS operate

Organized activity by which people measure and represent geographic phenomena then transform these representations into other forms while interacting with social structures (Chrisman 1999).

Definition of GISc

Geographical information science (GISc) refers to the study of problems arising from the handling of spatial information in GI Systems, i.e. it entails finding solutions to problems associated with the use of GIS (Goodchild 1992).

Definition of GISoc

Geographic Information and Society (GISoc) refers to the social context and the impact of GIS (Sheppard 1995). GISoc is concerned with core issues such as the degree to which GIS empowers or disadvantages particular groups.

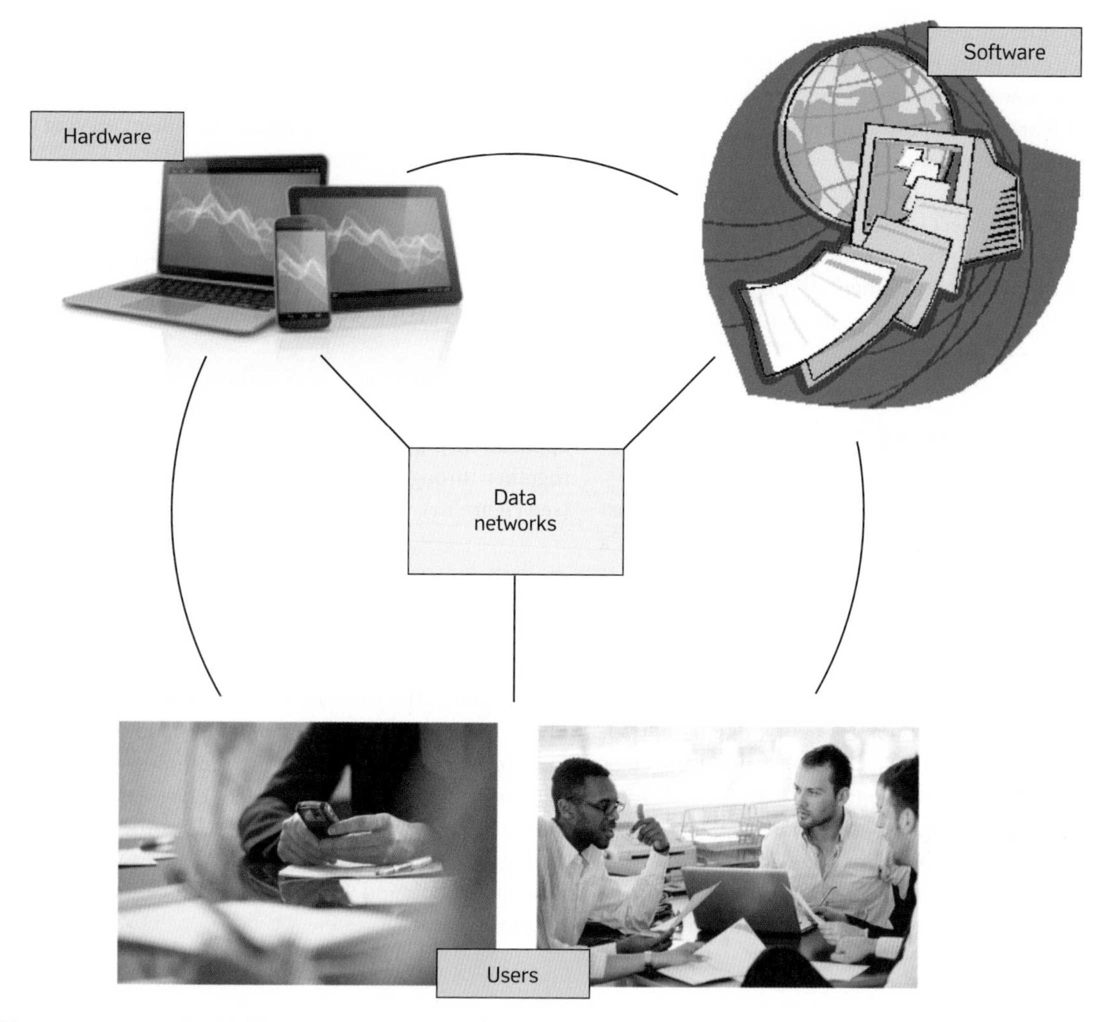

Figure 1.1 Key components of a GIS

The systems used 20 years ago were based on local powerful computing devices (known as **thick client systems**) and it was on these that most or all of the data processing and analysis took place. Since then, advances in data transfer capabilities, whether through local and global networks, fibre optics, satellite systems, or cloud computing, have transformed the IT environment, allowing users to be based at some distance from the main data processing. Today users can interact with a GIS through some keyboard, touchscreen, or pointer device that is in a different building, different city, or even different country as easily as with a stand-alone computer. Systems where the main data and software are stored on servers at a distance, with users accessing the GIS through locally loaded software on simple devices, are referred to as **thin client systems**. These are typical of many enterprise and web-based devices where the local client is really only used as a means of entering the operations and interactions of the user and for viewing the results.

The developments in **cloud computing** have impacted GIS as in other areas of computing with new software design supporting data collection, input, access, and analysis through a wide range of hardware devices using data and software held on distant databases. This is particularly useful where GIS users are away from any data access nodes or are mobile. It supports real-time GIS interaction from any location that has access to satellite or network systems, thus bringing a new space and time 'liberation' to users.

In addition to the main computer systems, various **peripheral** hardware devices may form part of the hardware

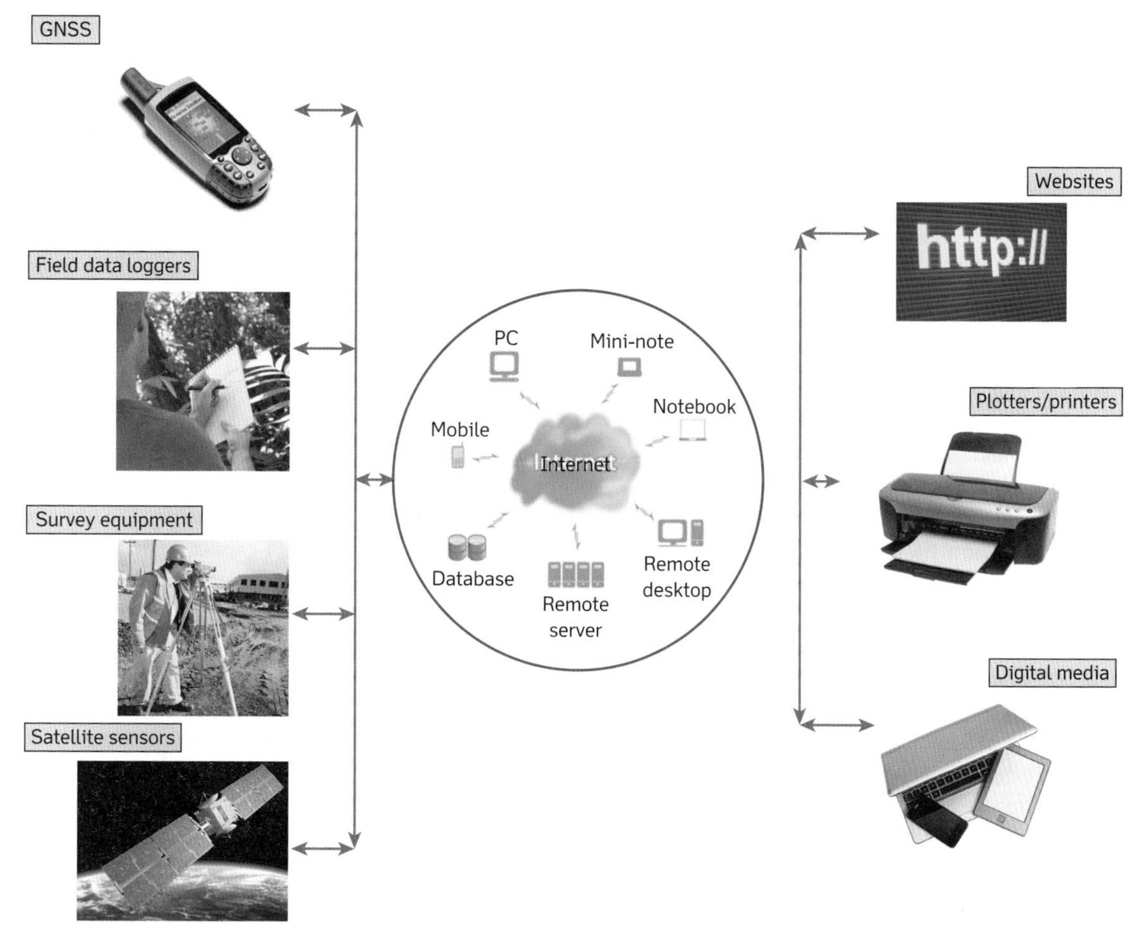

Figure 1.2 The major hardware components of a GIS

of the GIS, with devices for inputting or outputting data most used. These might include mobile devices that enable you to collect data and conduct **field observations**, or **digitizers** or **scanners** for conversion of legacy paper-based data into digital form, and **plotters** and printers for outputting in hard copy form.

GIS software

The range in GIS software systems is as variable as the hardware it runs, but their functionality is centred on four common types of actions (Figure 1.3):

a) data input and verification

b) data storage and database management

c) data output and presentation

d) data **transformations**, analysis, and modelling.

Data input (Figure 1.4) covers many possibilities, ranging from using existing digital data sets, field observations, or output from different digital sensors through to converting paper maps or images into digital data sets. This is the foundation on which a geographic database is built (explored in Chapter 4) and on which all subsequent analysis and output steps are based.

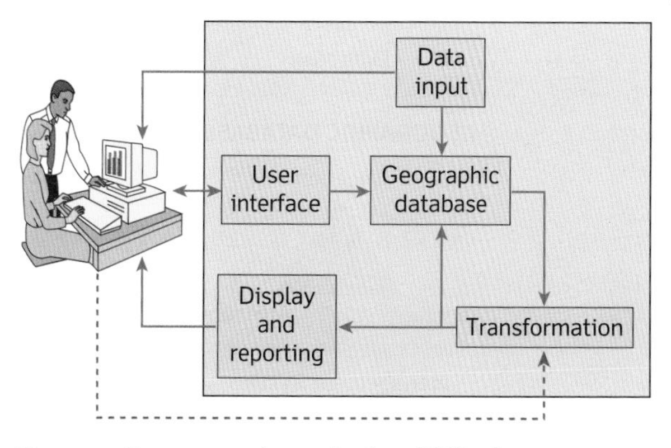

Figure 1.3 Four common types of action of GIS software

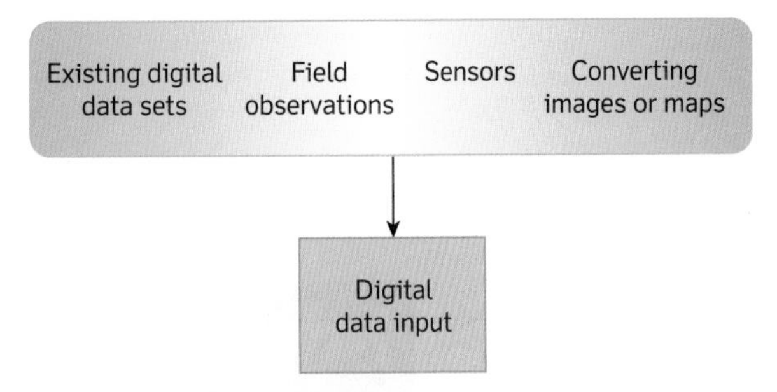

Figure 1.4 Data collection and input to a GIS

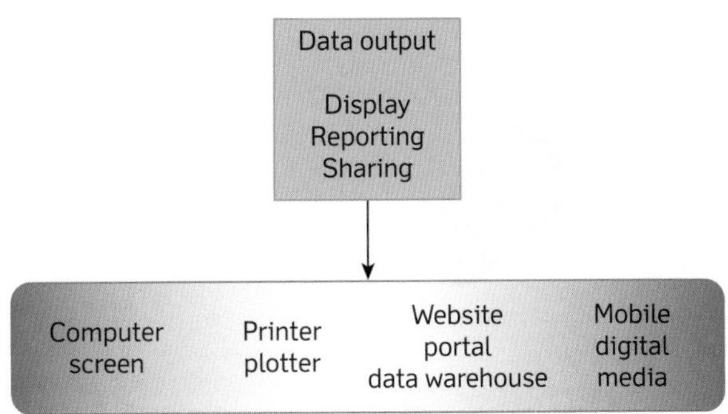

Figure 1.6 Data outputs from a GIS

Data storage and database management (Figure 1.5) concern the way in which data about the location, linkages (**topology**), and values/information (**attributes**) about geographical elements are structured and organized, both with respect to the way they must be handled in the computer and how they are perceived by the users of the system. The computer program used to organize the database is known as a **database management system (DBMS)** and the structuring of relative complexity of spatial data will be based on formal **data models** (which are discussed in Chapters 2 and 3).

Data output and presentation (Figure 1.6) concern the ways the data are displayed and how the results of analyses are reported to the users. Data may be presented as animations, maps, tables, or figures (graphs and charts). Methods of display and reporting are discussed in Chapter 5.

Data transformations (Figure 1.7) embrace two classes of operation: (a) transformations needed to remove errors from the data or to bring them up to date or to match them to other data sets, and (b) the large array of analysis methods that may be applied to the data in order to achieve answers to the questions asked of the GIS (Box 1.3). Transformations can operate on the spatial, topological, and the non-spatial aspects of the data, either separately or in combination. Many of these transformations, such as those associated with scale changing, transforming data to new **map projections**, logical retrieval of data, and calculation of areas and perimeters, are of such a general nature that one should expect to find them in every kind of GIS in one form or another. Other kinds of manipulation may be extremely application-specific, and their incorporation into a particular GIS may be only to satisfy the particular users of that system. The kinds of transformation methods available, their optimum use, the ways in which sets of simple transformations may be combined in order to achieve certain types of geographical or spatial modelling, and an understanding of when certain methods may be appropriate or not, are the major subject of this book.

While all GIS offer these four types of functionality, the nature and capabilities of the software is more varied than ever before. Commercial-based GIS systems from a few main vendors have formed the backbone of usage over the last three decades and are still widely used today, representing a multibillion-dollar business. More

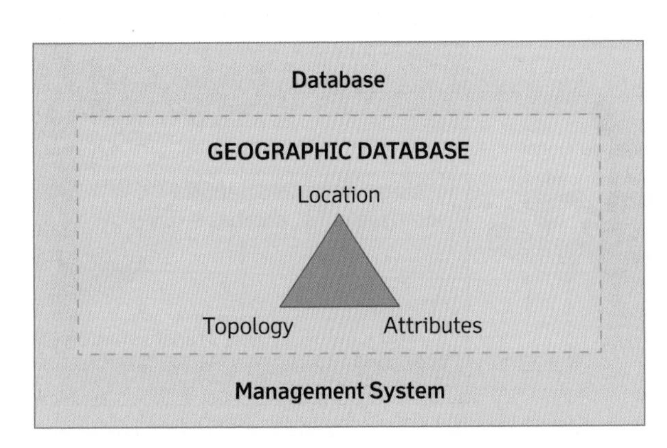

Figure 1.5 The components of the geographic database

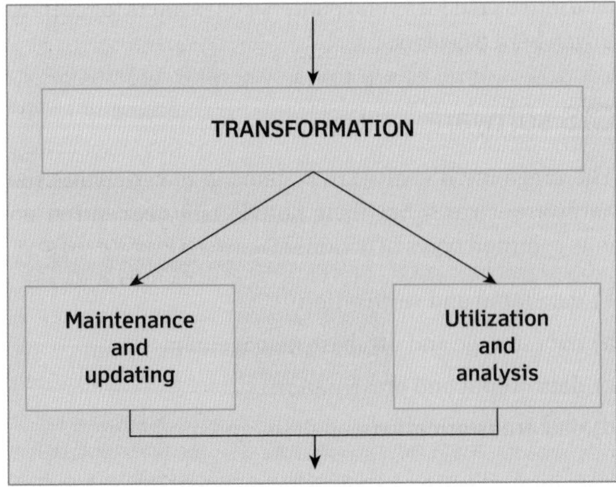

Figure 1.7 Data transformations in a GIS

Box 1.3 Basic requirements for a GIS

a) Show the locations of entities of type A.
b) Show the location of entity A in relation to place B.
c) Count the number of occurrences of entity type A within distance D of entity type B.
d) Evaluate function *f* at position X.
e) Compute the size of B (area, perimeter, count of inclusions).
f) Determine the result of intersecting or overlaying various kinds of spatial data.
g) Determine the path of least cost, resistance, or distance along the ground from X to Y over a network or a continuous surface.
h) List the attributes of entities located at points X_1, X_2.
i) Determine which entities are next to entities having certain combinations of attributes.
j) Reclassify or recolour entities having certain combinations of attributes.
k) Knowing the value of Z at points $X_1 X_2 \ldots X_n$, predict the value of z at points $Y_1, Y_2 \ldots Y_m$.
l) Use numerical methods to derive new attributes from existing attributes or new entities from existing entities.
m) Using the digital database as a model of the real world, simulate the effect of process P over time T for a given scenario S.

modules and functionality, standardized data transfer formats, and user interfaces have been added to these systems over the years, but the tried and tested system structures remain familiar. Reflecting the developments in the increasing range of hardware devices being used by GIS operations, the software now runs across different **operating system** environments including Linux, Mac OS X, Windows, and Android and many are now available in forms that support network functioning across server and client/server architecture. Today GIS support a growing number of data formats and functionalities as new and increasingly flexible formats for information become more widely used (see Chapter 3). Additional software modules have also been developed to support the media-rich possibilities for data collection and output of modern devices, particularly in the mobile arena.

The growth of web-based GIS has been one of the greatest changes in the last decade, exploiting advances in cloud computing, advances in data transfer, and interoperability standards and XML technologies. These offer users greater mobility and the capacity to interact with GIS anywhere that has Internet access. The functionality offered varies greatly from simple web-querying and mapping accessed through commonly used browsers to more comprehensive spatial data location, and data analysis and integration capabilities. Most of the new web-based solutions comply with **Open GIS Consortium** (an international body for setting standards in GIS and spatial data) specifications and protocols for 'web map services' or 'web feature services', ensuring the GIS software is more interoperable across the web, with wireless and location-based services, and with mainstream IT software.

A further change in the nature of GIS in recent years has been the growing availability and use of open-source software (see Steiniger and Hunter 2013 for a useful overview). GIS programs, sourced through the Internet, are available for all the major operating systems, including Android, and offer varying degrees of functionality—most provide file interchange capabilities and support map-making and simple analysis. Some, for example, offer the ability to modify and add new code or **apps**, allowing users to customize their systems. Much of this GIS software is free to download and use, offering a low-cost entry point, and is increasingly being adopted by businesses, government departments, in developing countries, as well as individuals and research environments.

GIS communities

All aspects of dealing with geographical information involve interactions with people, as Figure 1.8 illustrates. Place is no longer the domain of geographers, geodesists, or earth scientists, and the GIS user community is now worldwide, involving professional specialists alongside everyday users. There are many different groups that make up the GIS community with each playing an important role in realizing the benefits of working with these systems.

Behind the scenes are different groups of GIS developers responsible variously for developing the software, for the day-to-day running of installed systems, for generating new data sets, and for developing services (Longley 2010). These groups of skilled professionals are involved in many different activities ranging from developing new software systems or databases, ensuring existing software is functional and secure, developing **scripts** or apps, or customizing software to provide particular operations required by the users. These communities of professionals may be a continent or hemisphere away from the users, but connected through digital information networks.

The second group are the GIS user or practitioner communities which have expanded rapidly and been transformed in the last decade with the revolutions in

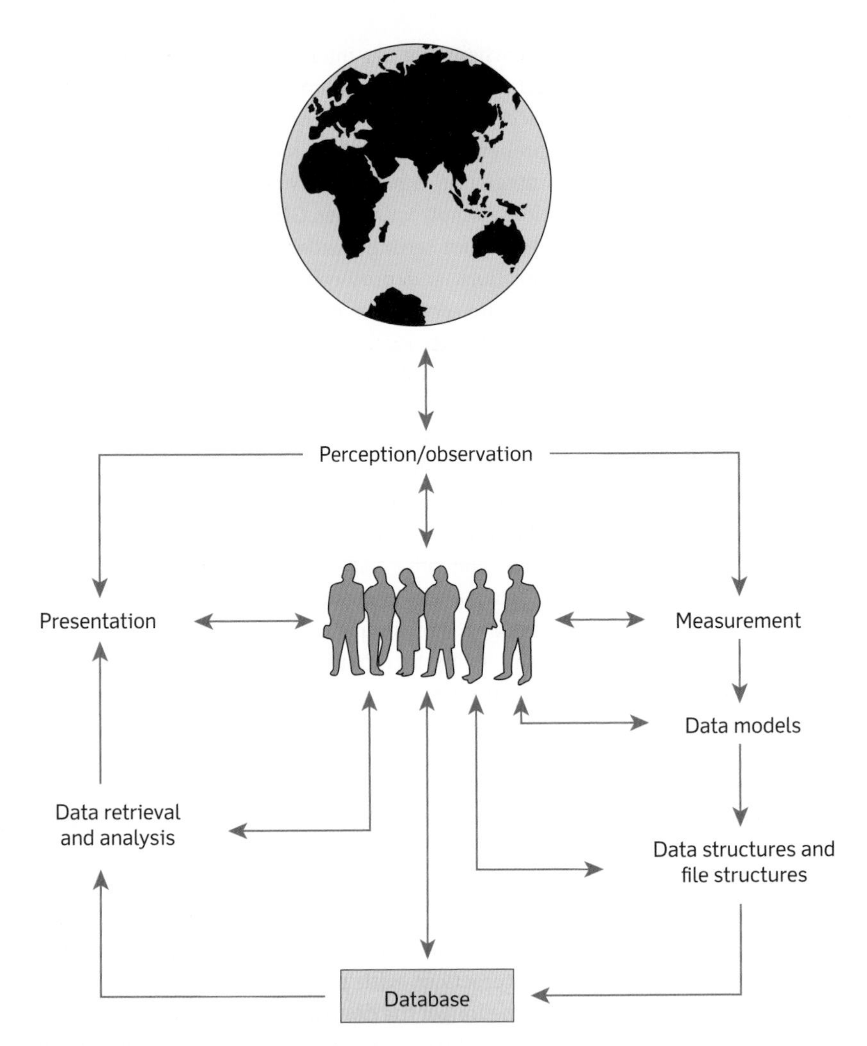

Figure 1.8 People are involved in all aspects of GIS

IT allied with the rapid expansion in availability of geographical information. These user communities span the globe, demographics, communities, disciplines, businesses, and governments. Those undertaking simple web-based mapping and querying may often not be aware that there is a formal name for the technology they are using, with their usage focused primarily on spatial browsing. Basic querying and mapping capabilities through embedded apps ensure digital maps are part of a plethora of web-based interactions. Easy-to-use interfaces are also allowing users to interact with the system directly, supporting the growth of participatory online map-making communities—such as the OpenStreetMap project—with many being '**spatial browsers**' and '**citizen cartographers**' (Crampton 2009; Newman 2012). Other user groups include those that use GIS every day in professional operational environments, for instance for mission-critical responses, managing human or natural resources, urban planning, local government, or in the private sector. For others, GIS is used in a more sporadic fashion, depending on the needs of particular projects or operations.

The GIS user and developer communities have been supported by a rapidly expanding information infrastructure of dedicated books, blogs, journals and magazines, conferences and exhibitions, and professional organizations. The Internet has transformed this exchange in the last decade, with the web bringing information, practices, techniques, and new ideas to the user community anywhere at any time. Blogs, for example, whether from individuals or from organizations, are an increasingly important means of finding out new thoughts, ideas, and experiences, particularly from the practitioner community.

1.4 Geographical information science and systems: through history

Since the earliest civilizations, knowing where resources are, such as animals, water, forests, metals, and then crops, have been fundamental to human existence. The depiction of animals in pictorial cave arts is one of the first forms of spatial data recording found. Through subsequent societies, cartographers, navigators, geographers, and surveyors have recorded and developed images of spatial data ranging in scale from the local to attempts at the global. The developments of surveying and map-making under Greek, Roman, Arab, and then Renaissance thinkers have provided the basic tools and techniques behind our spatial data today. By the seventeenth century, skilled cartographers such as **Mercator** had demonstrated that not only did the use of a mathematical projection system and an accurate set of coordinates improve the reliability of the measurement and location of areas of land, but the registration of spatial phenomena through an agreed standard provided a model of the distribution of natural phenomena and human settlements that was invaluable for navigation, route finding, and military strategy (Hodgkiss 1981). By the eighteenth century, the European states had reached a state of organization where many governments realized the value of systematic mapping of their lands.

These developments led to the creation of a Geographical Information Society through the establishment of **national mapping authorities (NMAs)**, whose mandate was to produce cadastral and **topographical maps** of whole countries. These highly disciplined institutes, often military in background, have continued to this day to render the spatial distribution of the features of the earth's surface, or topography, into map form. The fundamental tools of their trade—**geodetical surveying**, **photogrammetry**, and **cartography**—provided a powerful range of tools for accurately recording and representing the location and characteristics of well-defined natural and anthropomorphic phenomena.

The first maps of the spatial distribution of rocks or soil, of plant communities or people, were qualitative, reflecting the first aim of many surveys, which was inventory—observing, classifying, and recording what was there. Qualitative methods of classification and mapping were unavoidable given the huge quantities of complex data that most environmental surveys generate. The basic units of geographic information used in this mapping reflected the work of the first modern cartographers, who represented real-world objects or administrative units by accurately drawn **point** and **line** symbols that were chosen to illustrate their most important features. The attributes of areas were indicated by uniform colours or shading, though gradually varying shading was sometimes used to indicate the change in certain properties of the surface, such as the steepness of slopes, the depth of a lake, or the variation in land cover from forest to marsh (see Robinson et al. 1995 for in-depth coverage of cartography). The inclusion or exclusion of things was at the behest of the patron of the mapping exercise, whether a government, business, or private citizen (see, for example, Harley 1988, 1989; Dodge et al. 2011)

The printing technology of etching on copper plates, available from the seventeenth century onwards, reinforced the use of a geographical symbolism that relies on well-defined, crisp delineation. Today, geographical information is still often centred on the location of, and interactions between, well-defined objects in space like trees in a city park, houses on a street, aeroplanes en route to destinations, or administrative units like the *Dépendances* of France (Figure 1.9).

During the nineteenth century, the advancements in the scientific study of the earth required different kinds of attributes to be mapped. The study of the earth and its natural resources—**geophysics**, **geodesy**, geology, **geomorphology**, soil science, ecology, meteorology, and land cover—involved surveys to record data on phenomena that varied continuously from place to place, requiring huge amounts of data to provide accurate insight about their differences. Quantitative description was not only hindered by the volume of data that needed to be observed, but also by insufficient resources for making these quantitative observations. Further, there was a lack of appropriate mathematical tools for describing spatial variation quantitatively. The result was that surveyors often reduced their findings to sets of 'representative' **taxonomic classes** which were drawn on a map using different colours or shading (Figure 1.10). The use of this type of map to represent what are in reality continuously varying and complex phenomena artificially forced people to divide the space over which these phenomena occur into sets of 'things' that could then be mapped in the same way as much more well-defined objects such as trees, houses, or fenced paddocks.

The first developments in appropriate mathematics for mapping in a more representative manner the continuous natural variations in phenomena such as elevation above sea level or air pressure shown on a meteorological chart came in the 1930s and 1940s in parallel with developments in statistical methods and **time series**

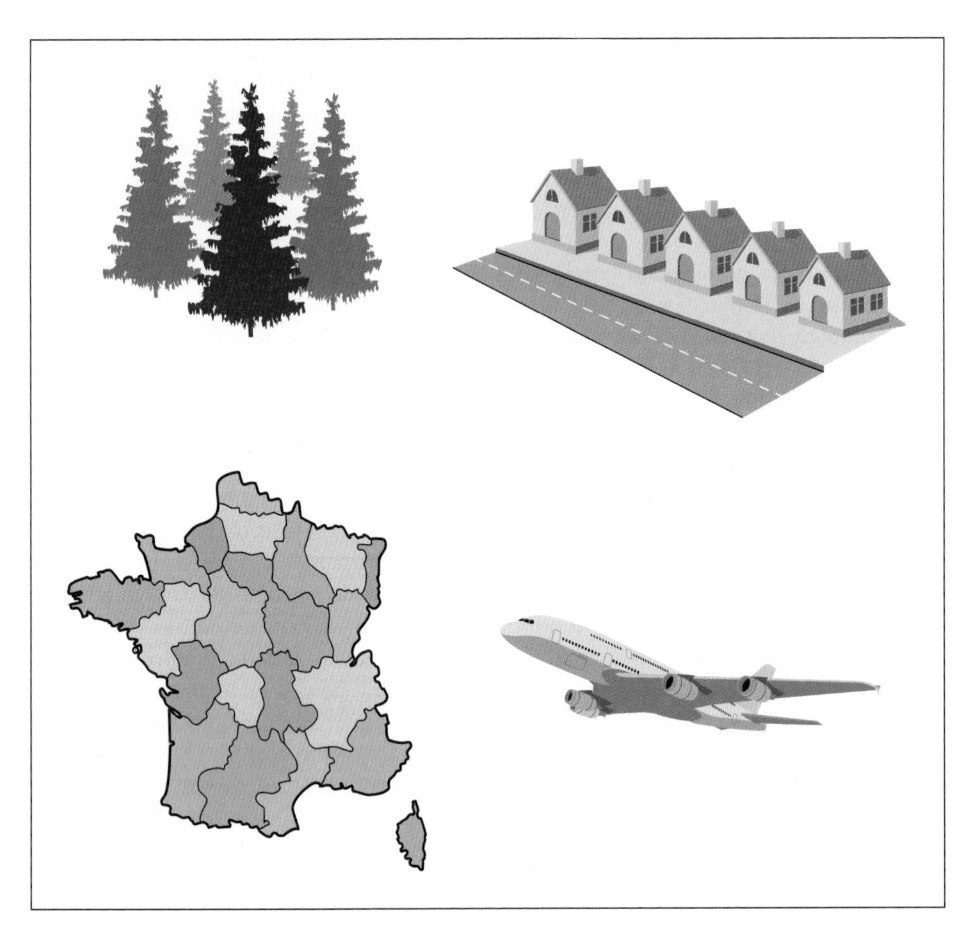

Figure 1.9 Geographical information is still often centred on the location of trees in a forest, houses on a street, aeroplanes en route to destinations, or administrative units like the *Dépendances* of France

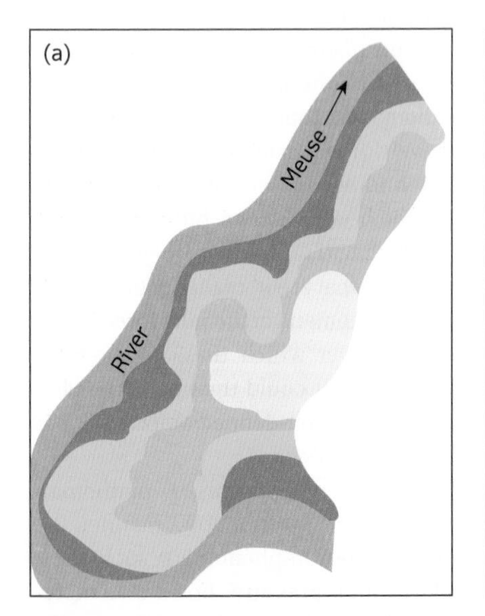

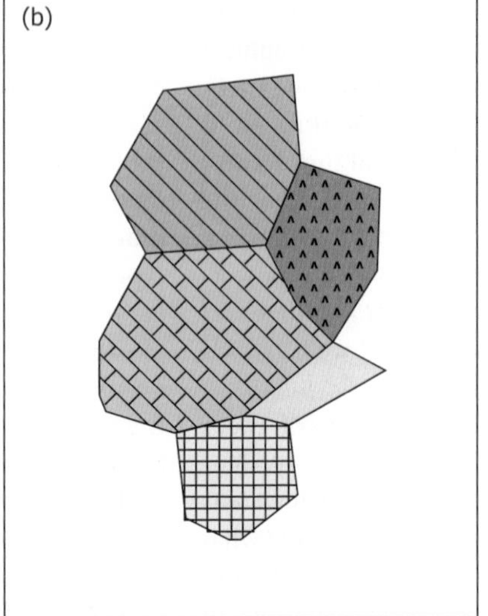

Figure 1.10 Conceptual models and representations of spatial phenomena as (a) continuously varying over an area such as elevation, or as (b) distinct entities such as forest types

analysis. Effective practical progress was completely blocked, however, by the lack of adequate computing tools. It is only since the 1960s, with the availability of digital computers, that both the conceptual methods and the actual potential for quantitative thematic mapping and spatial analysis have been able to blossom.

Mapping was further supported by the advent of **aerial photography**, but more especially images from different **satellite sensors**. The digital data from these **remote sensing systems** are not in the familiar form of points, lines, and areas representing the already recognized and classified features of the earth's surface, but are encoded in picture elements—**pixels**—cells in a two-dimensional matrix that contain merely a number indicating the strength of reflected **electromagnetic radiation** in a given band or radar signal. New tools were needed to turn these streams of numbers into pictures and to identify meaningful patterns. Moreover, the images need to be located with respect to a formal location grid; otherwise the information they carry cannot be related to a definite place. The need arose for a close interaction between remote sensing, earth-bound survey, and cartography—this has been made possible by the powerful spatial information-handling and mapping capabilities of GIS.

During the 1960s and 1970s there were new trends in the ways data on soil and landscape systems were being used for resource assessment, land evaluation, and planning. Realizing that different aspects of the earth's surface do not function independently from each other, and at the time having insufficient means to deal with huge amounts of disparate data, people attempted to evaluate them in an integrated, multidisciplinary way. The 'gestalt' method (Hills 1961; Hopkins 1977) classified land surfaces into assumed 'naturally occurring' environmental units or building blocks that could be recognized, described, and mapped in terms of the total interaction of the attributes under study. Within these 'natural units' there was supposed to be a recognizable, unique, and interdependent combination of the environmental characteristics of landform, geology, soil, vegetation, and water.

Early on, planners and landscape architects, particularly in the United States of America, realized that data from several monodisciplinary resource surveys could be combined and integrated simply by overlaying transparent copies of the different resource maps on a light table, and looking for the places where the boundaries on the several maps coincided. One of the best-known exponents of this simple technique was the American landscape architect Ian McHarg (McHarg 1969). In 1963, another American architect and city planner, Howard T. Fisher, elaborated Edgar M. Horwood's idea of using the computer to make simple maps by printing statistical values on a grid of plain paper (Sheehan 1979). Fisher's program **SYMAP**, short for SYnagraphic MAPping system (the name has its origin in the Greek word *synagein*, meaning to bring together), includes a set of modules for analysing data, and manipulating them to produce **choropleth** or **isoline** interpolations, with the results displayed in many ways using the overprinting of line-printer characters to produce suitable grey scales.

Fisher became director of the Harvard Graduate School of Design's Laboratory for Computer Graphics, and SYMAP was the first in a line of mapping programs that were produced by an enthusiastic, internationally well-known, and able staff. Among these programs were the grid cell (or raster) mapping programs GRID and IMGRID that allowed the user to do in the computer what McHarg had done with transparent overlays. Naturally, the Harvard group was not alone in this new field and many other workers developed programs with similar capabilities (e.g. Duffield and Coppock 1975; Steiner and Matt 1972, to name but two). Initially, none of these overlay programs supported users undertaking analysis beyond that of McHarg; they merely speeded up the process and made it reproducible. However, users soon began to realize that with little extra programming effort they could do other kinds of spatial and logical analysis on mapped data that were helpful for planning studies (e.g. Steinitz and Brown 1981) or ecological analysis; previously these computations had been extremely difficult to do by hand.

SYMAP, GRID, IMGRID, GEOMAP, MAP, and many of the other relatively simple programs were designed for quick and cheap analysis of gridded data, and their results could only be displayed by using crude line-printer graphics, with many cartographers challenging whether the results they produced were maps. Cartographers had themselves begun to adopt computer techniques in the 1960s, but these were largely limited to aids for automating the drafting and preparation of masters for printed maps. For traditional cartography, the new computer technology did not change fundamental attitudes to map-making—the high-quality paper map remained both the principal data store and the end product. However, by 1977 the experience of using computers in map-making had advanced so far that Rhind (1977) was able to present many cogent reasons for using computers in cartography.

By the late 1970s and early 1980s, major strides in using and developing computer-assisted cartography were made by government and private agencies in several nations, involving considerable investment in computer programs and systems (e.g. Tomlinson et al. 1976; Teicholz and Berry 1983). However, the introduction of computer-assisted cartography did not immediately lead to a direct saving in costs as had been hoped. The acquisition and development of the new tools was often very expensive; computer hardware was extremely expensive; there was a shortage of trained staff and many organizations were reluctant or unable to introduce new working practices. Many purchasers of expensive systems were forced to hire programming staff to adapt a particular system to their needs.

At the same time there were many parallel developments in automated data capture, data analysis, and presentation in several broadly related fields such as cadastral and topographical mapping, thematic cartography, civil engineering, geology, geography, hydrology, spatial statistics, soil science, surveying and photogrammetry, rural and urban planning, utility networks, and remote sensing and image analysis. Military applications overlapped and even dominated several of these fields. Consequently there was much duplication of effort and a multiplication of discipline-specific jargon for different applications in different lands. This multiplicity of effort in several initially separate, but closely related, fields has resulted in the emergence of the general purpose GIS.

During the 1990s there were several important technical and organizational developments that greatly assisted the wider application and appreciation of GIS. There was a growing awareness among many more people of the possibilities and importance of being able to manipulate large amounts of spatial information efficiently. At the same time, computer and network technologies were changing beyond recognition, providing vast processing power and data storage capacity at modest prices and increasingly fast transfers of data across the globe. This rapid expansion in GIS use was also facilitated by standardization in many areas including data transfer, interfaces, and for mathematically defining the earth as a whole. The use of technology for measuring where we were, **global navigation satellite systems (GNSS)** (colloquially called GPS), transformed the ability to undertake field surveys, enabling data collection at a more rapid pace in areas of study or for features not collected by NMAs. There were still areas limiting development in GIS, including those centred on the basic data models used in defining spatial data in modern GIS (described in Chapters 2 and 3), which were little different from those used in previous decades. These have restricted the ability of GIS users to represent changes in the time dimension as well as those over a space.

In the last decade, the opening up of geographical data to a wide user base, the explosion in the use of social media, along with a capacity to modify and contribute data to environments has transformed the way GIS are used (Sui and Goodchild 2012). The widespread availability of web-based mapping and its use by non-experts has resulted in the formation of what has been termed **neogeography** (Turner 2006). This capacity to modify and update data sources, the democratization of information, and links to the concept of Web 2.0 are just some of the areas of an ever-increasing academic debate (Lin and Batty 2009a; Graham 2009; Hudson-Smith and Crooks 2009; Warf and Sui 2010; Connors et al. 2013; Haklay 2013; Wilson and Graham 2013).

1.5 Geographical information systems today

The main result of more than 40 years of technical development is that GIS have become a worldwide and widely used phenomenon. These systems are used in many different fields that early pioneers would not have thought possible, becoming a component in mainstream operations in many areas of business and services and contributing to the development of new research insights. The following examples give some idea of the range of applications.

Urban planning in a fast-growing city: Dubai Municipality's GIS Centre

Dubai is known as one of the fastest growing cities, boasting the tallest building and the second busiest international airport in the world (Figure 1.11a). The pace of urban growth is continuing with new roads and railways, housing developments, office blocks, tourist resorts, port expansions, and even a new airport. Managing this rapid urban expansion and the required services that go with this is challenging, but Dubai Municipality's work is aided by its GIS Centre, formed in 2001, which collects and holds large databases on what is where (Figure 1.11b).

There were many challenges in developing such a centre, not least of which was how to integrate the existing large databases of the different departments of the Municipality and the private sector. The system also needed to be flexible and scalable to enable details of rapid urban development to be easily entered in the system as they became functional. On the user front, the system needed to reflect the demographics of the Dubai workforce, where expatriates as well as local Emiratis use this information, so it needed to be bilingual, supporting both Arabic and English.

The resulting GIS Centre is today a centralized but physically networked system bringing together in real time, data held in different databases to answer particular queries or to produce mapped information as needed. The development relied on the availability of fast network

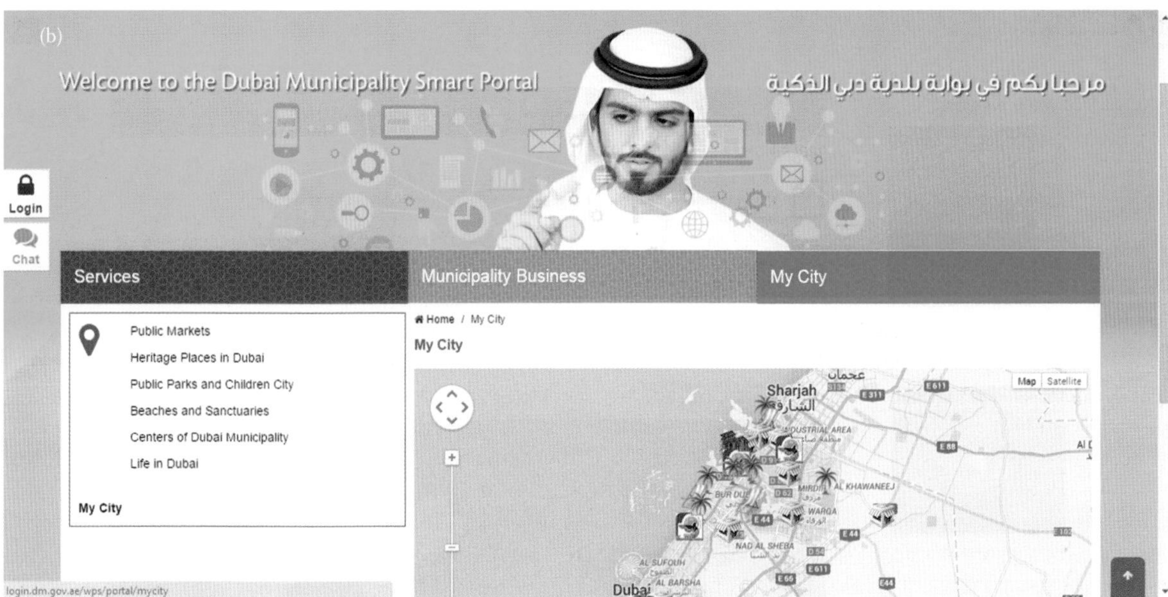

Figure 1.11 (a) Dubai skyline (image copyright Paul White) and (b) screenshot of Dubai Municipality's GIS web page (www.dm.gov.ae)

technologies, and the ability to integrate data across different computer operating systems and formats. It combined data collected by private sector developers with those from government departments.

The data sharing was facilitated by the passing of Dubai Law No. 6 for 2001, which established the GIS Centre as the only body responsible for protecting, maintaining, and classifying the data and for stating the technical guidelines for its interaction with the various public and private sector bodies. The law also required all collected data, whether aerial photographs, satellite pictures, three-dimensional digital models, topographical information, civic planning data, building legislation and information on addresses, or infrastructure and utility data, to be shared with the Centre as the developments took place.

The resulting mapped data are now part of the everyday work of the Municipality and are also used by many private, government, and semi-government agencies (Figure 1.11b). Data access security is managed through the granting of specific privileges to each user. Development of the GIS Centre has led to a greater return on investment on the many data sets collected, the initiation of more government e-services, and, at a more strategic level, enhanced integrated planning and management.

Food security is more than agricultural production: FEWS NET

Moving from the local to the global, this second example highlights the use of GIS to provide information and analysis of food insecurity among some of the world's most vulnerable people. The Famine Early Warning Systems Network (FEWS NET), created by United States of America International Development agency (USAID), integrates global spatial data collected by US government agencies (including those for space, weather, and water observations), national government ministries, and international partners for more than 30 of the world's most food-insecure countries. This involves continuous monitoring of weather, climate, agricultural production, prices, trade, and other factors, considered together with an understanding of local livelihoods and vulnerabilities. Using these data, FEWS NET researchers develop scenarios to forecast likely conditions in the coming days and anticipated changes over the next six to twelve months (see Figure 1.12). Access to these data on current and future conditions helps government decision-makers and relief agencies plan for food emergencies.

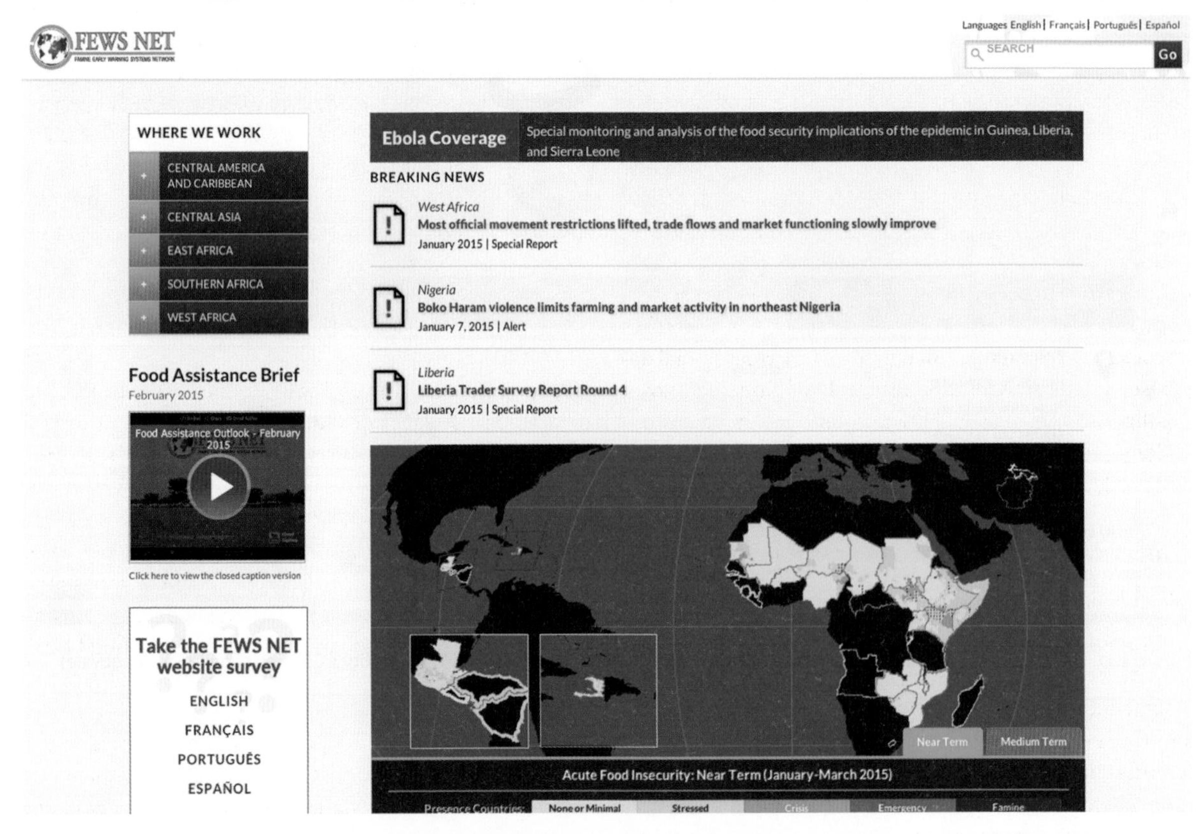

Figure 1.12 Screenshot of web page for USAID's Famine Early Warning Systems Network—FEWS NET provides important information on current conditions and future scenarios for 30 at-risk countries
(*Source:* USAID)

Managing rivers through participatory management: GIS bring information in new ways to local decision-makers

GIS and modelling have long played an important part in the planning, control, and attempted optimization of water resources. Mapping and modelling the inflows and outflows of a river contributes to solution building for water resource problems. Today there is emphasis in many countries on decentralized governance, moving decision-making from central authorities to those established at the river basin level. New local stakeholders are now included in this decision-making environment, bringing their own understanding of water flows, challenges, and solutions. This local knowledge, often referred to as **tacit** or **naïve knowledge**, may be expressed using words and expressions quite different to the formal language and jargon of engineering or water resources management. There is therefore a need to bring these different, but equally important, understandings together and a **participatory GIS** (PGIS) approach can be used to facilitate and support this inclusive decision-making (e.g. Dunn 2007; Cinderby et al. 2013).

The Rayong River Basin in Thailand is a case study example where a GIS was linked to the graphical dynamic modelling software system Stella as a platform to work directly with local village leaders, farmers, and other stakeholders. Both the GIS and Stella models showed the river system in a graphical form in which different colours and dynamic charts were used to highlight the different flows that would result from varying flow management scenarios, ensuring this was comprehensible to all the stakeholders. During the participatory process additional local knowledge was added, often by converting hand sketches or vocalized spatial referencing into the formal spatial database system of the PGIS. The study of Nonejuie (2010) highlighted the growing engagement in the decision-making press of local stakeholders, facilitated by the PGIS and modelling approaches (see Figure 1.13).

Gaining insight through showing spatial data differently

GIS have for a long time been an important tool for researchers in analysing geographical data to gain new insight into complex systems. Zoology is no exception, and this next example highlights how seeing things differently can help in developing understanding. For a long time, zoologists thought that animals living in the centre of their species' geographic range should be larger than those living on the edges. The logic was that conditions in the centre were better and produced healthier, larger individuals. In areas further from the core, habitat conditions were likely to be worse, survival less likely, and the quality of individuals therefore expected to be lower. But is this really true? To find out, a team of zoologists assembled a data set of 7 871 museum specimens of 25 species of carnivorous mammal, georeferenced by collection location (Meiri et al. 2009).

Ideally, the geodesic (geoid surface) distances between the specimens' location and the edge of their species' geographic range would be calculated. But in 2008 this wasn't feasible on the desktop: projected data had to be used. Rather than produce nearly 8 000 equidistant projections, each centred on one specimen (see Chapter 2 for more on map projections), a compromise projection needed to be found. R. Buckminster Fuller's Dymaxion projection was chosen. This is a projection of the globe onto an icosahedron, with each face of the icosahedron containing a centred gnomonic projection. Scale and conformal error are kept relatively low and universal by using this projection across the global domain. The Dymaxion also arranges coastlines so that none falls across a discontinuity in the projection, making it much easier to calculate distances between terrestrial objects (Figure 1.13).

These distance measurements revealed that mammal species did not seem to follow a general rule of 'bigger in the middle'. Instead, quality of local conditions seems to vary across the range, leading to idiosyncratic size variation—giving rise to examples such as the well-fed brown bears of Kodiak Island, Alaska, which are the world's joint largest terrestrial carnivores while being a peripheral population of the brown bear species as a whole (Meiri et al. 2009).

Mapping population change

GIS are useful for comparing differences over time. One of the main benefits of GIS is that they offer powerful tools for integrating diverse data sources. These may differ in terms of attributes or spatial features. However, for many types of data, such as population data for which there are censuses across different decades, the attributes recorded and the size and shape of zones may not be consistent across time if some property, such as population characteristics, is measured for several time points. Martin et al. (2002) describe some of the problems that users face if they wish to explore population change by comparing geographical population data for different time periods. The approaches can be used to develop comparable data sets which allow researchers to assess how the population has changed—where have employment levels increased the most? Is deprivation

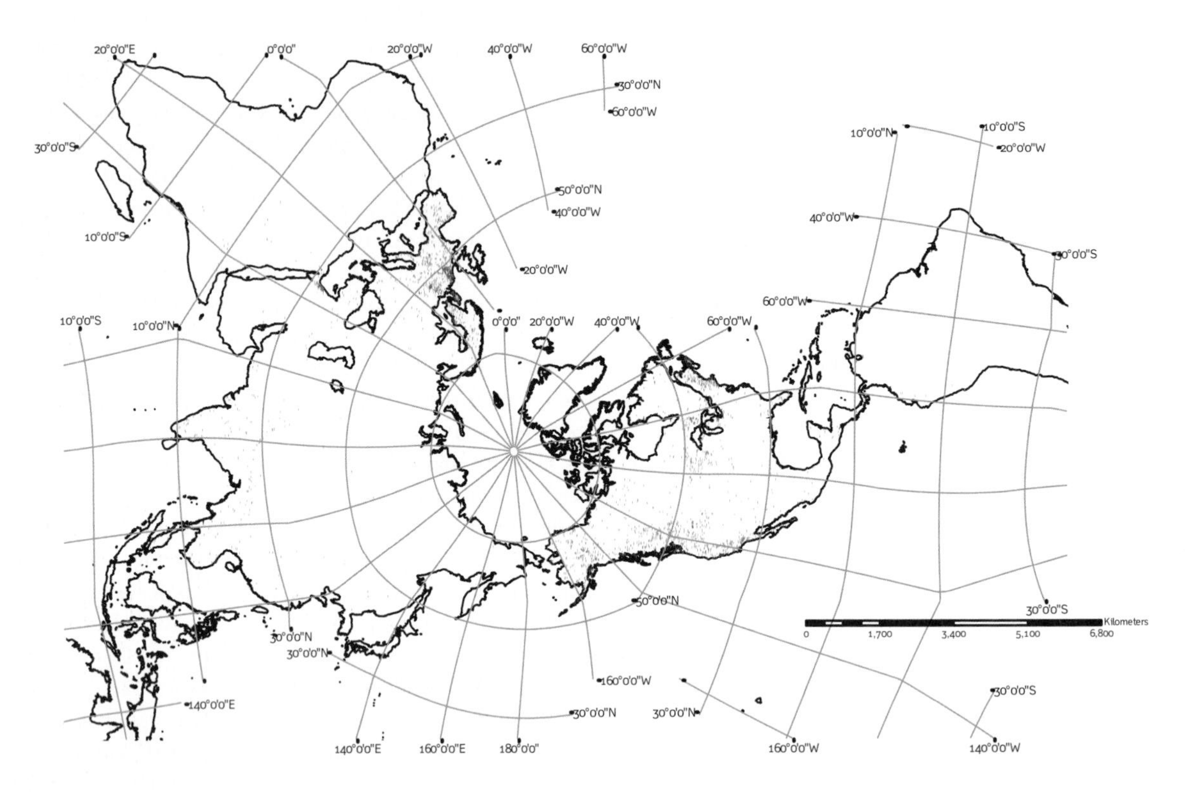

Figure 1.13 Visualizing distributions of 25 species of carnivorous mammals using Buckminster Fuller's Dymaxion projection (*Source:* Meiri et al. 2009)

growing more in urban areas than in rural areas? Is unemployment related to illness, and if so where is the relationship strongest and how is this changing over time? Before the advent of GIS, along with the development of digital population data sets, detailed examination of population distributions and how they change was not feasible.

Figure 1.14 shows area boundaries (**census wards**) for an area of England; the ward areas were not the same in 2001 and 2011, so to directly compare counts in the two sets of areas they must be made consistent. GIS have been used to convert population counts from one set of zones (for example, counties or wards) to another so that populations can be directly compared even though the data were not initially recorded for common areas.

1.6 Choosing to use GIS

These examples provide only the smallest of insights on the plethora of ways GIS are used today and the benefits this brings. The systems allow us to integrate all manner of geographical data; however, establishing a GIS and its database can be an onerous task, especially where the data are not easy to access or are in forms that are not readily input to the system. It is therefore important that before we start establishing a database and undertaking the analysis we ask ourselves a number of strategic questions (adapted from Longley et al. 2010) including:

- What value will the investment of time and money in establishing a GIS have for my research, our organization, the project to be undertaken, or the community it aims to serve?
- When will benefits be delivered, and does this fit my time horizons?
- Who will benefit from setting up the GIS and the results that follow?
- How much investment is needed, both initially and on an ongoing operational basis, and is this within the budget available?
- Who is going to deliver these benefits, and what resources are required—both internally and externally—to realize the expected benefits?

Figure 1.14 Changing census ward boundaries between 2001 and 2011
(*Source:* Chris Lloyd)

- What is the proven financial case—that is, does the investment in the GIS provide financial or other value to make it worthwhile?

1.7 The structure of this book

It is not the aim of this book to describe further the growth of awareness of the advantages of GIS among politicians, businessmen, academics, and resource managers, nor to explain how these tools can be efficiently incorporated in modern government or business. As with the first and second editions (Burrough 1986; Burrough and McDonnell 1998), the aim of this book is to explain the scientific and technical aspects of working with geographical information, so that users understand the general principles, opportunities, and pitfalls of recording, collecting, storing, retrieving, analysing, and presenting the information using GIS. The aim includes not only to outline the basic principles but to help the reader understand the limitations of the technology and what impact these errors in data at various stages of GIS database development may have on the results.

The old computer adage of *garbage in, garbage out!* holds true in GIS use. The very complexity of these systems makes it possible to input good data but to deliberately or unwittingly produce garbage which, thanks to multi-coloured high-resolution outputs or animated visualizations, looks like a high-quality product. At the same time, it is not difficult to conceal bad data through clever or high-quality presentation. The problem is not the technology, but one of the complexities of using spatial information (Monmonier 1993); with computers we can make bigger mistakes faster than ever.

This book provides a comprehensive, but of necessity limited, overview of the main aspects of handling spatial information in a GIS. In this chapter we present a brief history of development and the basic elements of GIS. In Chapter 2 we explore the problems of collecting and describing spatial data and the user's dilemma of needing to deal with both data about discrete **entities** in space and also with data about features while attribute values are ever changing over space in continuous **fields**. These problems raise questions such as how we should proceed when we are unsure which model to use, or which approaches are preferred by which disciplines. In Chapter 3 we explore ways in which the different paradigms introduced in Chapter 2 can be implemented technically in the computer, for storage, retrieval, and display.

The practical aspects of building a geographical database are described in Chapter 4. Chapter 5 considers some of the ways in which we can visualize or display geographic information. Indeed, visualization is the key objective of many users of GIS. Visualization is a powerful means of searching for patterns, and Chapter 6 takes this theme further by considering how we can summarize geographical data or, for example, assess if large or small data values are clustered in particular areas.

Chapter 7 is concerned with the analysis of discrete entities and it deals with topics such as measuring distance from objects, or overlaps between different features. Links are made to topics such as **multicriteria evaluation**, where locations are identified that fulfil a wide range of conditions (Malczewksi 2006). Chapters 8 and 9 introduce approaches for spatial **interpolation**. In brief, if we have sparse point samples (for example, measurements of an airborne pollutant), but we want to have values at all locations in our study area, then we can use an interpolation approach to generate gridded values. Chapter 8 focuses on key principles and describes a range of commonly used interpolation methods, while Chapter 9 introduces **geostatistical methods** which can help in exploring how a variable is spatially distributed, optimally mapping from sparse samples, and sampling design.

The analysis of continuous fields (often in the form of raster grids) is the focus of Chapter 10. This chapter outlines methods for processing images or gridded data and also introduces some key derivatives of **elevation**, such as **slope** and **aspect**. Chapter 11 builds on this latter theme by considering in detail **digital elevation models (DEMs)**, exploring methods for measuring topographic form, and for the analysis of surfaces through deriving **intervisibility**, **shaded relief**, and **irradiance maps**.

Space–time modelling and GIS is explored in Chapter 12 along with the ways in which errors can occur in the data and their effect on model results. Chapter 13 considers the theoretical and practical research ideas of **fuzzy logic** and continuous classification, and suggests that it is not always necessary to force spatial data into either one or the other paradigm—the real world is not made up of either crisp entities or continuous fields—but many phenomena have properties that can be dealt with by combining the best aspects of both. Previously it was thought impossible to deal with vague formulations of space in the computer, but Chapter 13 shows that this is far from being so. In fact, by refusing to be hidebound by the classical paradigms we find that we can often make more specific pronouncements about spatial relations than we can with conventional methods. This finding provides many new opportunities for useful research and applications that will take many more years to work out.

1.8 Summary

GIS are used in so many different areas of our lives, providing a means of developing new insight and informing decision-making from the individual through to the global level. While the systems build on a rich cartographic heritage, the information they produce today stretches far beyond the map. It is important to understand the fundamentals of GIS to not only comprehend how these systems work but to also be aware of their limits. It is a technology-based approach to understanding the geography of our world and this brings many advantages but also drawbacks that we should think about.

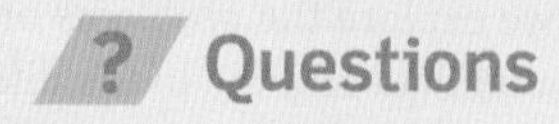

Questions

1. How do you use spatial information in your everyday life? How do you access it and where does the data come from?

2. How do governments collect and use data about you and your family? What would you like them to consider in holding this information on you?
3. What data should be freely available on the Internet and who should provide this?

Further reading

- Chrisman, N. R. (2003). ***Exploring Geographical Information Systems***. Wiley, Hoboken, NJ.
- Heywood, I., Cornelius, S., and Carver, S. (2011). ***An Introduction to Geographical Information Systems***. 4th edn. Pearson, Harlow.
- Lillesand, T. M., Kiefer, R. W., and Chipman, J. W. (2008). ***Remote Sensing and Image Interpretation***. 6th edn. Wiley, Hoboken, NJ.
- Lin, H. and Batty, M. (eds.) (2009). ***Virtual Geographic Environments***. Science Press, Beijing.
- Longley, P. A., Goodchild, M. F., Maguire, D. J., and Rhind, D. W. (2010). ***Geographic Information Systems and Science***. 3rd edn. Wiley, Hoboken, NJ.
- Monmonier, M. (1993). ***How to Lie with Maps***. University of Chicago Press, Chicago.
- Peng, Z.-H. and Tsou, M.-H. (2003). ***Internet GIS: Distributed Geographical Information Services for the Internet and Wireless Networks***. Wiley, Hoboken, NJ.
- Pickle, J. (1993). ***Ground Truth: The Social Implications of Geographic Information Systems***. Guildford Press, New York.
- Reitsma, F. (2012). Revisiting the 'Is GIScience a science?' debate (or quite possibly scientific gerrymandering). ***International Journal of Geographical Information Science***, 27 (2): 211–21, DOI:10.1080/13658816.2012.674529
- Wise, S. (2014). ***GIS fundamentals***. 2nd edn. CRC Press, Boca Raton, FL.

Spatial Data and their Models: Formal Abstractions of Reality

Digital technologies, social media, and the Internet have facilitated enormous changes in our access to digital geographical data, allowing many new uses and applications. Although in some uses these innovations have brought the data collector and the user closer together—and indeed they can be the same person—for most there is a distance, and often a conceptual disconnect, between the original data collector and the many data users. This raises important questions about how the geographical world is viewed, defined, and ultimately recorded by all of us. This has always been part of understanding and interpreting geographical data and is predicated on theoretical ideas on what phenomena exist and how they may be observed and recorded in a computer.

The basic theme in this chapter is that we need to adopt various structuring models when we portray the real world in a computerized system or analyse it, as we need to simplify its inherent complexity. We use these models to simplify, as well as understand, how others have simplified the diverse world before our eyes. However, since we all perceive the world differently, and have varying different purposes for using a GIS, it is a pragmatic choice when building a GIS to adopt existing data models and implement a standardized approach. It is important to examine the theoretical ideas behind the conceptual geographical models, as all the data we use in a GIS will have been schematized using these ideas.

Understanding this is important as the adoption of a particular model, whether the user realizes it consciously or not, will influence the type of data that may be used to describe the phenomena as well as the spatial analysis that may be undertaken. This material also gives an essential background to the following chapters, because of course we do not store real-world phenomena in the computer, only representations of them based on these formalized models.

Learning objectives

By the end of this chapter, you will:

- **understand how the properties, location/spatial geometry, and relationships of geographical data may defined**
- **understand why and how we conceptualize and formalize data models to describe geographical phenomena**

- **describe the differences in adopting entity conceptual models or field conceptual models in defining geographical phenomena**
- **show how the vector data model may be used, including how it defines features geometrically**
- **show how tessellations of continuous fields may be used to describe environments**
- **understand how changes in time and space, and differences in spatial and temporal scale affect the way geographical features might be represented.**

2.1 Fundamentals of geographic phenomena

Our complex, dynamic, multidimensional, multiscaled world contains infinite and unmanageable sets of phenomena. To bring some coherence to these phenomena, we need to abstract the objects of interest and define them in terms of their attributes (properties), geographical location (spatial coordinates or geometry), and relationships (topology, hierarchy, etc.).

What is present—attributes

Defining all the things that exist in the world and the relationships that are found between them is the subject of a field of study called **ontology**, an area of philosophy. With spatial data, the ontological interest is in defining 'geographical things' that exist (Smith and Mark 2001; Agarwal 2005). The development of schema for understanding this is important, as a common understanding allows information about them to be shared with different users. Distinctions may be made between:

a) something that exists before people are present (known as bona fide entities/boundaries), for example, a hill, a river, or a natural forest
b) something that exists because people are there (known as fiat entities/boundaries), such as an administrative boundary, city limits, or gross domestic product.

Both of these types of entities are frequently used in GIS.

Defining 'something' is often difficult, involving **classification** or measurements by humans, or capturing a moment in time, with interpretations and instrumentations influencing the results and meaningfulness of the resulting data. How we understand an environment and break it into units reflects:

- the experience and cultural background of the observer
- the purpose, context, and scale of a study
- the observable nature of the object, and events
- the simplicity or complexity of the 'thing'.

For example, in classifying marine habitats, much will depend on the survey conditions at the time, the procedures used, and the experience of the surveyor. Classifying a habitat as sandy or sea grass or coral is often difficult where conditions such as visibility and natural heterogeneity limit surveys.

For many geographical things we can use already defined and standardized terms such as 'village', 'town', or 'city', and 'river', 'floodplain', 'estuary', or 'delta' as the fundamental building blocks. Many of these phenomena are recognized and described in dictionaries; defined and accepted classification schema bring clarity and a certain degree of **precision** to our understanding of each characterized entity, ensuring there is reduced ambiguity (cf. Blackie 2010; Thomas and Goudie 2000; Allaby 2013; Castree et al. 2012).

Problems arise when dissimilar schema but clear definitions are used in allocating the same phenomena to different classes. For example, there are a number of global ecological classification systems used by national or more local agencies to identify, describe, and map progressively smaller areas of land with increasingly uniform ecological features. These systems use associations of biotic and environmental factors, including climate, geology, topography, soils, hydrology, and vegetation, but their definitions and class names vary—which makes combining maps from different organizations problematic. Quite marked differences occur even if the same number of classes is used and the central concepts of the classes are similar.

Once 'something' is defined, the next step is to assign values that describe its properties, using various means such as **alphanumeric characters**, digital images, or sounds (see Figure 2.1). The most common way has been with alphanumeric characters, and different data types can be used such as **Boolean, nominal, ordinal, integer**, or **scalar** values (see Table 2.1). A phenomenon may be defined using a combination of data types so, for example, the town of Oxford can be described by its name (nominal data), its settlement type (nominal data), the size of its student population (integer data), average summer and winter temperatures (interval data), and the peak flows (scalar data) and nitrate levels in its rivers (ratio). The data types used not only influence how attributes are defined but also the kinds of operations that can be used in analysing and querying them.

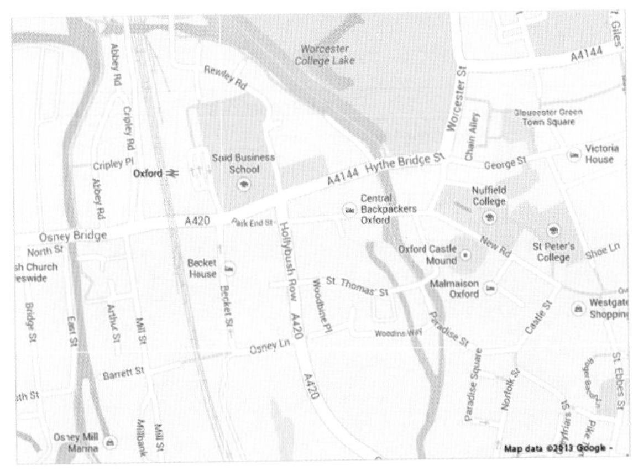

(a)

(b)

Figure 2.1 The city of Oxford described using (a) alphanumerics and (b) digital image (© Google, 2013)
(*Source*: Google Earth)

Where it is—location and geometry of what is there

Defining where something is involves detailing both its location in a space and its spatial geometry. One of two systems, known as **absolute space** or **relative space**, may be used to define a location: these are illustrated in Figure 2.2a and b.

With the Newtonian absolute notion of space, all objects are thought of as located within a huge container with positions defined using a georeferencing system, usually a Cartesian coordinate system. These systems are based on local or world coordinate systems, defined using standardized systems of **ellipsoids**, projections, and coordinates that give an approximation of the form of the earth (a **spheroid**), often portrayed as a flat surface (as detailed in Section 2.2).

In surveys of natural environments, geographical positions are usually recorded in terms of latitude/longitude or with respect to a **national grid**. Large-scale mapping—for instance of parcel boundaries for **land registry systems** or the locations of infrastructure and utilities (roads, electricity and telephone cables, etc.)—is usually referenced in metres to a local mapping agency grid.

The relative system of representing space in many ways reflects our human cognitive processes of positioning things, as locations are defined in terms of spatial relationships with other existing features rather than actual metric distances and coordinate systems. With relative-based thinking, important parts of definitions involve terms that reflect **contiguity**, **connectivity**, **proximity**, and distance, so 'something' might be described as 'joined to A' or 'next to B' or 'to the east of C'. Common examples of models and maps which use this form of locational referencing include those for infrastructure systems such as roads and pipelines, aboriginal rock paintings, and plans of underground train networks.

There are marked differences between relative and absolute systems of defining spaces. Absolute spatial referencing is the foundation of most mapping, and is commonly used in GIS. However, there are many examples of geographical data that are collected using relative referencing, so the resulting data points need

Table 2.1 Data types

Data type	Permitted values	Permitted operations
Boolean	0 or 1	Logical and indicator operations: Truth versus Falsehood
Nominal	Any names	Logical operations, classification, and identification
Ordinal	Numbers from 0 to ∞	Logical and ranking operations, comparisons of magnitude
Integer	Whole numbers from −∞ to +∞	Logical operations, integer arithmetic
Scalar	Real numbers (with decimals) from −∞ to +∞	All logical and numerical operations
Topological	Whole numbers	Indicate links between entities

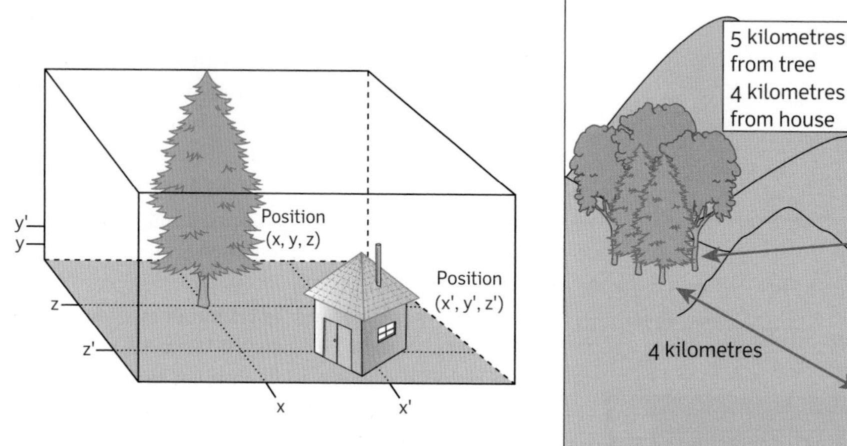

Figure 2.2 (a) Absolute and (b) relative concepts of space

to be converted to absolute positions before they can be input into a GIS. When interviewing people during field surveys, for example, they may describe the state of the river, the landscape, or their town in geographical terms that are based on relative concepts of space. Translating these understandings of space into absolute values requires absolute, defined geographical referencing to be added to the key phenomena in the landscape. Absolute spatial referencing is predominant in geographical studies, so we will now explore this in more detail.

2.2 Exploring absolute georeferencing systems

The common frame of reference used in absolute referencing systems has a number of components, depending on the scale of work and the nature of the data. When the study site covers a large area, the curved nature of the earth's surface needs to be accounted for and so a **global referencing system** is used. For more local studies, a planar, flat surface may be assumed, and a **local referencing system** adopted. The first important step is to define the shape of the earth, using ellipsoids and geoids, which are used as the basis for most mapping and sensing studies.

Ellipsoids, geoids, and datum

The earth's shape and surface is complex. Indeed, it is far more complex than the perfect sphere that is popularly used to portray it, with its topographic surface just one of the many-scaled undulations. The observation and measurement of the shape and size of the earth is a branch of applied mathematics known as geodesy, which has been practised since the early civilizations of the Greeks and Egyptians. Today the use of ever more sophisticated satellite and ground-based techniques has meant that the definition of the earth's shape is constantly being refined, and the resulting base references provide both a **horizontal datum** and a **vertical datum** from which location measurements may be made.

Given that the earth is flattened at the poles, and bulges at the Equator, the geometrical shape of an ellipsoid of revolution—derived by rotating an ellipse around its shorter (semi-minor) axis—is used as the base shape for defining locations. It is defined dimensionally in terms of its **semi-major axis**, designated as a (the equatorial radius), its **semi-minor axis**, known as b (the polar radius), and a value for **flattening** (f) which indicates how near to a sphere the ellipsoid is (see Figure 2.3). The ellipsoid that best approximates its shape varies from region to region, with many countries having their own national reference ellipsoid reflecting local conditions,

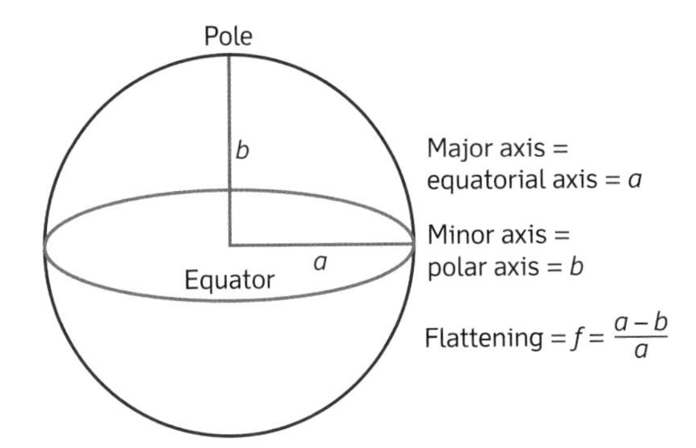

Figure 2.3 Ellipsoids are defined in terms of their semi-major and semi-minor axes, and a flattening value

as Table 2.2 shows. These form the horizontal datum from which all subsequent specific measurements of positions on the earth's surface are made. Different local versions of datum may be used—the same position can have many different geographic coordinates, so it is important to know which one of these has been used to create the reference.

While the mathematically derived smooth approximation of an ellipse provides a much-used reference frame for the earth's surface, some studies require location definitions which more accurately reflect the earth's physical shape. Geoids are the second main reference frames that are used; they have irregular shaped surfaces generated from gravity measurements (see Figure 2.4). The geoid is an **equipotential surface** of the earth's gravity field which best fits global mean sea level. This means it is essentially the hypothetical shape of the earth, where all gravity is equal, which coincides with mean sea level—and how it would look if the waters were extended out over land areas.

Vertical datum, the surface of zero elevation from which all other points are measured, can be based on a geoid or locally defined base level. Sea level measurements made at coastal gauging stations over a long period of time are used by national mapping agencies to define what is referred to as mean sea or base level. This averaging helps to remove the variations caused by the changing gravitational forces from the sun and moon. In the UK, for example, Ordnance Datum Newlyn (ODN) is the usual definition from which 'height above sea level' is defined, and is based on measurements at a site in the south-west of the country. For the Australian Height Datum (AHD), mean sea level at 30 tide gauges from around the Australian coastline were assigned a value of 0.000 m AHD, and a surface generated between them. While in the US, the North American Vertical Datum of 1988 (NAVD 88) is the vertical control datum established in 1991 by the minimum-constraint adjustment of the Canadian–Mexican–US levelling observations. It is fixed at the height of the primary tidal benchmark, referenced to the new International Great Lakes Datum of 1985 local mean sea level height value, at Father Point/Rimouski, Quebec, in Canada (National Geodetic Survey 2013).

With different base reference ellipsoids, geoids, and datums being defined by individual countries, this can mean that neighbouring country maps may not join exactly. To aid cross-boundary studies, long-term data sets from global navigation satellite system (GNSS)[1] and other space-based and ground observations have been used to develop and refine international reference ellipsoid frames such as the International Terrestrial Reference Frame (ITRF) and **World Geodetic System 84 (WGS84)**. The latter is one of the most commonly used references; it is defined not only in terms of an ellipsoid, angular velocity and the earth mass parameters, but values are also given which define a detailed gravity model of the earth. These latter values are important for determining the orbits of GPS navigation systems, which are fundamental to many forms of geographical data gathering.

[1] GNSS is the generic term used to refer to these systems, but in colloquial language 'GPS', based on the USA's Global Positioning Systems, is ubiquitously used.

Table 2.2 Reference ellipsoids used in many parts of the world

Ellipsoid name	Equatorial radius (a) (m)	Flattening $1/f$	Region covered
Airy 1830	6 377 563.0	299.325	Great Britain
Everest 1830	6 377 276.3	300.800	India, Pakistan, Thailand
Bessel 1841	6 377 397.2	299.153	Central Europe
NAD83 epoch 2010.00	6 378 206.4	294.980	North America
Australian 1965	6 378 160.0	298.250	Australia
World Geodetic System 1984 (WGS 84)	6 378 137.0	298.257	International, and basis of GPS

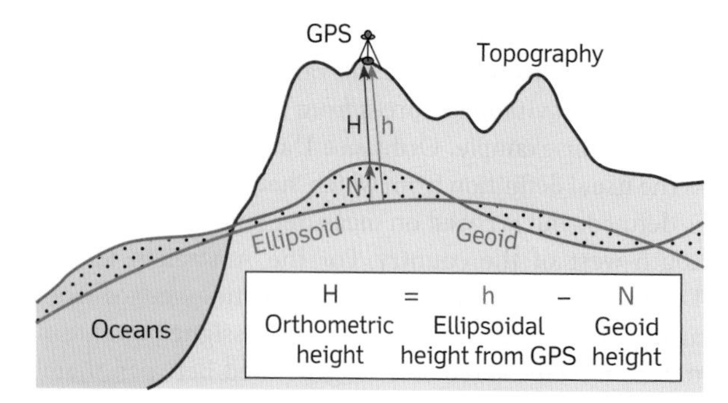

Figure 2.4 Geoids are irregular shaped surfaces generated from gravity measurements

Map projections

Working with an ellipsoid is difficult, so most studies are based on a flat planar surface representation of the earth. To achieve the conversion from three dimensions to two, one of three main mathematical transformations known as **planar** (**azimuthal**), **cylindrical**, and **conical projections** are used. The names describe the form of the surface used in the projection, as shown in Figure 2.5 (Maling 1992). Projections are centred on a point (**azimuthally**) or on one or more lines of origin (cylindrical and conical), and it is only at these positions that the scale of the map is true. The aspect that the cylinder, cone, or plane makes with the ellipsoid may be varied according to the purpose of the map. For example, a cylindrical projection centred tangentially to the equator (referred to as 'normal') is useful for world maps concentrating on the major land masses of the world, but in this projection the poles are distorted from points to lines so it is not suitable for studies in these areas.

In converting the ellipsoid to a plane, certain geometric aspects of the map—such as distance, true direction, shape, and area—are preserved or deformed to various degrees depending on the projection. Terms such as

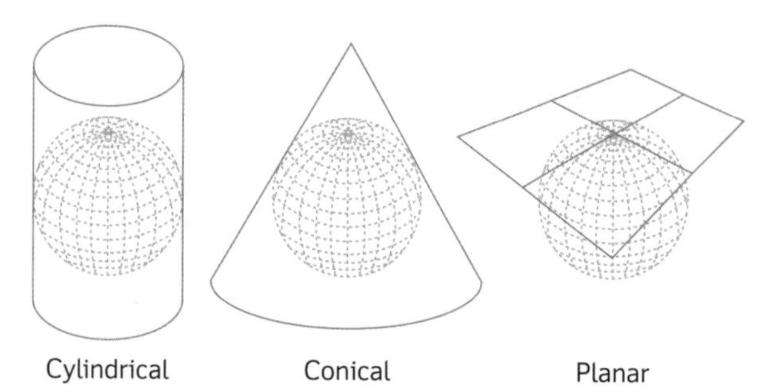

Figure 2.5 Planar, cylindrical and conical projections are used to convert the globe onto a flat surface

'**equal-area**' (preserving area proportions), '**conformal**' (preserving bearing), or '**equidistant**' (preserving distance) are used to describe the different projections. The resulting maps lead to quite different views of the world, with variations in the distances, shapes, and other related information of the land areas shown.

It is very important for users to understand the impact of the projection of a particular map as it affects its possible use, integration with other data, and interpretation. The best projection to use depends on the purpose of the work and the location of the site on the surface of the earth. Studies involving an analysis of changing land use obviously need to be based on maps where area is preserved, whereas those used in navigation need direction to be true. For example, cartographers have found that cylindrical projections are best for lands between the tropics, conical projections are best for temperate latitudes, and azimuthal projections are best for the polar areas. The standard grid for polar areas north of 84° north, and south of 80° south, is the Universal Polar Stereographic (UPS) grid.

The most widely used general projection is called the **Universal Transverse Mercator** (UTM), developed by the US Army in the late 1940s, which is now a world standard for topographic mapping and digital data exchange. It is centred tangentially on lines of longitude (the meridians), and so is referred to as transverse. Other arrangements of the cone, plane, or cylinder include the oblique (at an angle other than 90° to the Equator or a meridian) and the secant, which 'cuts' through the ellipsoid.

Coordinate systems

The actual location of an entity on the earth may be defined using one of a number of possible referencing systems. The referencing system used will depend on the scale of the study and where in the world the positions are. The most usual and convenient coordinate system used in global referencing is that of **geographical coordinates**, oriented conventionally north–south (longitude) and east–west (latitude), and measurements are given relative to the Origin (defined as the intersection of the Equator (0° latitude) and **Prime Meridian** (0° longitude) at Greenwich). Locations are defined using either negative or positive digital degrees or degrees/minutes/seconds values east/west and north/south, respectively, of this point, as shown in Figure 2.6a.

For smaller areas, **rectangular coordinates** are often used on a planar projected map. These are based on a straight perpendicular line grid and a position is given in terms of a displacement in the x (easting) and y (northing) directions from a locally defined origin. This origin is

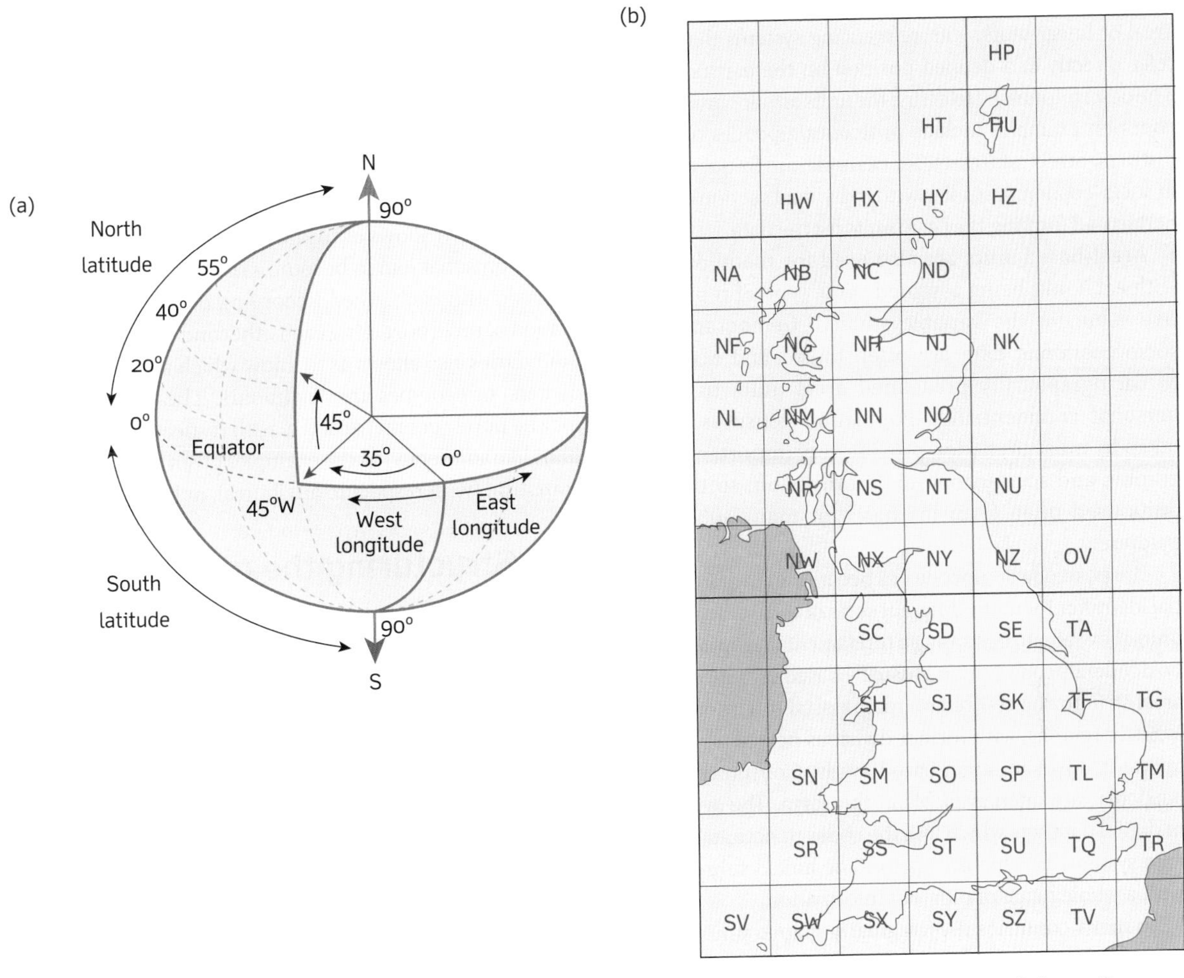

Figure 2.6 A location may be defined in different coordinate terms, such as (a) geographical or (b) planar, such as the Ordnance Survey National Grid

usually chosen to be the most south-western bottom corner of the grid, so all values are positive. Measurements may be defined in any standard distance metrics used to measure the earth, with metres being the most frequently used today. Precision in the location definition depends on the fineness of the division of the resulting squares.

The planar coordinate system of the UK is known as the Ordnance Survey National Grid, which uses a grid on a transverse Mercator projection of the area. The system uses a combination of letters (to define the basic 1 km grid square unit) and numbers (the division within the squares) to define a location.

The United States State Plane System was established during the 1930s and comprises more than 120 different zones across the different states; California, for example, has six zones, Alaska ten, and Florida has three. Different map projections, either the Lambert Conical or the UTM, are used in defining each zone depending on whether the shape of the state is predominantly east–west or north–south. These reference systems define the base mapping for local and state government geographical data and bring ease of use as they are based on **Cartesian coordinates** rather than geographical coordinates. The main difficulties arise when data need to be combined from different zones, especially across state lines. Various software packages are available online and in GIS to support the conversion and linking of data sets from different zones.

Discrete entity referencing systems

An alternative method to the continuous coordinate systems described so far, which is used for certain geographical data, is based on predefined regularly or irregularly shaped areal or linear units, with referencing systems that do not refer directly to a defined position on the earth's surface. The descriptors used to define the units are alphanumeric—common examples include postcodes/zipcodes or census districts. Street addresses, for example, are frequently used in socio-economic studies with individual sections or sides of the road forming the basic units for the data.

Areal-based units are also used for many data sets, with each unit being assigned some type of representative value for the variable studied. **Demographic** and socio-economic data are often linked and accredited to cartographically predefined areal units to protect personal confidentiality. Population censuses, undertaken by most countries, provide very important demographic and socio-economic information, so the basic units used often form the basis for many subsequent studies.

The postcode or zipcode has become an important spatial identifier for many different sorts of data, with the areal units they refer to becoming a fundamental geography. The postcode/zipcode system is usually based on a hierarchical areal structuring centred on main postal towns or administrative regions, with further divisions nested within this, taking it down to street-based levels. Consumer data in particular are often available in this form. The availability of digital data sets which link the codes to national coordinate systems has enabled them to be linked subsequently with a whole range of different information.

In many countries the irregularly shaped units for collecting administrative and other data are hierarchically structured so that regional and state-wide conclusions may be drawn, as well as more local ones. In England and Wales, for example, the census zone hierarchy includes local authority districts, middle super output areas, lower super output areas, and output areas (OAs). The basic OA is variable in areal size but contains on average approximately 100 households/300 persons. In the Australian Population and Housing Census system the country is subdivided into basic units of about 30 to 60 dwellings, known as mesh blocks. These may then be amalgamated into larger geographical areas, such as statistical areas of different levels, and states.

Georeferencing to these discrete areal units so that they can be registered to overlay with other GIS data involves allocating a representative locational identifier for each basic unit. Usually geographical coordinates are assigned to each unit. This might be the coordinates of the boundary line or a point at the **centroid** of the polygon. The United States and United Kingdom census authorities, for example, both provide full boundary data files for each census block unit. In other cases a value for the unit, often the centroid, is given (US Census 2013)

Topology

Topology is an important concept in geographical data as it captures characteristics such as the connectivity and contiguity of features with others. These spatial relationships do not change when the geometry is deformed, and this additional information supports more advanced analysis. For example, it allows us to know the connections of different streams that join to become a river, or the connections of roads with each other. Recording the contiguity information of units bordering each other and sharing common boundaries also allows us to know which soil types, house owners, or ecotypes are neighbours. These relationships may be purely geometrical (i.e. with respect to spatial relations like adjacency) as the examples given highlight, or hierarchical (with respect to attributes), or both (Figure 2.7).

2.3 Structuring the geographical world

Building from this introduction, it is now useful to consider in more detail how we actually conceptualize geographical data. The major steps involved in proceeding from human observations of the world—either directly or with the assistance of tools like digital sensors or aerial photographs—to digital representations are outlined in Box 2.1 and illustrated in Figure 2.8. The most important first step is that people observe the world and perceive phenomena that are either fixed or changing in space and/or time; their perceptions will influence all subsequent analysis. So success or failure in using

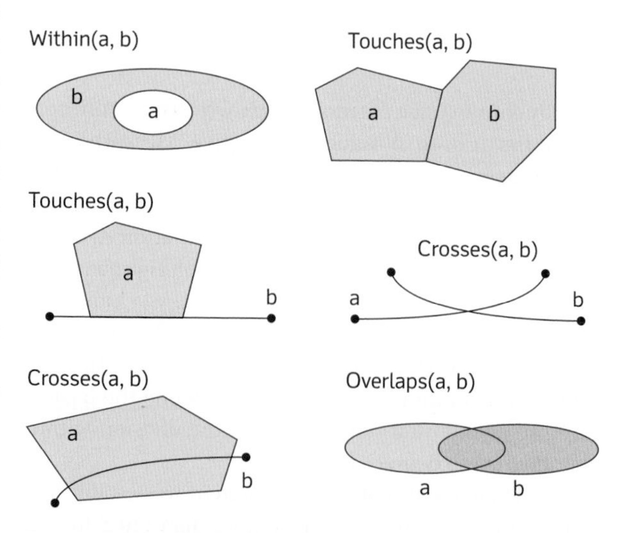

Figure 2.7 Topological relationships highlight connectedness and neighbourhoodness of geographical phenomena

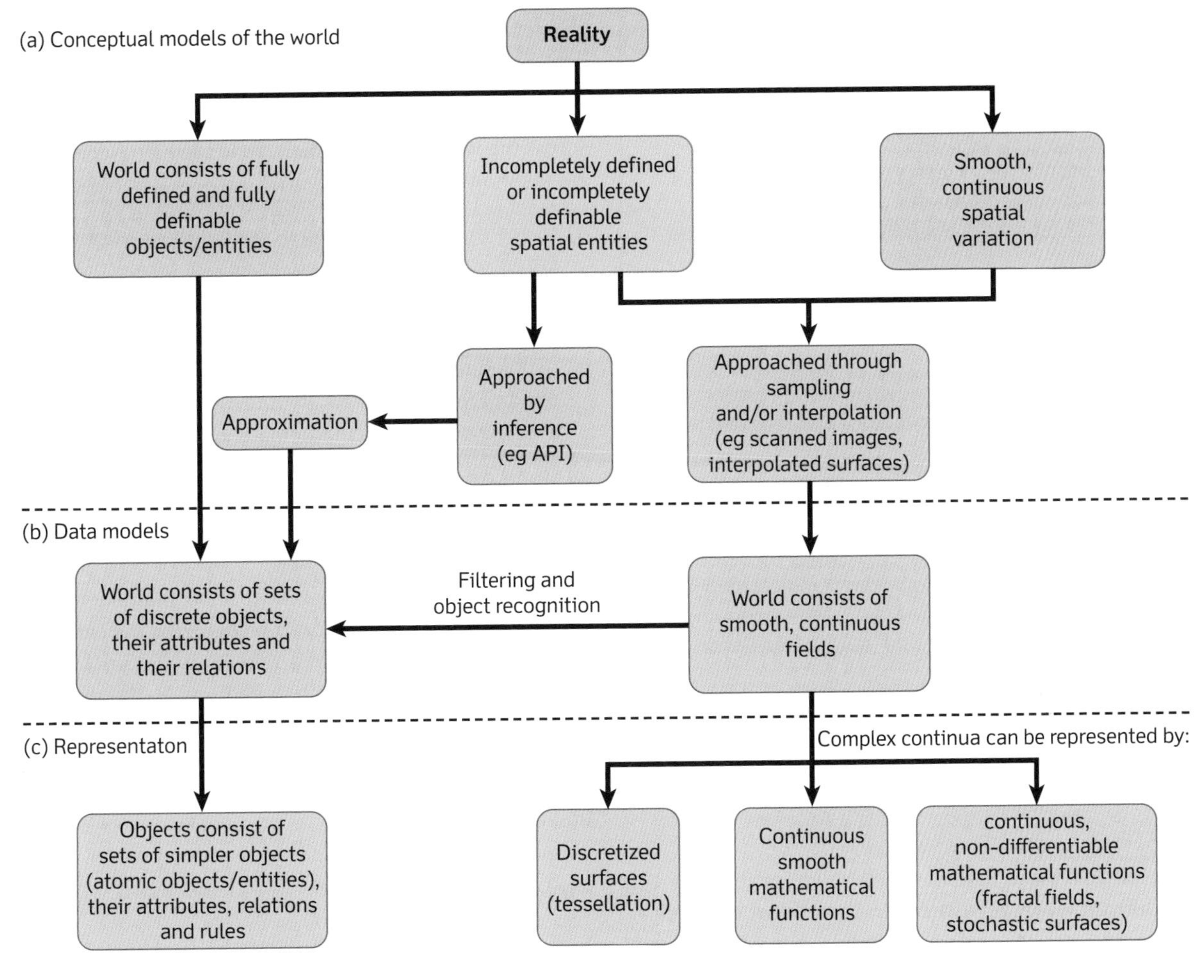

Figure 2.8 Steps in the process from observation of real-world phenomena to the creation of standardized data models

digital data in a GIS does not depend in the first instance on technology, but more on the appropriateness or otherwise of the conceptual models used to portray space and represent features and spatial interactions in this area.

Following Peuquet (1984), the creation of geographical data sets involves seven levels of structuring, development, and abstraction, going from the real world to the final information product from a GIS, as given in Box 2.1.

2.4 The human view of real-world geographical phenomena

Starting with the first level of abstraction, imagine that you are talking to someone on the telephone and they ask you to describe the view from your window. How would you depict the variations you see? Would you be likely to break down the landscape into units such as buildings, roads, fields, valleys, or hills? You would also probably use geographical prepositions to describe relationships between units and for referencing, such as 'beside', 'to the left of', or 'crossing'. You have, in fact, developed a structured abstraction of the landscape using relative concepts of space.

Your perceptions of the real world would be quite different if you were viewing the scene from on top of a hill or building, or if you had walked, driven, or flown over the environment. It would also be different if you were describing the scene captured by a photograph or video. Your experience of the environment, your cultural background, and that of the person you are describing the scene to, will influence your interpretation of the features observed and of what you decide to ignore.

Once the various geographical phenomena have been defined in terms of their characteristics or properties, describing the location and geometry, particularly

Box 2.1 The creation of analogue and digital spatial data sets involves seven levels of model development and abstraction

a) A human's view of reality (**conceptual model**). Human conceptualization turns this into an abstracted model of reality (creates an **analogue** model).

b) A formalization of the analogue abstraction without any conventions or restrictions on implementation (spatial data model).

c) A representation of the data model that reflects how the data need to be formulated and recorded in the computer (database model).

d) The representation of the data structure within the constraints of a file structure in the computer memory (physical computational model).

e) The handling of the digital geographical data is then structured by further accepted axioms and rules for handling the data (data manipulation model).

f) The display and presentation of data to people is structured by accepted rules and procedures (graphical model).

of explicit entities (such as 'hill', 'town', or 'beach'), can be difficult, especially where an exact form is not clear. Defining a house is relatively simple, but where exactly is the boundary of a city? A human might recognize the different soils, but where do you draw the boundary between them? For most human thoughts location is defined in relative terms, and the boundaries and geometry of features are more fuzzy than clearly defined.

2.5 Conceptual models of space: entities or fields

Is the geographic world a jig-saw puzzle of polygons, or a club-sandwich of data layers?

Couclelis 1992

This personal view of the geographical world may be formalized so that commonly used representations of space and spatial properties are used to share our understandings of what we see more unambiguously, effectively, and efficiently, especially given the vast amounts of geographical data being collected and exchanged today across the world. The two main structuring models used are (as shown in Figure 2.9):

a) to see the space as being occupied by *entities* which are described by their attributes or properties, and whose position can be mapped using a geometric co-ordinate system

b) to imagine that an attribute of interest occurs throughout an area and that its attributes vary over the space as some continuous mathematical function or *field*.

Entities

The most common view is that space is populated with 'objects' (entities). Defining and recognizing the entity (is it a house, a cable, a forest, a river, a mountain?) is the first step; listing its attributes, defining its boundaries and its location is the second. Entities are thus unique, non-divisible building blocks that have various properties. This is the most often used human conceptualization model.

Fields

In the continuous field approach, the simplest conceptual model represents geographical space in terms of continuous Cartesian coordinates in two or three dimensions (or four if time is included). The frame dividing up the space defines the scale of detail to be represented. The attribute is usually assumed to vary smoothly and continuously over that area, and the values of the phenomena (e.g. air pressure, temperature, elevation above sea level, clay content of the soil) and their spatial variations are considered first. Each individual cell is a unique item and it is only when there are remarkable clusters of like attribute values in geographical space or time (as with hurricanes, mountain peaks, or 'significant events') will these be recognized as zones of something—a thing (e.g. the Matterhorn, the Gulf Stream, or a clay layer rich in a particular element).

Choosing between an entity model or a continuous field approach can be difficult when the entities may also be seen as sets of extreme attribute values clustered in geographical space. Should one recognize Switzerland, for example, as a land of individual mountain entities (Mont Blanc, Eiger, or Matterhorn) or as a land in which the attribute 'elevation' demonstrates extreme variation?

Figure 2.9 Entity and field portrayals of geographical phenomena represent different ways of representing the real world with (a) the points, lines and polygons that define the individual entities in the centre of Washington DC, overlaying a satellite image of the area (Copyright Google Imagery, 2015). (b) Elevation data at a 50 m resolution for a section of the River Nile at Qena, Egypt, derived from NASA's Shuttle Radar Topography Mission (Source: United States Geololgical Survey http://srtm.usgs.gov/)

In practice, a pragmatic solution based on the aims of the user of the database must be made.

The choice of conceptual model thus determines how we first represent the geographical phenomena, but it also directly influences how information could be derived in subsequent modelling and analysis. Opting for an entity approach to mountain peaks will provide an excellent basis for a system that records who climbed the mountain and when, but it will not provide information for computing the slopes of its sides. Choosing a continuous representation allows the calculation of slopes as the first derivative of the surface (see Chapter 10), but does not give names to parts of the surface such as peaks or valleys.

As a gross oversimplification, the choice of an entity or a field approach may also depend on the scientific or technical discipline of the observer. Disciplines that focus on understanding spatial processes in the natural environment may be more likely to use the continuous field approach, while those in which work centres on human-constructed phenomena and interactions are more likely to view an area as a series of distinct units.

2.6 Geographical data models and geographical data primitives

The problem with conceptual models is that there are as many different models as there are people, as our different experiences mean we all view the world slightly differently. There are also times when we need an agreed definition such as a legal property boundary. We therefore use the formalized structures of geographical data models to ensure our abstractions are transferable and comprehensible to others. These vector and raster structures support and constrain our definition of the geographical phenomena we wish to capture. The result is a world of smooth, continuous tessellating units, or a set of discrete point, line, or polygon objects, each with their own attributes and relations.

In the next step, geographical data models are used as the formalized equivalent of the conceptual models. They structure how spatial data is broken up into parts for analysis and communication, and assume that phenomena can be uniquely identified, that attributes can be measured or specified, and that their geographical location can be registered through coordinates or other discrete descriptors.

The fundamental primitive of the geographical data model is therefore the tuple: (x, y, z, t, U), where phenomenon U (e.g. crop type, elevation, or town) is located at space-time coordinates (x, y, z, t) (position on the earth's surface at a particular time).

As with the conceptual models, two main geographical data models are used:

- vector data model of entities
- tessellations of fields.

Vector data model of entities

Most human phenomena (houses, land parcels, administrative units, roads, cables, pipelines, agricultural fields in western agriculture) can be handled best using the entity approach. The simplest and most frequently used

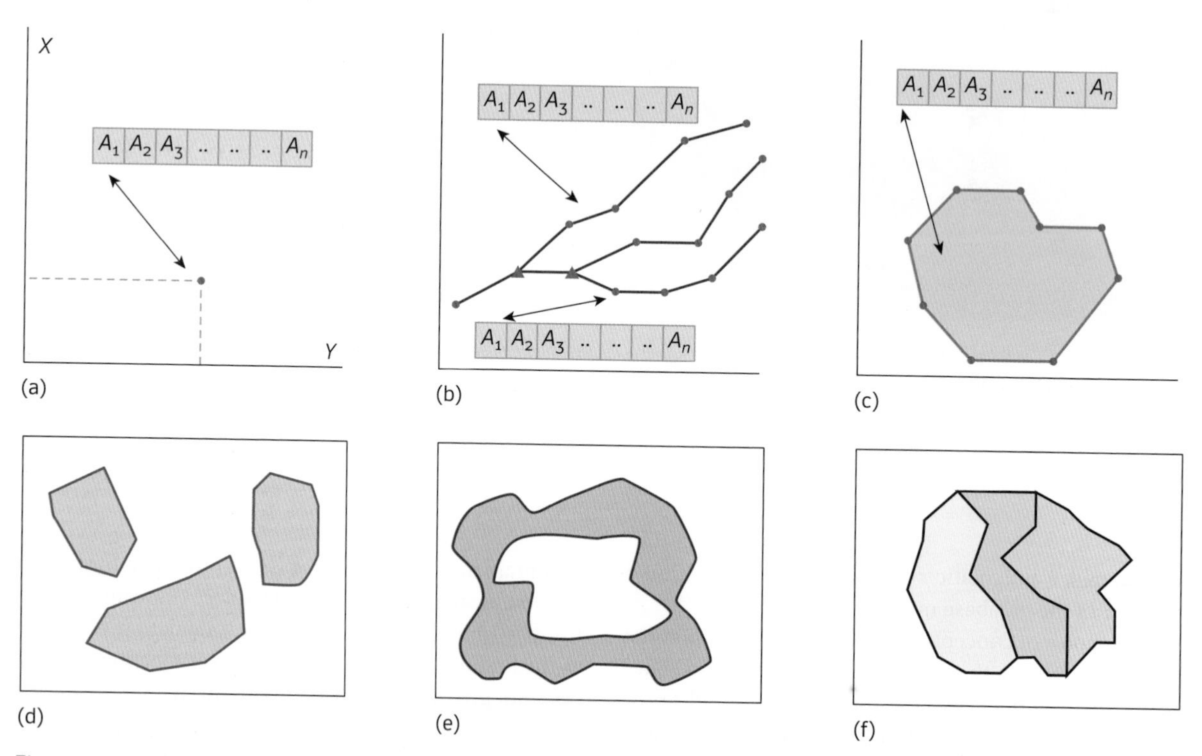

Figure 2.10 The fundamental geographical primitives: (a) point, (b) line, and (c) polygon, where $A_{1,2}$ is the value of different attributes such as tree species and tree height, and various polygon types: (d) discrete, (e) island, and (f) contiguous

data model of reality is of a basic spatial entity that is further specified by its attributes and its geographical location (see Table 2.3).

This can be further subdivided according to one of the three basic geographical data primitives, namely a '**point**', a '**line**', or an 'area' (which is most usually known as a '**polygon**' in GIS), which are shown in Figure 2.10. These are the fundamental units and they may take various forms depending on the nature of the information included in their definition, as summarized in Table 2.1 and illustrated in Figure 2.10a, b, and c. The series of discrete entity-defined point, line, or polygon units are geographically referenced by Cartesian coordinates; some further details of, for instance, connectedness, or what is near to an entity, can also be included.

Simple points, lines, and polygons

Simple point, line, and polygon entities are essentially static representations of phenomena in terms of XY coordinates. They are supposed to be unchanging, and do not contain any information about temporal or spatial variability. A point entity implies that the geographical extents of the object are limited to a location that can be specified by one set of XY coordinates at the abstraction level of resolution.

A line entity implies that the geographical extents of the object may be adequately represented by sets of XY coordinate pairs that define a connected path through space, but one that has no true width unless specified in terms of an attached attribute. A line adequately represents a road at national level; at street level it becomes an area of paving and the line representation is unrealistic. A telephone cable, on the other hand, can be represented as a line at most practical levels of resolution used in GIS.

For a polygon, the simplest definition is that it is a homogeneous representation of a two-dimensional space. The polygon can be represented in terms of the XY coordinates of its boundary, or in terms of the set of XY coordinates that are enclosed by that boundary. Polygons may contain holes, they generally have direct neighbours, and different polygons with the same characteristics can occur at different locations (see Figure 2.10d, e, and f).

If the boundary can be clearly identified, it can be specified in terms of a linked list of a limited number of XY coordinates; the size of the set of included XY coordinates depends on the level of spatial resolution (detail) that is used. Very often it is assumed that the level of detail is given by the number of decimal places to which the coordinates or the boundary are specified, which means that encoding using boundaries is more efficient than listing all coordinates that are inside the boundary envelope.

The resolution of the data will influence how an entity is represented. A town could be represented by a point entity at a continental level, with increases in the level of resolution revealing internal structure in the phenomenon (in the case of a town, subdistricts, suburbs, streets, houses, street lights, traffic signs) which may be important for some users such as urban planners and crime analysts.

Complex points, lines, polygons, and objects with functionality

More complex definitions of points, lines, and polygons can be used to capture the internal structure of an entity; these definitions may be functional or descriptive. An important addition is the inclusion of topology information. This is independent of the coordinate system and allows more complex relationships to be established between neighbouring objects. The main components of a topologically defined spatial database are illustrated in Figure 2.11.

In recent years, more highly structured ways of encapsulating entity data have been possible through the **object-oriented** approach, in which databases and programming tools are able to capture each object as the fundamental unit (for example McKinney and Cai 2002; Brown et al. 2005). Nested hierarchies and functional relations can be defined between related groups of entities that together form a single unit at a higher aggregation level. Object-orientation is described more fully in the context of data structures in Chapter 3.

Continuous data fields

For continuous data it is often useful to subdivide the field into a series of discrete spatial units, as given in Table 2.3 and shown in Figure 2.12. The tessellation can be based on regularly or irregularly shaped units depending on the nature of the phenomena and the analysis to be undertaken. It can be based on squares, triangles, or any other shape that tessellates. For example, changes in elevation may be represented using regular grid squares or irregular triangles with their shapes reflecting the nature of the terrain. The resulting tessellation is taken as a reasonable approximation of reality, at the level of resolution under consideration, and it is assumed that the operations such as differentiability that can be applied to continuous mathematical functions also apply to these discretized approximations. These different representations permit different types of analysis and modelling, as will be discussed in Chapters 10 and 11.

The most used is the regular tessellation or regular grid, known as the **raster data model**. The two-dimensional geometric surface is divided into pixels, whose size

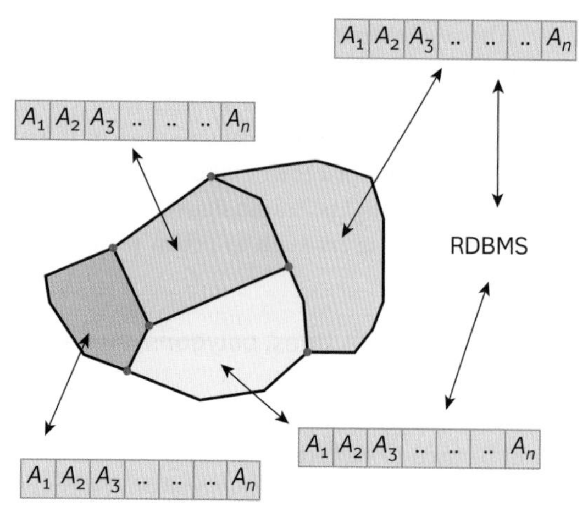

Vector implementation of entity data model with topology and attributes for each polygon

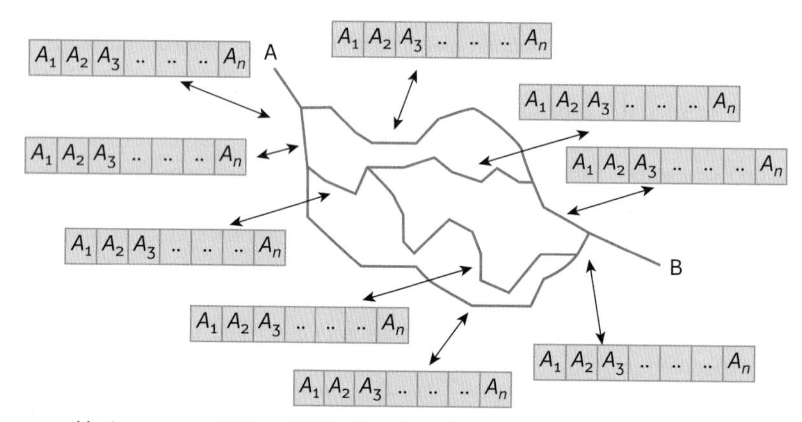

Vector representation of linked topology for modelling networks

Figure 2.11 The encoding of exact objects (entities) in different data models: (a) vector representation of crisp polygons; (b) vector representation of linked lines

is determined by the resolution required to represent the variation of an attribute for a given purpose. For each grid cell, a value is recorded for the phenomena such as plant type, soil class, or land cover. The size of the grid cell relative to the area on the ground—in other words the granularity—will determine how the phenomena are defined, with low-resolution data being most used for coarsely defined classes.

Table 2.3 Discrete data models for spatial data

Vector representation of exact entities	Tessellations of continuous fields
Non-topological structures (loose points and lines: 'spaghetti')	Regular triangular, square, or hexagonal grid Square pixels are known as a raster
Simple topology with linked lines—e.g. a drainage net or utility infrastructure	Irregular tessellation such as Thiessen polygons
Complex topology with linked lines and nested structures—e.g. linked polygons	Triangular irregular nets (TIN)
Complex topology of object orientation with internal structures and relations	Finite elements
Nested regular cells/quadtrees	Irregular nesting

When grid cells are used to represent the change in a continuously varying phenomenon, each will have a different attribute value; the variations between cells will be assumed to be mathematically continuous so that differential calculus may be used to compute local averages, rates of change, and so on. Each grid cell may be thought of as a separate entity that differs from vector polygons only in terms of its regular form and implicit rather than explicit delineation (Tobler 1995). Regular grid cells often form the basis for dynamic modelling (see Chapter 12) so capturing and representing environmental variations using this structure is often a first step in this form of analysis.

Although the regular grid is most often used to represent static phenomena, it can easily be adapted to deal with dynamic change. Changes over time may be recorded in separate layers of grid cells, one for each time step; so the change from the static to the dynamic data model only requires the basic structure to be repeated for each time step. Time, like space, is assumed to be discretized in this model. Regular tessellations may also be nested to provide more spatial detail through the use of linear and regional quadtrees and other nested structures (see Chapter 3).

Lateral changes over space may also be handled easily by the regular grid because of its approximation to a differentiable continuous surface. The flow of materials through space may be computed using **finite difference modelling** because the constant geometry of the cells means that by simple subtraction and addition, first and second order derivatives may be easily calculated. These modelling capabilities are increasingly available as additional modules for off-the-shelf GIS software.

With a continuously changing variable it is possible to use mathematical differentiation techniques to derive more information, such as a value for the rate of change of that entity. For example, attributes such as slope (rate of change of elevation with distance), aspect, plan, and profile curvature may be generated from triangular and grid cell representations of elevation. Chapters 9 and 10 give more details of this.

One of the commonly used tessellation structures is irregular triangles, which have long been used in land surveying. They are based on the principle of triangulation, in which the continuous surface of the land is approximated by a mesh of triangles whose apices or nodes are given by measured 'spot heights' at carefully located trigonometric points. A major advantage of this approach is that the density of the mesh can be easily adjusted according to the degree of variation in the surface—areas with little variation can be represented adequately by a few triangles, while areas with large variation require more. This effectively supports a variable resolution when storing the data.

The triangular surface can also easily accommodate variations in form as shown in three-dimensional representations of landform or other surfaces, or climate systems. These data models are essentially static representations of the hypsometric surface, which is supposed to be unchanging over time.

Triangular meshes are also used to represent continuous variation in the dynamic modelling of groundwater flows, surface water movement, wind fields, and so on. In this form they provide a structure for **finite element modelling**; the differences in attribute values between adjacent triangular cells that result from the flow of water or dissolved materials are modelled using differential calculus. The **triangular irregular network** or **TIN** is a data structure (see Chapter 3) used to model continuous surfaces in terms of a simplified polygonal vector schema (Figure 2.12b).

Representing three-dimensional entities and continuous fields

Representing three-dimensional data is more complex, as phenomena may be solids having the properties of volume, surface area, and mass. The main data models

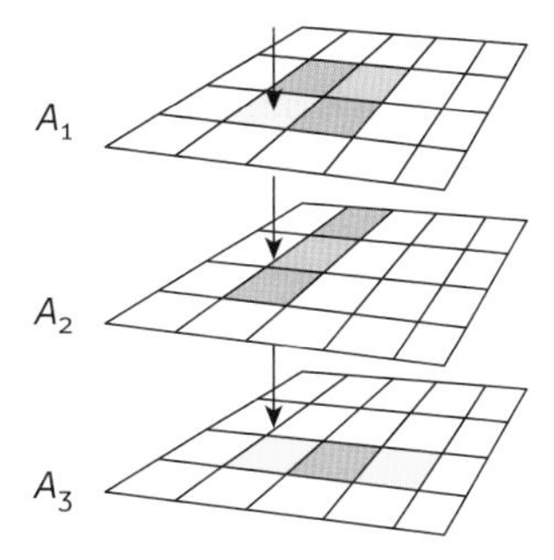

Regular tessellation with square cells (raster) with a separate layer for each attribute

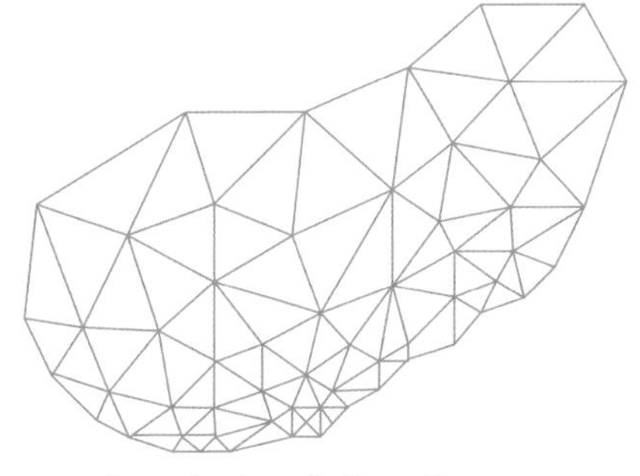

Irregular tessellation with triangular elements

Figure 2.12 The encoding of continuous fields in different data models: (a) raster model of continuous fields; (b) Delauney triangulation network of a continuous field

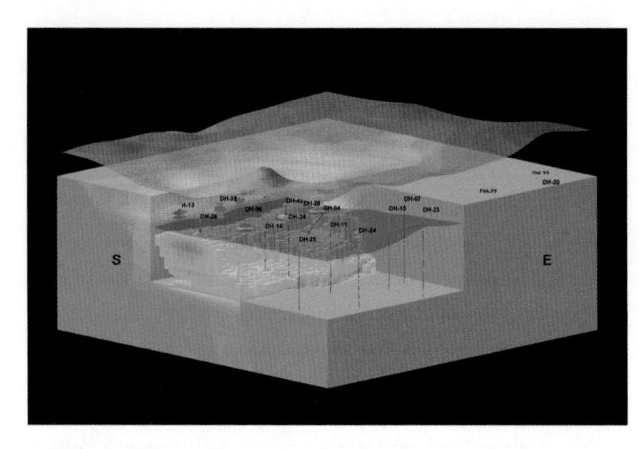

Figure 2.13 Three-dimensional spatial data models for a regular-shaped voxel
(Copyright 2014, RockWare, Inc., Golden, Colorado, USA)

used for three-dimensional data are extended vector and raster model equivalents.

The raster solution, known as the **voxel** (volume element), is based on the subdivision of the three-dimensional space into units (see Figure 2.13). The union of adjoining cells, where the cell is a primitive regular or irregular shape, defines solid objects. Some applications use rectangular or parallel-piped tessellations as an alternative to regular cube discretization.

With the vector equivalent, each entity is defined in terms of X, Y, and Z coordinate space using points, lines, and solids. The faces, edges, and vertices of their bounding surface and their topological relations are explicitly defined in the GIS, supporting the modelling of complex shapes. Attribute data are linked to the defined objects as in the two-dimensional model.

2.7 Overlap between the two geographic data models

The basic units of discretization in the regular tessellation of continuous space may also be used to provide a geometrical reference for the simple data units of points, lines, and areas. A vector 'point' can be represented by a single cell; a vector 'line' by a set of contiguous cells one cell wide having the same attribute value; a vector polygon by a set of contiguous cells having the same attribute value. Vector representation used to be often preferred over tessellations such as raster because regular grid cells may lose spatial details, though this is becoming less of a problem with the increased power of computers and greater storage space. This means raster data can now be stored at sufficiently high resolution that the discretization produces only small locational or attribute error.

This effective equivalence between the vector and raster models of space sometimes causes confusion about the nature of the phenomena being represented. For example, elevation can be represented as continuous fields of elevation using triangular or square tessellated cells, or as a point grid, or as **isolines** or contours which are sets of XY coordinates linking sites of equal attribute value. Contours are useful ways of representing attribute values on paper maps and computer screens, but they are less efficient for handling continuous spatial variation in numerical models of spatial interactions. Contour envelopes may be treated as simple polygons or as closed lines in terms of the entity approach, but they are merely artefacts of a terrain representation, not outlines of real-world 'objects'.

The overlap between the two approaches is often greatest when we deal with 'artificial' phenomena that do not exist as real-world features such as soil mapping units, or land use, or land cover units. The classic approach of conventional mapping is to define classes of soil, land use, land cover, etc. and then to identify areas of land (entities) that correspond to these classes. These areas can be represented by vector boundaries enclosing polygons or by sets of contiguous raster cells with the same value. This kind of representation is known as a **choropleth map**, because it contains zones of equal value. It may also be known as a **chorochromatic map** because each zone is displayed using a single colour or shading.

An alternative approach to representing such artificial (i.e. imposed) entities as land use or soil classes, is to postulate that land use or soil are continuous variables, not entities, but the geographical surface is made up of zones where the attributes have the same value (the polygons) and zones where the attribute values change abruptly (boundaries). This approach stems from the need to extract entities and homogeneous zones from grid-based data such as remotely sensed images.

Though the variation of these kinds of attributes may be thought of as being continuous in space (because every cell has a value for soil or land use, including classes for 'no soil' or 'waste land'), the surface is not continuous in terms of differential calculus. This means that mathematical operations like the calculation of slopes should not be applied to data that cannot be approximated by a continuous mathematical function.

In both two and three dimensions we can think of pixels (voxels) or polygons as units that can be treated as a series of open systems with regular or irregular form. The state of each cell (the local system) is determined by the value of its attributes; attribute values can be changed by operations that refer only to the cell in question, or that use information from other cells that are in one way or another part of the cell's surroundings. The difference between this approach for cells and the approach for polygons is not great (Tobler 1995); only the variable geometry of the vector representation of entities means that computing neighbourhood interactions is much more complex than with regular cell structures.

2.8 Representation changes with scale—granularity, generalization, and hierarchies

The spatial scale at which a study takes place influences how phenomena are defined. For example, a town is shown as a point on a coarse-resolution map but as a complex-shaped area on a large-scale one. Accounting for and capturing the differences in detail possible at the different scales—referred to as different levels of **granularity**—requires varying sized building blocks to be used. At coarse levels of granularity, a low level of detail is given using approximations for geometric shapes, whereas at fine grain small units with a high level of detail are used. For example, a river might be defined in the vector model using a line, but for fine granularity mapping it may be represented as a thin, irregular-shaped polygon with associated channel features such as berms shown as smaller multisided areas or as points within it (as shown in Figure 2.14). In the field model, the river might be represented by single contiguous grid cells at coarse granularity, but as varying numbers in fine detail representing the changing shape and dimensions of the channel.

Moving between different scales influences the accuracy of the representation relative to the phenomena on the ground. Changing to a coarser granularity means many of the complex details are simplified, merged, or aggregated while others need to be exaggerated or enhanced to maintain their distinctive characteristics. In the encoded data these differences are referred to as generalization, and they lead to changes in both the geometry and the attributes of the phenomena at different scales.

Sometimes, the way in which a feature varies over different scales can be preserved through the use of **hierarchical structuring**, with phenomena grouped or divided into units at different levels of spatial resolution. Hierarchies are established which explicitly link the same phenomena across the different spatial-scaled definitions. For example, a river may at one level be divided into individual streams and tributaries, while at a finer level units are used for the reaches, or shorter sections, to capture the in-channel form and habits. The various levels are linked to define the river, but users may query the data at different levels depending on the detail required. This form of structuring might yield multiple representations of the same phenomena with different

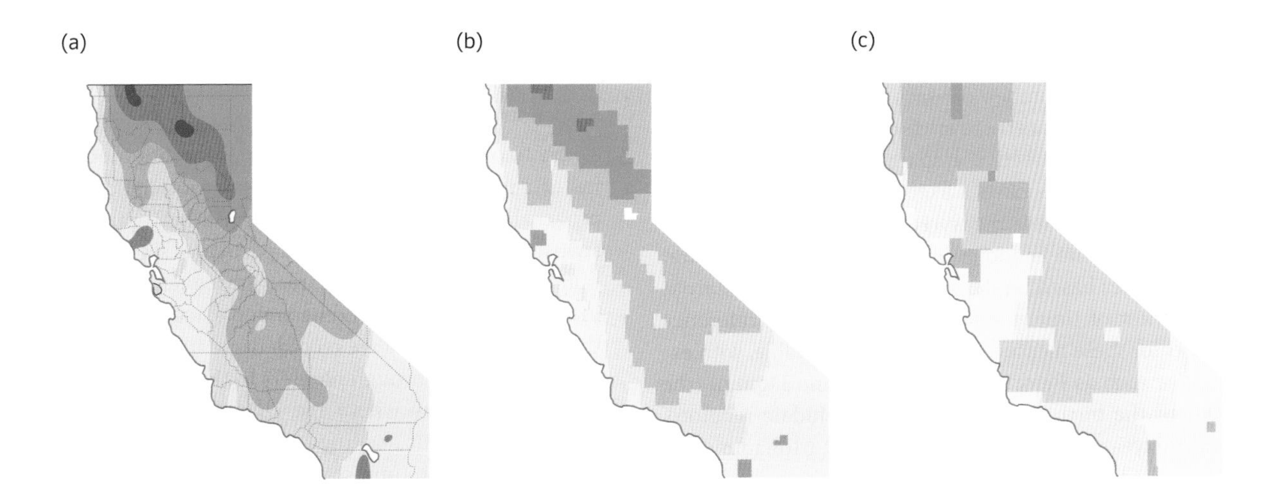

Figure 2.14 Moving between different levels of resolution brings varying levels of detail in the representation of phenomena

levels of detail recorded for the feature at each level of the hierarchy.

Hierarchies are also central to many definitions of spatial phenomena based on their attributes. The most common systems used for taxonomic classifications of soils, plants, or animals employ a nested approach where the units at the apex of the hierarchy contain and are composed of all lower division units. Thus an area or point will be in different classes based on increasingly generalized attribute definitions in the classes higher up the taxonomy. The alternative, non-nested hierarchies relax the requirement for containment of lower by higher units and are seen as a more appropriate structure where macroscopic properties of the phenomena may emerge.

2.9 Representing changes in time with geographic data models

So far we have considered a static view of geographic data—that phenomena with certain attribute values are located at particular positions. However, many geographical objects, particularly in the natural world, are continuously changing over space and time, either in location or shape, or new attribute values occurring, such as nitrate levels in a river, air temperature, and soil salinity. Some changes come about as the result of an event which has a duration, such as changes in the boundaries of a beach in an area of active erosion after a storm event, or they may result from a sudden event such as a change in the ownership of a house following a sale, or the planting of a crop grown in a field in a given season. With rapid advances in real-time location data, space-time data modelling is ever more crucial if the richness of the new information is to be exploited (Janelle 2012).

There are a number of ways to conceptualize and measure time, just as there are for space. The Newtonian view of time is directly comparable to its concepts of space, in that there exists a fixed dimension along which events or objects are measured, using units ranging in granularity from millions of years through solar and moon cycles based on days and months, to the international unit of time, the second. The starting reference points for these measurements are also variable, with examples such as Before Present (BP), Anno Domini (the year of the birth of Christ in Christo-Judaic religion), and the year of the birth of the Prophet Mohammed in Islam used in various calendars.

A different conceptual framework is that time itself does not exist but that it is a construct used by humans to define the dynamics of things and events in relation to others. This concept is, however, little used in most geographical data records and it is absolute terms that are almost exclusively used.

The basic tuple for geographical data which includes the time dimension is *x, y, attribute, time.* As time changes, t_1, t_2, etc. variations for any one or combination of the following may occur:

- location values
- attribute values
- topology between entities.

To capture this dynamism in geographical data, various spatio-temporal data models have been put forward by researchers over the last two decades. The following are the most common ideas, which have been used for both field and entity representations of space (shown in Figure 2.13) (Langran 1992; Frihida 2002; Worboys 2005; Bothwell and Yuan 2012; Janelle 2012; Long and Nelson 2012):

1. One of the most enduring frameworks is based on the concepts of a space-time cube in which each item follows a path in these three dimensions based on what state it is in and where it is at a particular time. The cube can be used for both entity-based models and with continuously changing field data such as atmospheric pressure.
2. The most commonly adopted framework in GIS today is of a series of sequential time shots in which the nature of an object is recorded and it is time-stamped. The frequency of the timing of the different records might be at regular or irregular intervals. Some of the drawbacks of this model are that there is redundancy in the records if the state of the phenomena remains the same, and there is no explicit definition of the relationship or degree of change between two different snapshots.
3. To try to overcome the limits of the snapshot model, raster and vector versions of an event-oriented approach have been explored. In this, a base map or snapshot of the initial state of a particular set of phenomena are recorded for an area. Subsequent changes in the state of these phenomena in particular parts of the geographical area are recorded and time-stamped and linked back through pointers to the original base map. These show where the change has occurred and when, without needing to record the state of the whole area at each time-stamped event.
4. Some of the most complex spatio-temporal modelling is used to capture the movement and interaction

of an entity or agent across a continuously changing landscape. Two types of scenario might occur. The first is where the agent moving over a landscape might be affected by it, but the environment remains unchanged. The second would be where the landscape is dynamic and interacts with the agent and is affected by it. In data modelling, this involves both a fixed entity that changes location or state over time and a continuous field whose state either remains constant or changes over time. This requires an event or snapshot approach in which interactions and reactions are captured between two spatial data sets. Usually a fixed time-stepped approach is adopted along with complex modelling, involving specialized agent-based spatial modelling software.

The field of research of spatio-temporal data modelling continues to be explored with ideas centred on objects and other models for unifying concepts (Goodchild 2007b) (discussed in more detail in Chapter 3). The addition of time, and with it dynamics, in processes and in the state of phenomena brings many complications that have not been resolved into commercial GIS systems to date. The implications for spatial database models are explored in Chapter 3.

2.10 Data modelling and spatial analysis

From the examination of geographical data so far, it becomes apparent that there are direct links between the data model, the fundamental axioms, the data type used to represent a geographical phenomenon, and the kinds of analysis that can be carried out with it. The following different situations help to illustrate this:

1. If the location and form of the entity is unchanging and needs to be known accurately, but the attributes can be changed to reflect differences in its state caused by inputs of new data or output from a numerical model, then the vector representation of the entity model is appropriate. This is the most common situation in conventional GIS.
2. If the attributes are fixed, but the entity may change form or shape but not position, as for instance if a lake dries up, then a vector model requires a redefinition of the boundary every time the area of the lake changes. A raster model of a continuous field, however, would treat the variation of the water surface as a response surface to a driving process so that the extents of the lake could be followed continuously.
3. If the attributes can vary and the entity can change position but not form, or its separate parts are linked together, the behaviour can be well described using a hierarchical approach where information can be passed from one level of the model to another.
4. If no clear entities can be discerned, then it is often preferable to treat the phenomenon as a discretized, continuous field.

2.11 Examples of the use of data models

Given the different possibilities for data models, collectors and users of geographical data need to make decisions on the choice of model to be adopted every day. The following nine, not necessarily independent, questions concerning geographical data are of fundamental importance when choosing data models, and database approaches for any given application:

1. Is the real-world situation/phenomenon under study simple or complex?
2. Are the kinds of entities used to describe the situation/phenomena detailed or generalized?
3. Is the data type used to record attributes Boolean, nominal, ordinal, integer, scalar, or topological?
4. Do the entities in the database represent objects that can be described exactly, or are these objects complex and possibly somewhat vague? Are their properties exact, deterministic, or stochastic?
5. Do the database entities represent discrete physical things or continuous fields?
6. Are the attributes of the database entities obtained by complete enumeration or by sampling?
7. Will the database be used for descriptive, administrative, or analytical purposes?
8. Will the users require logical, empirical, or process-based models to derive new information from the database and hence make inferences about the real world?
9. Is the process under consideration static or dynamic?

The following examples illustrate the importance of an understanding of these ideas.

Cadastre

The main aim of the cadastre or land registry is to provide a record of the division and ownership of land. The important issues are the location, area, and extent of the land in question and its attributes (such as the name and address of the owner), the address of the parcel in question, and information about transactions and legal matters. In this case the exact entity (vector) model works well, using nominal, integer, and scalar data types to record the attributes and scalar data types for the coordinates.

Given that land ownership is fundamental to many legal and economic activities, these registries are often highly organized. Parcel boundaries are surveyed with high accuracy to reduce the chance of disputes, so the assumption of a data model in which the parcel is bound by infinitely thin lines is a good approximation to what the land registry is trying to achieve (see Figure 2.15). The coordinates of these boundaries are located accurately with respect to a national or local reference and the attributes of the entity in the database are simply the properties associated with the parcel. An essential aspect of the polygon representation is that adjacent parcels may share boundaries. This saves double representation in the database (see Chapter 3) and links the boundaries into a topologically sound polygon net which can handle both adjacency and inclusions.

Water supply networks

Given that the water supply network for most countries is hidden below ground, information on the location and state of the pipe and pump infrastructure is crucial to the supplier. Three aspects of these networks are important when recording these phenomena:

1. the attributes of a given network (how much it carries, what kind of pipe, additional information on materials used, age, name of contractor who installed it, and so on)
2. the location of the network (so that people digging in the street will not damage it and so it can be found quickly when needing repair)
3. information on how different parts of the network are connected together.

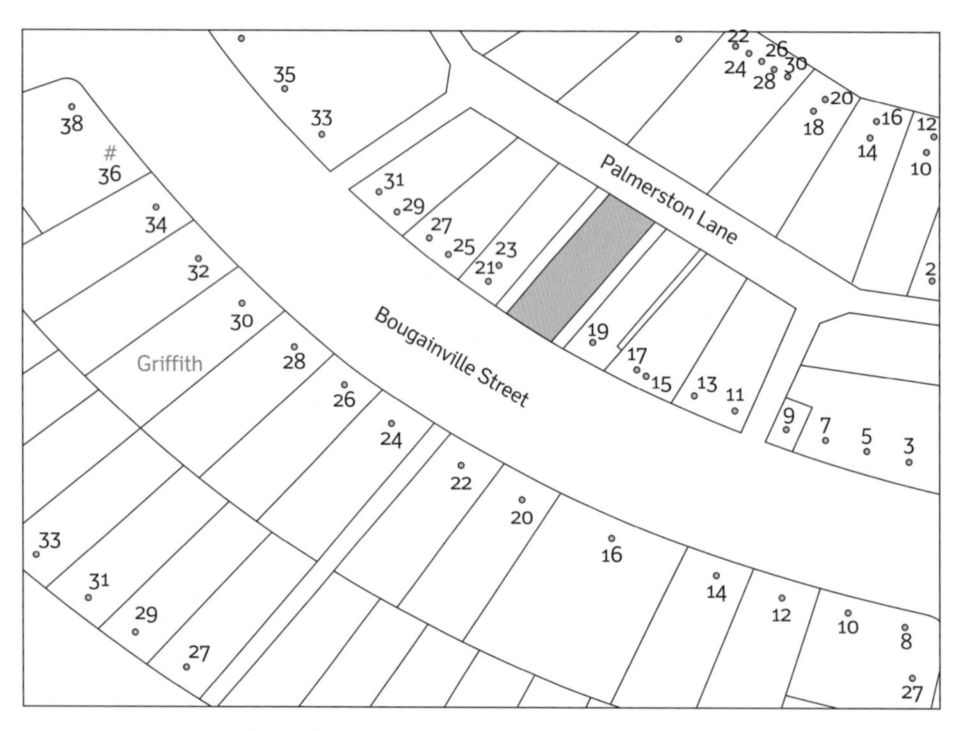

Figure 2.15 Map of parcel boundaries for part of Canberra, Australian Capital Territory, at a scale of 1:1 250 (© PSMA Australia)

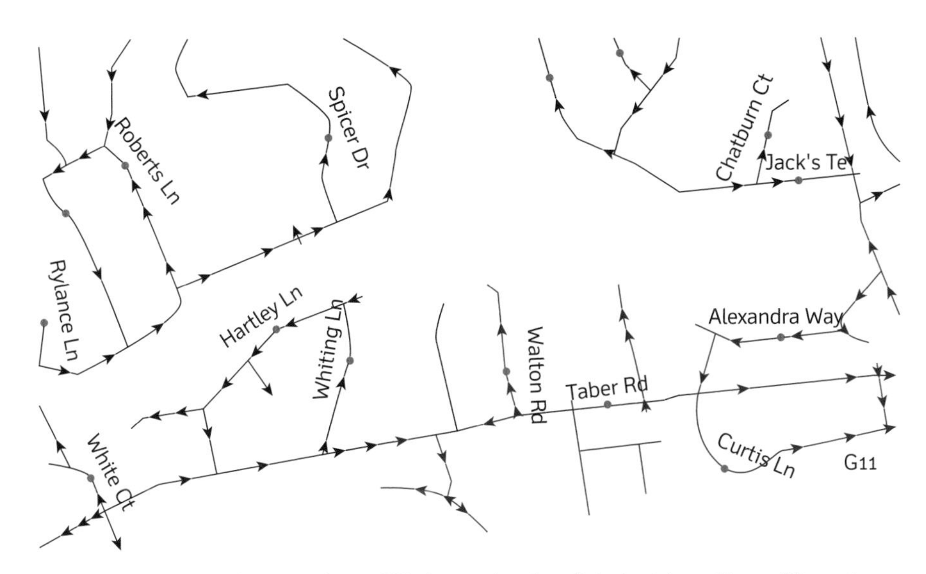

Figure 2.16 Water pipeline records in a GIS, showing topology linked points and lines of the system

Clearly all these requirements can be incorporated in a data model of topologically connected lines (entities) that are described by attributes (Figure 2.16), using a variety of data types from alphanumeric for the unique reference for the section of pipe to images of the items.

Land cover databases

National and international governments are interested in the division of the landscape according to classes of land cover—urban areas, arable crops, grassland, forest, water bodies, coasts, and mountains—for many aspects of urban and rural planning. This can be difficult, as different people or organizations might have different reasons for allocating land to different classes, even if the same number of classes is used and the central concepts of the classes are similar. Creating a data model for such an application requires several steps. Firstly it is necessary to define exactly what is characterized in each of the classes to avoid any ambiguity, overlap, or gaps. Secondly one needs to decide how to recognize them, and thirdly, one must choose a survey methodology (such as a point sample survey or a remote sensing scanner in a satellite) to collect surrogate data which are then interpreted to produce the result desired.

The simplest data model assumes that the classes are crisp and mutually exclusive and that there is a direct relation between the class and its location on the ground. If this is acceptable, then one can use the simple polygon primitives as a model for each occurrence of each class. The result is the well-known choropleth, or, more correctly, chorochromatic map. The data types used will range from nominal (for recording names of classes) to scalar (for computing and recording areas).

If we use remotely sensed data to identify land cover we automatically work with a gridded tessellation of continuous space, because that is how the satellite scanner works. The resolution of our spatial information is limited by the spatial resolution of the scanner. Unlike the case with sampling, we do have complete cover of the area so the information present in each pixel is of equal quality (although cloud cover and other factors may have an impact), which is not the case with interpolation. The major problem with identifying land cover with remotely sensed data is to convert the measurements of reflected radiation for each pixel into a prediction that a given land cover class dominates that cell. Obviously the success of the quality of the data depends on the quality of the classification process. Figure 2.17 shows a land cover map for an area of Tunisia derived from NASA's Landsat Thematic Mapper (TM) data.

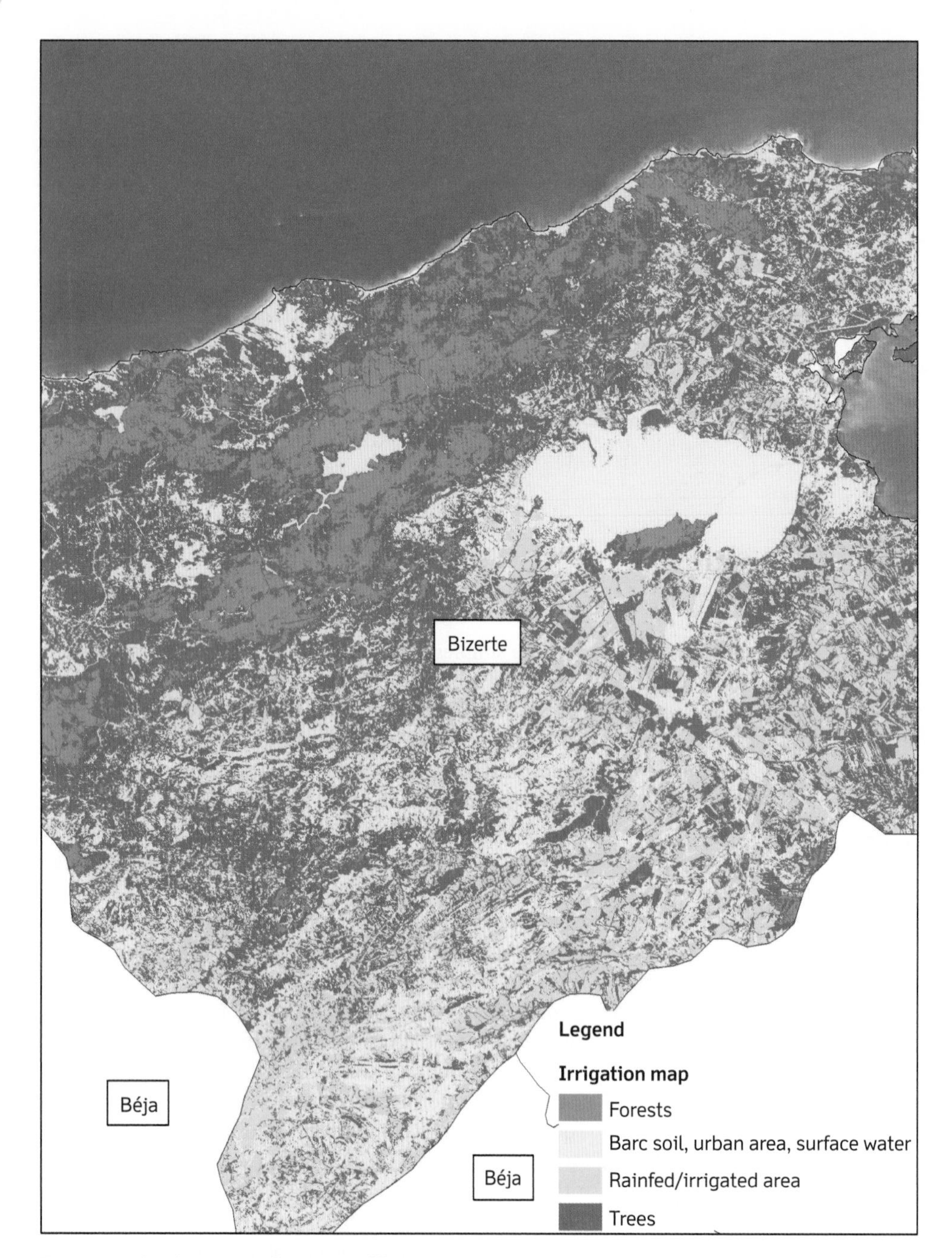

Figure 2.17 Land cover map for an area of Tunisia
(*Source*: International Center for Biosaline Agriculture, ICBA Dubai)

2.12 Summary

The steps needed to go from a perception of reality to a set of data models that can be used in a computerized information system are summarized in Figure 2.8. In some applications, the decision to opt for an entity-based approach or a field-based approach may be clear-cut. In others it may be a matter of opinion depending on the aims of the user. Both the entity and tessellation models assume that the phenomena can be specified exactly in terms of both their attributes and spatial position. In practice, there will be some situations where it is clear which is the more acceptable representation of reality, but there will be many others where this is uncertain and we must choose pragmatically based on the most common data type in an analysis or by the analysis to be undertaken.

In most GIS, all locational data and attribute values are deemed to be exact. Everything is supposed to be known so there is no room for uncertainty. Very often, uncertainty occurs not because we are unable to cope with statistical uncertainty, but rather because it costs too much to collect or process the data needed to give us the information about the error bands that should be associated with each attribute for each data unit. But in principle there is no reason why information on quality cannot be added to the data.

In essence, the intellectual level of the simple, crisp entity models of spatial phenomena is little different from that of children's plastic building blocks. These toys obey the basic axioms of information systems, including the one which says that it is possible to create a wide variety of derived objects by combining various blocks in different ways. Logically this is no different from combining sets of points, lines, and areas from a GIS to make a new map. Given enough bricks, one can build houses, recreate landscapes, or even construct life-sized models of animals like giraffes and elephants.

And this is the point. The giraffe built out of plastic blocks can be the size, the colour, and the shape of a real giraffe, but the model does not, and cannot, have the functions of a giraffe. It cannot walk, eat, sleep, procreate, or breathe because the basic units it is built out of (the blocks or database elements) are not capable of supporting these functions. While this is a trivial example, the same point can be made for many database units that are used to supply geographical data to drive analytical or process-oriented models. No amount of data processing can provide true functionality unless the basic units of the data models have been properly selected.

? Questions

1. Which data model would you choose in the following applications?
   a) Managing a road transport information system
   b) Organizing a night out with friends
   c) Controlling an emergency unit
   d) Monitoring an airborne pollution incident
   e) Understanding the impact of deforestation on a river system
   f) Predicting the incidence of famine
2. How do you think time and space can be represented in the same data model?
3. What map projections would you choose, and why, in studying the following?
   a) Transboundary rivers
   b) Global population growth in the last 20 years
   c) Changes in city areal extents in your country over time
   d) Flight distances from Los Angeles to major cities around the world

Further reading

▼ Date, C. J. (1995). ***An Introduction to Database Systems***. 6th edn. Addison-Wesley, Reading, MA.

▼ Heywood, I., Cornelius, S., and Carver, S. (2011). ***An introduction to Geographical Information Systems***. 4th edn. Pearson, Harlow.

▼ Hodgkiss, A. G. (1981). ***Understanding Maps***. Dawson, Folkstone.

▼ Hofmann-Wellenhof, B., Lichtenegger, H., and Wasle, E. (2008). ***GNSS—Global Navigation Satellite Systems: GPS, GLONASS, Galileo, and More***. Springer Verlag, Vienna.

▼ Kennedy, M. (1996). ***The Global Positioning System and GIS: An Introduction***. Ann Arbor Press, Chelsea, MI.

▼ Lillesand, T. M., Kiefer, R. W., and Chipman, J. W. (2008). ***Remote Sensing and Image Interpretation***. 6th edn. Wiley, Hoboken, NJ.

▼ Longley, P. A., Goodchild, M. F., Maguire, D. J., and Rhind, D. W. (2010). ***Geographic Information Systems and Science***. 3rd edn. Wiley, Hoboken, NJ.

▼ Mather, P. and Koch, M. (2011). ***Computer Processing of Remotely-Sensed Images: An Introduction***. 4th edn. Wiley-Blackwell, Oxford.

▼ Monmonier, M. (1993). ***How to Lie with Maps***. University of Chicago Press, Chicago.

▼ Snyder, J. P. (1997). ***Flattening the Earth: Two Thousand Years of Map Projections***. University of Chicago Press, Chicago.

▼ Wise, S. (2014). ***GIS Fundamentals***. 2nd edn. CRC Press, Boca Raton, FL.

▼ Worboys, M. F. and Duckham, M. (2004). ***GIS: A Computing Perspective***. CRC Press, Boca Raton, FL.

Geographical Data in the Computer

This chapter describes ways in which spatial data may be efficiently coded into a computer to support the operations of a GIS. Computers encode data and carry out instructions using a series of switches that exist in one of two states: 'on' or 'off'. Therefore geographical data need to be converted into discrete records using these switches to represent the location, presence or absence, type etc. of the phenomenon. In the computer, the data need to be organized to allow efficient accessing, retrieval, and manipulation. Systems for organizing data in the computer range from simple lists or indexed files, through to highly structured databases based on hierarchical, network, relational, and object-oriented schemata. The various combinations impose limitations on the data representation and handling, which affect the ways in which the data can be used for analysis and modelling. There is further organization of the way geographical data are held in a computer, with the adoption of either vector or raster data structures. In this chapter, we explore the implications of particular choices for GIS users.

Learning objectives

By the end of this chapter, you will:

- **understand the basics of how digital data are recorded and stored in a computer**
- **be familiar with different database management systems and how these influence the way geographical data are structured for computer storage and retrieval**
- **understand how the vector data model codes geographical data into points, lines, and polygons, and how this is stored in a computer**
- **have learned about the processes involved in recording geographical data using a raster data model and of the various techniques possible for reducing the computer storage space needed.**

3.1 Geographical data and computers

The conversion of geographical data into structured inputs for computer-based GIS software is an important step as this then forms the basis of further interrogation, analysis, and modelling operations, which is the power behind these systems. The computer provides a greater range of possibilities for recording, storing, retrieving, analysing, and displaying spatial data than a conventional map. Whereas the conventional paper map was a general purpose database created through a survey at a given point in time, the computer can select many different views of the data for any given purpose. The computer allows us to add and retrieve data, and to compute new information from existing data, so a computerized spatial database need not remain static, but may be used to model changes in spatial phenomena over time. These possibilities raise a number of questions about the nature of space, of spatial data, and how spatial phenomena should be described in terms of conceptual models.

For most users of GIS the aim is to extend the ability to handle spatial data in such a way that our view of the world is not limited by the constraints of a printed map or table of data, and we are able to extract subsets of spatial data for specific purposes, or link information on spatial phenomena to the processes governing their appearance and distribution. A computer provides the means to interact with data in ways that were impossible with printed information. Data can be changed, retrieved, recomputed, and displayed not just as a reflection of new information that has been collected, but also in response to numerical models of spatial and temporal processes. The GIS provides both an archive of spatial data in digital form and a tool for exploring the interactions between process and pattern in spatial and temporal phenomena.

When geographical data are entered into a computer it would be very convenient if the GIS recognized the same conceptual models that the user is accustomed to (as explained in Chapter 2). This would allow the interactions between the user and the computer to be both possible and intuitive. However, human perception of space is frequently not the most efficient way to structure a computer database, and does not account for the physical requirements of storing and repeatedly using digital information. Computers for handling geographical data therefore need to be programmed to represent the phenomenological structures in an appropriate manner, as well as allowing the information to be written and stored on magnetic devices in an accessible way.

As shown in Chapter 2, geographical data are more complicated than many other kinds of data handled routinely by modern information systems, such as in spreadsheets and databases, because they refer to the position, attributes, and the internal and external topological relationships of the phenomena recorded. It is these latter two that distinguish them from the administrative records handled by data-processing systems used for banking, libraries, airline bookings, or medical records. Therefore, as a result of the nature, volume, and complexity of geographical data, a number of logical computer schemata have been developed for efficient storage, updating, and retrieval. These schemata are a continuance of the model/structure development processes discussed in Chapter 2 and are stages (d) and (e) of Box 2.1.

3.2 Overview of data in computers

It is useful to first consider how computers represent and store data, as this influences the structuring involved. People represent numbers and do arithmetic using the decimal system with base 10. Each position in the written form of a number indicates how many units of the power of 10 correspond to that position. So the first column to the left of the decimal represents multiples of unity (10^0), the second multiples of 10 (10^1), the third multiples of 100 (10^2), and so on.

We could equally represent numbers using different base systems such as base 2 or base 16. For base 2, the first column represents unity, the second multiples of two (2^1), the third multiples of four (2^2), and the fourth multiples of eight (2^3). The resulting binary number would look different, but the count would be the same (as shown in Box 3.1). Computers use a base 2 counting system, reflecting the fact that data are stored in the form of arrays of switches that have only two states–they are 'on' or they are 'off'—and the numbers 1 and 0 code these states respectively. Therefore computers code data and do computations using **binary arithmetic**. Box 3.1 demonstrates the binary system and arithmetic.

As with decimal numbers, the data in the binary system are represented using a series of 'columns' and in a computer each column is used to represent a switch. The switches, which are really small magnetized areas on memory or storage devices, are grouped in packets of eight known as a **byte**. Subsequently data records and **words** are amalgamated together in the commonly used computer file.

With a byte of eight bits we can represent 256 numbers from 0 to (2^8 – 1)—i.e. from 0 to 255. For example, the sequence of eight switches 11111111_2 means:

$$2^7*1 + 2^6*1 + 2^5*1 + 2^4*1 + 2^3*1 + 2^2*1 + 2^1*1 + 2^0*1 = 255_{10}$$

Remember that any positive number raised to a zero power equals 1. The equivalent in base 10 would be:

$$10^2*2 + 10^1*5 + 10^0*5 = 255_{10}$$

and in base 16 would be:

$$16^1*15 + 16^0*15 = 255_{10}$$

If we combine 2 bytes in a 16-bit word it is possible to code numbers from 0 to 65 535. However, it is also useful to be able to code positive and negative numbers, so only the first 15 bits (counting from the right) are used for

Box 3.1 Binary numbers and arithmetic

The binary system is in base 2, where numbers count using 1 and 0. As with ordinary numbers (the decimal system), the data are represented using a series of 'columns' where the first counts in units of 2^0 (i.e. 0 or 1), the second counts in units of 2^1 (0 or 2), the third counts in units of 2^2 (0 or 4), and so on.

The binary sequence 0010 means $2^3 * 0 + 2^2 * 0 + 2^1 * 1 + 2^0 * 0 = 2_{10}$

The binary sequence 1100 means $2^3 * 1 + 2^2 * 1 + 2^1 * 0 + 2^0 * 0 = 12_{10}$

Binary arithmetic is similar to decimal arithmetic but there are some differences.

Addition and subtraction are similar:

$$14_{10} - 3_{10} = 11_{10} \text{ is } 1110_2 - 0011_2 = 1011_2$$

Multiplication proceeds by adding numbers together:

2310 * 410 is carried out as 23 + 23 + 23 + 23 = 92

The binary equivalent is $10111_2 + 10111_2 + 10111_2 + 10111_2 = 1011100_2$

Division proceeds by subtracting numbers (the reverse of the example above):

$$1011100_2 / 10111_2 = 100_2 \text{ (i.e. } 4_{10}\text{)}$$

Multiplying by 2 is quickly achieved by shifting the columns one place to the left:

2* 0101 is 1010

Division by 2 is done by shifting the columns one place to the right:

1110/2 = 0111

Of course, if the rightmost bit is a '1', division results in it being removed from the word so that simply multiplying by 2 (the left shift) again does not give the number you started with unless other checks are added.

coding the number and the 16th bit (2^{15}) is used to determine the sign. This means that with 16 bits we can code numbers from −32 767 to +32 767. A number lacking a fractional component is called an **integer**, so numbers in the range −32 767 to +32 767 are called 16-bit integers.

Frequently it is necessary to code numbers that are larger or smaller than −32 767 to +32 767, or numbers that have fractional components, so computer words with more bits are needed. Real numbers with positive and negative values and decimals require 32-bit or even 64-bit words (so-called **double precision**) in which some of the bits are reserved for the decimal part (the part smaller than 1) and the rest are used for larger numbers. This method of coding numbers means that the accuracy with which any number may be coded is a function of the number of bits used. This means numbers are not coded exactly, which may lead to rounding errors when the user attempts to work with numbers that are larger than those allowed. Chapter 12 discusses the problems of errors in GIS in more detail.

3.3 Database structures: data organization in the computer

Before considering in detail the ways in which geographical data may be stored efficiently in the computer, we must first consider the general issues of organizing data for optimal storage and access. A little knowledge on the basics of data-modelling and data-structuring methods will help in understanding how GIS work and their limitations and advantages.

Examples of the types of questions posed and the tasks performed by the database underlying a GIS include:

- Where are all occurrences of woodland? (requiring querying of attribute values in the land cover data field then reporting the spatial coordinates of these points/polygons either as a table or more usually as values on a map display).
- What is the total area of irrigated agriculture within 5 km of a point? (involving the generation of a subset of data based on their geographical reference values, then from that the definition of a group based on their attribute values for land cover, and finally the addition of the areas of the individual parcels).

These types of queries are much more complex than typical operations on non-spatial data, requiring the aggregation of data from multiple data files held in the database. Spatial databases contain many files with data on related aspects of the same entities, or data on entities that, because of their spatial proximity or connectivity, have to be linked or grouped together. The following section presents only a brief introduction but alerts readers to the ideas of how their data may be organized and physically structured in a computer. This is usually hidden from the users by interfaces which prompt them for data entry but do not show how this is then stored in a data file, and the file in turn organized in a database system.

Data file system tasks of ordering, sorting, and retrieval have evolved over time, and today computer programs designed to store and manage large amounts of data, called database management systems (DBMS), are central in GIS software. The basic unit of a file, the **data record** or **tuple**, contains all the relevant information for each entity, and depending on the kinds of data collected the data records may all be of the same length, or may be of variable length. Given the importance of efficient data retrieval to so many government and business enterprises, DBMS such as ORACLE, IBM's DB2, and Microsoft's SQL server are an important part of their IT systems. Open-source systems such as MySQL, Hadoop, MongoDB, and PostgreSQL have grown in the last few years, with many offering different data-structuring schema.

The aim of the DBMS is to make data quickly available to a multitude of users while still maintaining its integrity, to protect the data against deletion and corruption, and to facilitate the addition, removal, and updating of data as necessary. DBMS use many methods for efficiently organizing, storing, and retrieving data, but today most systems are based on one of three fundamental database models (relational, NoSQL, and object-oriented) with each providing a series of constructs which are used in organizing and describing the information. Accessing and querying the data is usually through a high-level application programming interface (API) or a query language such as **SQL (Structured Query Language)**. User interfaces support easy-to-use interrogation of databases, while high-level programming interfaces allow the database to be linked directly to application programs such as GIS.

Geographical data, however, present a special case for database modelling as each data item has both spatial and attribute information associated with it. To understand how this can be achieved in GIS, it is necessary to examine the fundamental principles behind the different database structures, and to see how they can be used for spatial information that has been recorded using either the exact entity or the continuous field data models.

The relational database model

Relational database models have been used in GIS for over 30 years and are now established as major tools for handling the attributes of spatial entities in many well-known commercial systems. In its simplest form, the relational database structure stores data in simple records, known as tuples, which are sets of fields each containing an attribute; tuples are grouped together in two-dimensional tables, known as relations (as shown in Figure 3.1). All records have identification codes that are used as unique keys to identify the records in each file (M, I, II, and a, b, c ... in Figure 3.1). Each table of rows and columns or relations is usually a separate file.

Data are extracted from a relational database by the user defining the relation that is appropriate for the query. This relation is not necessarily already present in the existing files, so the controlling program uses the methods of relational algebra to construct the new tables. These rules are often encoded in SQL—the Structured Query Language—(see Date 1995). Figure 3.1a shows that the relational structure of a simple map also contains much redundancy, and methods known as normalization are used to create more efficient coding. For example, addressing the lines as sets of straight line segments (**arcs**) simplifies the relational structure without loss of information (Figure 3.1b).

Relational databases have the great advantage that their structure is very flexible and can meet the demands of all queries that can be formulated using the rules of Boolean logic and of mathematical operations. This allows different kinds of data to be searched, combined, and compared. Addition or removal of data is easy too, because this just involves adding or removing a tuple, or even a whole table (as shown in Figure 3.2). Querying across different relational tables is made by joining them through common fields (shown in Figure 3.2). This is good for situations where all records have the same number of attributes and there is no natural hierarchy. However, where the relationships between tables are complex and a number of joins are needed, operations take longer.

There are some difficulties with using relational database management systems (RDBMS). The structuring of data across a series of tables does not fit all data, especially when the relations between various data groups are hierarchical. There can also be difficulties when the need to alter or add new data requires a different structure in the RDBMS. Further limits come in scaling across multiple horizontal platforms which, given the exponential increase in new geographical data available through social media (discussed in detail in Chapter 4), becomes ever more pertinent. However, these limits notwithstanding, RDBMS are by far the most used database systems used in current GIS and their proven robustness in so many applications will ensure they continue to be a commonly used database structure.

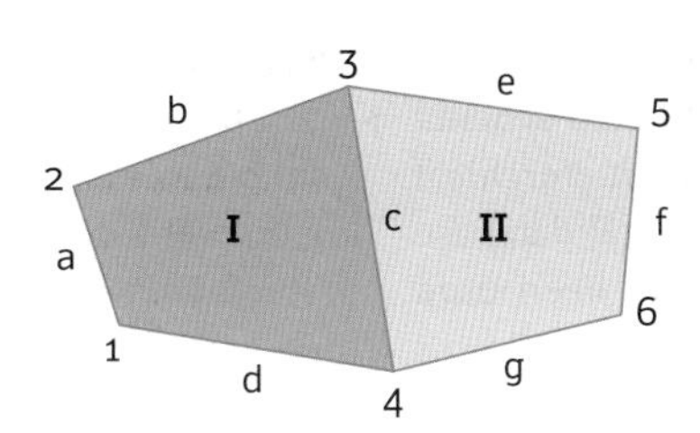

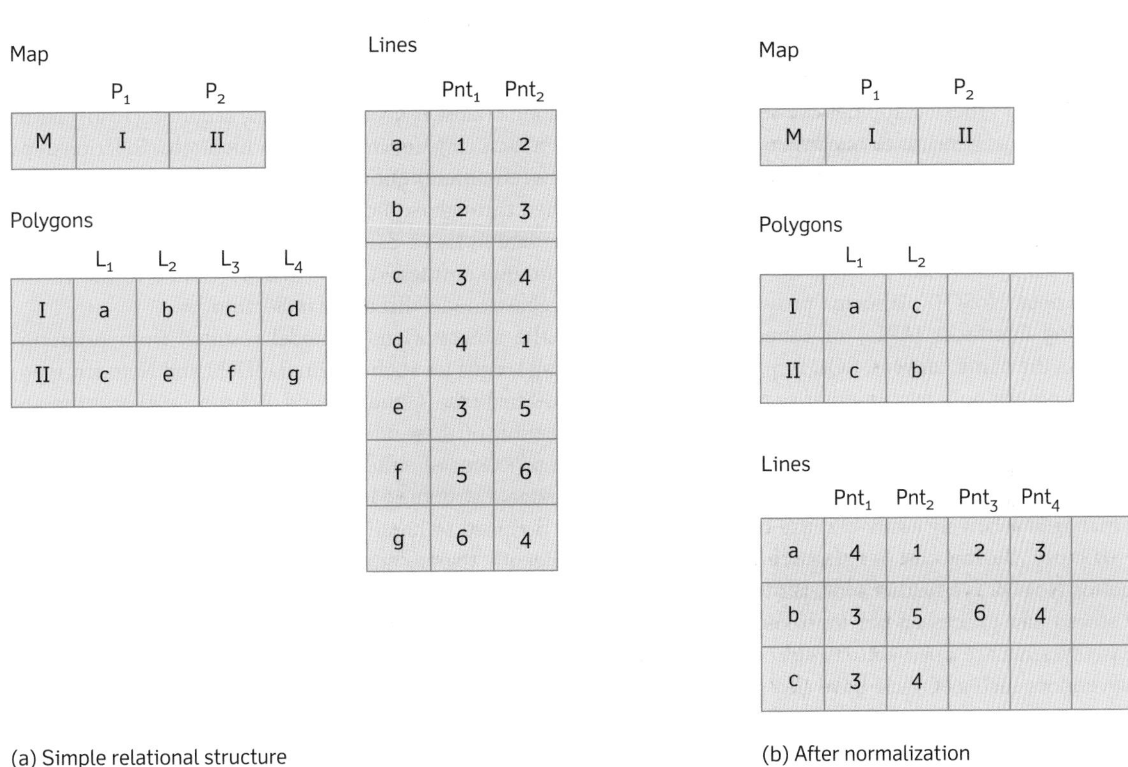

Figure 3.1 Relational organization of the vector data, (a) using simple structure and (b) after normalization used to reduce redundancy

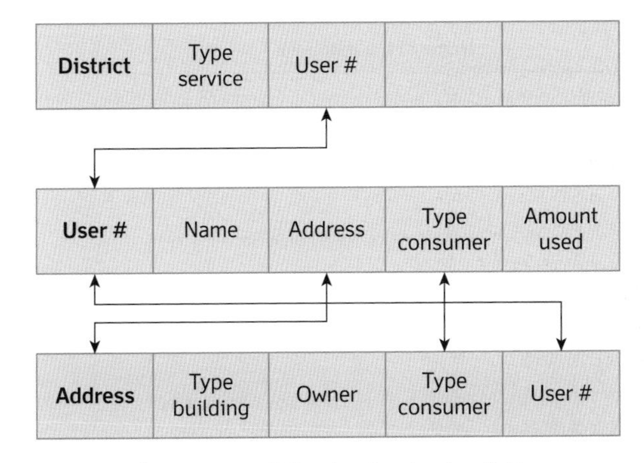

Figure 3.2 Querying in a relational system involves linking across tables—this example shows how, in a utility consumer database, details of building type can be obtained for a particular user

NoSQL database management system

In the last decade, in response to the burgeoning data storage volumes of the Big Data era, a broad group of more flexible database systems, known as NoSQL (often interpreted as Not Only SQL), are being used across a broad range of applications, including behind GIS.Their particular advantage is in applications that require quick retrieval operations or where databases are joined across multiple platforms, often including cloud computing infrastructure. This type of database is used by companies such as Facebook, Twitter, and Amazon in one form or another. NoSQL systems have been used in a number of GIS, particularly web-based systems, with the more rigidly defined BigTable-based NoSQL databases used in Google Earth (Chang et al. 2006).

These database systems store virtually any structure that is needed in a data element, with no fixed schema and no joins, and can accommodate nested hierarchical configurations which are difficult in relational database systems. The different structures that have been used are broadly classified into four main groups, as Table 3.1 shows. Key-values structures—the structures most used in GIS so far—are centred on the development of a hash table, where there is a unique key and a pointer to a particular item of data and where it is stored in the database for later retrieval. These mappings are usually accompanied by cache mechanisms to maximize performance.

In terms of querying and data-retrieval operations, NoSQL databases support little advanced functionality except retrieval, so many of the queries typically undertaken in an RDBMS are not possible. To develop querying, the various NoSQL platforms provide **application programming interfaces (APIs)** or other interfaces to the data. As the name conveys, SQL is not used and the systems do not support relation queries relations across records.

There are many advantages to these databases resulting from their high flexibility, which enables changes to be managed relatively easily because of this lack of a rigid structure. The low cost per gigabyte or transaction/second for NoSQL is a further plus, allowing more data to be stored and processed per unit cost. Many of the database systems are open-source, although major database vendors such as Oracle have also produced their own versions such as Oracle NoSQL. The main limits result from the restrictions on the possibilities for querying and so NoSQL databases are often used in conjunction with a RDBMS instead of replacing it.

The object-oriented database model

As the name suggests, object-oriented database models organize the data around the actual entities as opposed to the functions being processed. In the relational structure, each entity is defined in terms of its data records and the logical relations that can be elucidated between the attributes and their values. In object-oriented databases, data are defined in terms of a series of unique objects which are organized into groups of similar phenomena (known as object classes) according to any natural structuring, for instance in a hierarchy. Relationships between different objects and different classes are established through explicit links (as shown in Figure 3.3). As the data in these databases need to be clearly definable as unique entities, it is particularly suited to the vector database model discussed in Section 3.5.

The object-oriented model resulted from programming languages such as Simula (Dahl and Nygaard 1966) and Smalltalk (Goldberg and Robson 1983) and the application of these ideas to databases was stimulated by the problems of redundancy and sequential search in the relational structure. In GIS their use has been stimulated by the need to handle complex spatial entities more intelligently than as simple point, line, polygon primitives, and also by the problems of database modifications when analysis operations like polygon overlay are carried out. The approach is being applied increasingly in a number of fields although there are many different formalizations of the concept (Frihida et al. 2002; McKinney and Cai 2002).

The characteristics of an object may be described in the database in terms of its attributes (called its state) as well as a set of procedures which describe its behaviour

Table 3.1 Comparison of database systems

Categories of NoSQL database	Description	Name of the database
Document-oriented	Data are stored as documents. An example format may be like: FirstName="Arun", Address="St. Xavier's Road", Spouse=[{Name:"Kiran"}], Children=[{Name:"Rihit", Age:8}]	CouchDB, Jackrabbit, MongoDB, OrientDB, SimpleDB,Terrastore, etc.
XML database	Data are stored in XML format.	BaseX, eXist, MarkLogic Server, etc.
Graph databases	Data are stored as a collection of nodes and edges, where nodes are analogous to objects in a programming language. Nodes are connected using edges (type attached to them) optimized for just this sort of problem.	AllegroGraph, DEX, Neo4j, FlockDB, Sones GraphDB, etc.
Key-value store	In the key-value store category of NoSQL databases, a user can store data in a schema-less way. A key may be strings, hashes, lists, sets, or sorted sets, and values are stored against these keys.	Oracle NoSql, Cassandra, Riak, Redis, memcached, BigTable, etc.

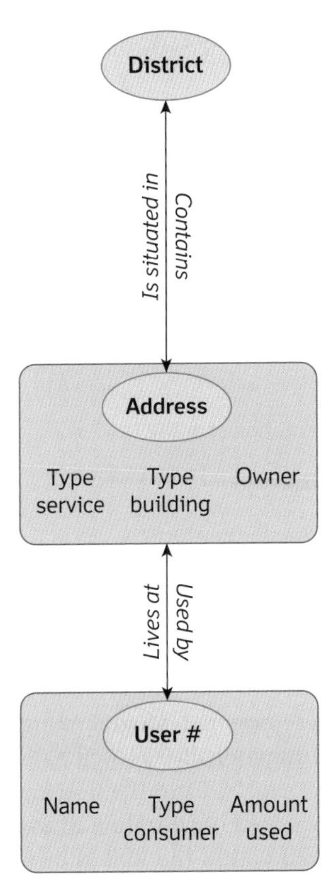

Figure 3.3 The object-oriented database system for the utility consumer database of Figure 3.2

(called operations or methods). Part of the attribute definition will describe the geometric nature of the object (point, line, polygon, or cell); more than one geometric type may be used to reflect the differences in shape found at different spatial scales.

These data are encapsulated within an object which is defined by a unique identifier in the database. This remains the same whatever changes are made to the values that describe their characteristics. For example, a building 'object' might change over time in terms of structure (or use) but its unique identifier will remain the same. The relational model of a bicycle is a list of parts that offers flexibility in how they may be brought together; the object-oriented model links the parts so that their function in relation to each other and the behaviour of the object is clearly expressed.

The structuring of objects in the database is established using pointers which refer directly to the unique identifiers. The pointers show various relationships and hierarchies and link the classes and instances within them. Where hierarchies are established, forming general, sub, and super classes, various defined states or methods are passed down through the system of inheritance. This means that efficiencies may be made both in characterizing the attributes of objects and in retrieving them from the database.

Figures 3.2 and 3.3 contrast a relational database with that of an object-oriented approach for storing a water utility company's data. In the relational approach each main table (District, User #, Address) is linked by data that are repeated from one table to the next. In the object-oriented approach, 'District', 'Address', and 'User #' are designated as 'objects', with defined relations between them such as 'contains', 'is situated in', 'used by', 'lives at', etc. All data are held once only, and the directed pointers serve not only for rapid retrieval of data, but also for passing commands up and down the database hierarchy. For example, it is much easier with the object-oriented approach to find all users of a given type in a given district in order to adjust their billing arrangements.

Once the data have been encapsulated as an object in the database, the only way to change or to query them is to send a message to carry out one of its operations. The types of querying possible depend on the operations that have been used to define the objects. The response of the object to a message will depend on its state, and the same message may bring about a different reaction when received by different objects or in a different context; this is termed polymorphism.

The structuring of the database into a series of self-contained, fundamental units brings with it both problems and possibilities and it has proved attractive to certain GIS users as it offers a way of modelling the semantics and processes of the real world in a more integrated, intuitive manner than is possible in relational systems. People working with human-made objects, such as utility companies, have found that these systems provide an approach which suits the data types they use and the querying capabilities needed. However, for continuously changing environments, it is very difficult to break down continuous spatial fields into separate units. How do you break up a hill into a series of distinct objects that may be used in analysis and modelling? The choice of boundary is often subjective.

To date, the implementation of object-oriented databases in GIS has been limited. The problem is that there are few generic object-oriented database products available to act as an engine to support GIS functionality, e.g. Smallworld based on Smalltalk.

Hybrid databases: linking geometric representation to attributes

The availability of commercial DBMS, particularly RDBMS, greatly eased the work of GIS system designers as they were able to apply ready developed and tested systems to their data-handling needs. These databases allowed designers to divide the problems of spatial data management into two parts. The first part was how to represent the geometry and topology of the spatial objects—should this be done using vector or raster data structures? Today it is also possible to store advanced topology outside of the geometric files using separate tables such as origin–destination matrices. The second part was how to handle the attributes of the spatial objects, which may be done through a standard commercial DBMS, with new possibilities today also from some free non-commercial DBMS. The resulting hybrid structures (sometimes referred to as geo-relational models) have a number of distinct advantages:

a) Attribute data need not be stored with the spatial database but may be kept anywhere on the system, or even online via a network.

b) Attribute data can be expanded, accessed, deleted, or updated without having to modify the spatial database.

c) They ensure that new developments in DBMS are incorporated as standard.

d) Data structures may be defined in standard ways using data dictionaries: data can be retrieved using general methods such as query languages that are independent of the DBMS.

e) Keeping the attribute data in a DBMS does not interfere with the basic principles of layers in a GIS.

f) Attributes in a DBMS can be linked to spatial units that may be represented in a wide variety of ways.

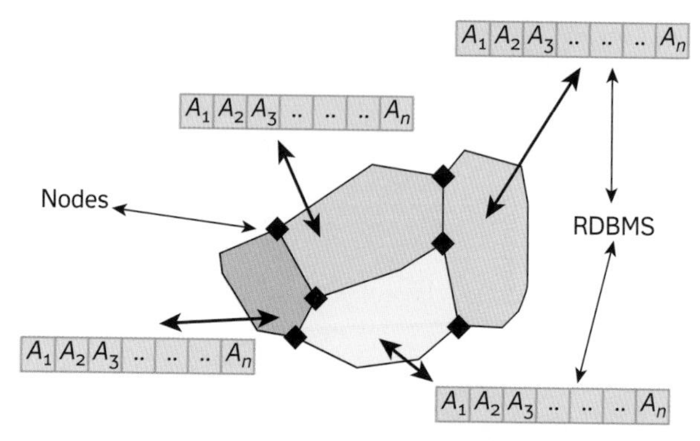

Figure 3.4 Hybrid relational database structure for a polygon with topology and entities

1. *Arc-node RDBMS.* This is probably the most used database system, in which the full vector arc-node topology is used to describe networks of lines and polygon boundaries as explained above. Each spatial unit is identified by a unique number or code, and it may be placed in a chosen layer or overlay. The attributes of the spatial unit are stored as records in relational tables that may be handled by the RDBMS. Some of the topological information and attributes such as coordinates, neighbours, areas, coordinates of minimum bounding rectangles, and similar data may also be stored in tables in the RDBMS. Figure 3.4 shows a typical example of such a data structure.

One complication with this type of system is the creation of new spatial entities and their associated attributes when layers are intersected, because this means building new sets of records and links in the RDBMS.

2. *Object RDBMS.* In recent years, with the object-oriented approach the advantages of both systems have been capitalized on and used in organizing both raster and vector data structures in the same GIS. In these systems the various geometric and attribute data are stored in relational tables and object-oriented programming languages provide analytical functionality as well as a graphical object-based interface to the data. The systems allow the benefits of object-oriented organization of geographical data to be exploited within the well-known relational database environment. An example of this is PostGIS, which adds support for geographic objects to the PostgreSQL object-relational database. In effect, PostGIS 'spatially enables' the PostgreSQL server, allowing it to be used as a backend spatial database for GIS, much like Oracle's Spatial extension.

3. *Compact raster RDBMS.* If the spatial objects are represented by *sets of pixels* instead of topologically linked lines, then the raster equivalent to the above may be created. When most of the spatial data refer to thematic units that are internally homogeneous, such as choropleth map units, then raster compaction methods of run-length codes may be used to save space.

3.4 Coding the basic data models for input to the computer

As we explored in Chapter 2, it is necessary to use discrete building blocks of geographical data that are capable of representing entities and continuous fields, as well as being able to represent exact or inexact attributes, location, and relations. The main spatial units used in representing the data were shown to be the vector (point, line, and polygon) and raster (pixel) primitives; the different data types associated with each kind of data model were also highlighted. In the database, the geographical data are often structured using layers (also referred to as **overlays** or **feature planes**) which divide the phenomena into intuitively useful different groups.

To organize their storage in the computer, these basic building blocks may be formalized into logical schemata (known as data structures), which are used for representing both the entity and continuous data models. Table 3.2 summarizes the properties of the different data structures with respect to entity locational data, attribute storage, and topology, and Table 3.3 shows the relative advantages and disadvantages of the two data structures. Information formulated according to one geographical data model does not necessarily have to be organized in the computer using a data structure of the same name. Raster data structures may be used for representing data formulated according to the vector data model, and vice versa.

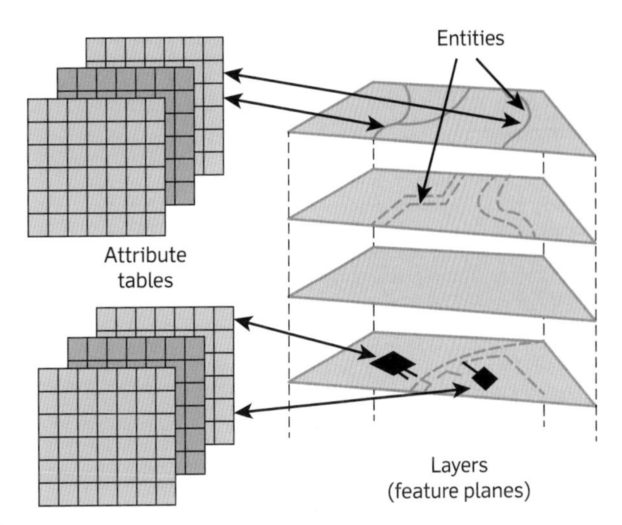

Figure 3.5 Feature plane layers in vector data structures

3.5 Points, lines, and areas: vector data structures

A data structure that uses points, lines, or polygons to describe geographical phenomena is known as a **vector data structure**. Vector units are characterized by the fact that their geographical location may be independently and very precisely defined, as may be their topological relationships. They are usually homogeneous, containing no internal variation (unless they are composed of simpler units), so that their attributes refer to the whole unit.

A vector database is built up from what the user perceives as a number of feature planes which are used to separate different classes of phenomena (shown in Figure 3.5). The units are represented as crisp world objects using a coordinate space that is assumed to be continuous, not quantized as with the raster structure, allowing all positions, lengths, and dimensions to be defined precisely. In fact this is not exactly possible because of the limitations of the length of a computer word on the exact representation of a coordinate and because all display devices have basic step sizes or resolutions. Besides the assumption of mathematically exact coordinates, vector methods of data storage use implicit relations that

Table 3.2 Characteristics of basic spatial entities in vector and raster mode

Basic unit type	Locational data	Attributes held as	Topology
Vector units			
Points	*X*,*Y*,(*Z*)	records	implicit
Nodes	*X*,*Y*,(*Z*)	records	explicit
Simple line	*N*[*X*,*Y*,*Z*]	records	implicit
Complex line (arc)	*N*[*X*,*Y*,*Z*]	*N*[records]	explicit
Polyhedron (solid body)	*M*[lines]	records	explicit
Raster units			
Grid cell (pixel)	row, column	single value	implicit
Line	*N*[pixels]	single values or pointer to records	implicit
Polygon	*M*[pixels]	single values or pointer to records	implicit
Voxel	row, column, layer	single values or pointer to records	implicit

Table 3.3 The relative merits of vector and raster data structures

Vector data structures	Raster data structures
Advantages Good representation of entity data models. Compact data structure. Topology can be described explicitly—therefore good for network analysis. Coordinate transformation and rubber sheeting is easy. Accurate graphic representation at all scales. Retrieval, updating, and generalization of graphics and attributes are possible.	**Advantages** Simple data structures. Location-specific manipulation of attribute data is easy. Many kinds of spatial analysis and filtering may be used. Mathematical modelling is easy because all spatial entities have a simple, regular shape. The technology is cheap. Many forms of data are available.
Disadvantages Complex data structures. Combining several polygon networks by intersection and overlay is difficult and requires considerable computer power. Display and plotting may be time-consuming and expensive, particularly for high-quality drawing, colouring, and shading. Spatial analysis within basic units such as polygons is impossible without extra data because they are considered to be internally homogeneous. Simulation modelling of processes of spatial interaction over paths not defined by explicit topology is more difficult than with raster structures because each spatial entity has a different shape and form.	**Disadvantages** Creates large data volumes. Using large grid cells to reduce data volumes reduces spatial resolution, resulting in loss of information and an inability to recognize phenomenologically defined structures. Crude raster maps are aesthetically less pleasing, though graphic elegance is becoming much less of a problem. Coordinate transformations are difficult and time-consuming unless special algorithms and hardware are used, and even then may result in loss of information or distortion of grid cell shape.

allow complex data to be stored in a minimum of space. There is, however, no single, preferred method. This section describes a range of vector structures used in GIS for the storage of points, lines, and areas.

Point entities

Point entities may be considered to embrace all geographical and graphical entities that are positioned by a single XY coordinate pair. In addition to XY coordinates, other data must be stored to indicate what kind of 'point' it is and any other information associated with it. Figure 3.6 illustrates a possible data structure for 'point' entities.

Line entities

Line entities may be defined as all linear features built up of straight-line segments made up of two or more coordinates. The simplest line requires the storage of a 'begin' and an 'end' point (two XY coordinate pairs). An arc is a set of n XY coordinate pairs describing a continuous complex line. The shorter the line segments, and the larger the number of XY coordinate pairs, the more closely the chain will approximate a complex curve.

Networks

Simple lines and chains carry no inherent spatial information about connectivity, which might be required for road and transport or drainage network analyses, for example. To achieve representation of more complex phenomena such as a linear network, it is necessary to add topological pointers to the data structure. The pointer structure is often built up with the help of nodes. Figure 3.7 illustrates the sort of data structure that would be necessary to establish connectivity between all branches of a network. Besides carrying pointers to the arcs, the nodes would probably also carry data records indicating the angle at which each chain joins the node, thereby fully defining the topology of the network. This simple linkage structure incorporates some data redundancy because the coordinates at each node are recorded a total of $(n + 1)$ times, where n is the number of chains joining a node. Attributes attached to the lines (indicated by black dots in Figure 3.7) can be used to select preferred routes.

Polygons

Polygons may be represented in various ways in a vector database. As many kinds of spatial data are linked to polygons, the way in which these entities may be represented and manipulated has received considerable attention. The aim of a polygon data structure is to describe the topological properties of areas (that is, their shapes, neighbours, and hierarchy) in such a way that the associated attributes of these basic spatial building blocks may be displayed and manipulated as thematic map data. Many forms of geographical analyses require the data structure to be able to record the neighbours

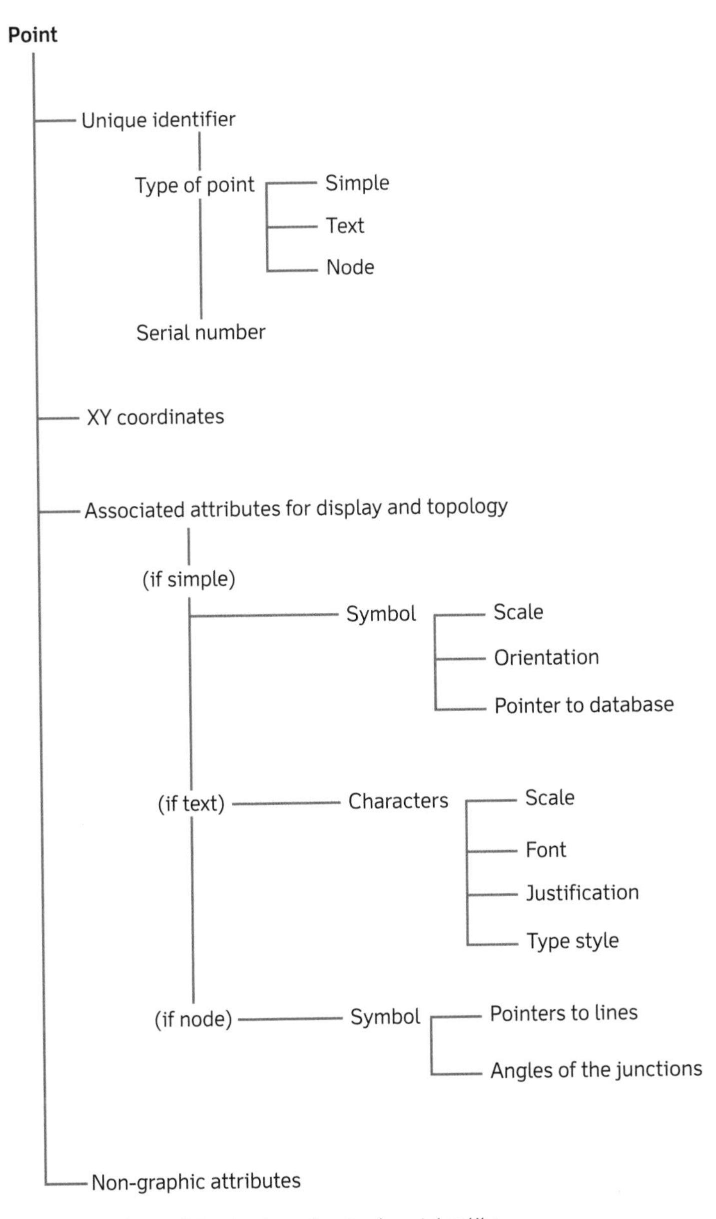

Figure 3.6 Vector data structure of a simple point entity

of each polygon in the same way that the network required connectivity. Each of the component polygons on a map will have a unique shape, perimeter, and area, with no single, standard basic unit as is the case in raster systems.

Simple polygons

The simplest way to represent a polygon is as an extension of the simple chain—i.e. to represent each polygon as a set of XY coordinates on the boundary (as shown in Figure 3.8). The names used to tell the user what each polygon is are then held as a set of simple text entities. While this method has the advantage of simplicity, it has many disadvantages: (a) lines between adjacent polygons must be recorded and stored twice (this can lead to serious errors in mismatching, giving rise to slivers and gaps along the common boundary); (b) there is no neighbourhood information; (c) islands are impossible to encode except as purely graphical constructions; and (d) there are no easy ways to check if the topology of the boundary is correct, whether it is incomplete ('dead-end'), or makes topologically inadmissible loops ('weird polygons') (see Figure 3.9).

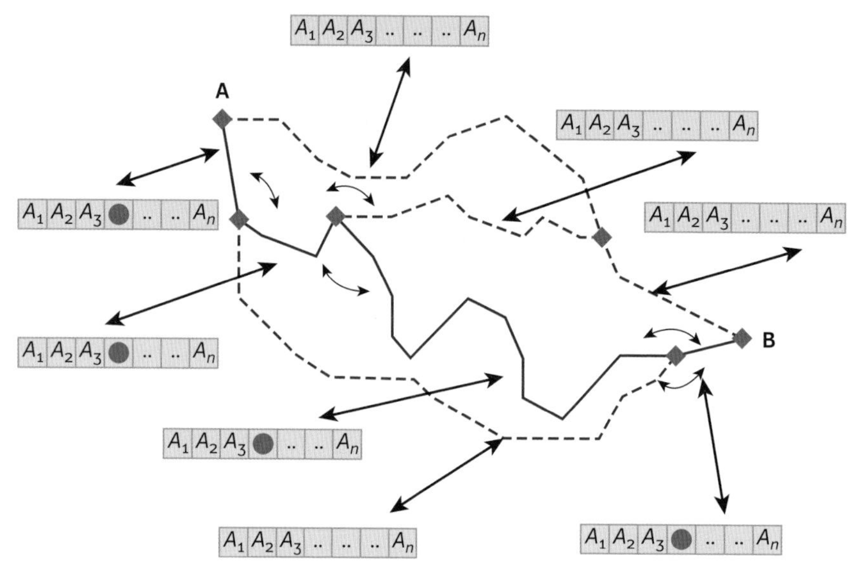

Figure 3.7 Hybrid data structure for network analysis

Polygons with point dictionaries

With this representation, all coordinate pairs are numbered sequentially and are referenced by a dictionary that records which points are associated with each polygon (Figure 3.10a). The point dictionary database has the advantage that boundaries between adjacent polygons are unique, but the problem of neighbourhood functions still exists. Also, the structure does not easily allow boundaries between adjacent polygons to be suppressed or dissolved if a renumbering or reclassification should result in them both being allocated to the same class. The problem of island polygons still exists, as do the problems of checking for weird polygons and dead ends. As with simple polygons, polygons may be used with chain dictionaries (Figure 3.10b). This has the advantage that over-defined chains (resulting from continuous or stream digitizing) may be reduced in size by weeding algorithms (see Chapter 4) without having to modify the dictionary. Polygon attributes are linked by pointers to data tables (Figure 3.10c).

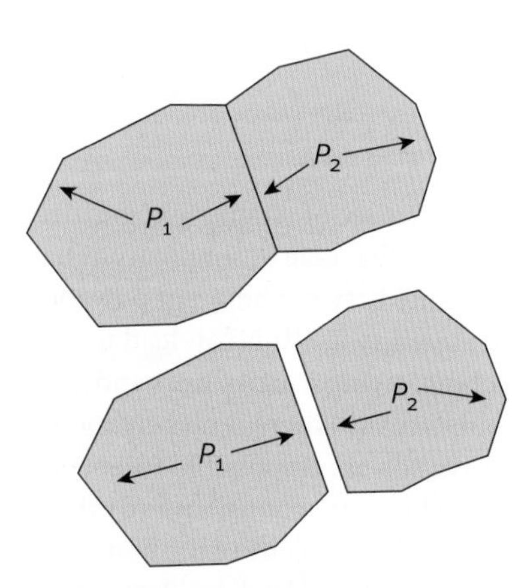

Figure 3.8 Without topology the database cannot distinguish between polygons that share boundary lines (top) or that are truly separate entities (bottom)

Polygon systems with explicit topological structures

Islands and neighbours may only be properly handled by incorporating explicit topological relationships into the data structure. Building the topological structure is an iterative process involving explicitly creating the links during data input or using software routines within GIS to form them. The software usually supports automatic checks for weird polygons and dead ends, and automated or semi-automated association of non-spatial attributes with the resulting polygons.

A topologically sound polygon network as shown in Figure 3.11 has the following advantages:

a) The polygon network is fully integrated and is free from gaps and slivers and excessive amounts of redundant coordinates.

b) All polygons, arcs, and associated attributes are part of an interlinked unit so that all kinds of neighbourhood analyses are possible. Note that the system just

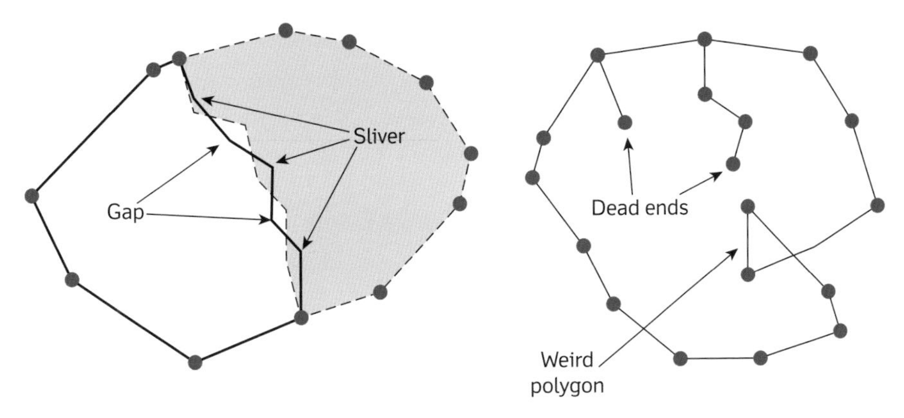

Figure 3.9 Topological errors in a polygon net

described also allows the arcs to have non-graphic attributes associated with them.

c) The number of continent–island nestings is unlimited.

d) The locational accuracy of the database is limited only by the accuracy of the digitizer and the length of the computer word.

Editing and updating the polygon net

Vector polygon networks may be edited by moving the coordinates of individual points and nodes, by changing the polygon attributes, and by cutting out or adding sections of lines or even whole polygons. Changing coordinates or associated attributes requires no modification to the topology, but modifying the network by cutting out or adding lines and polygons requires local recalculation of topology and rebuilding the database. Consequently, these kinds of data structures are not efficient for spatial patterns that are constantly changing.

Special purpose vector data structures—the triangular irregular network

An important, and much used, vector polygon structure is the triangular irregular network (TIN). It is built from joining known point values into a series of triangles based on a **Delaunay triangulation**. The triangulation allows a variable density and distribution of points to be used which reflects the changes in attribute values within an area, as shown in Figure 3.12a. The structure model regards the nodes of the network as primary units. The topological relations are built into the database by constructing pointers from each node to each of its neighbouring nodes. The neighbour list is sorted clockwise around each node, starting at north. A dummy node on the 'reverse side' of the topological sphere onto which the TIN is projected represents the world outside the area modelled by the TIN. This dummy node assists with describing the topology of the border points and simplifies their processing.

Figure 3.12b shows a part of the network data structure (three nodes and two triangles) used to define a

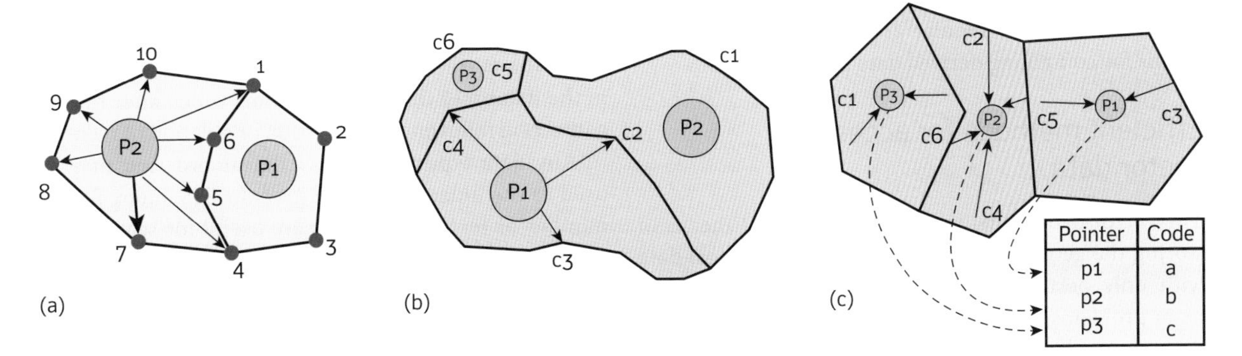

Figure 3.10 Three different ways of incorporating simple topology in polygon nets

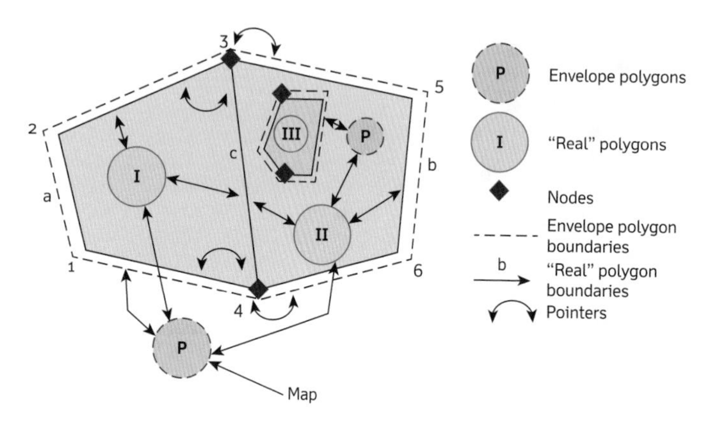

Figure 3.11 Full topological structure of a polygon map

TIN. The database consists of three sets of records called a node list, a pointer list, and a trilist (triangle list). The node list consists of records identifying each node and containing its coordinates, the number of neighbouring nodes, and the start location of the identifiers of these neighbouring nodes in the pointer list. Nodes on the edge of the area have a dummy pointer set to 32 000 to indicate that they border onto the outside world.

The node list and pointer list contain all the essential altitude information and linkages, so they are sufficient for many applications, such as slope mapping and hill shading. For associating other attributes with the triangles, it is necessary to be able to reference the triangles directly. Using a trilist to associate each directed edge with the triangle to its right establishes this. In Figure 3.12b, triangle T2 is associated with three directed edges held in the pointer list, namely from node 1 to 2, from node 2 to 3, and from node 3 to 1.

Nodes located in areas of greatest change help to reduce errors in the derived landforms. In digital elevation models, for example, values along ridges and valleys help to ensure that the resulting elevation surface does not have anomalies such as rivers which flow upstream. By using these irregularly spaced points, TINs avoid the redundancies of the regular grid and provide efficient means for computing derived data such as slopes.

Efficient methods for accessing vector data

The vector data model is a relatively efficient means of storing the geometric information of geographical data with only pertinent coordinate values recorded. The main problems have been associated with accessing the data, particularly where a number of data layers have been integrated. Various methods for improving this have been devised, which all use some form of indexing. Methods such as the KD-tree, the binary space partitioning tree, or **R-tree** use spatial location or bounding areas to divide up the data into indexed groups.

Database indexing using B-trees and R-trees

The importance of indexing a database to speed up querying of large or complex databases, as in GIS applications, is all too well understood by anyone experiencing long waits for results to a query. Structures have therefore been developed which give hierarchical indexes of indexes so that searching is more efficient and directed. They are known as multilevel indexes and are particularly useful with vector data structures where much of the topological and attribute data are held in index files.

The **B-tree** structure provides a multilevel index which is structured using 'internal nodes' and 'leaf nodes', which are analogous respectively to the branches and leaves of a tree. In the example shown in Figure 3.13 a file of street name data is indexed alphabetically. The first index splits the data into groups of letters a–f, g–m, n–s, t–z, which are listed along with pointers to the data in the respective nodes. The next level of the index splits these groups into finer divisions. When a search is initiated it is therefore limited at each stage, thus saving time. The B-tree structure adjusts to dynamic changes in the databases through algorithms which alter the organization following the insertion or deletion of records.

Any data type which has a linear ordering may be used as the index field in a B-tree. In GIS, index files of number or text string values are useful for searches on attribute records. However, this structure does not address the spatial (two-dimensional) nature of many of the queries of a GIS.

There are various alternatives to the B-tree model which allow geometrical properties of the database to

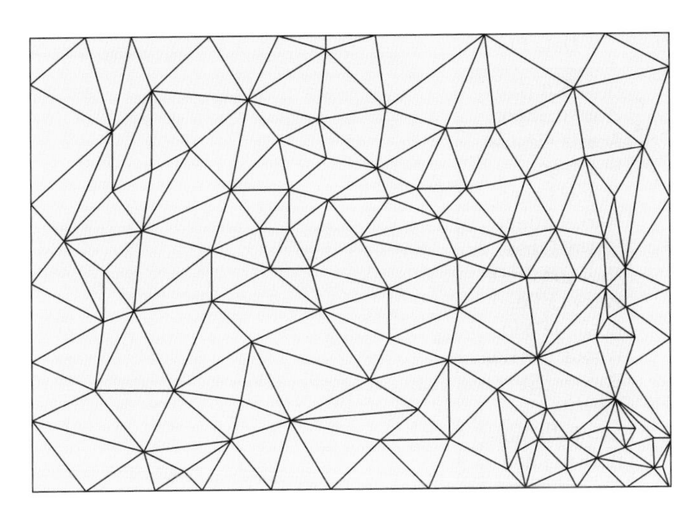

(a) Triangular irregular network based on a Delaunay triangulation

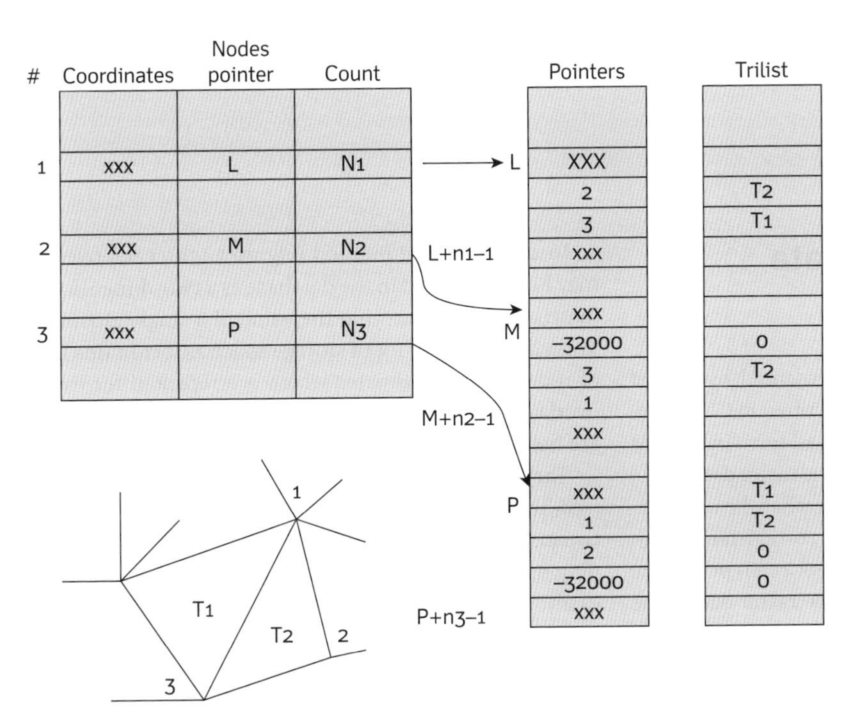

(b) Data structure of a TIN (detail)

Figure 3.12 Triangular irregular networks based on the (a) Delaunay triangulation provide a (b) vector data model of a continuous field

be included in structuring an index. For example, the R-tree divides space into a series of boxes, known as **minimal bounding rectangles (MBRs)** (Manolopoulos et al. 2010). Figure 3.14 shows a series of hospital locations and the three MBRs used to divide the space. Nodes in the index represent these rectangles, and a search for a particular hospital location would be directed first by an algorithm to one of these. A hierarchical structure of different rectangle sizes of MBRs may be set up using a tree-shape structure for the index. The search algorithm checks which large rectangle an entity is contained in and then follows the branches of the tree down various levels until the conditions of the query are met.

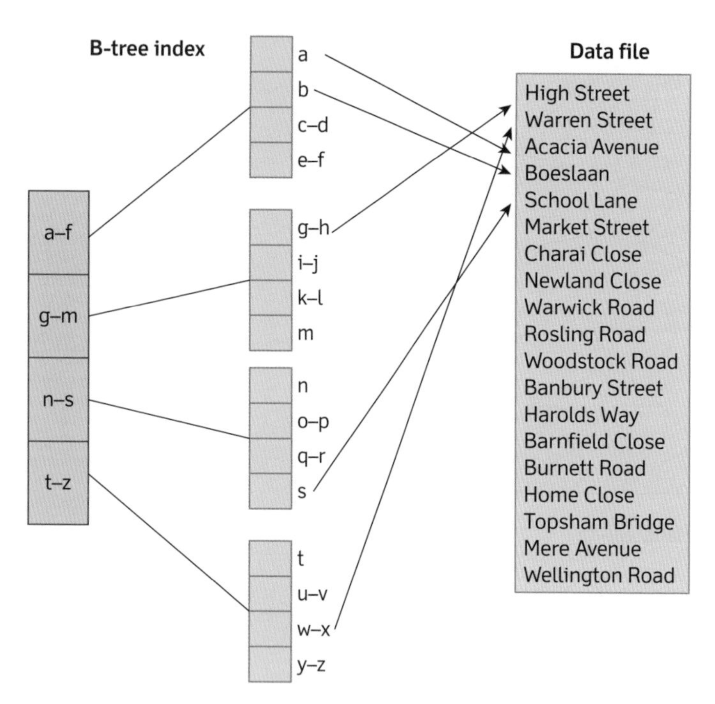

Figure 3.13 Data indexing using B-trees

3.6 Grid cells: raster data structures

Spatial phenomena may also be represented by sets of regular shaped tessellated units as described in Chapter 2. The simplest and mostly commonly used form in a GIS is the tessellated regular grid, known as the raster data structure. A raster database is built up from what the user perceives as a number of layers which are co-registered, with the location of entities defined by direct reference to the grid cells. The resolution or scale of the raster data is the relation between the pixel size and the area the cell represents on the ground. Whereas in vector data structures the topology between different units is explicitly recorded through database pointers, in raster databases this is only implicitly coded through the attribute values in the pixels. Relational database systems influence the main structuring used.

In simple raster structures where each cell on each overlay is assumed to be an independent unit in the database (i.e. a one-to-one relation between data value, pixel, and location), each cell is identified by a coordinate pair and a set of attribute values for each overlay, as shown in Figure 3.15a. This is a very data-hungry structure with no data held on cell size or display symbols, and no compression techniques used to reduce storage demands.

In an alternative method (Figure 3.15b), each overlay may be represented in the database as a two-dimensional matrix of points carrying the value of a single attribute. This still requires a lot of storage space as it contains lists of redundant coordinates which are repeated for every overlay, and again no data on cell size or display symbols are held.

The hierarchical structures shown in Figure 3.15b and c (see Tomlin 1983) establish a many-to-one relation between attribute values and the set of points in the mapping unit so uniform areas (polygons) may be addressed easily. Recoding or changing variables is made easy as it requires rewriting only one number per mapping unit per overlay, as opposed to all cell values with the previous two structures. This structuring also allows run-length code and quadtree data compression techniques to be used to reduce storage demands and is efficient in handling structured data. The main disadvantage of this structure is that it is clumsy for highly variable field data.

With the fourth structure, each overlay is stored as a separate file with a general header containing information such as map projection, cell size, numbers of rows and columns, and data type; this is followed by a simple list of values which are ordered according to the sequence of rows and columns (Figure 3.15d). This is obviously more efficient, as coordinate values are not stored for each cell and generic geometry and display

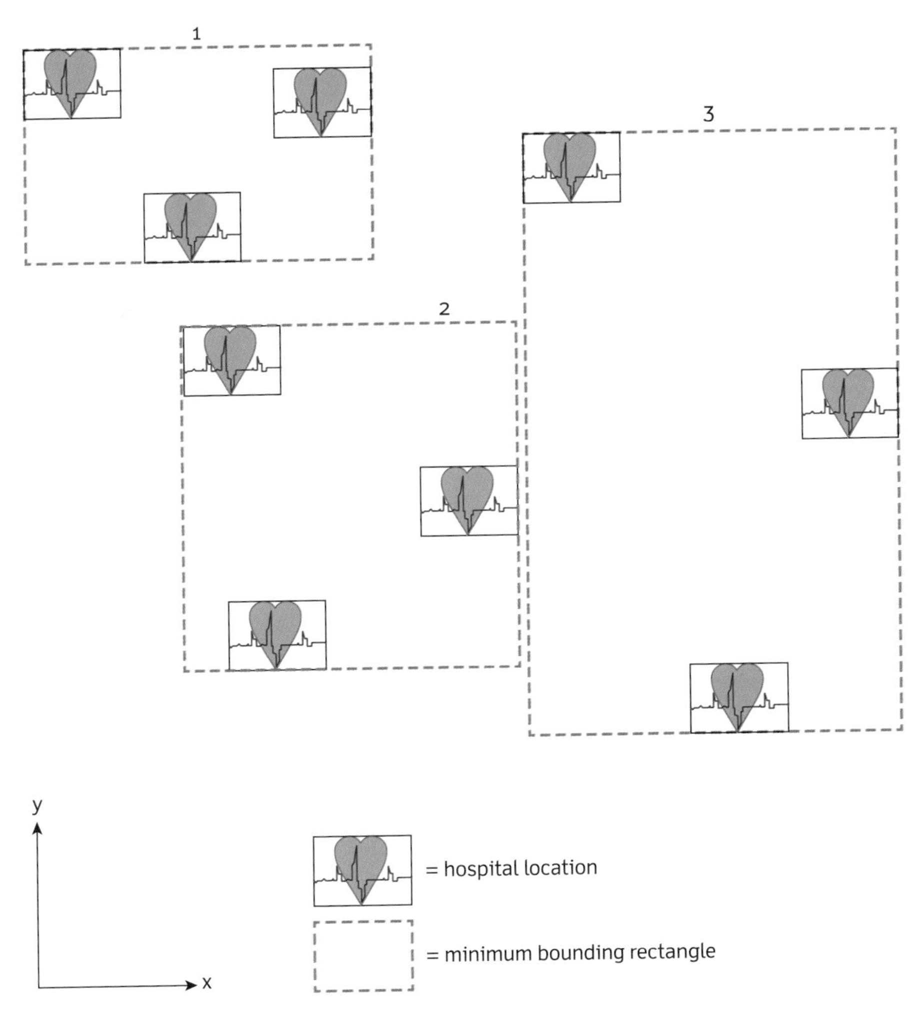

Figure 3.14 Data indexing using minimum bounding rectangles

values are written in the header of the overlay. PCRaster (Wesseling et al. 1996) uses this structure.

Compact methods for storing raster data

Given that in basic raster data structures each cell has a unique value and it takes a total of n rows × m columns to encode each overlay, plus general information on map projection, grid origin, grid size, and data type, the raster data structure requires ample storage space. The most storage-hungry data layer is one where the data type is a scalar and each cell contains a real number, as in the case of altitude matrices for digital elevation models and other continuous surfaces, perhaps obtained from interpolation. Other data types will require less space because they can code their data in fewer bits and compact methods, particularly for raster images; methods such as **quadtrees**, **run-length codes**, **block codes**, and **chain codes** have been used to great effect in this.

When raster structures are used to represent lines or areas in which the pixels everywhere have the same value, it is possible to gain considerable savings in the storage requirements for the raster data, providing of course that the data structures are properly designed. The structures described in Figure 3.15a and b may use array coordinates to reduce the actual quantity of numbers stored (with the limitation that all spatial operations must be carried out in terms of array row and column numbers). These systems do not encode the data in the form of a one-to-many relation between mapping unit value and cell coordinates, so compact methods for encoding cannot be used.

The third structure given in Figure 3.15c references the sets of points per region (or mapping unit) and allows a

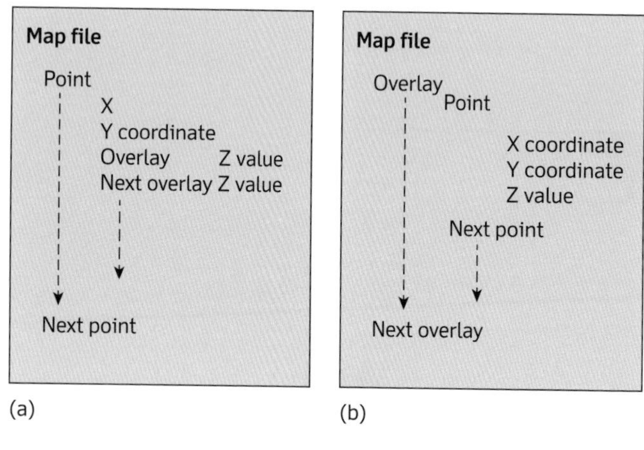

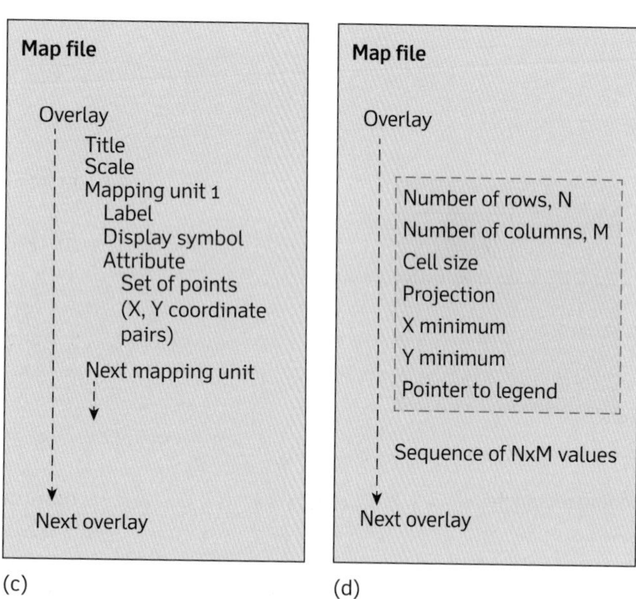

Figure 3.15 Four different ways of creating a raster data structure: (a) each cell is referenced directly; (b) each overlay is referenced directly; (c) each mapping unit is referenced directly; (d) each overlay is a separate file with a general header

variety of methods of compact storage to be used. There are four main ways in which the spatial data for a mapping unit or polygon may be stored more economically: chain codes, run-length codes, block codes, and quadtrees.

Chain codes

Consider Figure 3.16; the boundary of region A may be given in terms of its origin and a sequence of unit vectors in the cardinal directions. These directions can be numbered (East=0, North=1, West=2, South=3). For example, if we start at cell row=10, column =1, the boundary of the region is coded clockwise by:

$$0, 1, 0^2, 3, 0^2, 1, 0, 3, 0, 1, 0^3, 3^2, 2, 3^3, 0^2, 1, 0^5, 3^2, 2^2, 3, 2^3, 3, 2^3, 1, 2^2, 1, 2^2, 1, 2^2, 1, 2^2, 1^3$$

Chain codes can be stored using integer data types and therefore provide a very compact way of storing a region representation; they allow certain operations such as estimation of areas and perimeters, or detection of sharp turns and concavities to be carried out easily. Overlay operations such as union and intersection (see Section 7.7) are difficult to perform with chain codes without returning to a full grid representation. Another disadvantage is the redundancy introduced because all boundaries between regions must be stored twice.

Run-length codes

Run-length codes allow the points in each mapping unit to be stored from left to right per row for each class encountered in terms of a begin cell, an end cell, and an

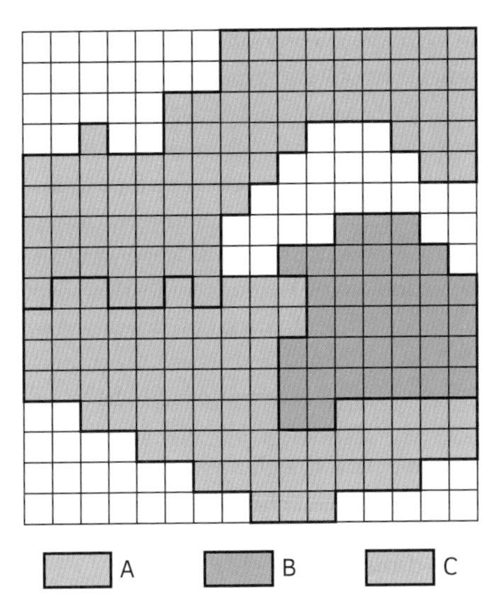

Figure 3.16 A simple raster map

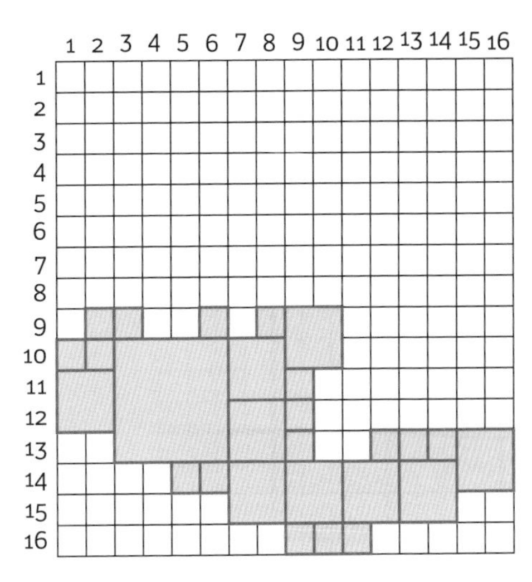

Figure 3.17 Medial axis transformation of polygon A in Figure 3.16

attribute. The integer data type is usually all that is required, and even smaller integer representations at 8-bit word level could suffice for some applications.

For the region A shown in Figure 3.16, the codes would be as follows:

Row 9	2,3 6,6 8,10
Row 10	1,10
Row 11	1,9
Row 12	1,9
Row 13	3,9 12,16
Row 14	5,16
Row 15	7,14
Row 16	9,11.

In this example, the 69 cells of region A have been completely coded by 22 numbers, giving a considerable reduction in the space needed to store the data.

Block codes

The idea of run-length codes can be extended to two dimensions by using square blocks to tile the area to be mapped. Figure 3.17 shows how this can be done for region A of the raster map in Figure 3.16. The data structure consists of just three numbers, the origin (the centre or bottom left) and radius of each square. This is called a medial axis transformation or MAT (Rosenfeld 1980). Region A may be stored by 17 unit squares +9 4-squares + 1 16-square. Given that two coordinates are needed for each square, the region may be stored using 57 numbers (54 for coordinates and three for cell size). Clearly, the larger the square that may be fitted in any given region and the simpler the boundary, the more efficient block coding becomes. Both run-length and block codes are clearly most efficient for large, simple shapes and least so for small, complicated areas that are only a few times larger than the basic cell. Medial axis transformation is used by fax machines to reduce the size of the images for transmission; this method also has advantages for performing union and intersection of regions and for detecting shape properties such as elongatedness (Rosenfeld 1980).

Quadtrees and binary trees

A problem with regular grids is that the minimum resolution of the data is limited by the size of the basic grid cell. Binary trees and quadtrees provide an approach to addressing successively finer levels of spatial detail with, in principle, an infinite set of levels (Samet 1990), although a minimum cell resolution still exists, at which the highest level of spatial detail is recorded for some areas. By using variable resolution structures like quadtrees it is possible to avoid storing finer resolutions of data in areas of little change, where this is not needed. However, it should be remembered that they do not increase the data resolution itself, as this is only possible through the systems and sensors used in the data-collection exercise.

The most efficient methods of compact representation of space are based on successive, hierarchical division of a $2^n \times 2^n$ array. If the division occurs by dividing the area into half each time, the method is known as a binary tree. If, however, the region is tiled by subdividing the area step by step into quadrants, it is known as a quadtree—this is the most commonly used form of hierarchically subdividing spatial data. In both cases the lowest limit of division is the single pixel.

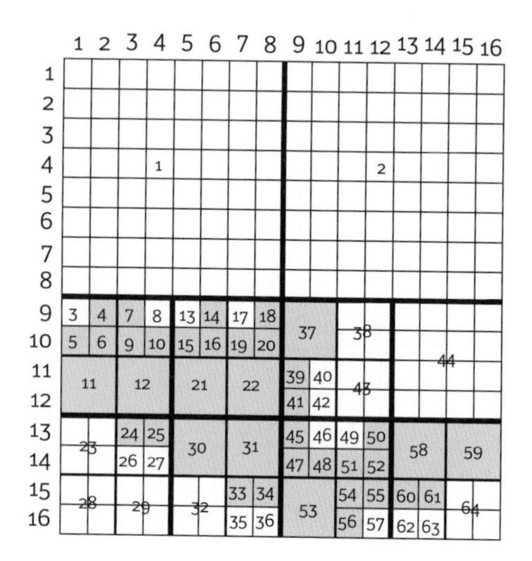

Figure 3.18 Quadtree encoding of polygon A in Figure 3.16

Figure 3.18 shows the successive division of the square region A (Figure 3.16) into quadrant blocks. This structure may be described by a tree of degree 4, known as a quadtree, as shown in Figure 3.19. The entire array of $2^n \times 2^n$ cell values starts from the root node of the tree, and the height of the tree is at most n levels. Each node has four children, respectively the NW, NE, SW, and SE quadrants. Leaf nodes correspond to those quadrants for which no further subdivision is necessary.

Quadtrees have many interesting advantages and drawbacks compared to other methods of raster representation (Gahegan 1989; Mark et al. 1989; Kothuri et al. 2002). In brief, standard region properties may be easily and efficiently computed on the compressed data, without requiring it to be expanded to recreate the explicit raster. Quadtrees are 'variable resolution' structures, in which detail is represented only when necessary or available, without requiring excessive storage for parts where detail is lacking or when a simpler representation is adequate for purpose (see Figure 3.20). One of the drawbacks of using quadtrees is that the tree representation is data frame-dependent and not translation-invariant—so two regions of identical shape and size may have quite different quadtrees, depending on how the square extent is defined and where the primary subdivisions are drawn. The quadtree also has to be rebuilt if a region changes, or is reprojected. Shape analysis and pattern recognition are not straightforward, which is a problem for objects that move or change over time.

Orderings of two-dimensional space

Various ordering methods for two-dimensional space have been used with quadtrees and other pixel addressing systems which, in effect, simplify the task of spatial referencing of a two-dimensional space by reducing this down to a one-dimensional sequence of cell addresses. This simplifies algorithms used in GIS and exploits various list data structures that are available in computer science, so reducing demands on computer disk storage and memory (Abel and Mark 1990). The orderings define pathways or directions through the two- (or by extension, three-) dimensional gridded space as shown in Figure 3.21; the row orderings simply number the cells in a square matrix of grid cells in a row-by-row sequence (Figure 3.21a and 3.21b). The various paths pass through all pixels in the space but have different aims in terms of total or unit length or in how they link neighbouring cells in the sequence. The

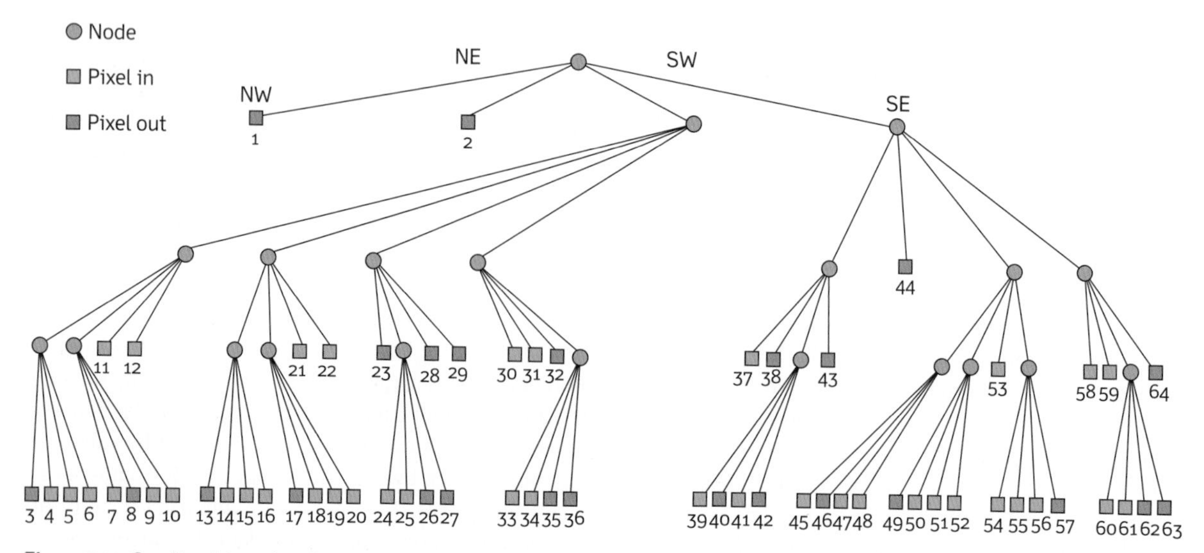

Figure 3.19 Quadtree hierarchy of polygon A in Figure 3.16

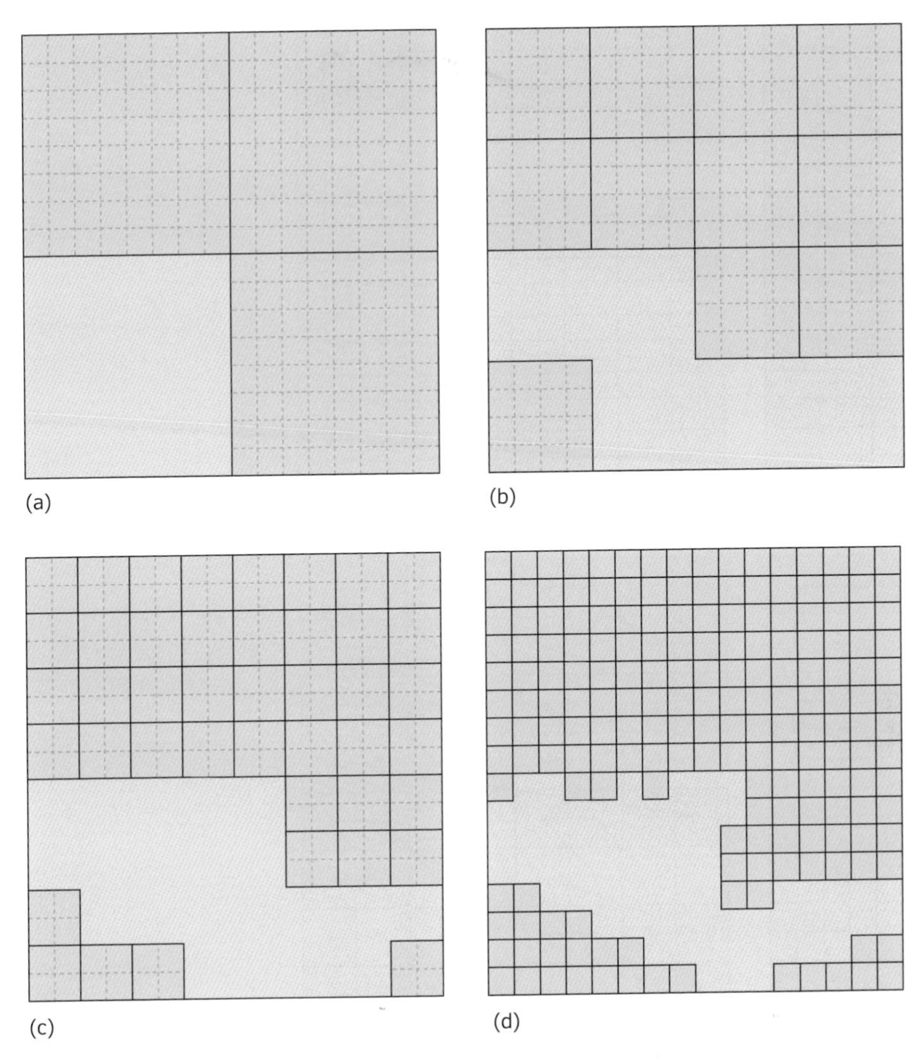

Figure 3.20 Quadtree hierarchies permit display at different levels of resolution

most commonly used orderings are **Peano–Hilbert** and **Morton orderings**, in which the pixels are indexed according to special recursive sequences based on subquadrants, as shown in Figures 3.21c and 3.21d.

The similarity between the Morton ordering and the quadtree division of data has been used to store and retrieve the two-dimensional locations of blocks of pixels of various sizes as a one-dimensional list in a computer file. This brings about considerable savings in time when querying the larger spatial databases that are increasingly common today.

Compact raster structures and data analysis

Compact raster structures are efficient for data analysis when the spatial unit (pixels, lines, or polygons) are exact, static entities. Quadtrees with sufficient levels of nesting can provide as fine a spatial resolution of bounded entities as vector methods of coding. However, compact raster structures provide few advantages for handling the continuous fields encountered in remotely sensed images, digital elevation models, interpolated surfaces, and numerical spatial modelling because each grid cell takes a different value. They are also of little value for modelling movement or change. Arithmetic operations or Boolean logic can be applied on data sets that have been compressed and stored as quadtrees, and the result can be written out as a quadtree without the need to expand each of the data sets back to uncompressed rasters. However, when data are stored in block codes or run-length codes, the compressed data files have first to be expanded back to the full, simple raster format so that cell-by-cell computations can be carried out.

Figure 3.21 Alternative two-dimensional orderings of data: (a) row, (b) row-prime, (c) Peano–Hilbert, (d) Morton

3.7 GIS and time

Developments in translating the spatio-temporal models described in Chapter 2 into working database systems for GIS have been limited so far. Most commercially available temporal GIS available today have harnessed RDBMS and object RDBMS technology and introduced means through which the temporal dimension can be recorded as part of a data entry. The most common way has been through time-stamping entries so that location and attribute values are recorded for a particular start point/duration, with subsequent changes becoming a replacement for, or being held in addition to, the initial record. There are problems with redundancy in the database, especially in RDBMS, and where the spatial values of a vector entity change.

Although taking the theoretical ideas of spatio-temporal models into working commercial systems has been limited to date, significant advances have been made in the visualization of the dynamics of an area. Software is often included as part of commercial GIS to develop animation that shows changes over time, whether raster or vector, and the resulting moving images provide insightful views of spatio-temporal dynamics. Changes in forest cover, movements along transportation systems, and individuals' interactions in an urban environment can be shown and further understanding developed.

3.8 Summary

Converting geographical information into digital data involves further structuring to encode it into binary configurations. These digital data are stored, manipulated, secured, and queried in database management systems which act as both the vault and the workhorse of the GIS. The choice of database system will influence how the geographical data are structured. While relational database systems, with structuring based on tables, still dominate most commercial GIS, others based on object-oriented and NoSQL are being used by some organizations as they offer users greater flexibility. The design of the database is crucial, and for the main database system types this takes place during the initial stages of creation and is hard to change once established.

With geographical data, both the location and attribute information need to be brought together and many GIS use hybrid DBMS configurations of two databases (one for spatial and one for attribute data) which make use of the advantages of each system. Further demands in data structuring and storage result from the spatial models used to represent the geographical data. The vector and raster data models require different structuring of the location information within the specific database management system used by the GIS, with relational databases currently used for storing both vector and raster data and object-oriented databases predominantly used to store vector data. This chapter has described various methods for speeding up data access and compression in GIS, including indexing, quadtrees, and two-dimensional orderings.

In these days of increasingly large spatial data sets and user expectations, quick access to information is important, so choosing the most appropriate spatial data model to encode data and selecting a GIS with an efficient database management system are vital parts of overall project design and implementation.

Questions

1. How do the limits of computer encoding affect the spatial data structures currently available in GIS?
2. Why are database management systems so important?
3. What data models are used in your mobile phone apps, and in your main web-based activities? Do they support spatial data types?
4. How does the data structure influence what and how you can do something in GIS?
5. Why do you think data accessing and compression methods are so important in today's spatial data-handling systems? What are the main benefits they bring?

Further reading

- Chrisman, N. R. (2003). ***Exploring Geographical Information Systems***. Wiley, Hoboken, NJ.
- Date, C. J. (1995). ***An Introduction to Database Systems***. 6th edn. Addison-Wesley, Reading, MA.
- Longley, P. A., Goodchild, M. F., Maguire, D. J., and Rhind, D. W. (2010). ***Geographic Information Systems and Science***. 3rd edn. Wiley, Hoboken, NJ.
- O'Sullivan, D. and Unwin, D. J. (2010). ***Geographic Information Analysis***. 2nd edn. Wiley, Hoboken, NJ.
- Tomlin, C. D. (2012). ***GISystems and Cartographic Modeling***. ESRI Press, Redlands, CA.

Summary

Questions

Further reading

Data Input and Verification

This chapter describes the exacting task of building a GIS database of spatial entities. The many different forms of raw geographical data, such as paper maps, satellite images, and Internet databases, need to be in digital form, registered to a geographical coordinate system (as described in Chapter 2), and then input into the structures and database management systems of the GIS to be used (as described in Chapter 3). A geographical database can be created using existing digital data, by extracting information from hard copy maps, or through a new digital survey. With the proliferation of many different forms of digital geographical data, standard exchange formats are important—these are described later in this chapter. The final sections of the chapter consider the importance of data about the data being used: that is, metadata. With the growing use of already collected information, users need to know how, when, and by whom it was collected.

Learning objectives

By the end of this chapter, you will:

- **understand the basics of how geographical data are captured digitally**
- **be familiar with different sources of geographical data and the methods used in collecting them**
- **be aware of the different data providers and of the impact the Internet has had on the supply of data**
- **know about the processes involved in converting from hard copy geographical data (paper maps, photographs, etc.) to digital forms.**

4.1 Creating a digital database

The last chapter explained how geographical data are structured in a database. In this chapter we explore how the many sources of geographical data are input into these databases. Creating a spatial database involves the abstraction of often complex entities into a computer format and a common geographical referencing system, to support interrogations and modelling across the different data sets. Given the complexity of the data and the variation in data sources, this is no mean task. Because raw geographical data are available in many different forms, such as paper maps, satellite images, or Internet databases, spatial databases can be built in several, not mutually exclusive, ways. These are, as summarized in the introduction (and shown in Figure 4.1):

a) acquire already existing digital data

b) convert existing hard copy data such as maps and images into digital form

c) carry out one's own digital survey.

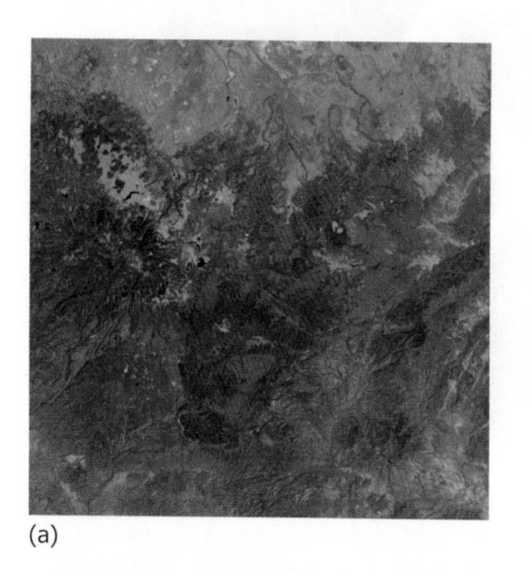

(a)

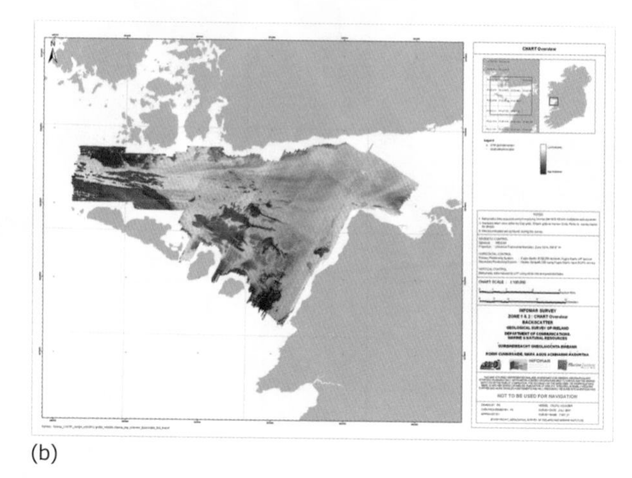

(b)

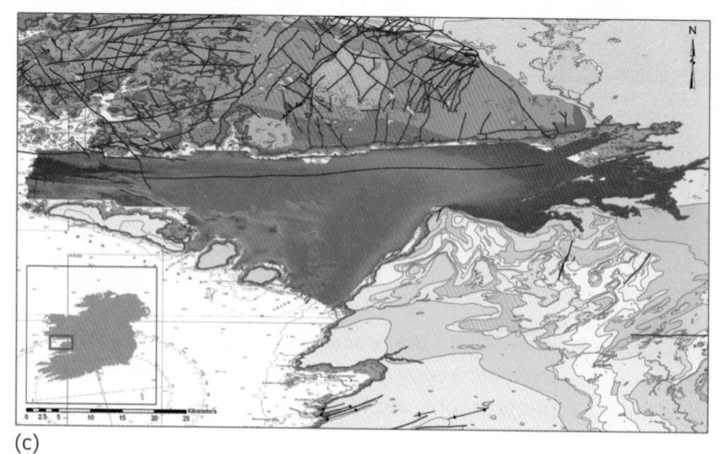

(c)

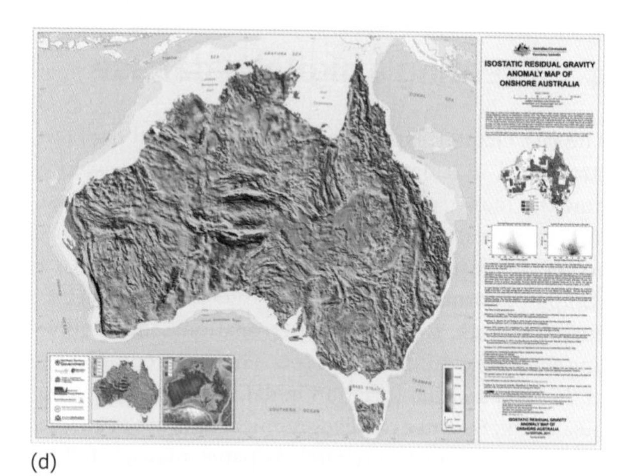

(d)

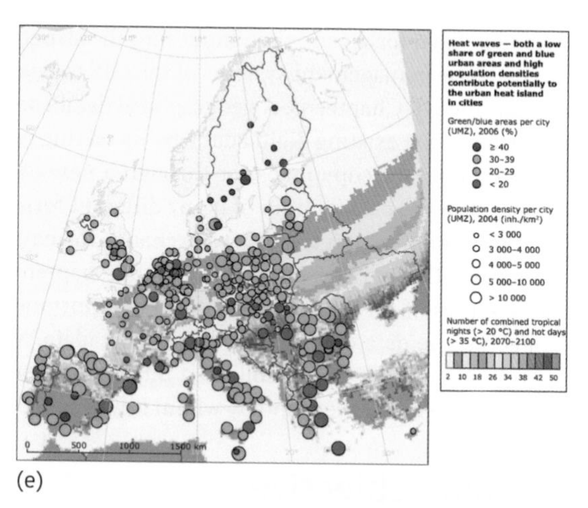

(e)

Figure 4.1 Different forms and sources of geographical data may be input to a GIS: (a) satellite image from NASA's Landsat 7 showing wildfires in Arizona in June 2011 (image courtesy of United States Geological Survey); (b) echo sounder imagery of Galway Bay, Ireland (image courtesy of Geological Survey of Ireland); (c) geology and LiDAR data of Galway Bay, Ireland (image courtesy of Geological Survey of Ireland); (d) Isostatic residual gravity anomaly map of onshore Australia (© Commonwealth of Australia (Geoscience Australia) 2013. This product is released under the Creative Commons Attribution 3.0 Australia Licence http://creativecommons.org/licenses/by/3.0/au/deed.en); (e) map of heat waves and urban centres in Europe http://www.eea.europa.eu/legal/copyright. Copyright holder: European Environment Agency (EEA)).

(*Source*: (a) USGS (b) © Copyright, Geological Survey of Ireland and Marine Institute (c) © Copyright, Geological Survey of Ireland (d) © Commonwealth of Australia 2009/CC-BY (e) European Environment)

Developments in the provision of data already in digital format have eased the whole process of database creation for many users. However, this access brings with it its own problems for the unwary. It is now, more than ever, important for users to be aware of the differences in the various spatial and attribute referencing systems used and the data-collection techniques.

4.2 Sources of geographical data

Geographical data have been an important information input to societies for many thousands of years. Early civilizations were concerned with the location of food, people, and landscape phenomena, as well as expressing their perceptions and theoretical ideas of the earth. Today, geographical data are collected on a whole gamut of subjects across a wide range of spatial scales and using many different techniques.

Maps

For many people the most common source of geographical data is a paper or digital map, which provides a graphical representation of the distribution of spatial phenomena. These powerful information sources have been used since the early civilizations of the Egyptians, Greeks, Romans, and Arabs, although the great age of map-making began with the geographical discoveries of the Renaissance period. By the seventeenth century, skilled cartographers such as Mercator had demonstrated that the use of a mathematical projection system and an accurate set of coordinates could increase the reliability of the measurement and location of areas of land (see Section 2.2 for more detail). The registration of spatial phenomena through an agreed standard also provided a model of the distribution of natural phenomena and human settlements that was invaluable for navigation, route finding, and military strategy.

Today many different organizations and individuals produce maps and they are available in digital as well as paper form. Maps serve a multitude of functions but have in common the principal feature of being scaled abstractions of the real world. They all have some type of georeferencing system so the user can identify where a phenomenon is found, and various graphical devices such as colour, lines, and shading, with their associated legends, for the attribute referencing system (described in Section 2.2). The general purpose topographic map, available at a range of different scales, showing many types of key features such as roads, housing, elevation, land cover, rivers, etc., is used principally as a reference device and often produced by national mapping authorities (NMAs). Government bodies also usually produce hydrographic and land registration mapping. Mapping conventions are discussed in Chapter 5.

Many subject-specific maps show the distribution of phenomena such as rock types, soil series, land use, voting habits, or vulnerability to hurricanes (see Figure 4.2 for examples), and their scope and use have grown rapidly with the access to GIS and digital data-collection devices. The result is a plethora of maps now available on subjects as wide-ranging as land cover to voting patterns, and from access to water to global telecommunications traffic.

Survey data collection

The traditional means of collecting geographical data is by ground or field surveys to record sample values at known locations, using instruments ranging from questionnaires to field instruments, through to automatic data loggers linked to radio and satellite telemetry which ensure that collected data can be downloaded directly to a computer without human presence. The Internet and social media are also being actively used to collect data, changing the interactions between the people collecting and providing the data. The results of these surveys are usually recorded in terms of a series of point location and attribute values, and these may later used to create areal mapped information.

The collection of surveys has been greatly assisted by the availability of global navigation satellite systems (GNSS) technology in many mobile devices (see Figure 4.3). GNSS are able to record the location data anywhere on the earth's surface (see Box 4.1 for more details). For individual users GNSS have been liberating, allowing the collection of raw data to be georeferenced accurately in the field without the difficulties of paper maps, and their affordability and inclusion in so many mobile devices—including cameras, tablets, and telephones as well as surveying and sensing equipment—ensures they are widely used.

This equipment is of particular advantage when surveying the natural environment or when working in countries where maps are unavailable or are out of date. Many hundreds of points may be collected in a day and the systems are useful at sea as well as on the land. With social media, geographical data are being recorded without users necessarily realizing this, as locations are

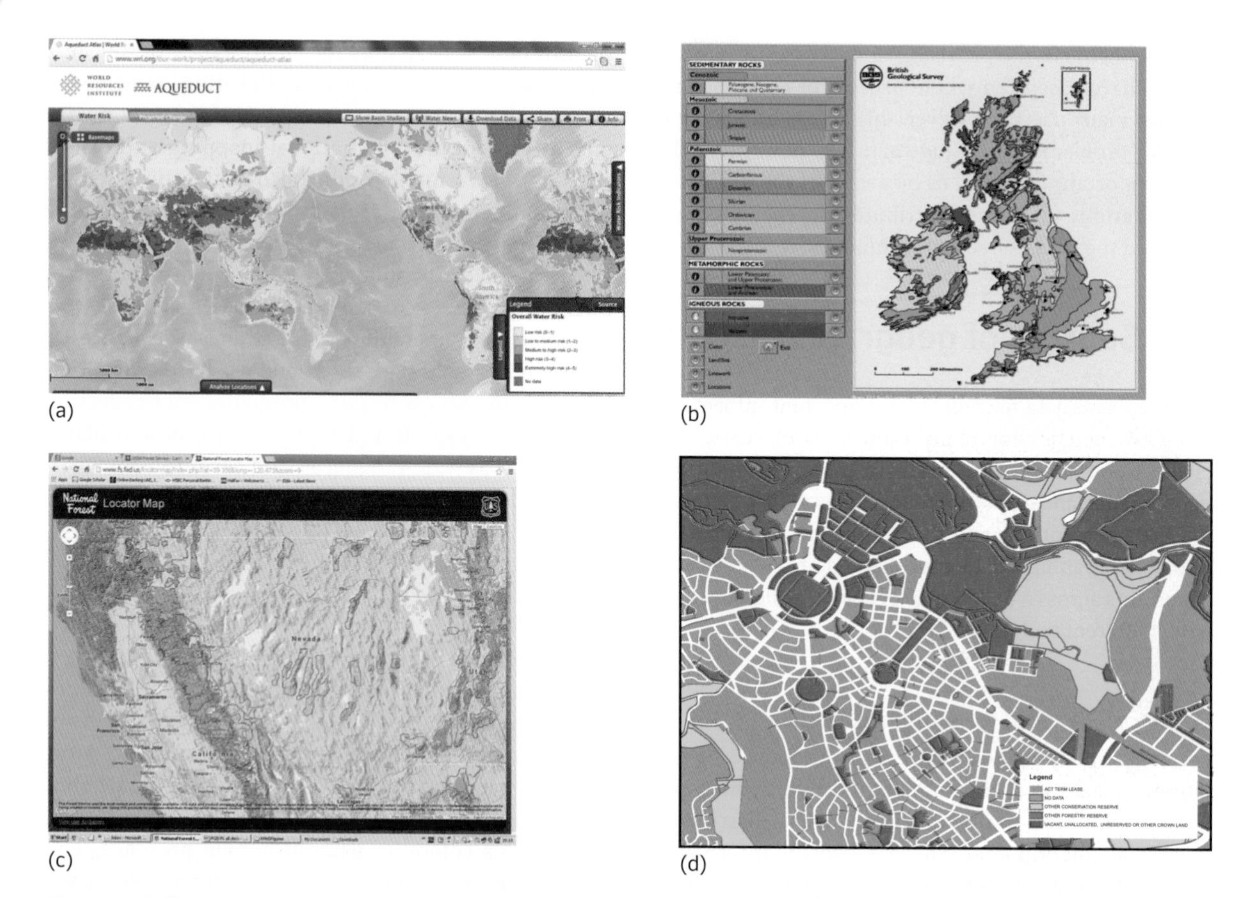

Figure 4.2 Different maps using various scales and sources: (a) screenshot of global water risk map from Aqueduct; (b) geological map of Great Britain and Ireland (contains British Geological Survey materials © NERC 2013); (c) screenshot of US Forestry Service forestry locator; (d) map of land tenures for part of Canberra, Australia Capital Territory at a scale of 1:20 000 (copyright PSMA Australia)

(*Source*: (a) World Resource Institute (b) Reproduced with the permission of the British Geological Survey ©NERC. All rights Reserved (c) US Forest Service (d) Copyright PSMA Australia)

being placed through interaction with a digital map or GNSS.

On a national scale GPS are now the main surveying system, with numerous NMAs shifting their topographic control network to be based on them. The National GNSS network in the UK, for example, consists of around 30 active stations (permanently installed, continuously operating) and about 900 passive stations (user-accessible ground monuments). This network will allow GPS users to carry out positioning in the European GPS coordinate system, ETRS89, with no additional costs. It fulfils the role of the traditional survey control network with higher accuracy.

Remotely sensed images

Remotely sensed images of the earth's surface such as those seen readily on Google Earth and daily on weather forecasts are an important source of geographical information. Satellite images are currently used in GIS and remote sensing systems to provide background information for thematic data. Analysis of the spectral information in the satellite images can also supply useful and timely information about land use, vegetation, soil conditions, hydrology, and so on. Image analysis techniques are used to help identify and classify geographical objects from the spectral information in the images. Very similar analysis techniques can also be used with spectral data obtained from airborne sensors of non-visible radiation (e.g. radar, infra-red), which may give finer spatial resolution than the satellite images. Remote sensing is often used as an affordable means of obtaining information about temporal changes on the earth's surface. Remotely sensed images and their classified products are usually available as digital databases.

(a) (b) (c)

Figure 4.3 Satellite navigation is embedded within many devices such as (a) mobile telephones, (b) portable route navigation systems, (c) watches

Box 4.1 Georeferencing with GNSS/GPS

The development of global positioning systems (GPS)—a US Department of Defense-owned navigation system based on signals received from satellite transmitters—transformed the way we are able to determine our location in the world. This NAVSTAR (sometimes expanded to Navigation Satellite Timing and Ranging) GPS satellite group is a constellation of 24 satellites (referred to as space vehicles or SVs), each following a path with 55° rotation at an orbit 20 200 km from the earth, giving one revolution in 12 hours. The Russian GLONASS system is also still operational, with plans for European (Galileo), Chinese (Beidou), and Japanese systems (Quasi Zenith Satellite System or QZSS) to become globally functional in the next few years. They are collectively referred to as global navigation satellite systems (GNSS).

GNSS receiver instruments are able to define the geographical location and altitude, to varying degrees of accuracy, for most places on the earth's surface using triangulation geometry based on signals emitted by constellations of satellites. Problems are found in using GNSS at high altitudes and in deep valleys. The position is determined by timing how long it takes for a radio signal from the satellite to reach the GNSS receiver; three or more different satellite transmitters are needed to give precision. All GNSS generate identical digital codes ('pseudo random codes') every millisecond so the time taken to travel to the receiver can be deduced by comparing the time the satellite generated the code and the time the receiver receives it.

Once these distances have been calculated, it is then possible to determine a position on the earth's surface using the principles of trilateration: using the measurements of distance from at least three satellites, and applying some basic trigonometry, a point in three-dimensional space may be defined. GPS receivers generate this location based on the WGS84 geodetic datum and height above ellipsoid vertical datum. Most GPS receivers have numerous preset datum and coordinate systems to choose from in the software settings for display, and selecting the appropriate one is critical for activities that use both paper maps and GPS.

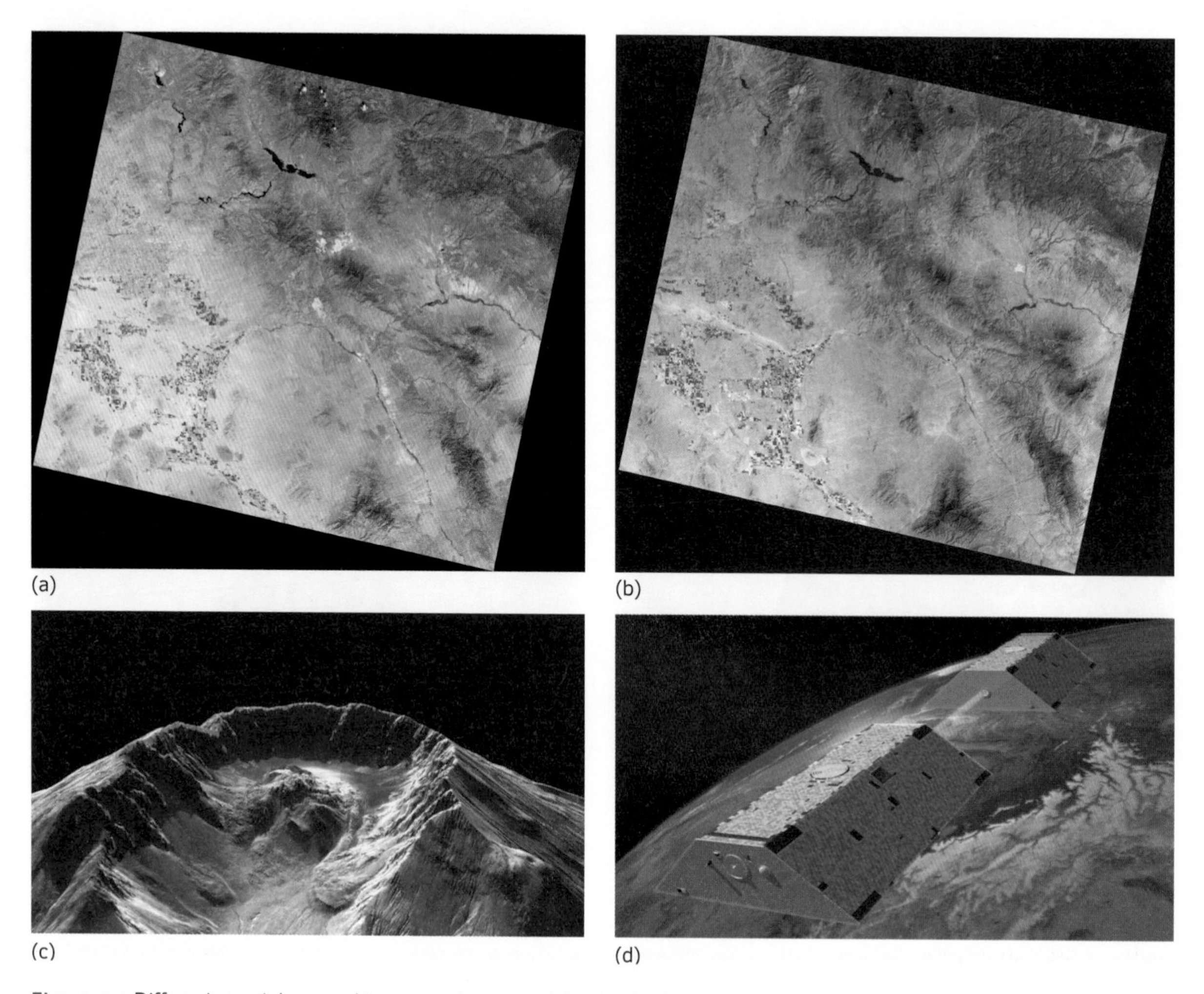

(a) (b) (c) (d)

Figure 4.4 Different remotely sensed images and sensors: (a) natural colour image of Phoenix, Arizona area acquired on 31 March 2013 from Landsat 8 (source: NASA's Earth Observatory); (b) thermal infrared image from Landsat 8 of same area as (a), also acquired on 31 March 2013 from Landsat Data Continuity Mission (source: NASA's Earth Observatory); (c) scientists from the United States Geological Survey (USGS) and NASA surveyed topographic change at Mount St Helens using Light Detection and Ranging (LiDAR) technology (*source*: NASA's Earth Observatory); (d) gravity-based GRACE sensor (image courtesy of NASA/JPL-Caltech)

The many different sensors, mounted on numerous aircraft and satellite platforms, measure different wavelengths (spectral) of electromagnetic radiation. Some wavelengths measured are reflected or absorbed radiation from the sun; others are emitted such heat energy (as shown in Figure 4.4a, b, and c) (Lillesand et al. 2008; Kubiak and Dzieszko 2012). Some sensors, such as those in the radar waveband, both generate and measure reflected energy, while systems such as NASA's Gravity Recovery and Climate Experiment (GRACE) system relies on sensing of gravity fields (Rodell et al. 2011) (Figure 4.4d).

The data collection is available from both national space agencies and commercial businesses, and data may be provided free or on a fee-based system to users, usually through a web portal. These data are usually corrected for geometric and radiometric distortions before distribution. With archives of satellite imagery now available from the last 40 years, satellite images may also be used in studies of changes, in processes such as land cover change, and urbanization taking place over long time spans.

Aerial photographs sensed using optical devices mounted on aircraft are still used in detailed mapping studies, with features distinguished through the changes of tone, pattern, and texture as well as size, shape, shadows, site, and association (Lillesand et al. 2008). In other

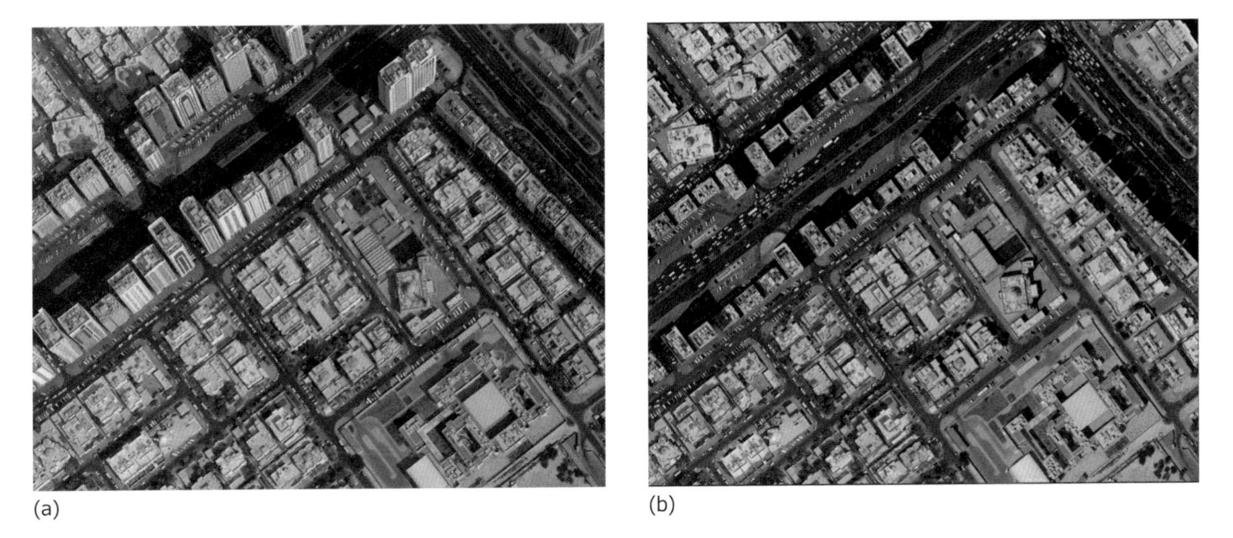

(a) (b)

Figure 4.5 Abu Dhabi: (a) raw aerial photo and (b) orthophoto, where any distortions caused by the tilt of the camera or topography of the land have been removed

systems based on **LiDAR** (Laser Detection and Ranging) and radar maps may be derived, showing surface roughness, with the results used to identify past and present landscapes and acting processes (see Chapter 11 for more detail).

Photogrammetric techniques may also be used to gain additional elevation detail from aerial photographs and satellite images. Using digital photogrammetric workstations, the three-dimensional nature of an area may be captured and these techniques provide a major source of data for the mapping of geology, soil, vegetation, or land cover as well as human urban environments (Chapter 11 provides more information on this topic).

The main problems associated with remotely sensed data as a data source stem from the following:

a) Data availability is limited by factors such as cloud cover, haze, and surface vegetation cover.
b) Their cost (which is highly variable between the different government and commercial supplying organizations).
c) Specialized image-processing systems are needed for preprocessing the sensor data before the information is ready for use in a GIS.

Where these limits are overcome, remote sensing systems provide invaluable data inputs for GIS, often in a readily transferable format.

Orthophotos

Orthophotos are a hybrid between maps and aerial photographs; they can provide a valuable source of highly detailed data on land use and land cover as well as information on utilities, municipal administration, and environmental studies (see Figure 4.5). Although they have a similar appearance to aerial photographs, the images have been corrected for geometrical distortion, and registered to a map projection/coordinate system using mathematical correction methods based on information obtained from ground control points and other data sources such as terrain elevation differences from digital terrain models (see Chapter 11 for more detail). A photomosaic is made by merging the corrected digital images from adjacent photographs and adjusting them for differences in greyscale intensity or colour balance.

Topographic information on, for example, administrative boundaries and other features can be added from the existing digital databases or digitized by hand (discussed in Section 4.5). Orthophotos are particularly useful in areas where the map coverage is outdated or unavailable, and they provide a relatively cheap and easy-to-interpret form of spatial data.

Three-dimensional data

Three-dimensional data are used in the modelling and analysis of many environmental systems, such as geological, hydrological, and atmospheric, in applications ranging from weather forecasting to hazardous waste disposal. The data are defined in terms of x, y, and z in the spatial dimensions with attribute values representing the phenomena. This type of data should not be confused with two-dimensional surface data draped over an elevation representation of an area—these types of data are often referred to as 2½-dimensional, or 2.5D data.

Three-dimensional (3D) data are collected by numerous methods involving both direct and indirect measurements. For soil and geological studies, surveys often involve the collection of material from cores, sections, or bore holes. Seismic methods might also be used which give a series of cross-sectional depth readings that may be linked to give a 3D picture. Atmospheric vertical measurements are collected with the rise of devices such as aerosonde balloons. Developments in positioning and data-collection technology have benefited these areas enormously. Today offshore seismic surveys collect 3D data directly, based on typically 25 x 25 m voxels using ships with large booms.

The sampling framework for collecting 3D data is often difficult to devise, as variability in the phenomena is hard to assess from surface expressions alone. With geological phenomena, for example, a vertical sample may not be orthogonal to the axis of the main object studied. It is also difficult to know if the occurrence of a particular type of material is the repeated occurrence of a single object or if it is a number of different objects. Most surveys use a grid network as a basis and samples taken at cross-sectional points.

4.3 Geographical data collectors

There are a plethora of different organizations, agencies, and individuals collecting and disseminating geographical data, ranging from NMAs, natural resource survey institutes, and commercial organizations through to non-specialist interested individuals.

Government data collection

Government agencies continue to be the main collectors, providers, and users of geographical information; of all the data they collect, the majority has a location reference. Traditionally, government agencies have maintained the systematic collection of geographical data. Many countries have national or regional mapping agencies responsible for collecting systematic data on phenomena such as the nature of the terrain, natural resources, human settlements, and infrastructure. In many countries NMAs, which were or still are part of military organizations, have the mandate for defining and maintaining the map spheroids, projections, grid origins, orientation and spacing, reference sea levels, data measurement, and collection standards (see Section 2.2) which are then used by others, including commercial mapping companies, as well as being responsible for providing standard topographic map coverage. The data they collect tend to be general purpose and are employed by a broad range of users (see Figure 4.6

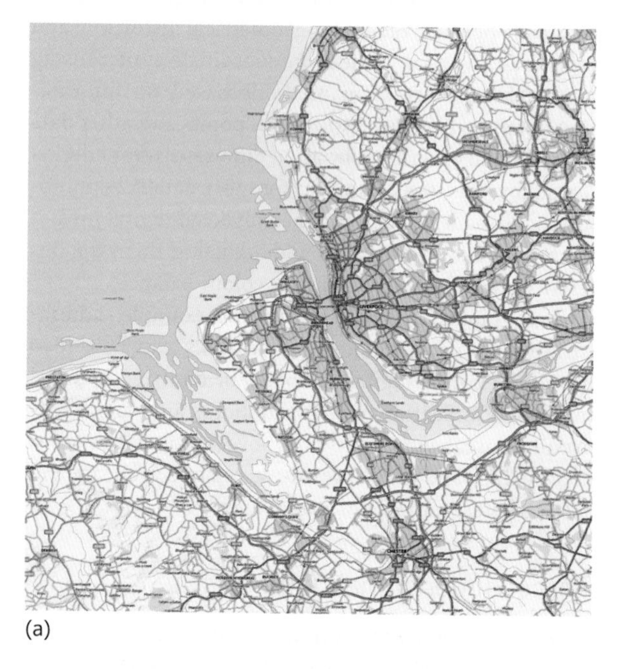

(a)

(b)

Figure 4.6 Ordnance Survey maps for the same area of north-west England at different scales: (a) OS District View data; (b) OS Metropolitan data sets (© Crown copyright/database right 2014)
(*Source*: Ordnance Survey)

for examples). These data are frequently now available digitally, although often at a cost, with NMAs now producing value-added products from their existing records.

Other surveys are carried out by more specialized government agencies that record variables such as land ownership, weather conditions, employment and journey to work patterns, soils and geology, rainfall and temperature, river flow and water quality, and information about people. The decadal census in many countries is one of the most important data-collection exercises and influences government public policy for many years. Systematic collection of remotely sensed data is provided by government space centres and private remote sensing corporations of countries or regions including Europe (ESA), USA (NASA), Canada (CSA-ASC), Russia (FKA), India (ISRO), and Japan (JAXA).

The main types of geographical data available from government agencies include:

- topographic maps at a wide range of scales
- administrative boundaries; property boundaries
- statistical data on people, including wide-ranging socio-economic variables
- data on land cover, land use, rocks, water, soil, atmosphere, biological activity, natural hazards, and disasters collected at a wide range of spatial and temporal levels of resolution
- in some countries, data on utilities (gas, water, electricity lines, cables) and the location of related infrastructure.

Geographical data collection by NGOs, private sector organizations, and others

Data are also generated by many non-governmental organizations (NGOs), commercial companies (e.g. private surveyors, civil engineers, and market researchers), political organizations, academic institutions, and individuals. Structured data are often collected for a specific purpose such as understanding consumer or voter thinking, evaluating mining possibilities, an environmental impact assessment, or an academic research study. The kinds of entities and attributes recorded may cover a myriad of subjects, and use various georeferencing systems ranging from geographical coordinates and planar coordinates through to census districts and postcode areas. The scale of the study and the observation methods used, along with the data classification and interpretation schema applied, are often quite unique to the survey, which limits their use in other applications. Access to the data is often restricted because of commercial or political interests.

Crowdsourced data

In recent years the role of individuals, who knowingly or unknowingly collect vast amounts of geographical data, has begun to be recognized, and is a largely untapped resource. Some data-generation exercises are structured using crowdsourcing of data—referred to by Goodchild (2007a) as 'people sensors'—by volunteer non-specialists who contribute actively to data-collection exercises, which was often previously the domain of a particular agent or agency (Ball 2010). It has been referred to as the democratization of geographical data collection, allowing anyone who is interested, whether communities or individuals, to contribute important information that reflects more their own (often local) insights, as opposed to those of officials who may be located remotely.

Crowdsourced geographical data, also called **volunteered geographic information (VGI)** (Goodchild 2007a), is often based on the use of mobile data collection enabled by GPS to capture the locational details, while attribute data detailing local knowledge are entered through alphanumeric keys. This may also include SMS text messages, digital images, non-specialist environmental sensor values, and many other devices which capture characteristics of an environment. Early successes have been gazetteer-type data where local data and images on where people live, collected by local non-specialist observations, have been brought together in projects such as OpenStreetMap, Platial, Google MyMaps, and GeoCommons, to provide a wealth of information. Many of these websites give free access to the collected data, providing unprecedented information to anyone who has web access, with geographical information from all over the world.

With the rapid increase in mobile phones with built-in GPS and SMS text messaging capacity, millions of members of the general public are carrying geographical data-collection devices with them in their pockets at all times. This is particularly important for measuring geographical characteristics in remote rural areas where it is difficult and expensive for technical experts to deploy sensors. Crowdsourcing has been used to collect data on many different variables from the state of the environment, city life, or the health of an individual, through to information provided about a hazard or disaster (Jiang and Lu 2012; special issue of *Transactions in GIS*, 16 (4); Connors et al. 2013; Dodge and Kitchin 2013).

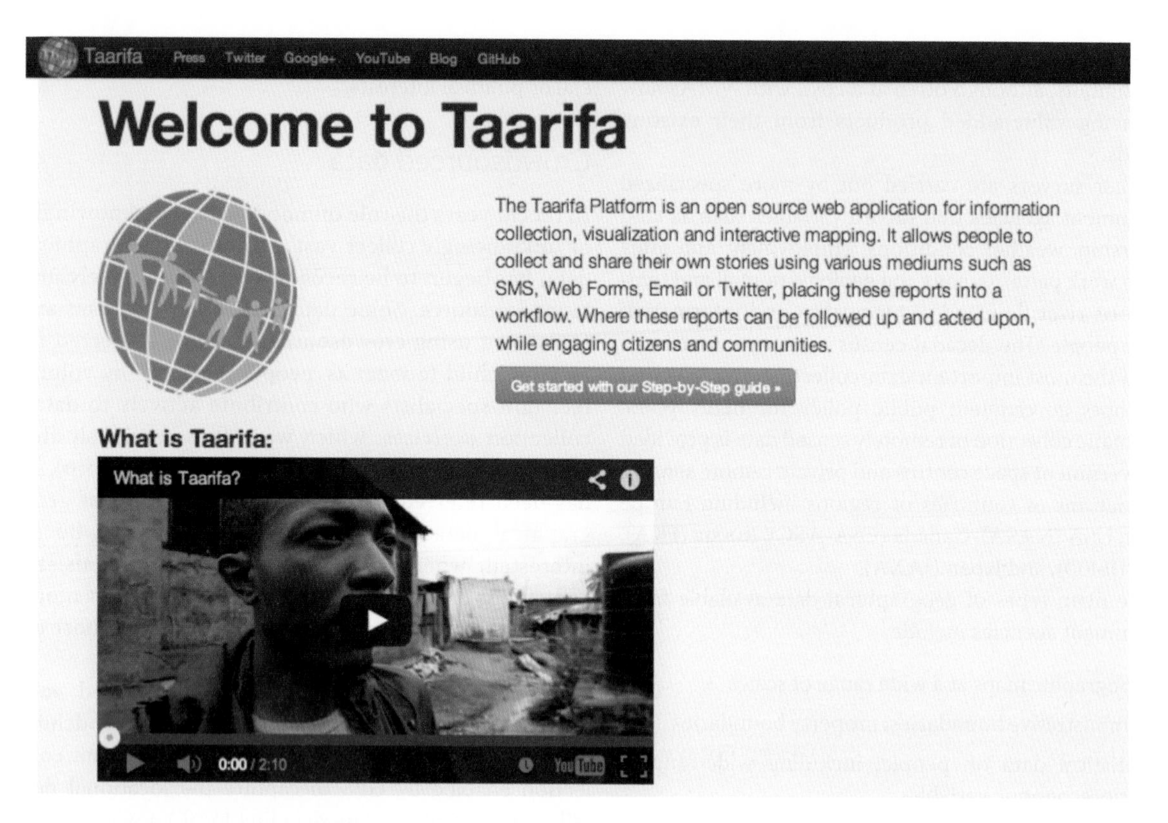

Figure 4.7 Screenshot of home page of Taarifa website
(*Source*: Taarifa 2013)

For example, the Taarifa Platform is an open-source web application for information collection, visualization, and interactive mapping, being used in Uganda, Ghana, and Tanzania (http://taarifa.org) (see Figure 4.7). It allows people to collect and share their own stories using various mediums such as SMS, web forms, email, or Twitter, and places these reports into a workflow, providing invaluable insight to local and national governments.

Often unknowingly, millions of social media users are generating their own geospatial data by adding location details and time stamps to their activity posts on Instagram, Facebook, Twitter, Flickr, etc. (Hollenstein and Purves 2010; Graham et al. 2013). Locational data are added through interaction with a digital map, by adding the name of a place, or through built-in GNSS. Although access to these geospatial data is often restricted because of privacy rights, where these data are available this represents never before possible insights into the habits, characteristics, thoughts, and current situations of users of social media. For example, they can give immediate data on the geographic nature of a crisis such as an earthquake, flood, or wildfire observed through the eyes of a non-specialist (Gelernter and Mushegian 2011; Vieweg et al. 2010).

This new source of data is particularly important in areas where little data exists and it can often be the only source of information for managers and policymakers.

The advantage of crowdsourcing is that vast amounts of data can be generated at a low cost, and that the data provided can offer a richer insight into some areas. There have, however, been increasing discussions on the reliability of the data (Fowler et al. 2013). The main problems identified are centred on:

- accuracy of data observed by non-specialists
- representativeness—one person's views do not reflect all, and not all citizens are equally able to access the Internet and use smartphones
- bias of the data collectors
- privacy issues and ethics in using individuals' data
- lack of quality information or **metadata**
- accuracy in geographical location and precision where GPS is not used.

4.4 Geographical data providers, metadata, and data exchange standards

Data providers make available geographical data in a variety of detail, formats, scales, and structures, for a price or for free. Traditionally data collectors and providers were the same relatively limited number of organizations. But with the Internet able to support fast transmission of data without the need for physical media, there are now a plethora of different possible suppliers involved, including those from the public or private sector and non-governmental bodies or civil society. Some collect and disseminate data, while others redistribute data gathered by others through web portals or spatial data warehouses (Bedard et al. 2009).

The legal side of data provision touches many areas, such as privacy and trade laws as well as ownership of information and copyright, data protection, liability, and intellectual property rights. Issues associated with ownership, intellectual property, and copyright have become more critical. The provision, distribution, and use of geographical data by parties outside of an organization is subject to various physical and legal restrictions, which are especially stringent where military agencies have been responsible for collecting the information. Needless to say, there are many differences in legal requirements across the world, and even for different types of organization within a country.

The development of geographical data portals (also called clearing houses) has made finding and accessing different data sets much easier. Portals have been established across the world: some based on specific themes such as agriculture, environment, and health, while others focus on a particular city, country, or region (see Figure 4.8). The organizations providing the portals

Figure 4.8 Screenshot of an example of a digital data portal

vary widely, ranging across the public and private sectors and non-governmental organizations (both local and international). They may offer data from one particular source or integrate data sets from different data collectors, so the information may be available at different scales and use different georeferencing systems. The portals themselves may be central repositories where all the data are stored or alternatively they may contain a series of Internet links to websites where the data can be accessed through the portal web interface. The data may be available for downloading to users' computers, or some simple web-based display and querying functions may be used. Licensing agreements may be required to access the data sets.

Although using existing data sets is attractive, attention must be paid to data compatibility when data from different suppliers are combined. There may be differences in projection, scale, base level, and description of attributes that could cause problems (this is discussed in further detail in Chapter 2) and it is important to know how and by whom the data were collected. Where data are used from a number of sources, and particularly where the area of study crosses administrative boundaries, difficulties in data integration are caused if the individual surveys use different geographical referencing systems, data classification, and sampling strategies. Users need to ensure that the data set's semantics correspond with those of their application and if it will ultimately be useful to them. They need to consider:

- the currency of the data
- the length of record
- the scale of the data
- the georeferencing system used
- the data-collection technique and sampling strategy used
- the quality of the data collected
- the data classification and interpolation methods used
- the size and shape of the individual mapping units.

In recent years, the development of national and international standards for passing on information about the data (metadata) and for physical transfer (spatial data transfer standards) of the data sets between dissimilar computing systems has been key to supporting the explosion in the transfer and use of data from remote providers by users.

Metadata—data about data

To address these problems and facilitate greater data sharing, metadata—data about the data—may be provided for each feature collection object. This typically records information on the data-collection exercise and includes elements in the following categories:

- details and accuracy of the georeferencing system
- definition and measurement of attributes
- data-collection details
- data accuracy/error details
- details of the collection agency
- data-processing information
- data lineage (previous sources and uses)
- status and distribution information.

With the growth in crowdsourced data, many of these details are difficult or impossible to provide. It may mean little to some potential data users to learn of the many different cameras, photographers, and dates involved in generating a particular set of information, yet understanding the provenance, veracity, and accuracy of the data has implications for how it may subsequently be used.

There are many possible formats for characterizing these elements of the metadata; different organizations and institutions develop their own standards for the information provided. For example, the United States Federal Government Data Committee have developed the Content Standard for Digital Geospatial Metadata (CSDGM), and put forward over 300 compound and data elements which may be used in the classification of spatial information, including entries such as 'altitude datum name', 'bearing resolution', and 'number of data bits'.

The main differences found between the different country and organizational standards are in the actual data elements specified and terminology used. To resolve any issues that might result from this, international standards have been developed, the most prevalent being those from the International Organization for Standardization (ISO), with ISO 19115:2003. This standard is gradually being endorsed and transitioned towards by national agencies responsible for metadata. The key elements are shown in Figure 4.9.

Geographical data exchange standards

On a more technical side, there is also a need for standards in the actual format of the data files to enable the data sets to be transferred between different computer systems. Dissimilar GIS and other spatial data-handling software may be used, so to transfer geographical data files from one to another, and to enable the various

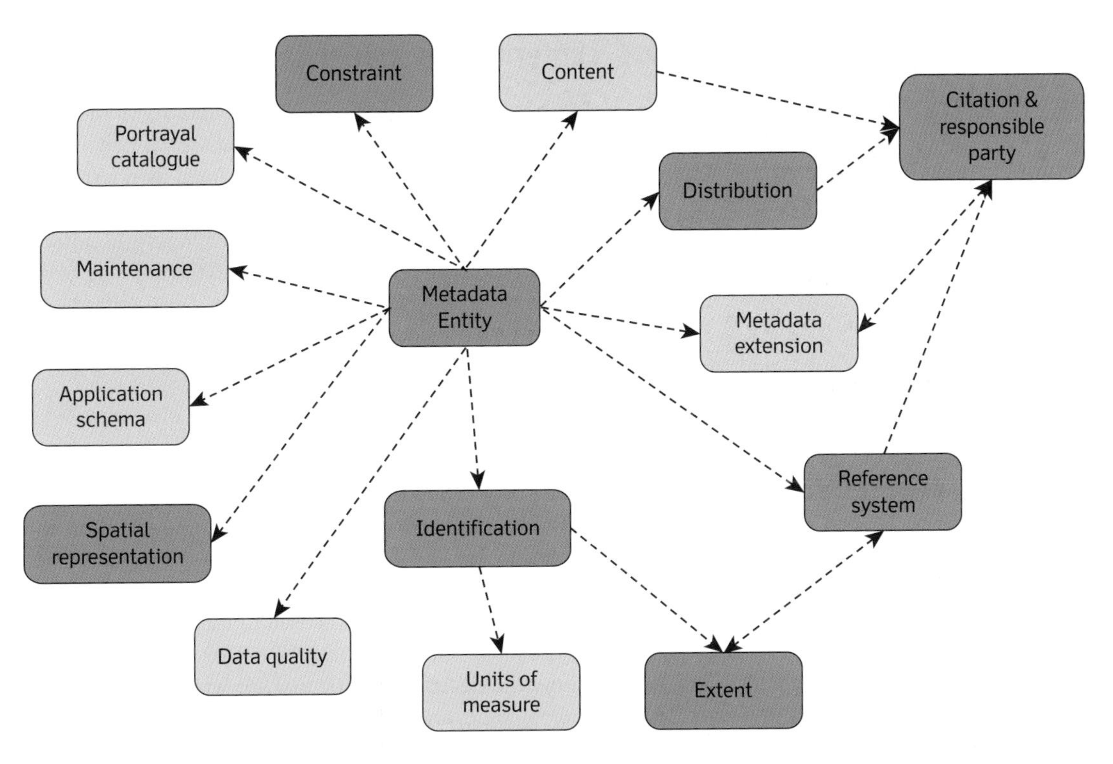

Figure 4.9 ISO 19115 standards for geospatial metadata: the blue boxes highlight the core metadata
(*Source*: Open Geospatial Consortium)

elements of the data to be recognized and used, there needs to be a common format that each is able to read. Geographical **data exchange standards** have been developed which capture all the elements of the data from the spatial data model, through the data and database structures, to the physical file structure.

There has been progress in open data exchange structures, with early developments centred on the main GIS vendors creating export files and functions that could import/translate and read export files created in different systems (such as .SHP and .DGX files).

Further developments have come in the last decade with national and/or international technical commissions defining various transfer standards, often as part of formalized overarching frameworks, to provide access for all users. Government agencies collect huge amounts of geographical data—ensuring these can be exchanged freely within their organizations has been a major driver leading to formalized systems, known as **spatial data infrastructures (SDIs)**. Examples include Canada's Geospatial Data Infrastructure (CGDI) and the USA's National Spatial Data Infrastructure (NSDI) (various examples including model development and challenges are given in Camara et al. 2006; Coleman and Webber 2008; Nedovic-Budic et al. 2011; Cooper et al. 2013; Hendriks et al. 2012). With the expansion of data-collection agencies, the private sector and sub-national governments are now also involved in the development of spatial data infrastructures (Harvey and Tulloch 2006; Rajabifard et al. 2006).

Internationally, efforts have been largely through the Open GIS Consortium, which has been leading the development of open standards and interoperability. It has been particularly active in supporting standards for accessing data over the Internet, leading to developments in GML, the geographic data exchange version of eXtensible Markup Language (XML). Mark-up languages use tags to associate a rule or give meaning to a set of information, with the tags defined in a separate schema, thus bringing flexibility to the data set definitions and enabling computer programs to readily interpret the XML packets. GML is the geographic data version, which is increasingly being adopted as the main method for describing geographical data on the Internet (detailed specification is given on the OGC website). To an increasing extent, national mapping agencies such as Britain's Ordnance Survey are adopting GML as the format for defining and exchanging their geographic information.

4.5 Creating digital data sets by manual input

If the data are not in digital form yet, various manual operations need to be undertaken to convert them. There are four main stages to this:

- entering the spatial data
- entering the attribute data
- spatial and attribute data verification and editing
- where necessary, linking the spatial to the attribute data.

Figures 4.10 and 4.11 summarize the processes for raster and vector data structures. The various database structures used in GIS also require the data to be input differently. The main differences are associated with the second and third stages of the process. With hybrid relational arc databases, the spatial and attribute data are stored separately in the GIS (see Chapter 3) and need to be linked prior to any analysis. With object-type relational databases, a similar procedure is needed as the spatial and attribute data are stored in different databases and the object identifier is used in linking the data. With object-oriented databases, the attribute and spatial data are linked through object and class definitions (see Chapter 3), so there is no need for a separate linking process to take place.

Entering the spatial data

With the entity model, geographical data are in the form of points, lines, or areas or pixels that are defined using a series of coordinates. These are obtained by referring to the geographical referencing system of the map or aerial photograph, or by overlaying a graticule or grid onto it. While it is possible to manually record the different coordinate values into an input mechanism, various hardware devices such as digitizers, scanners, and stereoplotters may be used to speed up the encoding of the X and Y coordinates.

Digitizers

A digitizer is an electronic or electromagnetic tablet on which a map or document is placed. Embedded in the tablet, or located directly under it, is a sensing device that can accurately locate the centre of a pointing device, known as a puck, which is used to trace the data points of the map. The most common types currently used are either the electrical-orthogonal fine wire grid or the electromagnetic, and they range in size from 30 × 30 cm to approximately 1.1 × 1.5 m. Positioning the cursor over a point on the map and pressing a button on it sends an electrical signal directly to the computer, indicating the cursor's coordinates with respect to the digitizer's frame of reference.

The principal aim of the digitizer is to input quickly and accurately the coordinates of points and bounding

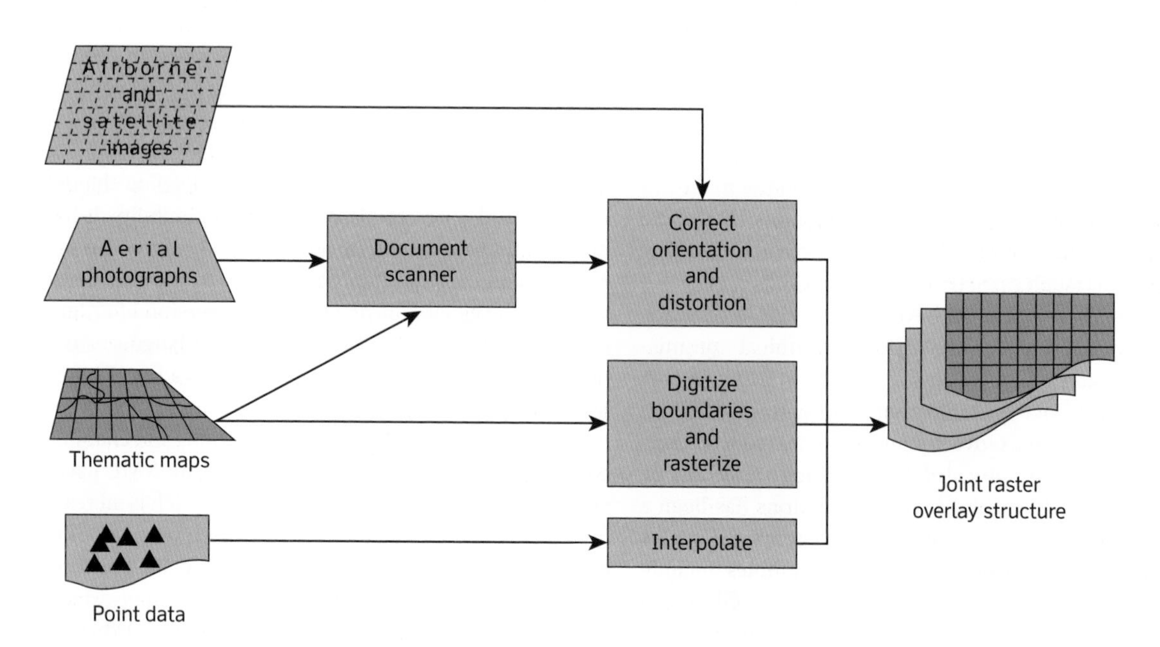

Figure 4.10 Manually capturing and processing of spatial data to build a raster database

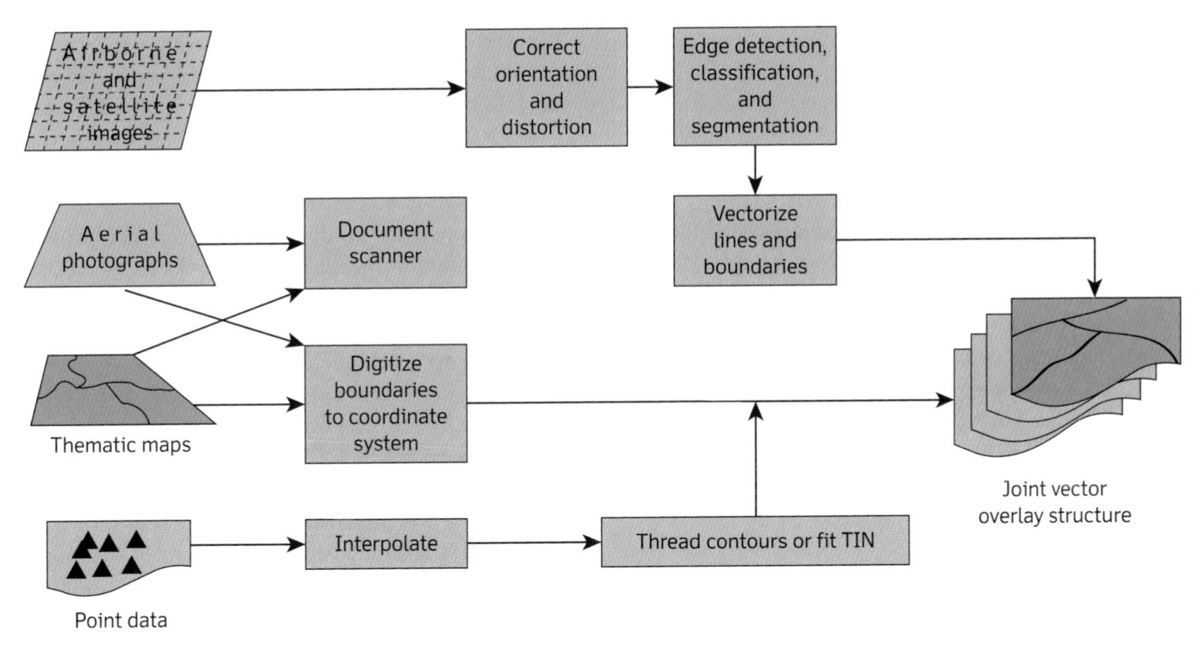

Figure 4.11 Manually capturing and processing of spatial data to build a vector database

lines. Lines or points are digitized by the operator moving the cursor along the feature and recording coordinates either by pushing a button on the puck, or it may be done automatically by the computer at a given time or distance interval. In spite of the development of modern table digitizers, digitizing is time-consuming and enervating work, a drudge. For many applications, the digitizer is a thing of the past as many types of data are collected and provided digitally.

Scanners

Scanners are devices for converting paper maps and photos into digital high-resolution, grid-based images which may be used directly or further processed to give vector representations. The resulting images are very similar to the raster images obtained by remote sensing, and the step size and the rate at which either the document or the light source moves control the cell or pixel size of the scanned image.

The scanned image will be far from perfect, as it will contain all the smudges and defects of the original map plus mistakes caused in areas where the map detail is complex. The digital image may then need interactive improvements to make it usable, with operations from various image-processing methods to remove excess data. The resulting scanned image may be vectorized (described in 'Converting data to vector or raster formats') or transformed into another kind of raster structure for direct input to the GIS. Matching, scale correction, and alignment may be controlled through known point locations, such as registration or fiducial marks which may be defined directly from the scanned image and processed automatically. Scanning, like digitizing, is most likely to be used to capture information from historic maps, since new data are likely to be collected digitally.

Analytical stereoplotters

A third type of technology used for capturing digital geographical data is a **stereoplotter**, which is used extensively in capturing continuous elevation data for digital elevation models and orthophotos. This photogrammetric instrument is used to record the levels and positions of terrain and entities directly from stereo pairs of aerial photographs (taken of the same area but from a slightly different viewing position). In recent developments, digital stereo images from satellite sensors, video recordings, and digital cameras have been used to generate elevation data using specialized photogrammetric algorithms in image-processing systems (see Chapter 11).

Converting data to vector or raster formats

Rasterization (**vector to raster conversion**) is the process of converting vector data into a grid of pixel values. This involves placing a grid over the map and then coding the pixels according to the occurrence or not of the phenomena. A vector line or polygon becomes a series of diagonally, horizontally, or vertically touching pixels. Boundaries are jagged and the number of pixels used in their representation is dependent on the grid resolution and the angle and thickness of the line relative to the underlying grid.

Vectorization (**raster to vector conversion**) is usually undertaken using specialist software which provides algorithms that convert arrays of pixels to line data. The process involves threading a line through the pixels of the scanned image using what are known as thinning algorithms. These reduce the pixelated lines to only one pixel thick. They are then linked from linear or areal units using automatic algorithms which scan and join neighbouring pixels of the same value, or through user-controlled operations.

Entering the attribute data

Attribute data are the characteristic properties of a spatial entity that are linked explicitly to the locational information in a GIS. For example, a road may be captured as a set of contiguous pixels, or as a line entity, and represented in the spatial part of the GIS. Linked to this information are attribute data that describe the type of road (e.g. motorway or track) and other characteristics such as the width, the type of surface, any specific traffic regulations, and the estimated number of vehicles per hour, etc. All this attribute data is recorded separately in the case of relational databases, or is inputted along with the spatial description in object-oriented databases (as described in Chapter 3).

Attribute data may come from many different sources including paper records, existing databases, and spreadsheets. They may be input into the GIS database either manually (though this is increasingly rare) or by importing the data using a standard transfer format, and can include alphanumeric characters as well as digital images or sound records. Where relational databases are used, an identifier is included in the attribute record and this is included in the attribute definition.

Data verification and editing

Once the spatial and attribute data have been entered it is important to check them for errors (possible inaccuracies, omissions, and other problems) before linking the spatial and the attribute data. Checks, by comparing scanned maps or other digital sources with the new data in the GIS, are then made for missing data, double representations, locational errors, or where the data are recorded using too many points.

These errors need to be addressed through various editing and updating functions that are supported directly by most GIS. Minor locational errors in a vector database may be corrected by moving the spatial entity using the screen cursor or by indicating their position on the digitizer tablet. New data may be added with the digitizer or keyboard. Some entity editing operations cannot be used in isolation, but must be followed by checks or operations to ensure the coherency of the database. For example, in networks in utility mapping (e.g. telephone lines) editing a single line into two branching lines requires the pointers indicating the flow of signals to be rebuilt. In polygon networks, if a line or part of a line is moved or changed, the polygon areas must be recomputed.

Where excess coordinates define a line, these may be removed using 'weeding' algorithms, the best known of which are by Douglas and Peucker (1973) and by Reumann and Witkam (1974). The visual appearance of the resulting set of straight line segments may be 'improved' by using **B-splines** (see Chapter 8 for further information).

Attribute values and spatial errors in raster data must be corrected by changing the values of the faulty cells. This may often be done by a simple command that digitizes the cell followed by inputting the correct attribute value. If large numbers of cells are 'wrong', the new information could be digitized and simply written over the existing values.

Linking spatial and attribute data

In GIS with separate geometric and attribute databases the final process in the manual capture of data involves linking the attribute and spatial databases through identifiers which are common to the records in both. For example a road may be given an identifier of A1 during the data input process; records in the attribute database referring to that road will also need to contain a value that identifies it as A1. The links are established for the first time and verified during this data-processing stage but are not immutable and are re-established each time the user calls particular parts of the data set. The linkage operation provides an ideal chance to verify the quality of both spatial and attribute data.

4.6 Data transformation and structuring

Following the data-acquisition process, whether manually or from a data supplier, various processing steps may still be needed, such as converting the data sets to the same coordinate system, joining different map sheets,

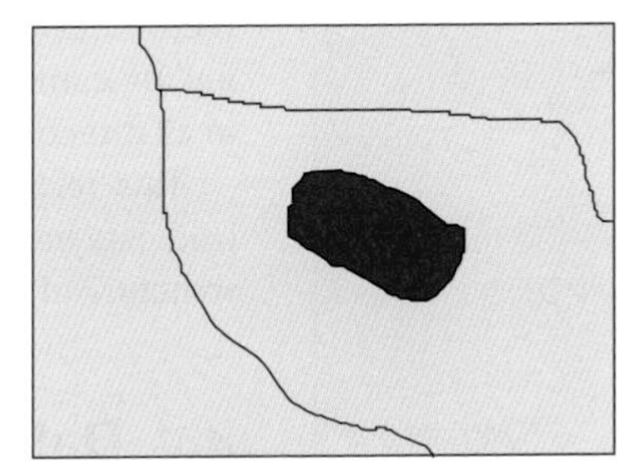

Figure 4.12 Edge-matching transforms of digital data

rasterization or vectorization, before the data sets may be used in any analysis.

Edge matching

In many studies the area of interest stretches across two or more images or map sheets. It is important for further analysis or modelling that the different files are joined to create a seamless database for the subject, using **edge matching** techniques. For vector structures, amalgamating the data is relatively easy for most of the entities in the files as they are self-contained, but for the ones whose boundaries cross maps, lines need to be joined and whole linear or polygonal objects created. The main problem in this processing is that the lines straddling the two map sheets rarely join perfectly, so first data points need to be repositioned so that they meet. Then the point, line, and/or polygon data need to be merged and the topology rebuilt to create the new complete object, and the common attribute value assigned to it (see Figure 4.12). Once this join and recreate operation has been performed for all the straddling entities, the common map sheet line is removed from the data.

Mosaicing

Frequently it is necessary to combine a number of digitally acquired images to create one large seamless file. This is called **mosaicing** and brings images together; this requires two processes:

1. All the images to be mosaiced must be georeferenced to the same coordinate system to ensure geometric integrity of the join.
2. Adjacent scenes that were scanned on separate days or are slightly different because of sun angle or atmospheric effects should be histogram matched to make the apparent distribution of brightness values in the images as close as possible. Matching the histograms of the images minimizes the brightness value variations across the join.

Histograms of images show the spread of digital number values from 0 to 255, and to achieve good results in mosaicing the two input images should have similar characteristics:

- The general shape of the histogram curves should be similar.
- Relative dark and light features in the image should be the same.
- For some applications, the spatial resolution of the data should be the same.

The relative distributions of land covers should be about the same, even when matching scenes that are quite dissimilar. If one image has clouds and the other does not, then the clouds should be 'removed' before matching the histograms.

Once these basic preprocessing steps are completed, the grids and values of the two or more images may be combined.

Coordinate and scaling transformations

An important step in the creation of an integrated GIS database is ensuring that the different data sets are referenced to the same coordinate system. Most GIS support

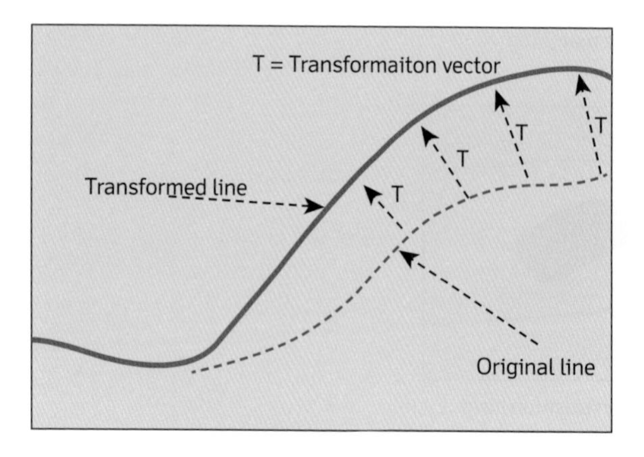

Figure 4.13 Transformation vectors for 'rubber sheeting'

the transformation of a wide range of coordinate systems which are used as part of the data file inputting process. The inbuilt programs use already established mathematical formulae to transform one to another using a combination of scaling, translating, and rotating operations (see Snyder 1987 for examples of formulae).

Where the transformation from one map system to another is not already programmed or where there are distortions in one relative to an accurate base map, more complex rotating and translating operations are needed. The digital map to be transformed is compared with a base map or GPS positions to identify some easily definable points. The software links the locations of unknown map space to their locations in measured map space (Figure 4.13). Mathematical transformations stretch and compress the original map until the linking vectors have shrunk to zero length and the tie points are registered with each other. It is then assumed that all the other points on the original map will have been relocated correctly. This process is known as 'rubber sheeting', 'warping', or 'conflation' because the original map is stretched in all directions like an elastic sheet to fit the other.

This technique cannot be used directly on rasterized data because of the rigidity of the fixed grid and the structure of the data.

4.7 Data quality

While every effort might have been made in the creation of a geographical database, it is important to understand that there will be limits to the accuracy of the data. This uncertainty influences how much the output results from any analysis or modelling exercises can be relied on (Heuvelink 1998; Crosetto and Tarantola 2001; Shi et al. 2004; Karssenberg and de Jong 2005). The quality of the data is affected by many factors, whether it is from primary field collection or laboratory analysis, derived from spatial interpolation methods or from GIS-derived secondary data (as highlighted in Box 4.2).

The starting point in any uncertainty assessment is therefore to consider the whole range of sources of measurement errors that may exist and their resulting influence on the modelling. For field information these can occur whatever the data-collection process, whether this is done by automatic loggers or telemetry, door-to-door surveys, or by social media returns. These might be random in occurrence and effect, perhaps as the result of changing weather conditions or human inaccuracies from one surveyor to another. Or they may

Box 4.2 Factors affecting the quality of spatial data

1. *Currency*

Are data up to date? Time series

2. *Completeness*

Areal coverage—is it partial or complete?

3. *Consistency*

Map scale; Standard descriptions; Relevance

4. *Accessibility*

Format; Copyright; Cost

5. *Accuracy and precision*

Density of observations; Positional accuracy; Attribute accuracy—qualitative and quantitative; Topological accuracy; Lineage—when collected, by whom, how?

6. *Sources of errors in data*

Data entry or output faults; Choice of the original data model; Natural variation and uncertainty in boundary location and topology; Observer bias; Processing; Numerical errors in the computer; Limitations of computer representations of numbers

7. *Sources of errors in derived data and in the results of modelling and analysis*

Problems associated with map overlay; Classification and generalization problems; Choice of analysis model; Misuse of logic; Error propagation; Method used for interpolation be systematic, resulting from biases in poorly calibrated sensors and other measuring devices.

With spatial data, a further important but unseen source of statistical uncertainty may be the number of points collected and their areal spread in any field collection exercise. This can be particularly problematic when data from different sources are combined in a GIS. Even if the data share the same grid, different sets of spatial data can be poorly matched when their spatial patterns have been sampled at different resolutions. If sample surveys for different attributes are tuned to different spatial correlation structures then it is difficult to obtain a sensible match for spatial modelling. The problem is not with GIS, but with the balance of spatial variation in natural and human phenomena. Analysis of indices of spatial co-variation such as the variogram (see Chapter 9) can be useful to ensure that different data are spatially compatible.

Mismatches can occur through:

a) each phenomenon being measured on a different support (area/volume)

b) each phenomenon having different intrinsic spatial variability

c) some phenomena being sampled directly while others are collected or classified using externally imposed spatial aggregation blocks that are inappropriate

d) the spatial variation of different phenomena being governed by processes that operate at different scales—to complicate matters, the same attribute can have different spatial correlation structures at different scales.

Data in a GIS will contain inaccuracies and uncertainty to varying degrees. It is important to acknowledge this even if it is not possible to change the data set. Most GIS do not provide analytical means to determine the effects of errors and uncertainty on the results of the operations undertaken, so further analysis needs to be undertaken, usually away from the system, using forms of statistical and probabilistic analysis. Chapter 12, for example, describes error propagation in modelling and methods that can be used for analysing the uncertainty in the output results.

4.8 Data updating

Geographical data are not inviolate for all time, and are subject to change. Few of these changes are so deterministic that they can be performed automatically. For example, political boundaries may change with the whims of parliament, land use and field boundaries may change as the result of reallotment, or soil boundaries change as a result of land improvement or degradation. If these changes in the landscape are not included in the database, then its credibility and integrity will be undermined; updating is therefore needed. The basic operations just described for editing are used to update the data, so new data are input, existing information relocated, and new or modified attribute values introduced. When changes to the spatial structuring are needed, the topology of the data set will need to be generated again.

Updating is rather more than just modifying an ageing database; it implies resurvey and processing new information. Some aspects of the earth's surface, such as rock types, change slowly, and important changes are few, so updating remains a small problem. However, there are other kinds of geographical data where it may be more cost-effective to resurvey completely every few years rather than attempt to update old databases. In comparison, updating the attribute data is trivial, provided the one-to-one links between attribute data records and the spatial entities remain unaltered.

4.9 Considering local tacit knowledge

The addition of different data sets into a GIS represents component parts of the knowledge we understand of the world we are studying. This knowledge, often based on formal scientific frameworks, is acknowledged to be complex, socially defined, subject to changes over space and time, and its content depends on the ontological frameworks used to both simplify and represent the complexity of the subject.

Forms of knowledge may be roughly categorized into two—explicit and tacit. The former is knowledge which can be articulated into formal language, including grammatical statements (words and numbers), mathematical expressions, or specifications, and is the form used extensively in GIS. For example, landscapes in a natural science framework are defined in terms of flow patterns, resources stocks, cycles, and yield. There is an implicit power that goes with using this form of knowledge, in that it is perceived as accurate, based on scientific principles, and so above emotional influence.

There are, however, other forms of knowledge that are often referred to as local, tacit, or naïve, which are acknowledged to be important in developing understanding yet are little used in GIS (McDonnell 2008). This form of knowledge is usually personal knowledge, embedded in individual experience, and often involves less tangible factors and sometimes subjective insights. For example, the information a farmer knows such as exactly where the soil is wet or more fertile or receives the most sun on the land is rarely if ever included in a GIS. The form of spatial referencing is often relative with a person's own mental map of an area defining the positions and interactions of key objects. While this knowledge does not conform to the scientific rigours of most data, it can often bring increased depth of understanding specific to a particular area.

This knowledge is often excluded from analysis or is converted from spoken or drawn information into data that conforms to the spatial data model of a GIS and alphanumeric defined attribute values (Elwood 2006b). In doing so, some of the richness of the knowledge is lost as this cannot be captured in this way. Social media use is helping to ensure more of this local knowledge is included through location-defined images or thought capture, but so far the ability to analyse this form of information is limited.

4.10 Summary

Given the enormous effort involved in creating databases in a GIS, it is important that they are as accurate and error-free as possible. Today, living in the era of 'Big Data', enormous amounts of information are being generated and made available, particularly through the Internet. This brings new insight and understanding to many fields of study and parts of the world where little data previously existed. However, there is a need for the user to think critically about the application the sourced data are being used for and to consider which data will provide the most insight. Unfortunately it is easy to introduce errors, inaccuracies, and lack of precision if the data used are not collected using the most suitable spatial scale, measuring systems, and currency. More importantly, in any given application, secondary data can account for over 90 per cent of the total data, and so there is a need to assess the quality of these data sets as well as design primary data-collection activities.

? Questions

1. What data sources would you use to assemble spatial data for the following applications?
   - i. Disaster relief from hurricanes
   - ii. Climate change adaptation in a small island state
   - iii. Managing the environmental impact of mining operations
   - iv. Increasing the number of people with access to improved water supplies
   - v. Maximizing retail success for a cake shop
   - vi. Choosing a property that is little affected by floods
2. What do you think are the important issues that should be considered when using data collected by someone else?
3. How would you design a crowdsourcing exercise to find out information on an issue of importance in your home town?
4. How do you collect spatial data in your everyday activities?

→ Further reading

- Chrisman, N. R. (2003). *Exploring Geographical Information Systems*. Wiley, Hoboken, NJ.
- Heywood, I., Cornelius, S., and Carver, S. (2011). *An Introduction to Geographical Information Systems*. 4th edn. Pearson, Harlow.
- Hodgkiss, A. G. (1981). *Understanding Maps*. Dawson, Folkstone.
- Hofmann-Wellenhof, B., Lichtenegger, H., and Wasle, E. (2008). *GNSS—Global Navigation Satellite Systems: GPS, GLONASS, Galileo, and More*. Springer Verlag, Vienna.
- Kennedy, M. (1996). *The Global Positioning System and GIS: An Introduction*. Ann Arbor Press, Chelsea, MI.
- Lillesand, T. M., Kiefer, R. W., and Chipman, J. W. (2008). *Remote Sensing and Image Interpretation*. 6th edn. Wiley, Hoboken, NJ.
- Mather, P. and Koch, M. (2011). *Computer Processing of Remotely-Sensed Images: An Introduction*. 4th edn. Wiley-Blackwell, Oxford.
- Monmonier, M. (1993). *How to Lie with Maps*. University of Chicago Press, Chicago.
- Nedovic-Budic, Z., Crompvoets, J., and Georgiadou, Y. (eds.) (2011). *Spatial Data Infrastructures in Context: North and South*. CRC Press, London.
- O'Sullivan, D. and Unwin, D. J. (2010). *Geographic Information Analysis*. 2nd edn. Wiley, Hoboken, NJ.
- Peng, Z.-H. and Tsou, M.-H. (2003). *Internet GIS: Distributed Geographical Information Services for the Internet and Wireless Networks*. Wiley, Hoboken, NJ.
- Rogerson, P. A. (2006) *Statistical Methods for Geography: A Student's Guide*. 2nd edn. SAGE Publications, London.
- Skidmore, A. K. (2002). *Environmental Modelling with GIS and Remote Sensing* (Geographic Information Systems Workshop). CRC Press, Boca Raton, FL.
- Wise, S. (2014). *GIS Fundamentals*. 2nd edn. CRC Press, Boca Raton, FL.

Further reading

Visualization

Maps, whether paper or digital, are encountered in a wide diversity of contexts. But the meaning of maps and the principles behind their creation are by no means simple. All users of GIS software are likely to want to visualize their data in some way—to enable them to identify patterns, to help decide how to proceed with an analysis, or simply to create a map which can be used by different individuals or groups. Visualization and presentation of geographical data is a complex topic, and a multitude of issues must be considered in order to visualize data in appropriate ways, or to generate maps which are easy to use and interpret. This chapter considers some key issues, including mapping point locations, categorization of variables, and cartographic mapping principles, as well as visualization of surfaces, among other themes. A wide range of examples are presented to illustrate some of the available modes of visualization, as well as particular topics which it is important to consider when visualizing or presenting data in a GIS.

Learning objectives

By the end of this chapter, you will:

- **understand some of the key ways in which we can present and visualize geographical information**
- **be able to apply your knowledge to create maps which convey information effectively**
- **have developed knowledge of how visualization approaches have evolved in the last decade.**

The next six chapters of this book introduce a range of methods for the analysis of geographical data. These include methods that assist with identifying spatial patterns or for identifying areas which meet several different criteria. But the first stage of any application of geographical data is likely to consist of simply looking at the data. Visualization provides a means of assessing major patterns, and simple inspection of maps is likely to be an important first step in using the data to solve problems. Indeed, visual examination of maps is often a key objective in itself and many users of GIS are interested predominantly in map creation. Maps (if well designed) allow us to ask 'what?' and 'where?' questions. For example, questions such as 'what is the land use at this location?', 'where are pollution levels high?', or 'where is the density of sites greatest?' may be addressed by looking at maps. In many applications, maps may be used to guide the analysis steps which follow.

Visualization can be divided into **ephemeral** and **permanent outputs**. Most visualization in GIS falls under the first heading. In the second case, cartographic principles of map design are important, and it is usually necessary to include on maps information on **spatial orientation**, **scale**, and **map symbology**. Maps may be produced as hard output—that is, paper maps. Alternatively, maps may be generated and disseminated digitally—many users access geographical maps online and may never require a hard copy. Wood (1994) defines (a) **cartographic communication**—a map is designed for a particular purpose and has a specific message (e.g. a map showing the characteristics of surface geology in an area) and (b) **cartographic visualization**—the message is not known and there is no specific purpose. Wood argues that only the latter assists with the discovery of patterns and relationships in spatial data. These two categories thus mirror what we have called, respectively, presentation and visualization.

Kraak (2008) considers three roles for visualization. Firstly, visualization allows us to present spatial information and address questions about what something is, where it is, and what belongs together. **Cartographic principles** may be followed to develop a map which shows the features of interest. Secondly, visualization may be used for analysis. For example, questions about the best site for a given purpose could be answered by visually overlaying two data sets. In this case, users must be able to access and manipulate the data. Thirdly, visualization can be used to help identify the data set which is most useful for a particular purpose. Kraak (2008) cites the example of a set of remotely sensed images acquired at different time periods. In this case, a user may need to decide which of the images displays patterns which are related to a particular problem. Returning to our two-part categorization above, the first of these three themes corresponds to presentation and the second and third to visualization.

This chapter focuses on visualization or presentation of data for a particular time point. For example, our map may make use of information on the transport network in a city and on the population of that city as recorded in a census survey. Many applications which make use of spatial data are real-time applications—as an example, an ambulance station or taxi company may track the location of their vehicles using GPS and thus the visualization is 'live'. The main focus here is on access to, and visualization of, data using a conventional GIS environment (whether using locally installed software or a web-based GIS). For many users of GIS, personal digital assistants (PDAs) or smartphones provide a key means of accessing geographical information. Users of PDAs often construct purpose-specific (often single-use) maps (for example a map showing all restaurants within 2 km of where they are standing). While the main focus here is on desktop GIS, many of the principles discussed in this chapter are also relevant for mobile GIS-type applications.

The following sections consider the mapping of point events, selection of categories (relevant to points, lines, area features, or continuous fields), cartographic principles, and a variety of other themes relating to visualization of geographical data.

5.1 Mapping points

Modes of visualization depend on the form of the data we want to visualize. That is, objects (frequently represented using vectors) may be visualized in quite different ways to raster grids. The simplest mappable object is a point, which comprises information on its location in two dimensions—there is an X coordinate and a Y coordinate. A set of points which relate to a particular set of events (e.g. locations of trees or people with a particular disease) is termed a **point pattern**. In many cases, the events consist of locations and there are no values (e.g. a measurement of some property) attached. Where values are attached (e.g. tree locations with a measure of foliage density), then this is called a **marked point pattern**. An example of a point pattern—the locations of redwood trees—is given in Figure 5.1. Point samples (e.g. measurements of precipitation at particular locations) are not discrete events, and so a set of samples of this kind does *not* comprise a point pattern. However, whether point data are point patterns or sample locations,

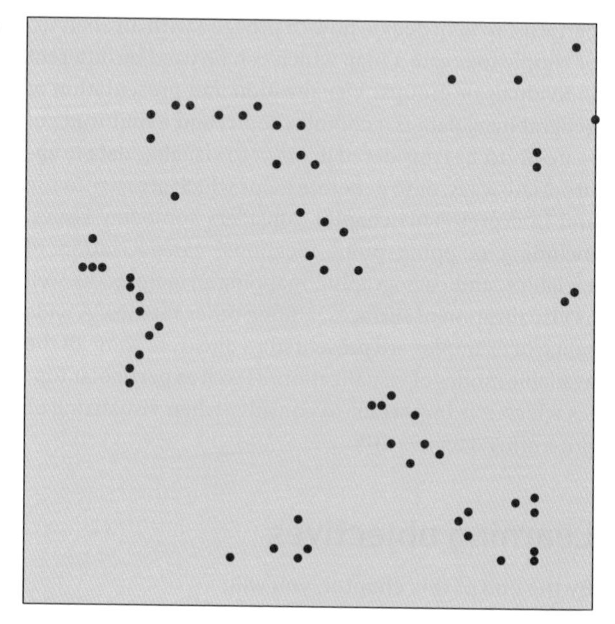

Figure 5.1 An example of a point pattern—the locations of redwood trees

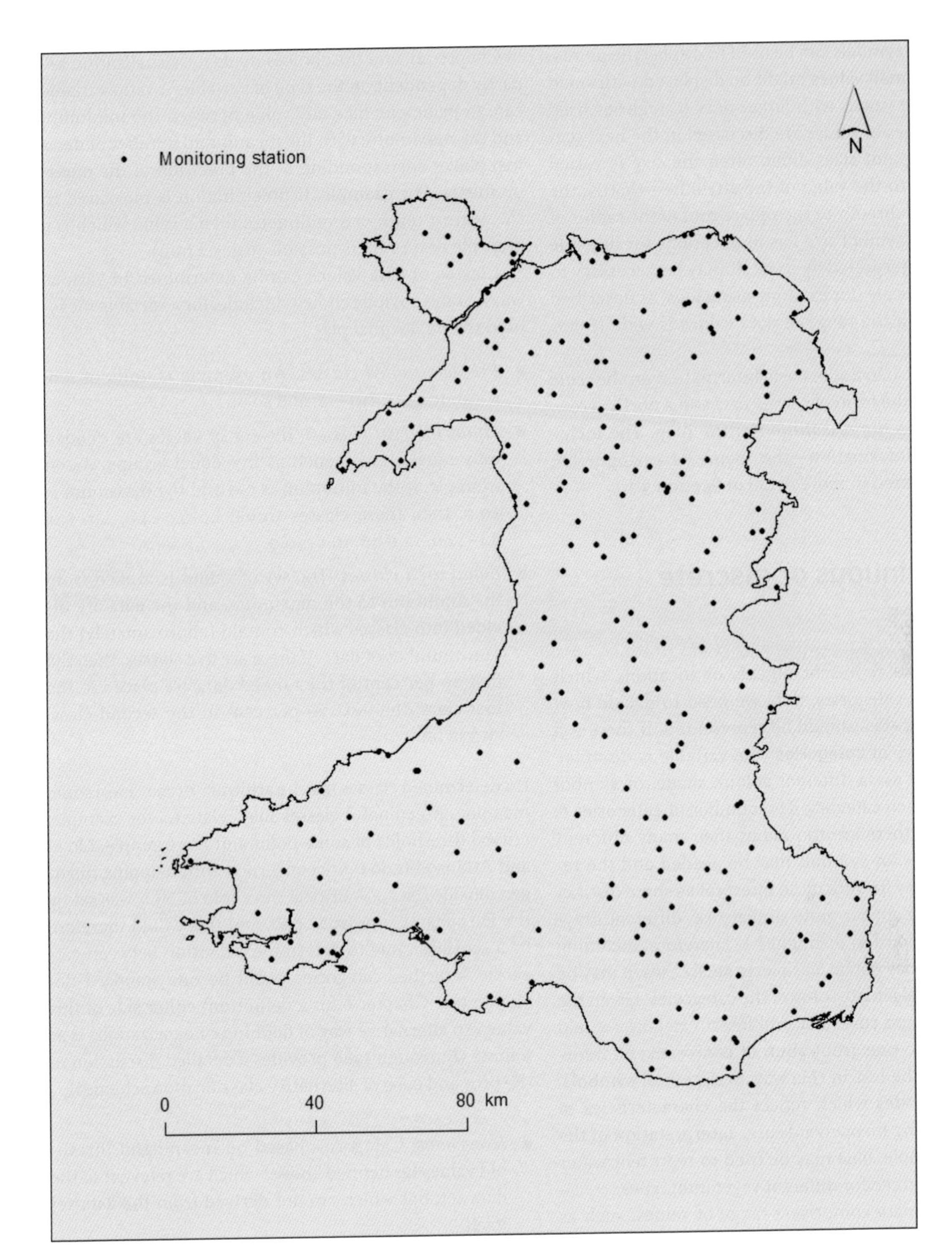

Figure 5.2 Precipitation monitoring stations in Wales for the month of January 2006; $n = 244$

similar visualization strategies might be employed. In both cases, a simple map of the point locations is likely to be generated.

Figure 5.2 shows a map of precipitation monitoring stations in Wales for the month of January 2006. In this case the points are samples of a continuous property and they are not, therefore, a point pattern. Each sample has a value attached to it, but these are ignored in the map—the focus is on spatial location. If there are overlapping points or the values attached to points are of interest, then shades or symbols can be used to display the points. For example, small values might be displayed with small dots, and large values with larger dots (categories used to display ranges of values are discussed in the next section). Where point sizes differ, often the size is varied in proportion to the value of the attribute—that is, the value is related directly to the square root of the radius of the circle. But symbol size has implications for how the symbols are interpreted by users. It may be necessary to transform the scale (for example using logs, as described in Chapter 6) if the range of data values is wide (Jones 1997).

Figure 5.2 also includes basic information on the *scale* of the map (in the form of a scale bar) and a north arrow, which indicates the *orientation* of the map. The inclusion of such information—the theme of cartographic design—is covered in more detail in Section 5.3.

5.2 Continuous or discrete categories

Where the data represent objects or locations which have values or categories, then we need to decide how these characteristics should be represented. If there is a specific number of **categories** (the variable is *discrete*), then we could use a different colour, shade, or symbol to represent each category. The number of categories is important—if there are too many, then many different colours, shades, or symbols may be needed and the resulting map may be difficult to interpret as some distinct categories may appear quite similar (e.g. different greyscales may be hard to distinguish). Therefore, including all categories may not be feasible. In such cases, it may be sensible to amalgamate some of the categories. Given the example of a land cover map, all forest categories could be merged into one group. But, of course, useful information might be lost in this way. Selection of symbols, colours, or shades which reflect the characteristics of the feature being mapped aids user interpretation of the map. For example, blue may be used to represent water and shades of green for different vegetation types.

Where the data comprise a range of values, such as from the minimum precipitation amount to the maximum precipitation amount, then the values may be split into groups. Any map which uses categories is, of course, a function of those categories, and it is essential that users consider how particular features may be emphasized by selecting different sets of categories; this theme is explored below. Chapter 6 discusses different types of variable, such as ratio or interval variables (as discussed in Chapter 2), and the choices made for visualization are partly dependent on the type of variable. A ratio variable can, in principle, take any value between the minimum and the maximum, with the meaningful number of decimal places corresponding to the precision of the measurements. For example, if precipitation is measured to the nearest tenth of a millimetre then a value which is a multiple of 0.1 mm is possible (e.g. 5.2 mm).

Classes of data values can be determined in various ways. Some obvious choices include, for a variable measured to one decimal place:

- *Pre-determined classes*: An example is units of ten: 0–10, 10.1–11, 11.1–12, and so on.
- *Equal interval classes*: The set of values are divided into equal classes, such as five equal groups. As an example, if the minimum is 0.0 and the maximum is 20.0, then these classes would be 0.0–4.0, 4.1–8.0, 8.1–12.0, 12.1–16, 16.1–20.0.
- *Equal area classes*: The set of values is ranked from the minimum to the maximum, and the data are divided into classes which contain (approximately) the same number of data. If there are five classes, then the first 20 per cent of the ranked data are placed in the first class, the next 20 per cent in the second class, and so on.

Predetermined classes may be arbitrary or may have some meaning. Meaningful classes may relate to, for example, critical thresholds of some pollutant. For example, Lloyd and Atkinson (2004) are concerned with mapping nitrogen dioxide (NO_2). A critical threshold of NO_2 was set by the European Union at 21 ppb, and this would therefore be a sensible value to determine a division between categories—further categories could be one standard deviation (see Chapter 6 for a definition) either side of this value. An alternative way of defining categorizations is as follows (Burrough 1986 provides a detailed discussion of the pros and cons of alternative classification schemes):

- *Exogenous*: Categories based on meaningful threshold values (as defined above) which are relevant to the data set, but which are not derived from the data set itself.
- *Arbitrary*: Categories with no prior rationale, often created following limited examination of the data.
- *Idiographic*: Categories derived from the data; these include percentile classes (equal area classes above).
- *Serial*: Categories which comprise a consistent numerical sequence; an example is equal subdivisions of the data range (equal interval classes above). These include:

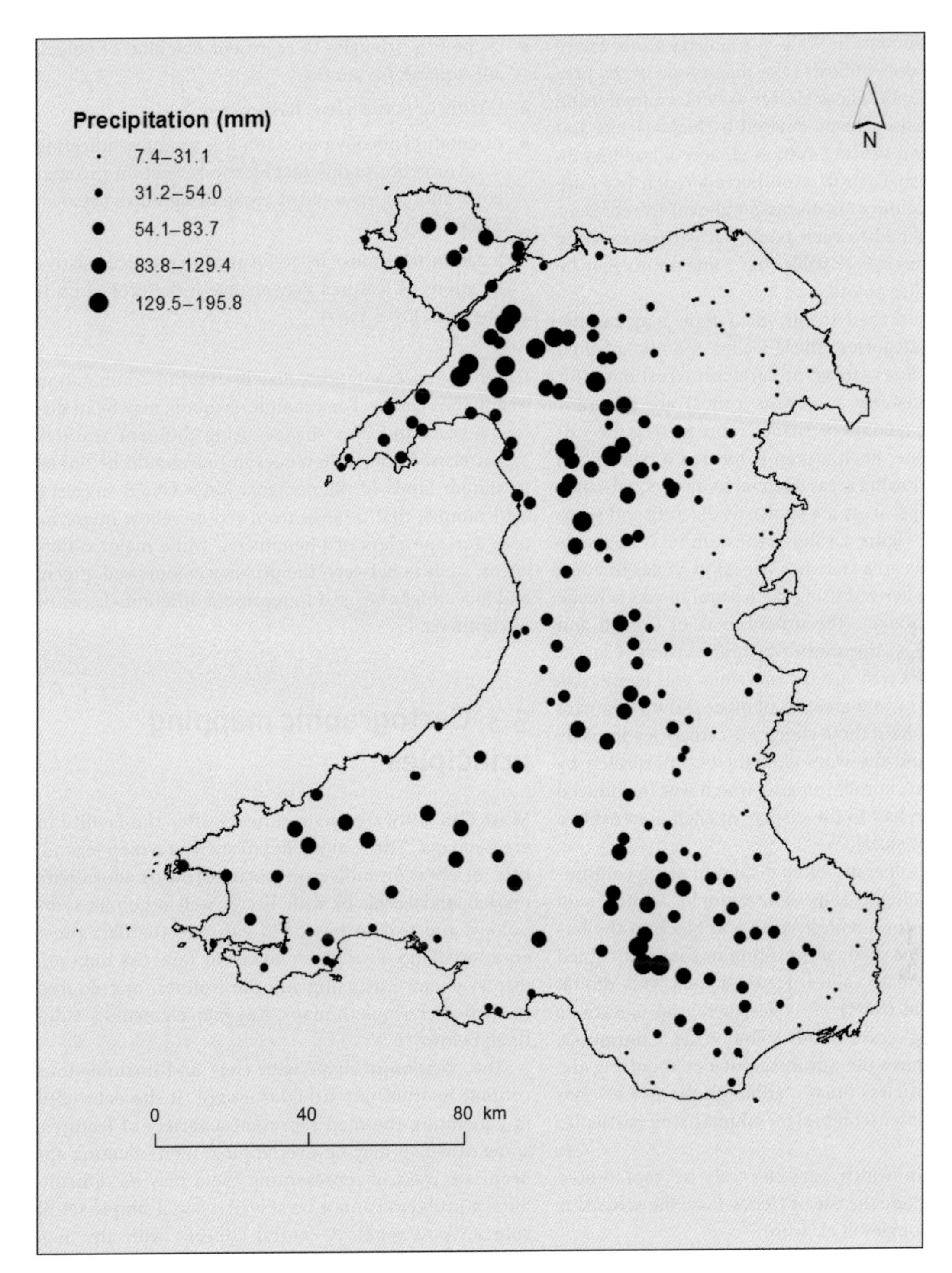

Figure 5.3 Precipitation amounts in Wales for the month of January 2006 using different-sized points

- normal percentiles—division of a normal distribution into classes with equal frequency
- classes defined as a proportion of the standard deviation, where the classes are centred on the mean
- equal arithmetic intervals (fixed class width)
- equal intervals on other scales such as reciprocal, trigonometric, geometric, or logarithmic scales.

In Figure 5.2, data on precipitation amounts in Wales were used as an example. Figure 5.3 shows a map of precipitation amounts in Wales for January 2006, where the size of the dots indicates the magnitude of the precipitation amounts. These classes are determined using the **natural breaks scheme** devised by Jenks (Jenks and Caspall 1971; and see de Smith et al. 2007), building on the work of Fisher (1958). Another approach is to use greyscales or colours (as discussed above) to represent the values attached to each point. Of course, in using these approaches it is possible that some points may be obscured by other points.

Attributes attached to any data type may be displayed using categories. These include points (as illustrated above), lines, areas, or surfaces. Areal data (for example, populations in census zones) are generally depicted using colours or shades representing the values in each zone. Such a map is termed a choropleth map—a map showing areas deemed to be of equal value. With these maps, areas are shown as discrete and separated by edges. Figure 5.4 shows the numbers of persons by ward (census area statistics zones) in Wales in 2001. The classes are derived using the natural breaks scheme of Jenks (see above). The urban areas of Cardiff and Swansea, both on the south coast, are visible. Clearly, selection of different groups of values will emphasize different features, and creators of maps should take care to understand how their choice of categories impacts on the map, and the ways it might be interpreted by users. The term chorochromatic, which was introduced in Chapter 2, refers to the display of each area using a single colour or shade.

Raster grid values are often displayed using continuous grey or colour scales—for example, white could be assigned to the smallest value and black to the largest value, with a gradual transition in shades assigned to the intermediate values. Figure 5.5 shows a **digital elevation model (DEM)** of Wales, with the elevations indicated using a continuous colour scale. Continuous colour scales have the advantage of not requiring the specification of class breaks, although the use of class breaks provides a useful tool for emphasizing particular features.

Key ways in which variables can be represented graphically include the use of (Jones 1997; the schema is expanded by Longley et al. 2010):

- hue (colours as defined by names, such as the colours of the rainbow)
- lightness (or value, referring to lightness or darkness of an area of pigment assigned to a symbol) (see Figure 5.5 for an example)
- size (see Figure 5.3 for an example)
- shape (e.g. triangles to represent one kind of object, and squares for another)
- texture or pattern (e.g. hashes or dots)
- orientation (an obvious example is symbols indicating wind direction at different locations; Section 5.10 suggests another example of commuting flows between places)
- location (the most basic form of depiction; relative locations of features may change if the projection is changed; Jones 1997).

Different representations may be used in combination to enhance clarity. For example, symbols may be of different sizes and also shaded using different textures or patterns. Minor differences in hue should be linked to similar kinds of phenomena. Jones (1997) suggests, for example, that a range from red to yellow might be used for one class of phenomena, while major differences, such as between the primary colours red, green, and blue, might be used to represent different classes of phenomena.

5.3 Cartographic mapping principles

Most GIS software environments offer the facility to create maps. These allow detail such as a map legend, title, orientation indicator (north arrow or coordinate marks), and a scale or scale bar, as well as colour, symbology, and text to be added easily to the data presentation. They also give choices in how the data are displayed, such as using smooth isolines, or coloured or shaded choropleth maps. Key map elements are defined below.

The creation of maps with clear and unambiguous content is often not straightforward. If the data used in generating the map represent a variety of features, some of which may be overlapping, then selecting appropriate ways of representing them may be difficult. Such a problem cannot be solved using a simple set of rules (Wood 1994). A central concern with any map is that users should understand what is shown on the map. The meaning of shades, colours, or symbols must be clear—their meanings are generally conveyed using a **legend**. It is also necessary that the area covered by the map is made apparent—this links to the spatial scale of the map. In addition, the *orientation* of the map should be indicated. By convention, north is usually at the top of a map, and is indicated by a north arrow. The

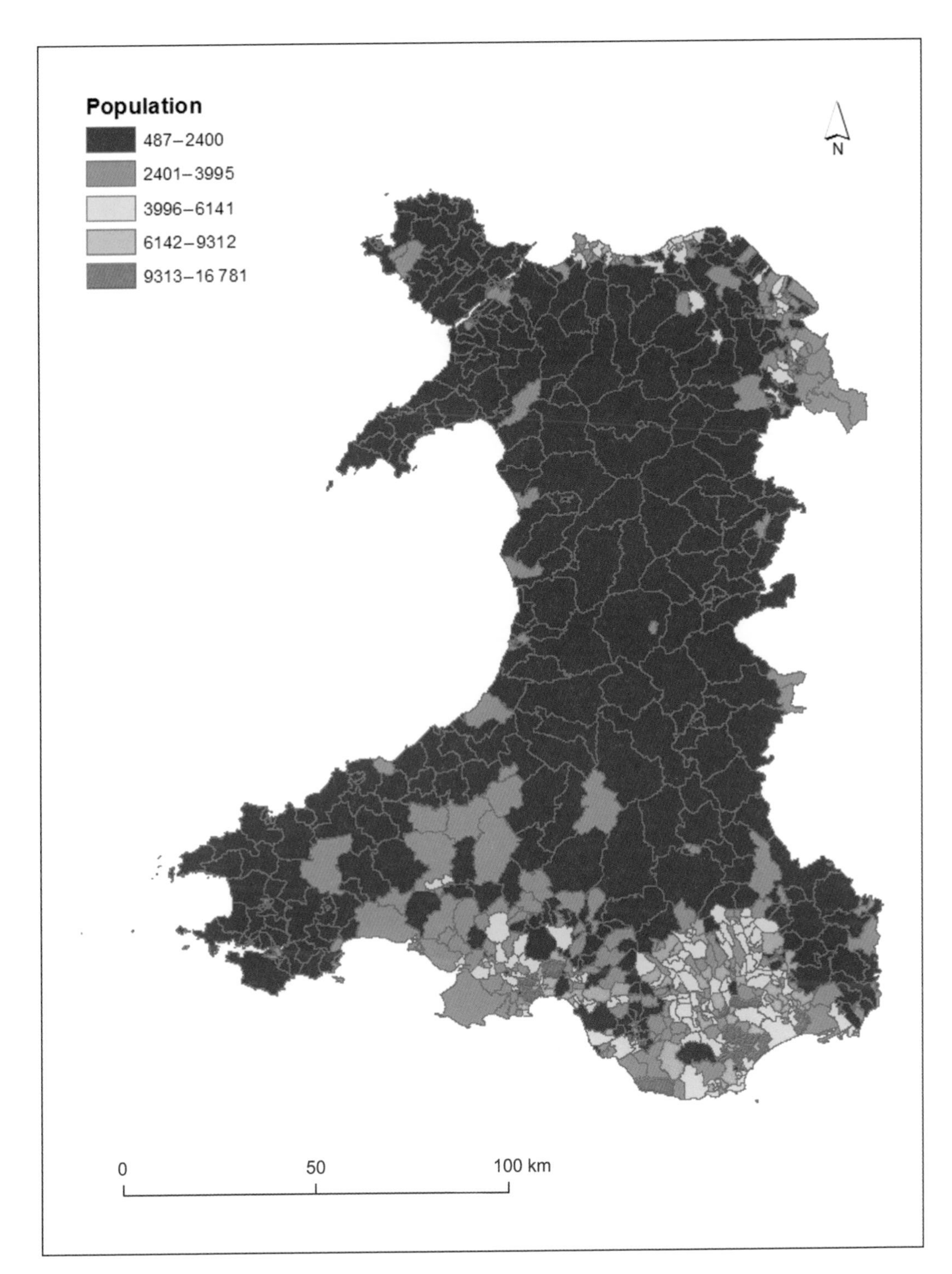

Figure 5.4 Numbers of persons by ward in Wales in 2001

details of the projection (see Chapter 2) used are generally noted on the map, or in the legend associated with the map. Other details which may be included on a map include information on survey dates and revisions to the map.

The maps in Figures 5.4 and 5.5 include a legend, north arrow, and scale bar. The selection of categories or symbols has already been discussed. The spatial scale of a map can be indicated in a variety of ways. In all cases, the objective is to indicate the relationship between the

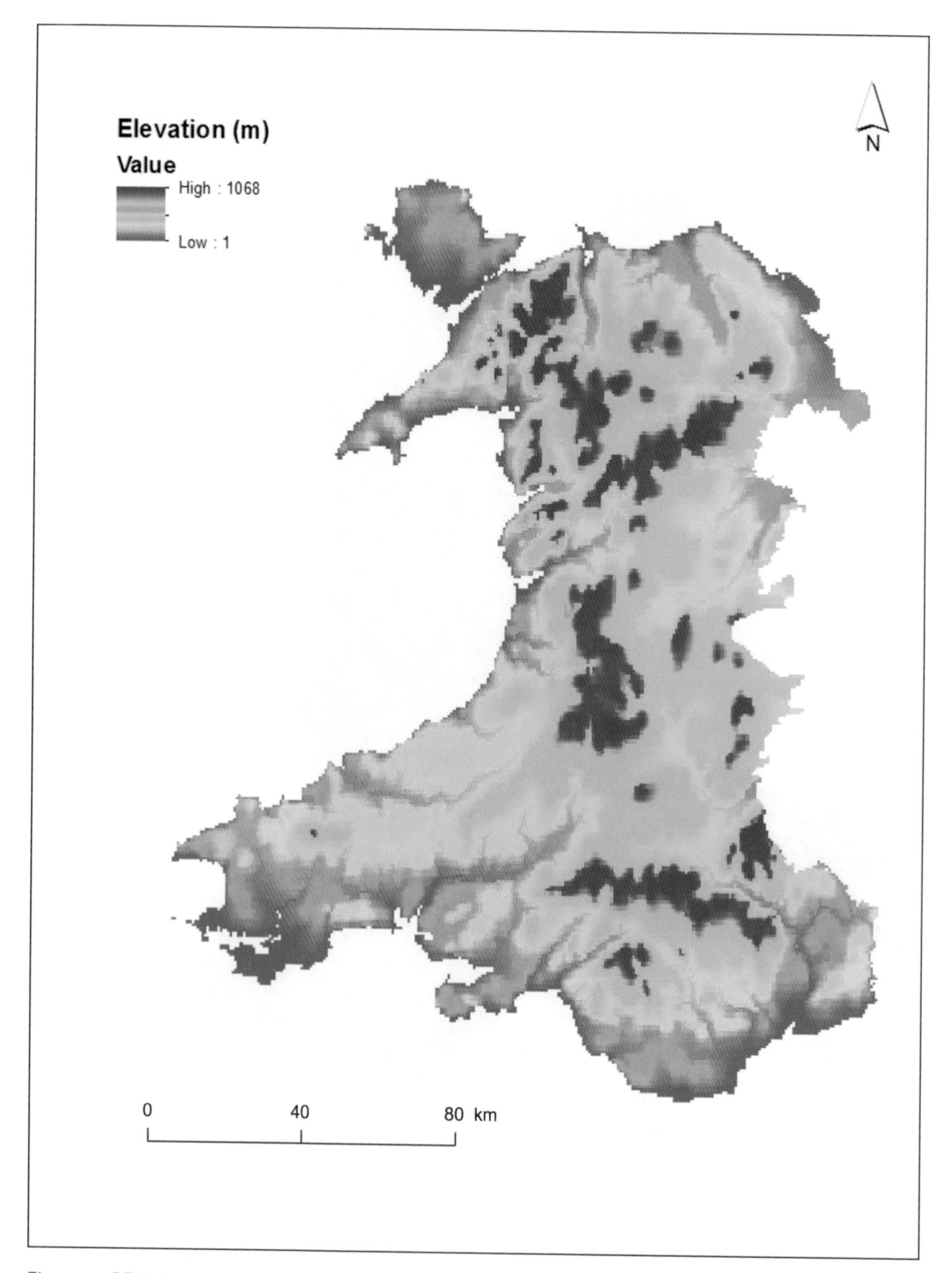

Figure 5.5 DEM of Wales

sizes of features on the map and their size in the real world. Figures 5.4 and 5.5, for example, include a scale bar which directly relates real-world distances to distances on the map. Grid lines of a fixed spacing could also be superimposed on a map to indicate the size of areas on the map (e.g. 1 km grid squares). Another common way of indicating map scale is to use a **representative fraction**.

A representative fraction is the ratio of distances on a map to distances in the same units in the real world.

A representative fraction of 1:10 000 indicates that one unit of measurement on the map—e.g. 1 cm—represents 10 000 of those same units on the ground. For a map with a representative fraction of 1:10 000, a distance of 1 mm in length on the map represents a distance of 10 m in the real world (for clarity, 10 000 mm = 10 m). A large-scale map (for example, a map with a representative fraction of 1:10 000) shows features with greater detail, while a small-scale map has a smaller representative fraction (for example, 1:1 000 000), and shows features with less detail. The terms 'large-scale' or 'small-scale' are not necessarily used consistently by specific individuals or organizations, and it is possible that what one organization terms a 'large-scale' map would be termed a 'small-scale' map by another. Figure 5.6 shows sections of maps for the same area of Cardiff using 1:10 000 and 1:250 000 maps, while Figure 5.7 shows parts of the same maps using the correct relative scale.

It is important to remember that the objects displayed on a map are often not represented at a scale which is consistent with the stated scale of the map. For example, if a map has a representative fraction of 1:10 000 and it shows a minor road using a line which is 1 mm thick then this suggests that the road is 10 m wide. The size of objects used to represent features is intended to make those features clear on the map, and the objects which appear on the map are often symbols rather than, for example, attempts to depict objects as they would appear from a given altitude. It is worth noting that maps do not strictly show a bird's eye view, as the scale of features tends to be constant across a map—this is not true if we look at a landscape from an aeroplane, as features directly below the plane are, conceptually, viewed at a different scale than features further away from the aeroplane. Another core theme in cartographic design is **generalization**. When a map is created, features are selected for inclusion in the map, and representations are chosen to show the selected features clearly. This process may involve a loss of information, both in terms of selecting only a subset of the available information for inclusion in the map and in terms of graphic abstraction, whereby the scale at which the map is produced does not allow representation of the features mapped at the same level of detail as the original survey. An example is the construction of a road map at a scale of 1:100 000 using information from a more detailed 1:10 000 map whereby only major roads are shown on the new map and the spatial detail of the shape of roads is reduced. Generalization is discussed in some detail by Jones (1997).

Robinson et al. (1995) is a standard text on the principles of cartographic map design, with detailed coverage of maps, mapping, and data sources, and core issues including scale and map projections. Harvey (2008)

(a)

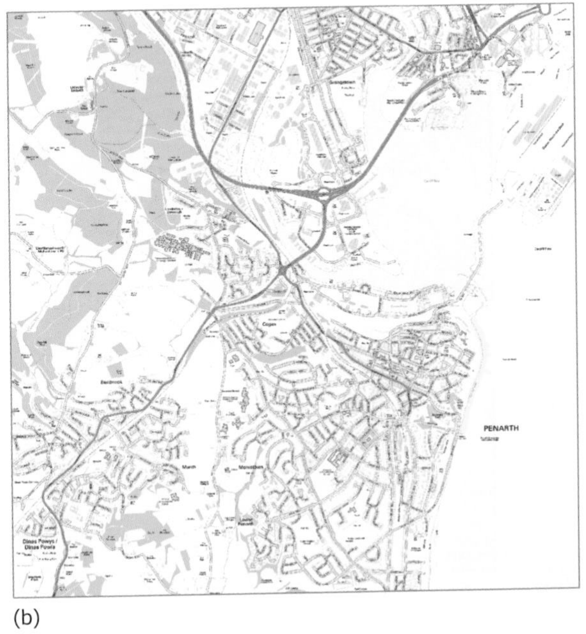

(b)

Figure 5.6 The same area of Cardiff represented using 1:10 000 (OS Streetview) and 1:250 000 maps
(Contains Ordnance Survey data © Crown copyright and database right 2012)

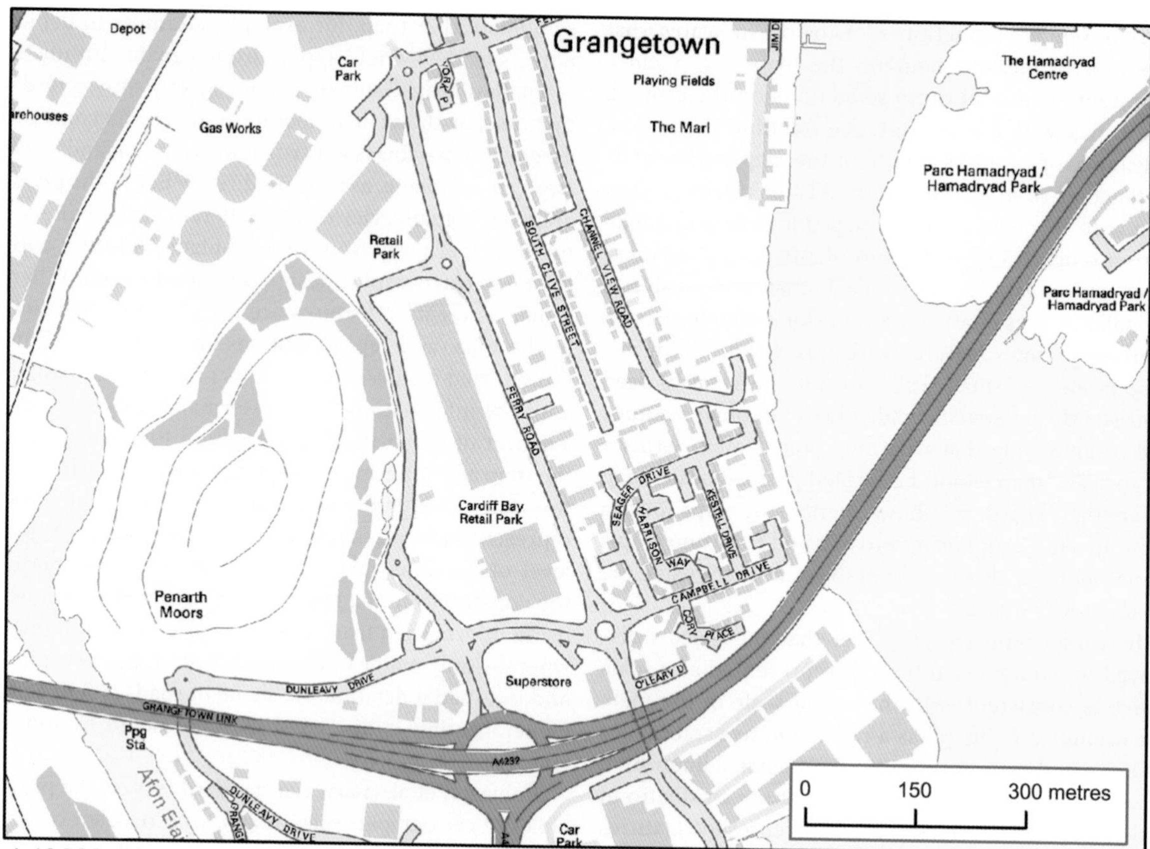

1:10000 (10 cm = 1 km on the ground)

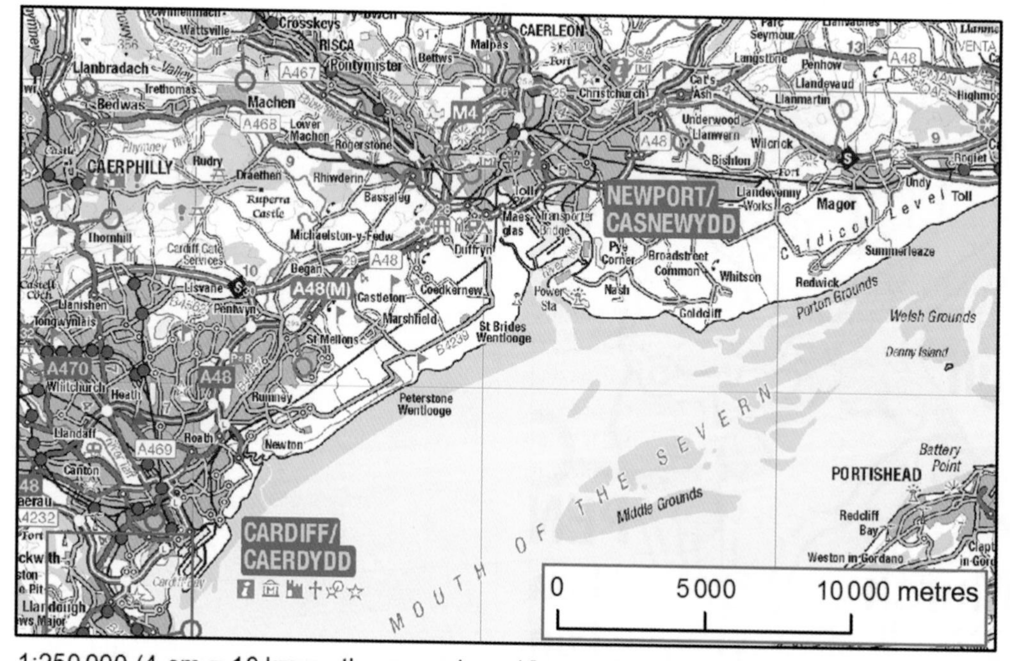

1:250000 (4 cm = 10 km on the ground, so 10 cm = 25 km on the ground)

Figure 5.7 Parts of the maps in Figure 5.6 shown to the correct relative scale

(Contains Ordnance Survey data © Crown copyright and database right 2012)

provides another detailed introduction to cartographic design principles.

5.4 Distorting space: cartograms

So far in this chapter we have assumed that distances between places are Euclidean—they are straight line cartographic distances. In some cases, it is useful to be able to distort distances to emphasize particular characteristics or meanings of the data. The outcome of the distortion of a spatial data set in this way in called a **cartogram**. A famous example is the map of the London Underground, familiar to anyone who has tried to navigate their way around that transport system. Henry (Harry) Beck, an Underground electrical draughtsman, was the creator of the Underground map which provides the basis of the map used today; the first version of the map was submitted to London Underground in 1931. In that map, the distances between places do not reflect the 'real' relative distances between stations. The map is a simplified (or generalized) version of the Underground network. The purpose of this simplification is to make it easier for travellers to use, and it is based on the realization that the connections between tube stations are more important than the distances between them. Many other maps of transport systems are based on similar principles.

Euclidean space can be distorted in a wide variety of ways. The well-known London Underground cartogram is a *line cartogram*: the lengths of lines are distorted. **Area cartograms** are also encountered in many different contexts. Figure 5.4 showed the numbers of persons by ward in Wales in 2001. Rather than only using colour, the areas of the zones used to report the data can be altered to reflect the sizes of the populations they contain. Figure 5.8 shows an area cartogram derived from the data shown in Figure 5.4—the zones with large populations are made (in relative terms) larger, while the zones with smaller populations are made smaller.[1] This has the effect of emphasizing the areas with larger populations and reducing the visual impact of areas with small populations. Area cartograms have been used widely in mapping population variables, and distortion as a function of population size or density is common in these cases. Note that most GIS packages do not allow the construction of cartograms and specialist software is required.

[1] This map was generated using the ScapeToad software http://scapetoad.choros.ch/.

5.5 Displaying multiple characteristics

Dorling (1994) applies multivariate cartograms. In this application, Chernoff Faces (Chernoff 1973) are utilized, in which the sizes of faces are made proportional to the electorate. The shape of each face and the eyes, nose, and mouth are each related to specific socio-economic variables such as the percentage of unemployed people or households which are rented from a public authority. This system allows for a maximum of 625 different faces, although only a small proportion of the possible permutations are observed. Such symbols are termed glyphs (from hieroglyphics) and they offer a powerful means of displaying complex multivariate data.

There are many instances in which multiple characteristics at a common set of locations might be usefully viewed at the same time, and the example of socio-economic variables derived from census data sets is one. Multivariate statistical analyses produce many outputs. Where spatial data are analysed using such methods and there is a need to visualize multiple outputs on one map, then glyphs may be useful. A simple scheme is used to display simultaneously two sets of statistical coefficients on one map by Lloyd and Lilley (2009), in an analysis of spatial distortion in a medieval map of Britain. In this case, different symbols represent different values of one coefficient, while different sizes and shades represent different values of the second coefficient. So the two together uniquely identify a combination of the two sets of coefficients. The number of dimensions represented in one map can be increased, and an example of a map summarizing three soil variables—percentage sand, silt, and clay—is given by Longley et al. (2010).

5.6 Visualization

Visualization as analysis

The term **geovisualization** (geographical visualization), often encountered in the GIS literature, relates to methods which can be used to explore geographical data through interactive visualization (see Dykes et al. 2005 for a detailed account and case studies). As noted previously, visualization of geographical data can be a means in itself, or it can be used to guide processing and analysis of the data. Visualization has a central role in spatial decision support systems, developed to help in identifying solutions to complex problems such as location selection (Densham and Armstrong 2003).

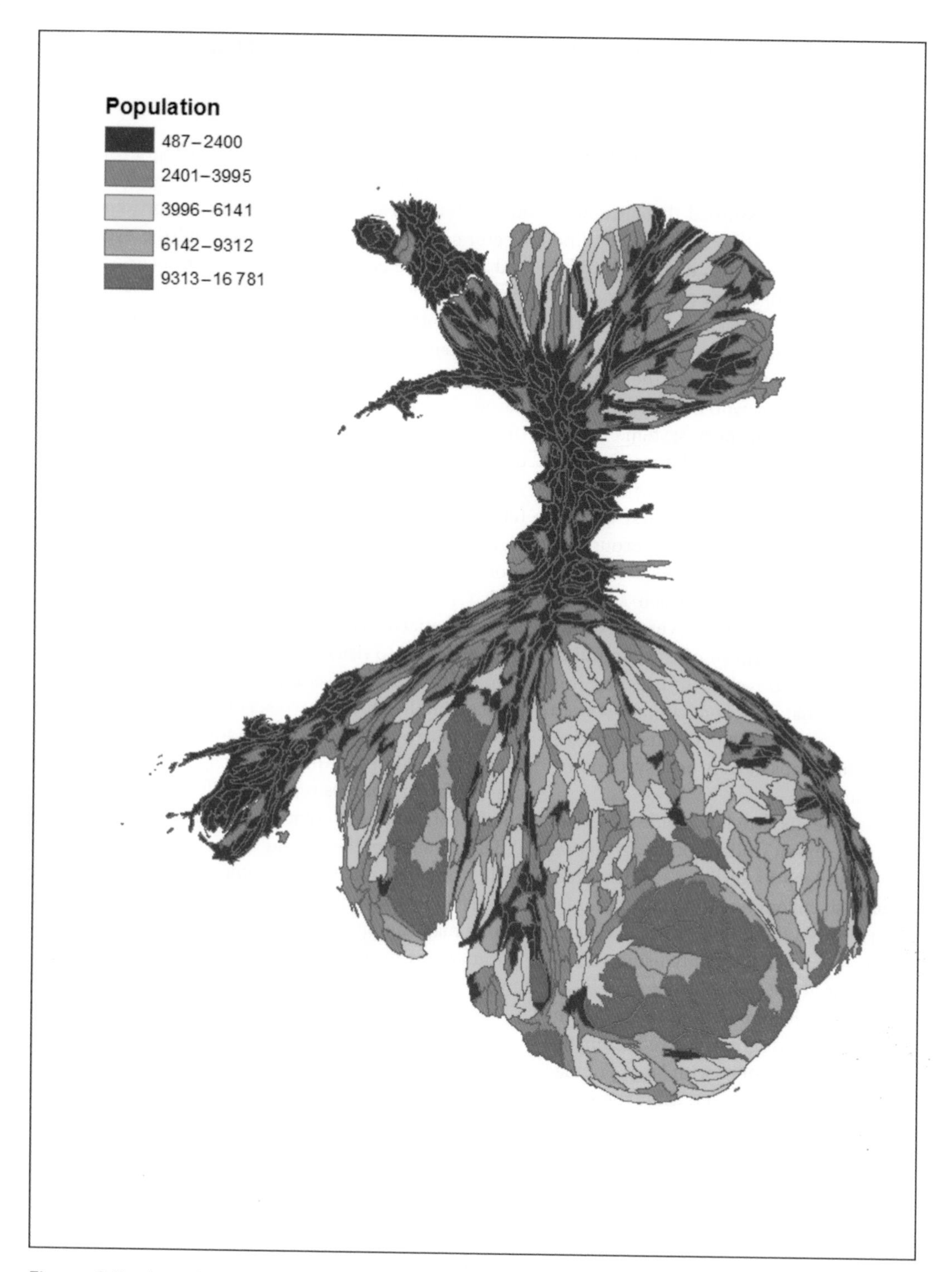

Figure 5.8 Numbers of persons by ward in Wales in 2001: cartogram based on zone populations

Appropriate visualizations of alternative options allow stakeholders to make informed decisions about these choices.

Often, there is an interest in confirming the visual impressions formed through examination of maps. For example, the population in an area may appear to be clustered by some characteristic—low unemployment rates may tend to occur only in specific areas—and the degree of clustering can be explored using statistical approaches, as discussed in Chapter 6. But visualization should always come first. Examination of maps based on different categorizations (where the data can be presented using alternative schemes, as outlined above) is essential to avoid forming views based on a categorization which may stress particular characteristics and (in visual terms) downplay others. Cartograms provide a useful alternative to direct visualization of choropleth data.

The visualization of continuous surfaces

Continuous surfaces are usually represented by images or lines. Image methods include regular and irregular grids and tessellations in which the variations in the value of the mapped attribute is indicated by graded zones of colour or grey levels, as discussed in Section 5.2. 'Two and a half D' representation is achieved by 'draping' the attribute surface over a continuous surface that represents the topography of the land (see below). In generating a '2.5D' view, users can adjust the angle of view (viewer inclination with respect to the surface), the viewing azimuth (direction), and viewing distance (how far away from the surface).

Line representation includes **isolines** (lines of equal value), vertical slices (profiles), and critical lines, such as ridges, stream courses, shorelines, and breaks of slope. Line methods and image methods can be combined to enhance perception.

The perspective, quasi-three-dimensional display or block diagram (otherwise known as **draping**) is a popular method of displaying surficial thematic information (e.g. land use) in relation to the landforms on which it occurs. The method can be used to provide visual impressions that are impossible to achieve with the usual two-dimensional map format. Examples of perspective displays are given in Figures 5.9, 5.10, and 5.11. Draped images are usually best produced in colour because the human eye cannot distinguish sufficient greyscales to make monochrome plots of thematic data fully comprehensible.

The combination of perspective plots and dynamic visualization is a powerful means of displaying the results of space–time models in GIS. In this section, the focus is on the use of DEMs or other continuous surfaces to visualize data. Alternative sources of data, such as building plans, can be used to generate more flexible, and more visually impressive, visualizations. In other words, while a DEM may include the heights of buildings and vegetation (for example), we can also use information on the heights of features other than that captured in the DEM. Box 5.1 summarizes one initiative which has used diverse data sources to create virtual cities.

Figure 5.9 OS VectorMap® district data (raster) for tile ST18 draped over DEM, vertical exaggeration of 3

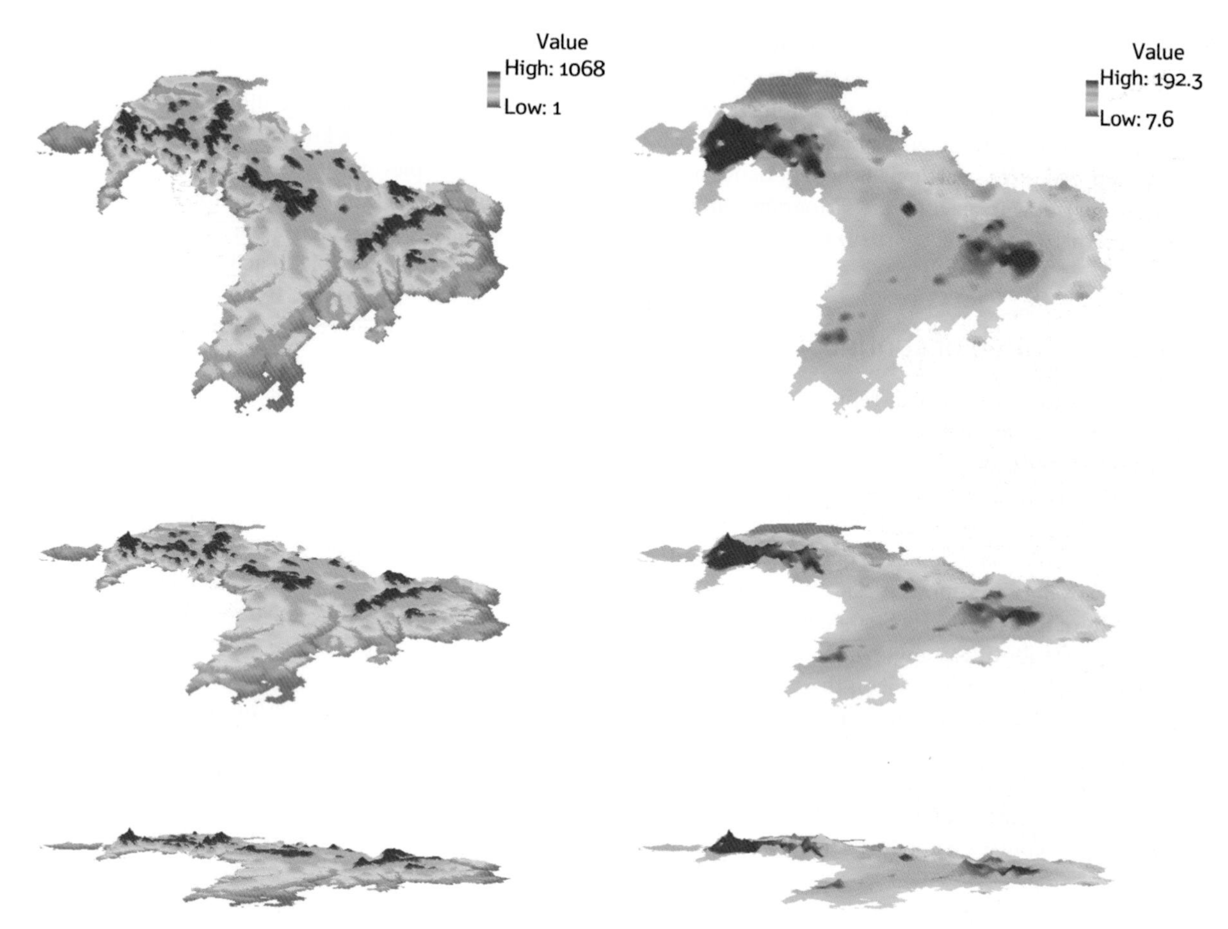

Figure 5.10 DEM of Wales, vertical exaggeration of 10, for three different viewing angles (units are metres)

Figure 5.11 Precipitation for January 2006 draped over DEM of Wales, vertical exaggeration of 10, for three different viewing angles (units are mm)

Box 5.1 Virtual cities

Virtual London

There are now many examples of digital or virtual cities, which use information on the size and shape of buildings to construct digital versions of urban areas. This is possible using remote sensing technologies such as LiDAR (see Chapter 11), or using building plans. The figure below shows an image of 'Virtual London', developed at the Centre for Advanced Spatial Analysis (CASA) at University College London.[2] Evans et al. (2012) discuss the background to, and development of, the Virtual London project. In Virtual London, the morphology of buildings is represented using height models derived using LiDAR and, where available, more detailed models (such as CAD (computer-aided design) models) of particular buildings. The visualization of selected buildings was enhanced through the use of sets of photographs for each building which were used to help guide the creation of building shapes and to texture the facades of the buildings.

The integration of imagery of buildings with height models is leading to ever more impressive digital models of urban environments; these seem likely to have a growing role in planning, as well as in simply giving a detailed visual impression of an area.

Such models open up the possibility of virtual reality environments (see Batty 2008) in which users with

[2] http://www.casa.ucl.ac.uk/

Image of Virtual London
(Courtesy of CASA, UCL; data supplied by Ordnance Survey and Infoterra)

headsets and other technologies can navigate virtual cities or other environments. Desktop virtual reality environments are opened up further through online 'virtual experiences' (Lin and Batty 2009b). Virtual geographic environments are the subject of a recent book which includes many case studies, as well as discussions about the development of, and key concepts behind, such environments (Lin and Batty 2009a).

3D models

Although surfaces (whether interpolated, or derived directly) show variation along three data axes, namely the x and y coordinates and the axis of the interpolated attribute, they are usually not considered as 3D representations. The term 'three-dimensional' is usually (and properly) reserved for situations in which an attribute varies continuously through a 3D spatial frame of reference—an example is the image of a 3D geological model in Figure 5.12. True 3D and 4D representation and visualization requires specialist software not usually found in standard GIS software environments.

Continuous volumes are usually represented by fence diagrams (slices through a 3D image), as shown by an example in Figure 5.13, or by 3D display of a crisped image of a zone of concentration, such as an ore body or a pollution plume. Pan et al. (2012) detail general approaches to 3D visualization of geological data. The methods of structural modelling which they outline are (a) modelling based on borehole data, (b) modelling based on parallel sections, (c) modelling based on parallel sections, and (d) interactive modelling based on multiple data sources. In turn, (a) entails viewing a column or columns of material, (b) refers to slices through a surface, and (c) commonly relates to a grid of cross-sections. In this way,

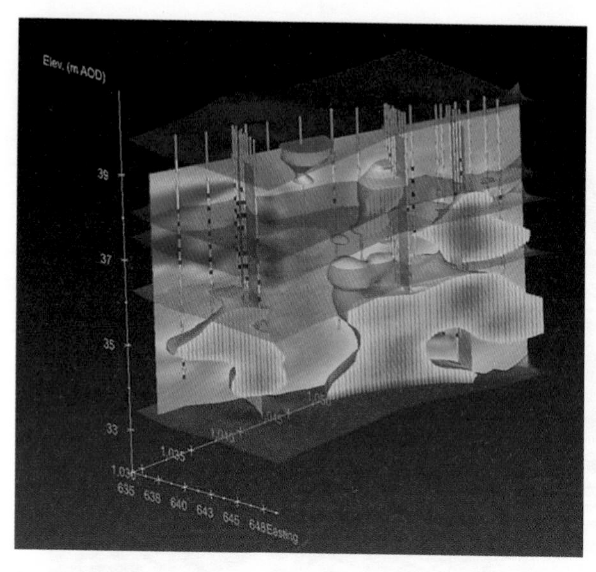

Figure 5.12 3D geology
(Courtesy of the British Geological Survey)

the x, y position of the slices can be viewed along with variation vertically (that is, geological features within sections). Approach (d), the authors argue, can provide the best results by utilizing multiple sources of information, although such approaches require significant human interaction and they cannot be fully automated. However, these approaches are not possible in standard GIS environments.

5.7 Non-cartographic output

This chapter has a focus on visualization of geographical information. In many GIS contexts, outputs may be non-geographical and may take the form of tables and charts which summarize attributes of the geographical information. Many statistical outputs are not mappable, or at least are not mapped, and the outputs of many analyses of geographical data are not maps. Vector data have geographical features and attributes which are linked—selecting a spatial feature also selects the corresponding table entry. Thus, as described in Chapter 3, queries on the database allow extraction of spatial features. Other approaches to feature selection include the use of linked displays. In 'linking and brushing', the selection of items in one display results in their being selected in another display (Monmonier 1990; Gahegan 2008). An example is a scatter plot where a subset of points is selected on the scatter plot, and the corresponding features are automatically selected in the map. So the location of a cluster of points in the scatter plot can be linked to the locations of the points, with subsequent benefits for interpretation.

Figure 5.13 3D geological fence diagram of Britain
(Courtesy of the British Geological Survey)

5.8 Dynamic visualization

Much GIS-based visualization is essentially static: we look at and use a map, or part of a map, for a single time period in a single window. Where data sets are available for multiple time points, as is the case for meteorological variables for example, then animation becomes a possibility. Other examples of the value of animation include viewing, in sequence, multiple data layers for an urban area (e.g. physical features, buildings, transport infrastructure), or the evolution of a river delta (Kraak 2008). Animations ('fly-throughs') of 3D topographic surfaces are another increasingly popular approach to animation using spatial data. In addition to a wealth of 'spatio-temporal' data sets derived through repeated monitoring (e.g. regular measurements at precipitation monitoring stations and series of remotely sensed images for parts of the surface of the earth), integration of multiple data sources also offers new opportunities for visualization and analysis. An example of this is the Population 24/7 project which seeks to generate, using multiple administrative sources, population surfaces (raster grids where the cells represent population counts or densities) for parts of the United Kingdom at specific times of day (Cockings et al. 2010). In this case, it is possible to generate animations which show how the population distribution changes across the day. Dynamic modelling is the subject of Chapter 12, and some elements of visualization are considered there.

5.9 Multimedia and GIS

As noted elsewhere in this book, GIS is a powerful unifying technology in that it is possible to integrate multiple diverse data sources. This principle can be extended beyond explicitly spatial features, and a wide range of objects can be linked to spatial locations. As examples, scanned documents, images, video clips (or, indeed, live camera imagery), or sound files can be linked to spatial objects to provide part of a spatial database on a particular theme or themes (see Chapter 4). The growing integration of diverse data sources in GIS environments may add depth to a user's experience of GIS and help to open up participation (Elwood 2011), as discussed in Section 5.11. Video clips may enhance a user's understanding of a landscape by allowing them to assess the visual impact of, for example, a scenic viewpoint. As pointed out by Kraak (2008), developments in virtual reality offer new possibilities in allowing users to more fully experience, in a virtual sense, particular environments such as the interiors of buildings.

5.10 Spatial interaction data: mapping movement

Spatial interaction data present particular challenges for visualization. An example of such data is derived from population censuses which ask where a resident lives at the time of the census and where they lived (for example) one year before. The resulting data are often outputted as a matrix of 'flows' showing every combination of zones, so that it is possible to identify how many people lived in zone x in 2000 and zone y in 2001. Showing moves between small areas may be difficult because there are so many possible combinations. Wood et al. (2010) address this issue and consider some approaches to the visualization of such data sources. Figure 5.14 gives an example visualization of a flow map—in this case, commuting flows

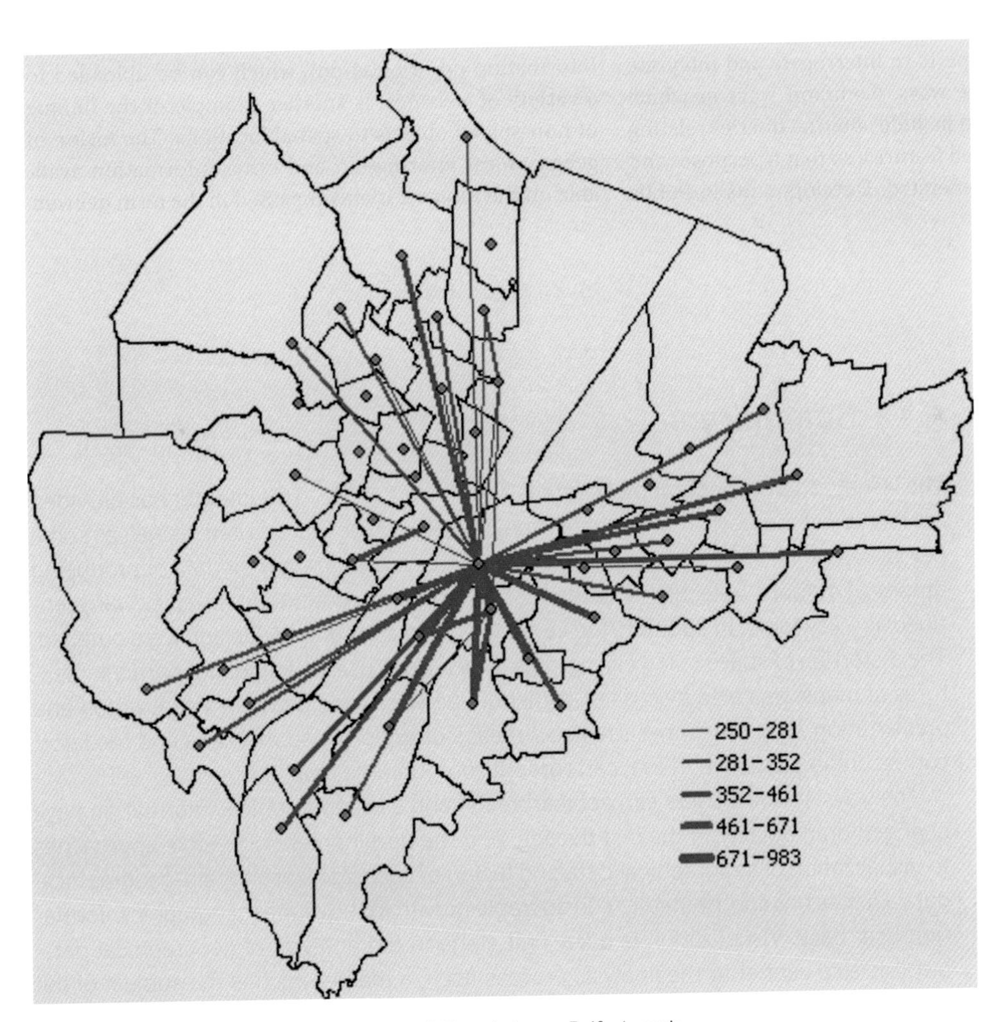

Figure 5.14 Large (≥250 people) travel-to-work flows between Belfast wards
(*Source*: 2001 Census, Output Area Boundaries, Crown copyright 2003)

between census wards in 2001 in the city of Belfast are shown using lines with thicknesses that relate to the numbers of people who live in one ward (one end of a line) and work in another (the other end of a line). Note the large commuting flows from the outskirts of the city into the centre. Where flows are directional—that is, they relate specifically to flows from one location to another—then arrows can be added to the flow lines to indicate the direction of flows.

5.11 Visualization and opening up access to data

New, more flexible, modes of geographical data visualization open up enhanced possibilities for public participation in GIS (see, for example, Krygier 2002). More generally, access to easy-to-use online GIS environments and allied technological frameworks, such as Google Earth™, enable non-experts to interrogate and integrate data sources in diverse ways. Common ways in which such resources are used include entering queries relating to locations and selected features, so that a purpose- and user-specific map is generated. Developments in **Public Participation GIS (PPGIS)** allow members of the general public to make more informed decisions about the planning process, for example, by allowing the flexible assessment of alternative scenarios. Also, initiatives such as the http://www.openstreetmap.org website, which aims to develop a worldwide freely accessible and modifiable geographic data resource, and the http://www.maptube.org website, which provides a free resource for visualizing, disseminating, and mashing mapped information (combining online mapping resources), are opening up the ways in which users can access, visualize, and modify geographical data (see Chapter 4).

The related concept of neogeography was introduced in Chapter 1. In addition to increased access to geographical data, recent research is adding value to non-spatial data sources. As an example, Gregory and Hardie (2011) outline an approach to the mapping of places mentioned in connection with particular terms in historic documents. **Geotagging** of photographs (often based on where the photographer was standing, and perhaps with information on orientation), which can be uploaded to a variety of websites, is another example of the linking of non-spatial objects to spatial locations. The fusion of geographical information and other information available on the Internet is encompassed in the term **geoweb**.

* 5.12 Summary

The visualization of geographical data is at the heart of GIS. This chapter has provided a summary of the ways in which geographical data can be represented, as well as some key issues which users of GIS should consider when visualizing their data or producing graphical outputs. Selection of categories to represent continuous variables was identified as a key topic. In addition, the key elements of cartographic design were outlined. The distortion of space using cartograms, which allow users to highlight particular features of maps, was also discussed. A distinction was made between visualization and presentation, with the former seen as a means of exploring spatial data and the latter corresponding to conventional cartographic output—the map.

The ways in which users can access, create, and use geographical information have changed dramatically in the last decade, and the development of flexible approaches to visualization of geographical data and fusion of geographical and non-geographical data sources has contributed to considerable growth in the digital geographic information user base. Visualization is a key first stage in the analysis of geographical data, but it is also central to the analysis process itself, a theme which is the subject of the next chapter.There are many books, book chapters, and journal papers which discuss approaches to visualization. These include the chapters cited above and more general texts such as the books by Heywood et al. (2011) and Longley et al. (2010).

? Questions

1. What are the most important aspects of a map for communicating information? What makes a good map?
2. How should you choose colours and greyscales for displaying data?
3. In what ways does '2.5D' visualization offer benefits over conventional 2D maps?
4. What are cartograms and how might they enhance the visualization of choropleth maps?
5. How can virtual cities enhance our experience of real cities?

→ Further reading

▼ Dykes, J., MacEachren, A. M., and Kraak, M.-J. (eds.) (2005). ***Exploring Geovisualization***. Elsevier, Amsterdam.

▼ Jones, C. (1997). ***Geographical Information Systems and Computer Cartography***. Longman, Harlow.

▼ Kraak, M.-J. (2008). Visualising spatial distributions. In P. A. Longley, M. F. Goodchild, D. J. Maguire, and D. W. Rhind (eds.) ***Geographical Information Systems: Principles, Techniques, Management, and Applications***. 2nd edn, abridged. Wiley, Hoboken, NJ, pp. 49–65.

▼ Lin, H. and Batty, M. (eds.) (2009). ***Virtual Geographic Environments***. Science Press, Beijing.

▼ Longley, P. A., Goodchild, M. F., Maguire, D. J., and Rhind, D. W. (2010). ***Geographic Information Systems and Science***. 3rd edn. Wiley, Hoboken, NJ.

▼ Robinson, A. H., Morrison, J. L., Muehrcke, P. C., Kimerling, A. J., and Guptill, S. C. (1995). ***Elements of Cartography***. 6th edn. Wiley, New York.

Exploring Geographical Data

Learning about the characteristics of data is an essential first step in any application of those data. As well as understanding the data lineage—where they come from, how they were collected, possible sources of error, and other factors—it is necessary to consider the characteristics of the values attached to the samples or the spatial configuration of these samples. Exploratory analyses are likely to be useful in their own right and are also likely to aid interpretation of the variables. In addition, provisional analyses of this type may help the user decide how best to go on to treat the data, or decide what methods might be suitable in further processing or analysis of the data. In this chapter, some key ways of extracting information from geographical data are outlined, from simple statistical summaries of the data, through to local methods for analysing relationships between multiple variables.

Learning objectives

By the end of this chapter, you will:

- **understand some key ways of summarizing numerical data**
- **understand key concepts such as spatial autocorrelation, and be able to interpret measures of this property**
- **be able to apply methods for exploring spatial relationships between variables**
- **know why point patterns are distinct from other types of data, and how to characterize these kinds of data.**

Having gone to the trouble of collecting data and building a spatial database, the next issue is how to use these data to provide information to answer questions about the real world. This involves a wide range of methods of data manipulation, from simple data retrieval and display to the creation and application of complex models for the analysis and comparison of different planning scenarios. In this chapter, summary statistical methods are outlined, along with methods for exploring spatial patterns and relationships between variables.

Geographical information systems provide a large range of analysis capabilities that can be used in many ways. These analytical capabilities will usually be organized in modular commands so that each kind of analysis can be performed separately, or combined with others

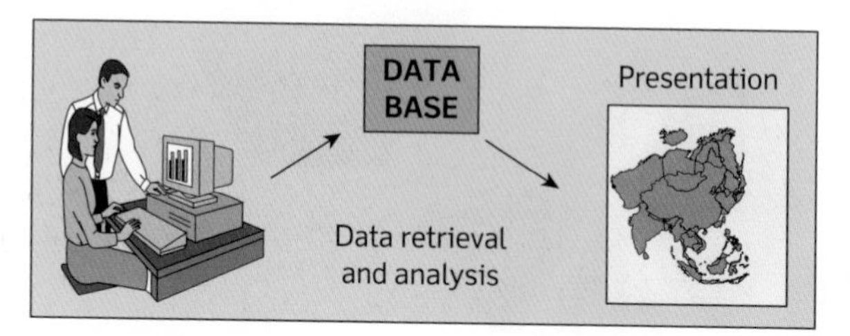

Figure 6.1 Data retrieval is the first step to visualizing information

to build **data analysis models**. The actual user interface can be provided through typed commands, using menus and buttons, or as statements in a user-written high-level programming language. The aim of this and the following chapters is not to instruct the reader on how to call up program modules (which are system-dependent) but to provide a proper understanding of the kinds of analytical functionality available and the tasks that can be accomplished.

The general problem of data analysis is stated in Figure 6.1 The user has a particular problem or query. The database contains information that can be used to answer the user's problem and will provide that answer in the form of a map, tables, or figures. To answer the query it is necessary to set up a formal set of data retrieval and analysis operations to recall the data, to compute new information, and to display the results. A key focus in this book is on how spatial data can be accessed, queried, and analysed. The present chapter covers summaries of numerical data, the analysis of spatial patterns, and of relationships between different attributes at particular locations. In the following chapter, the theme of analysis is extended to operations on discrete features of different forms—points, lines, and areas. Chapters 8 and 9 introduce methods for spatial interpolation (with a focus on the creation of surfaces), while Chapter 10 discusses the analysis of continuous fields. Chapter 11 describes digital elevation models and some approaches to their analysis. Together, these chapters outline many of the key GIS functions which can be used to extract information from our data and to address a multitude of questions about the world around us.

6.1 Summarizing and analysing spatial data

The previous chapter emphasized the value of visualization in exploring geographical data. This chapter outlines a variety of other approaches for extracting information from such data. GIS software includes functions for computing statistical summaries of data values and some packages include sophisticated tools for describing the spatial characteristics of data sets. Such tools are likely to be central to any application which makes use of geographical data, and this chapter outlines some key concepts and methods. Approaches detailed include simple statistical summaries (for example, the mean average of a set of values) and methods for assessing relationships between different variables. In addition, the chapter recognizes some characteristics of geographical data that affect how these data can be treated.

6.2 Statistical methods

Descriptive statistics

If we have a set of numeric values, it is often useful to summarize these values to get an idea of the characteristics of the data. Measures of central tendency, or **averages**, are a key means of summarizing data values. The mean average is the most widely used measure and is simply the sum of the data values divided by the number of values. The median is the middle value in a list of ordered values, and the mode is the most frequently occurring value. Examples of these three averages follow for a set of five data values: 2, 5, 7, 2, and 3:

Mean: $(2 + 5 + 7 + 2 + 3)/5 = 19/5 = 3.8$

Median: middle value of ranked values (2, 2, 3, 5, 7) = 3

Mode: most frequent value = 2

The mode can be used to summarize nominal or ordinal variables. Nominal variables have names (e.g. 'apartment', 'terraced house', 'detached house'), with no relative values, while ordinal variables have an inherent ordering (e.g. 'small', 'medium', 'large'). The median could be used to summarize ordinal data, but not nominal data, whereas the mean can only be used to summarize interval or ratio variables. With interval variables, the intervals between values are meaningful but the variable has no natural zero point—an example is temperature. Ratio variables do have a natural zero point—measurements of distance are an example. In much of this book, the examples make use of ratio data. Variable types are defined in Chapter 2.

The set of data values can be summarized using a **histogram**. This is a plot of the number of data values in a set of classes—for example, there may be seven values in the range 0 to 4.9, fifteen values in the range 5 to 9.9, and so on. The histogram represents the **distribution** of the variable. Many variables have a distribution which is called normal. A common example is the heights of adult humans in a population, where the majority have a height somewhere in the middle of the range and the proportions of (relatively) short and tall people are fairly similar to one another. With such distributions, the mean, median, and mode will be similar (and perhaps the same). Variation around the mean (that is, how different values tend to be from the mean) is described by the variance and its square root, the standard deviation. Box 6.1 outlines the basic principles of the mean and the variance.

To illustrate methods, this chapter makes use of a set of data representing precipitation in Wales for the month of January 2006. The elevations of each of the monitoring stations are also available, and these values are also used in the examples below. The locations of the monitoring stations were shown in Chapter 5. There

Box 6.1 Basic principles of means and variance

The meaning of the mean and variance of a population and a sample

Not all attributes can be measured exactly; there is a natural variation of values around the ***mean*** or average. In many cases this variation can be described by the ***bell curve***, or normal distribution, whose parameters are the mean μ and standard deviation σ. The width of the normal distribution is given by σ; 65 per cent of all values of a normally distributed population fall within $\mu \pm \sigma$, 95 per cent fall within $\mu \pm 2\sigma$ and 99 per cent fall within $\mu \pm 3\sigma$. Therefore, the larger the standard deviation of a population, the lower the ***precision*** that can be associated with the value of any sample of that population. In many cases our population of geographical entities can be regarded as almost infinite (think of all possible soil profiles, ground water observation wells, and even fast-food restaurants) so the data collected are only a ***sample***. We estimate μ and σ by computing the sample mean m and sample standard deviation s from n observations as:

$$m = 1/n\sum_{i}^{n} z_i$$

and

$$s = \left[1/n\sum_{i}^{n}(z_i - m)^2\right]^{0.5}$$

Obviously, it is useful to know s and to ensure that it is as small as possible to obtain the best precision.

The square of the standard deviation is called the ***variance*** σ^2. The variance is a very useful quantity because variances computed from different sources of variation can be added together to combine estimates of uncertainty from different sources. So if we have a map with p_k polygons, containing n_i observations per polygon, we can compute three variances, namely σ_T^2, the total variance of all observations, which is made up of σ_W^2, the ***between class variance***, and σ_W^2, the ***within class variance*** (pooled over all classes). These variances are estimated by our sample which computes:

$$S_T^2 = S_B^2 + S_W^2$$

The value of dividing the data according to the p_k polygons is indicated by the statistical ***significance*** of the variance ratio S_W^2 / S_B^2 for the appropriate ***degrees of freedom***. The degrees of freedom are given by $p_k - 1$ for S_B^2 and by $n - p_k$ for S_W^2 and can be found in statistical tables of F.

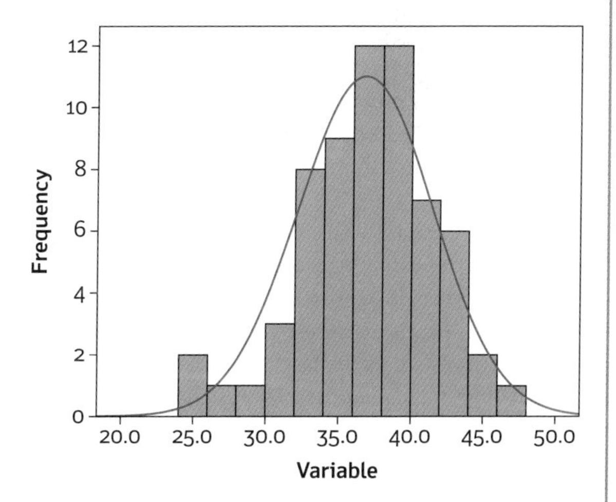

The standardized normal curve fitted to experimental data
Mean = 36.8, standard deviation = 4.65

The magnitude of the variation 'explained' by the soft data classification into p_k polygons is given by S_B^2 / S_T^2, which in regression (see 'Exploring relationships between variables') is called R^2; the larger R^2, the more of the variation is explained by the p_k classes, and the more likely that the mean values of each class m_k will be a good global predictor.

The variation within each polygon can be explored by subtracting the data values z_i from the polygon mean m_k in which they are situated.

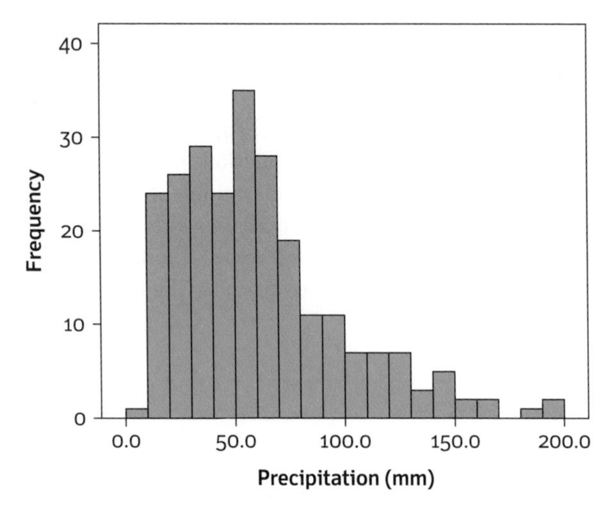

Figure 6.2 Histogram of precipitation amounts

are 244 observations, with a minimum precipitation amount of 7.4 mm, a maximum value of 195.8, a mean of 61.5 mm, and a standard deviation of 37.8 mm. The histogram of precipitation amounts is given in Figure 6.2. The large majority of values are below 100 mm and the histogram shows that there is a small proportion of large values (a tail of large values). The distribution can be described as **positively skewed.** If the form of the histogram was mirrored, and there was a tail of small values, then the distribution would be termed **negatively skewed.**

Where the data are non-normal it may be advisable to **transform** the data, for example by taking their logs or square roots. Positive skew can be reduced by logging the data, and logging is often conducted to make the data distribution closer to normal. Since the standard statistical approaches outlined in this chapter assume a normal distribution, the use of data with a skewed distribution may restrict the ways we can use or interpret the data. Therefore it is important to inspect the histogram of your data to ensure the distribution is not markedly skewed.

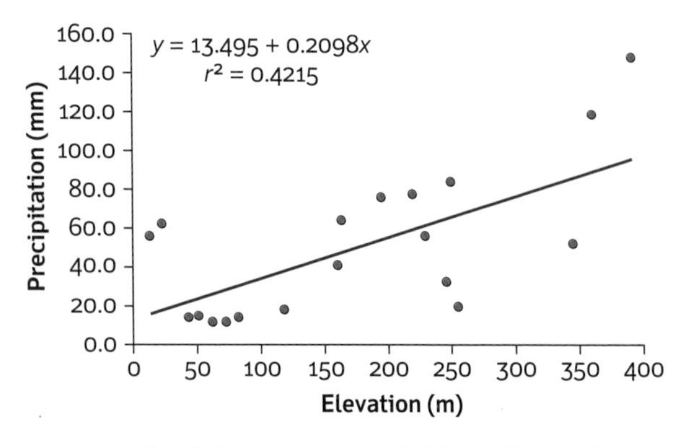

Figure 6.3 Plot of altitude against precipitation with line of best fit

Exploring relationships between variables

This section has discussed summaries of individual variables, but often we want to assess how different variables are related. For example, we may want to consider if census areas with high unemployment rates also tend to have low qualification rates. Relationships between variables can be assessed using regression and correlation. With these tools, it is possible to consider how variables are related and how strong the relationship between them is. **Regression** and **correlation** are illustrated here using an example which refers to measurements of precipitation amount and altitude. In some areas, precipitation is related to altitude and precipitation amounts tend to increase as altitude increases. If this is so, we can say that altitude and precipitation are positively related. Figure 6.3 shows a plot of altitude (on the x axis) and precipitation (on the y axis); the data are for a part of North Wales and they relate to precipitation recorded at 20 locations over the month of January 2006 (this is a subset of the data described above). The line which runs through the points is termed the **line of best fit**, and it is fitted using regression—the relevant procedure is described below. The equation which describes the line of best fit summarizes the relationship between the variables.

In this example, there are two variables: altitude and precipitation. The regression equation can be given by:

$$\hat{z}_i = \beta_0 + \beta_1 y_i \qquad 6.1$$

where z_i indicates the dependent variable (in this example, precipitation) at location i and $\hat{z}_i$ is the predicted value of z_i. The predicted value is obtained by estimating values for β_0 (termed the intercept) and β_1 (the slope; collectively, the intercept and slope are called the beta coefficients). In this example, $\beta_0 = 13.495$ and $\beta_1 = 0.2098$. The strength of the relationship between the variables can be measured using the correlation coefficient, r. The graph in Figure 6.3 shows r^2, the **coefficient of determination**. This value indicates the proportion of the variation in the data points explained by the line of best fit. Where the points lie exactly on this line $r^2 = 1$. Where r^2 is close to 0, the line does not represent the relationship between the points well. In this case, $r^2 = 0.4215$, and the line of best fit explains just over 42 per cent of the variation.

Once we have estimated the beta coefficients, we can predict z_i for any value of y_i (in this example, the elevation value at the location i). The intercept, β_0, indicates that the line of best fit crosses the vertical (y) axis, while the slope coefficient, β_1, indicates the nature of the relationship between the independent and dependent variables. Where the slope is negative, this indicates a negative relationship (in this example, this would mean that precipitation decreases as altitude increases) and a positive value indicates a positive relationship (precipitation increases as altitude increases). The beta coefficients can be estimated using a technique called **ordinary least squares**—the basic process is outlined in Box 6.2.

Box 6.2 Fitting a linear regression model and measuring correlation

The slope coefficient, β_1, is obtained with:

$$\beta_1 = \frac{\sum_{i=1}^{n}(y_i - \bar{y})(z_i - \bar{z})}{\sum_{i=1}^{n}(y_i - \bar{y})^2}$$

The intercept, β_0, is given by:

$$\beta_0 = \frac{\sum_{i=1}^{n} z_i - \beta_1 \sum_{i=1}^{n} y_i}{n} = \bar{z} - \beta_1 \bar{y}$$

The small set of data values given in Table 1 are used in this example. The slope is computed first. Initially, we compute the numerator of the equation:

$$\sum_{i=1}^{n}(y_i - \bar{y})(z_i - \bar{z})$$

In words, we take each value of **y** and subtract the mean value of **y**, then we take each value of **z** and subtract the mean value of **z**. The difference between each **y** value and its mean and each **z** value and its mean is then multiplied together. Once this has been done for all of the paired data values, the multiplied values are added together.

In this example, the mean value of **y** is 13.4 and the mean value of **z** is 26.

The sum of the multiplied differences in the table is 398. The denominator of the equation for the slope, $\sum_{i=1}^{n}(y_i - \bar{y})^2$, is used next. Following this, we take each value of **y**, subtract its mean, square this difference (column $(y_i - \bar{y})^2$ in the table), and all of the squared differences are summed, giving a value of 242.40.

The slope value is obtained by dividing the first summed value by the second summed value: 398/242.40 = 1.6418.

The intercept, β_0, is calculated with

$$\bar{z} - \beta_1 \bar{y} = 26 - 1.6418 \times 13.4 = 3.9983.$$

$\beta_1 \bar{y}$ means the two components are multiplied by one another.

The fitted line is shown in the graph. Values for β_0 and β_1 (and indeed for multiple independent variables) can also be obtained using matrix algebra. This approach is illustrated by Lloyd (2010b).

Table 1 Variable 1 (y) and variable 2 (z), differences from their mean, differences multiplied, and the square of the differences from the mean of variable 1. The last row (in italics) contains the means (first two columns) and sums (last two columns) of the column values.

Variable 1 (y_i)	Variable 2 (z_i)	$(y_i - \bar{y})$	$(z_i - \bar{z})$	$(y_i - \bar{y}) \times (z_i - \bar{z})$	$(y_i - \bar{y})^2$
12	20	−1.4	−6	8.40	1.96
14	30	0.6	4	2.40	0.36
15	32	1.6	6	9.60	2.56
6	12	−7.4	−14	103.60	54.76
22	38	8.6	12	103.20	73.96
10	18	−3.4	−8	27.20	11.56
20	38	6.6	12	79.20	43.56
10	24	−3.4	−2	6.80	11.56
8	18	−5.4	−8	43.20	29.16
13.4	*26*			*398.00*	*242.40*

(*continued*)

Box 6.2 *(continued)*

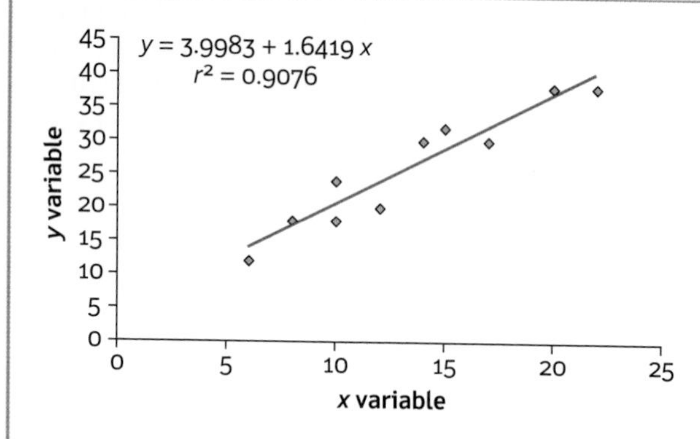

In this example, $r^2 = 0.9076$ and so the line of best fit explains 90.76 per cent of the variation. Therefore the line represents the relationship between the two variables well.

The correlation coefficient, r, is given by:

$$r = \frac{\sum_{i=1}^{n}(y_i - \bar{y})(z_i - \bar{z})}{\sqrt{\sum_{i=1}^{n}(y_i - \bar{y})^2}\sqrt{\sum_{i=1}^{n}(z_i - \bar{z})^2}}$$

The numerator is the same as for the estimation of β_1. The denominator is the square root of the sum of squared differences between **y** and its mean, multiplied by the square root of the sum of squared differences between **z** and its mean.

The differences between **y** and **z** and their means were given in Table 1. The squared differences and their sums are given Table 2 (values in column 3 are the same as those in the final column of Table 1).

Table 2 Variable 1 (y) and variable 2 (z), squared differences from their mean, and summed values

Variable 1 (y_i)	Variable 2 (z_i)	$(y_i - \bar{y})^2$	$(z_i - \bar{z})^2$
12	20	1.96	36
14	30	0.36	16
15	32	2.56	36
6	12	54.76	196
22	38	73.96	144
10	18	11.56	64
20	38	43.56	144
10	24	11.56	4
8	18	29.16	64
Sum		242.40	720

Given these values, the numerator is obtained with:

$$\sum_{i=1}^{n}(y_i - \bar{y})(z_i - \bar{z}) = 398 \text{ (as calculated earlier)}$$

and the denominator is obtained with:

$$\sqrt{\sum_{i=1}^{n}(y_i - \bar{y})^2}\sqrt{\sum_{i=1}^{n}(z_i - \bar{z})^2}$$
$$= \sqrt{242.40} \times \sqrt{720.00}$$
$$= 15.5692 \times 26.8328 = 417.7655$$

The correlation coefficient and coefficient of determination are then given with:

$$r = 398 / 417.7655 = 0.9529$$
$$r^2 = 0.9529^2 = 0.9076$$

A test of significance for the correlation coefficient can be given by:

$$t = \frac{r\sqrt{n-2}}{\sqrt{1-r^2}}$$

For the current example, this leads to:

$$t = \frac{0.9529\sqrt{9-2}}{\sqrt{1-0.9076}} = \frac{0.9529 \times 2.6458}{0.3040}$$
$$= \frac{2.5212}{0.3040} = 8.2939$$

To assess if **r** is significant, this figure is compared to the values contained in a **t**-table—this is a table which indicates the significance level associated with a particular **t** value for a given number of degrees of freedom, which are here given by **n** minus the number of variables (so, 9 − 2 = 7). Ebdon (1985) includes a **t** table, although statistical software packages compute significance levels automatically so manual assessment of a **t** table is therefore unlikely to be required. In tests of this kind, it is commonly the case that the 5 per cent significance level is used (the significance level, denoted by the Greek letter alpha, α, is 0.05, i.e. 5 per cent of 1). For 7 degrees of freedom and with $\alpha = 0.05$, the critical **t** value (for a two-tailed test—where we want to ascertain if the observed value is above or below some threshold) is 2.365. In this example, **t** = 8.347 and, because it exceeds the critical value of 2.365, we can say that **r** is significant at greater than the 0.05 level—there is a smaller than 5 per cent chance that the result has occurred by chance—and we can reject the null hypothesis that **r** is equal to zero.

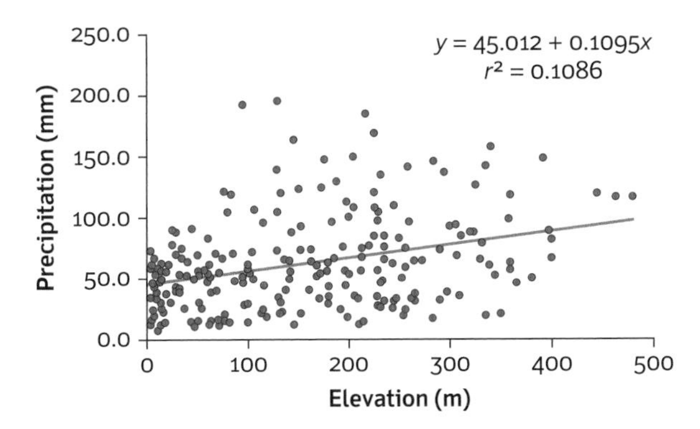

Figure 6.4 Plot of altitude against precipitation in Wales for January 2006, with line of best fit

Figure 6.4 gives a scatterplot of elevation against precipitation in Wales for January 2006. In this example, r^2 = 0.1086 and therefore the line of best fit explains only some 10 per cent of the variation in the scatterplot. Prior knowledge would suggest that elevation and precipitation would be quite strongly related in at least some parts of Wales, and a more sophisticated approach for exploring the relationship between variables, which may help to pick apart this relationship, is outlined later in this chapter.

6.3 Geographical data: problems and properties

Treating geographical data as though they are not spatially referenced is often unwise—this ignores the spatial context of the data and the special properties of such data. Many geographical variables exhibit what is termed **spatial dependence**. Obvious examples are altitude and precipitation—with both variables, neighbouring values tend to be similar while, on average, values at locations separated by large distances tend to be more different. An assumption of standard statistical tests is that the observations are independent, but if neighbouring values tend to be similar then the data values are described as spatially dependent and results from standard statistical tests will be biased. In Box 6.2, the *t* test for the correlation coefficient is outlined. This test assumes that the data are independent. If the data are spatially dependent this condition is violated, with the result that the associated significance levels are inflated and the test results may be unreliable. This topic is outside the remit of this book—Rogerson (2006) and Harris and Jarvis (2011) provide more information about the impact of spatial dependence.

6.4 Spatial autocorrelation

Following from the previous section, the term **autocorrelation** is important; it refers to the correlation of a variable with itself. For example, if we compare values of a variable at one location with values of that variable at other locations then we can assess autocorrelation of that variable. This is useful if, for example, we want to assess clustering. In this case, nearby values of the variable will tend to be similar—this is termed *positive* spatial autocorrelation. A variable which is positively spatially autocorrelated is spatially dependent. In contrast, where neighbouring values of a variable tend to be dissimilar, the variable is said to be *negatively* spatially autocorrelated and the values are spatially independent.

There is a variety of measures of spatial autocorrelation (and thus spatial dependence). These include the joins counts method, which summarizes the number of times neighbouring observations have the same value. As an example, a set of cells are coded B or W for black or white and we can summarize the spatial configuration of the cells by counting how many times immediate neighbours of 'B' cells are 'B' cells or 'W' cells, and how many times 'W' cells neighbour 'W' cells. If the BB and WW combinations (that is, joins) dominate, then this suggests positive autocorrelation. In contrast, a predominance of BW (with WB indicating the same thing) joins suggests negative autocorrelation.

One of the most widely applied measures of autocorrelation is the ***I* coefficient** developed by Moran (Moran 1950). The *I* coefficient measures covariation in a single variable measured at multiple locations. Taking an example, assume we are concerned with pH in soil, measured on a regular grid, and we want to consider to what extent neighbouring values of pH tend to be similar. First, we need to define a neighbourhood—with a regular grid, we could simply say that all grid cells which share edges or corners with other cells are neighbours of that cell. This is called 'queen contiguity' and it is illustrated in Figure 6.5 along with another neighbourhood definition, 'rook contiguity'. In both cases, arrows from the central cell point to neighbouring cells which are considered neighbours. The two terms come from the chess pieces which can make the corresponding movements. This principle extends to irregular zones (for example, census tracts or output areas), where queen

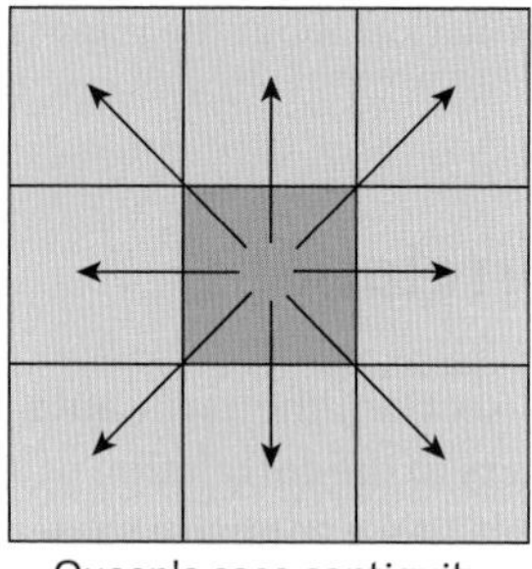

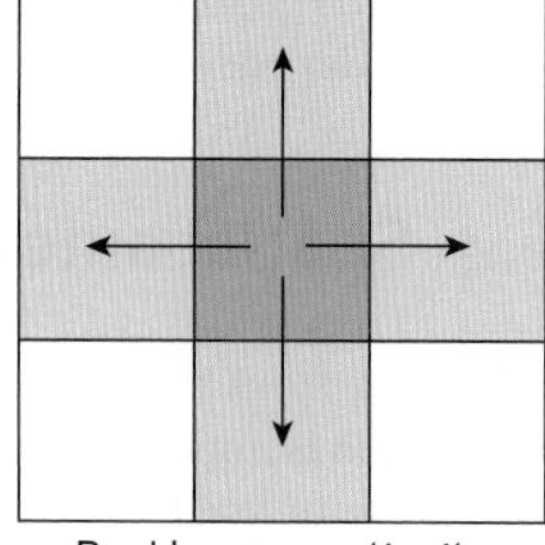

Figure 6.5 Queen and rook continuity for square cells: 3 x 3 pixels

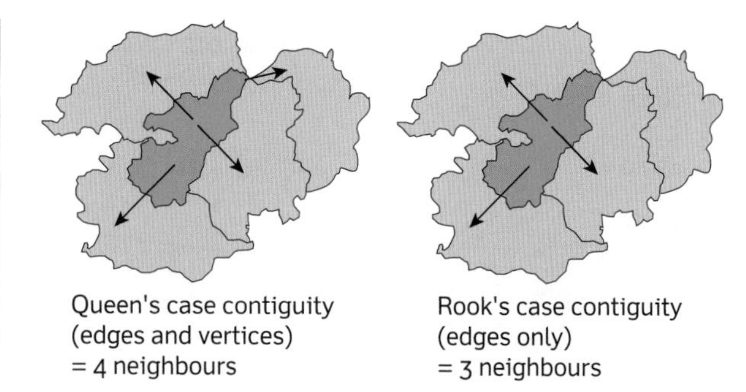

Figure 6.6 Queen and rook continuity for irregular zones

contiguity relates to shared edges *and* shared vertices of zones and rook contiguity relates to shared edges alone (see Figure 6.6 for examples of both).

Moran's *I* coefficient is given by:

$$I = \frac{n\sum_{i=1}^{n}\sum_{j=1}^{n} w_{ij}(y_i - \bar{y})(y_j - \bar{y})}{\left(\sum_{i=1}^{n}(y_i - \bar{y})^2\right)\left(\sum_{i=1}^{n}\sum_{j=1}^{n} w_{ij}\right)} \quad 6.2$$

The right-hand part of the top line of the equation (the numerator), $w_{ij}(y_i - \bar{y})(y_j - \bar{y})$, comprises the weights for paired data locations *i* and *j* multiplied by the co-variance between y_i and y_j—the mean is subtracted from each and the products multiplied. The sum of these covariances for all paired locations is multiplied by *n*, the number of observations. This is then divided by the sum of the squared differences between all of the data values and their mean average, multiplied by the sum of all of the weights. Worked example 6.1 shows a calculation using Moran's *I*. Moran's *I* is a *spatial* statistic and it allows us to measure how similar values are as a function of their spatial location. In other words, we can answer questions like 'how similar are the populations of neighbouring census areas?' Other measures of spatial autocorrelation are outlined in Chapter 9.

Worked example 6.1 Measuring spatial autocorrelation with Moran's *I*

The calculation of Moran's *I* is illustrated using a small grid of values:

25	24	23
20	22	21
18	19	17

Values that are next to one another along rows or columns (e.g. 25 and 24 or 22 and 21) will be counted as neighbours, as will those that are next to one another diagonally (e.g. 25 and 22). This is termed queen case contiguity.

First we will calculate $(y_i - \bar{y})(y_j - \bar{y})$—the difference of each value from the mean multiplied by the difference between each neighbouring value and the mean. For example, the value 25 (top left cell) minus the mean (21) is 4. One of the neighbours of this cell has the value 24 and its difference from the mean is 3. We then multiply the two differences together, giving 12 (see the furthest right column of Table 1). This is done for every cell and its neighbours, as shown in the table. The table includes only cells that are neighbours (so the weight in each case is one). The sum of the products $\sum_{i=1}^{n}\sum_{j=1}^{n} w_{ij}(y_i - \bar{y})(y_j - \bar{y})$ is 3.975.

Table 1 Value, value – mean (21), neighbour value, neighbour value – mean, product (two differences from mean multiplied together)

Value y_i	Value – mean $y_i - \bar{y}$	Neighbour value y_j	Neighbour – mean $y_j - \bar{y}$	Product $(y_i - \bar{y}) \times (y_j - \bar{y})$
25	4	24	3	12
	4	22	1	4
	4	20	−1	−4
20	−1	25	4	−4
	−1	18	−3	3
	−1	24	3	−3
	−1	22	1	−1
	−1	19	−2	2
18	−3	20	−1	3
	−3	22	1	−3
	−3	19	−2	6

Value y_i	Value – mean $y_i - \bar{y}$	Neighbour value y_j	Neighbour – mean $y_j - \bar{y}$	Product $(y_i - \bar{y}) \times (y_j - \bar{y})$
24	3	25	4	12
	3	23	2	6
	3	20	−1	−3
	3	22	1	3
	3	21	0	0
22	1	25	4	4
	1	20	−1	−1
	1	18	−3	−3
	1	24	3	3
	1	19	−2	−2
	1	23	2	2
	1	21	0	0
	1	17	−4	−4
19	−2	20	−1	2
	−2	22	1	−2
	−2	21	0	0
	−2	18	−3	6
	−2	17	−4	8
23	2	24	3	6
	2	22	1	2
	2	21	0	0
21	0	24	3	0
	0	22	1	0
	0	19	−2	0
	0	23	2	0
	0	17	−4	0
17	−4	22	1	−4
	−4	19	−2	8
	−4	21	0	0

Next we will calculate $(y_i - \bar{y})^2$, the squared difference between each value and the mean. The results are shown in Table 2. The sum of squared differences from the mean $\sum_{i=1}^{n}(y_i - \bar{y})^2$ is 60.

There are nine observations; $\sum_{i=1}^{n}\sum_{j=1}^{n} w_{ij}(y_i - \bar{y})(y_j - \bar{y}) = 58$, the sum of squared differences from the mean is 60, and there are 20 adjacencies (the number of rows in the first table is twice the number of adjacencies). As an example, the cells with values 25 and 24 are neighbours (they are adjacent to one another). Each adjacency (like all the others) is counted twice as we have 25 paired with 24 and 24 paired with 25. So the sum of the weights, $\sum_{i=1}^{n}\sum_{j=1}^{n} w_{ij}$ (the right-hand side of the denominator of equation 6.2), is 40.

For this example, Moran's I is computed with:

$$I = \frac{9 \times 58}{60 \times 40} = \frac{522}{2400} = 0.218$$

So, I indicates positive spatial autocorrelation in this case.

Table 2 Values, difference from the mean (21), and the squared differences

Value y_i	Difference $y_i - \bar{y}$	Squared difference $(y_i - \bar{y})^2$
25	4	16
20	−1	1
18	−3	9
24	3	9
22	1	1
19	−2	4
23	2	4
21	0	0
17	−4	16

In the example above, the immediate neighbours of zones were used to define spatial contiguity. Where the data are points rather than connected zones, distances are likely to be used instead to determine weights. Moran's I was measured using the Welsh precipitation data. In this case, the weights were computed using the inverse distance weighting scheme—in the example above, the weights for zones i and j were 1 if the zones were neighbours and 0 otherwise. With inverse distance weighting, the distance between points (or zone centroids, for example) are given by d_{ij} and the weight, the inverse distance, d_{ij}^{-2}, is calculated with $1/d_{ij}^2$. The value of Moran's I was 0.295, indicating that the precipitation amounts are spatially dependent. It is also possible to measure spatial autocorrelation locally so that the degree of similarity of values to neighbouring values is assessed (see Lloyd 2010b for an introduction to such methods).

6.5 Statistics and GIS

As many GIS do not provide more than simple statistical functions, the user who wishes to carry out statistical computations is advised to choose a system in which the attributes of spatial entities can be analysed using a standard statistical package. Linking a GIS to a spreadsheet program or standard statistical package such as SPSS or S-Plus is an easy way to provide a wide range of logical, arithmetical, and statistical operations for transforming the attribute data of entities held in a GIS without having to have these computational tools in the GIS. Both spreadsheets and statistical packages include useful graphics routines for plotting graphs, histograms, and other kinds of statistical charts. The use of hypertext facilities to link the database, graphs, and map displays is providing powerful exploratory data analysis tools for users to examine patterns in the probability distributions and correspondences in attribute data in relation to the spatial distribution of the entities to which they are attached.

Most statistical packages provide at least the following procedures for statistical data analysis:

- basic statistics—means, standard deviations, variances, skewness, kurtosis, maxima and minima, etc.
- non-parametric statistics—median, mode, upper and lower quartiles
- histograms, 2D and 3D scatter plots, box and whisker plots, stem and leaf plots
- univariate and multivariate analysis of variance
- linear regression and correlation
- principal components and factor analysis
- cluster analysis
- canonical analysis
- discriminant analysis.

6.6 Exploring spatial relations: geographically weighted regression

Earlier in the chapter, the basic principles of regression were outlined. Like the mean average and standard deviation, standard linear regression is a *global* and *aspatial* method. The regression equation represents average behaviour across the whole study area (in this sense it is the global relationship) and the approach does not take into account the spatial location of observations (it is aspatial). Where data are spatial, and so can be mapped, it may be useful to consider the spatial configuration of the data. It may be, for example, that neighbouring values of some variable explain a large proportion of the variation in the dependent variables.

A range of spatial regression approaches, such as spatial autoregressive models, exist (see Lloyd 2010b for a summary). **Geographically weighted regression** (GWR; Fotheringham et al. 2002) is an increasingly widely used approach which allows the fitting of local regression models so that coefficients (intercepts and slopes) can be obtained for specific locations. Firstly, a weighting function must be specified. In simple terms, this determines how much weight or influence is given to an observation. With GWR, observations more distant from a particular location receive smaller weights. In the case of bivariate regression, for example, the regression line will be closer to points on the scatter plot with larger weights (that is, closer locations) than it will be to points with smaller weights (which correspond to more distant locations). So the fitted model is likely to be different at all locations since the data values and possibly the configuration of the data may differ across the study area.

The general GWR procedure can be outlined as follows:

1. Go to a location **s** (for example, the centre of a census zone).
2. Assign weights to all observations as a function of distance from location **s**.
3. Fit the regression model using the weights assigned in stage 2.

So, once all locations have been visited, there will be a set of regression coefficients at all locations, which can be mapped to allow an assessment of how the relationship between the variables differs across the study area. A worked example of GWR is given by Lloyd (2010b). In addition, the geographically weighted correlation coefficient r (and therefore the coefficient of determination, r^2) can be computed for each location. Figure 6.7 shows the geographically weighted coefficient of determination for each sampling location in the Welsh precipitation data set. In this example, the weights, which determine how much influence each set of paired values (that is, elevation and precipitation amount) have in the locally fitted model, were based on an adaptive bandwidth—the 19 closest observations to each location (this number was determined using the Akaike Information Criterion—see Fotheringham et al. 2002) were given weights, with the observations closest to each location given the most weight or influence. In the example, for some regions the

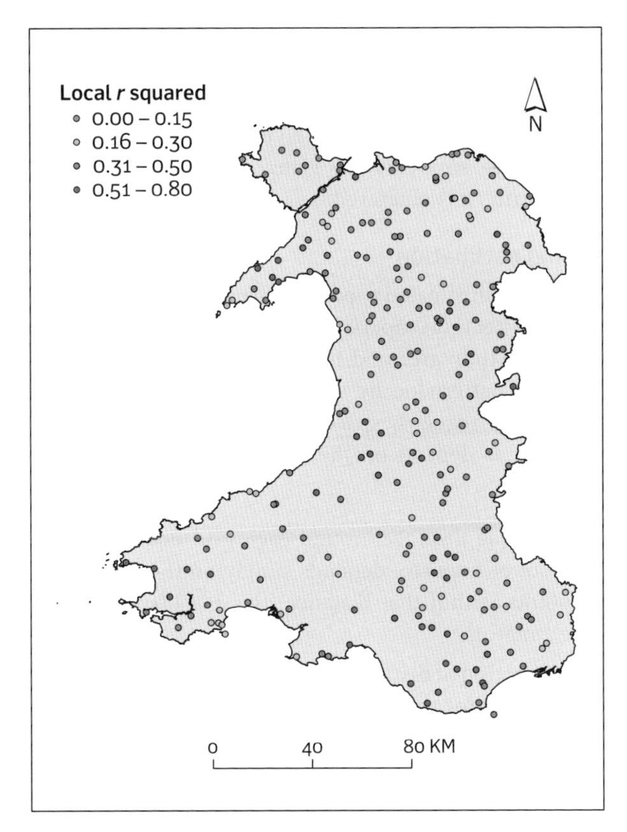

Figure 6.7 Geographically weighted r^2 for elevation against precipitation

relationship between elevation and precipitation amount is strong (the maximum local r^2 is 0.80), whereas in other areas the relationship is very weak. GWR allows the exploration of spatial variations in relationships, and it is increasingly widely applied in GIS contexts.

6.7 Point pattern analysis

The chapter has, so far, focused on spatial data with attached values. In this context, questions such as 'are values of a pollutant spatially clustered?' are relevant. In some applications, it is the location specifically which is of interest rather than the spatial distribution of observations according to their values. Analysis of point configurations is referred to as point pattern analysis. The point locations are termed events, and examples of **point patterns** (defined in Section 5.1) include trees, volcanoes, and people with a particular disease. Methods for analysing point patterns can be divided into two groups: (a) methods for exploring event intensity and (b) measures based on distances between events (Lloyd 2010b). This section gives a short summary of some key ways of analysing point patterns. The selected approaches are illustrated using a data set which is well known and widely applied in spatial analysis contexts. The data are the locations of 62 California redwoods and they are utilized by Diggle (2003). The original data (a subset of a larger data set) were placed in a square of 23 metres to a side, but they have been scaled to a unit square—that is, a square of one by one units. The redwoods data were shown in Figure 5.1 in Chapter 5.

Basic summaries of point patterns include the mean centre (defined by the mean of all of the x coordinates and the mean of all of the y coordinates). The average nearest neighbour distance is a commonly applied numerical summary. As its name suggests, this is the mean of the distances from each point to its nearest neighbour. We can calculate the distance, d, between points labelled i and j using Pythagoras' Theorem:

$$d_{ij} = \sqrt{(x_i - x_j)^2 + (y_i - y_j)^2} \tag{6.3}$$

Location i has the coordinates x_i, y_i and location j has the coordinates x_j, y_j. As an example, with a point x_i with location $x_i = 7.47$, $y_i = 13.40$ and a point x_j with location $x_j = 9.70$, $y_j = 10.50$, the distance between them is calculated as:

$$\sqrt{(7.47 - 9.70)^2 + (13.40 - 10.50)^2}$$
$$= \sqrt{4.97 + 8.41} = \sqrt{13.38} = 3.66$$

In the case of the redwoods data, the mean nearest neighbour distance is 0.0385 units. For point patterns which contain clusters of events, the mean nearest neighbour distance will be smaller than for point patterns where the events are more evenly distributed.

The measures discussed below provide various ways of assessing the degree of clustering or dispersion of a point pattern. To consider how clustered or how dispersed a point pattern is, a statistical test of some kind is needed. This is often achieved by considering whether an observed process is likely to be an outcome of some hypothesized process (for example, in effect, are the locations of the points due to some 'random' process or are the points a function of some process that causes them to cluster?). Models of **complete spatial randomness** (CSR) are used widely. For a CSR process the intensity is constant (that is, each location has an equal chance of containing an event) and events are independently distributed. There are various tests that can be applied to assess whether a point pattern corresponds to a CSR process. In other words, such tests support our visual interpretation of the point pattern.

0	0	0	0	0	0	0	1	1	1
0	0	3	2	3	0	0	0	0	1
0	0	1	0	2	1	0	0	2	0
0	1	1	0	1	2	0	0	0	0
0	5	2	0	0	3	0	0	0	2
0	1	3	0	0	0	0	0	0	0
0	1	0	0	0	0	4	0	0	0
0	0	0	0	0	0	2	2	0	0
0	0	0	0	1	0	0	3	3	0
0	0	0	1	2	1	0	0	3	0

Figure 6.8 Redwood locations: counts for 10 x 10 quadrat cells

Exploring event intensity

Mean intensity

Intensity can be referred to as the mean number of events per unit area at a particular location. The mean intensity of a point pattern is given by the number of events divided by the area of the study region. For the example of the redwood data, there are 62 observations and the study area has been rescaled to one by one unit, so the mean intensity is 62/1 = 62. If the original size of the study area is used (approximately 23 metres to a side = 529 square metres), then the mean intensity is 62/529 = 0.117 redwoods per square metre.

Quadrat analysis

A common way of assessing point intensity is to superimpose a grid over the point pattern and count the number of points that fall within each grid square. Various tests can be applied to these quadrat counts which enable the assessment of the degree of clustering in the point pattern. Figure 6.8 shows a 10 by 10-cell grid superimposed on the redwoods point pattern—the number of trees in each quadrat is indicated. The **variance/mean ratio** (VMR) provides a way to assess how far the point pattern is clustered or dispersed. Where the VMR takes a value of less than 1, this suggests a dispersed point pattern, while a value of greater than 1 suggests a clustered point pattern. In this example, there are 100 quadrats and 62 events, so the mean quadrat count is given by 62/100 = 0.62. The variance of the quadrat counts is 1.17 and so the VMR is given by 1.17/0.62 = 1.88. This figure supports the visual impression that this point pattern is clustered.

Kernel estimation

A more sophisticated approach than quadrat analysis entails (in effect) (a) superimposing a regular grid of points over the study area and (b) measuring the intensity of the point pattern locally within a neighbourhood around each of the regularly spaced points in (a). Two possible approaches depend on the way the neighbourhood is defined:

1. A simple distance around each grid node: the estimate of intensity depends simply on the number of points within that distance (this is called the naïve estimator).
2. A function of distance where the estimate of intensity accounts for the distance of each event from the grid node. As an example, there may be ten events around a grid node; if these events tend to be close to the grid node then the intensity estimate will be larger than if the events tend to be more distant (this is known as the kernel estimator).

So, if we want to estimate the intensity of points over an area we can simply calculate the intensity within a radius around the nodes of a grid. A kernel function is used to give larger weights to nearby events (points) than events that are more distant. A simple schematic example is given in Figure 6.9—here there are two sets of eight points, each in a circle of the same size. Using the naïve estimator, the intensity estimates for the centres of the two circles would be the same. With a kernel estimator, the estimated intensity for the centre of the circle on the right would be greater than for the circle on the left, as the points in the right-hand circle cluster around the

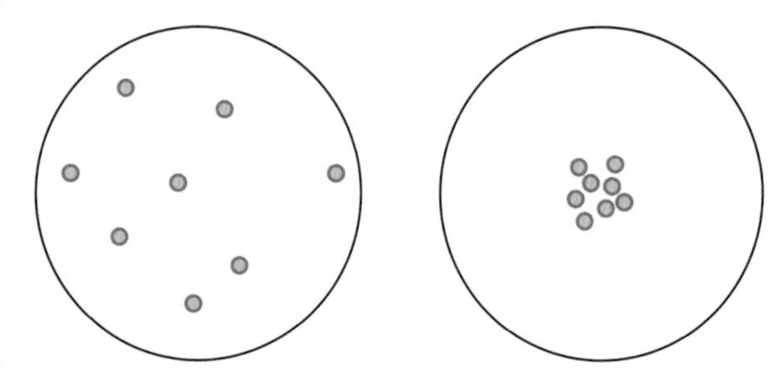

Figure 6.9 Point events within a circle

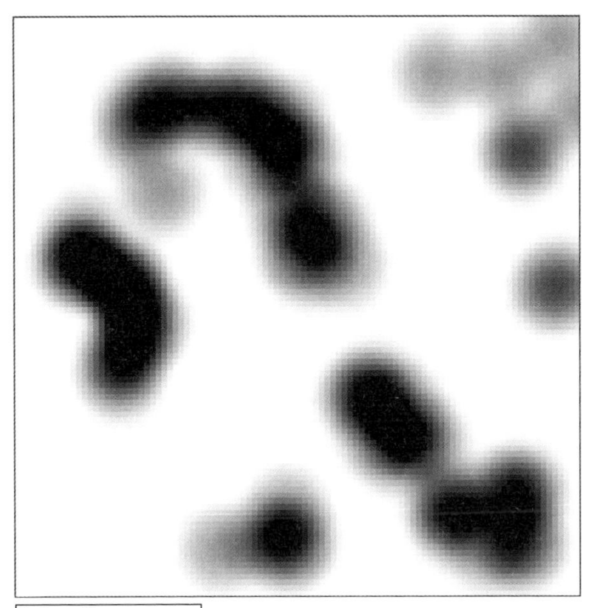

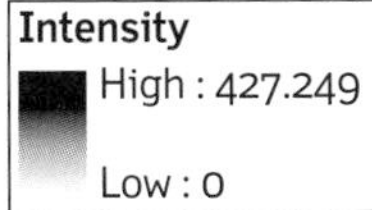

Figure 6.10 Intensity estimates for redwood locations

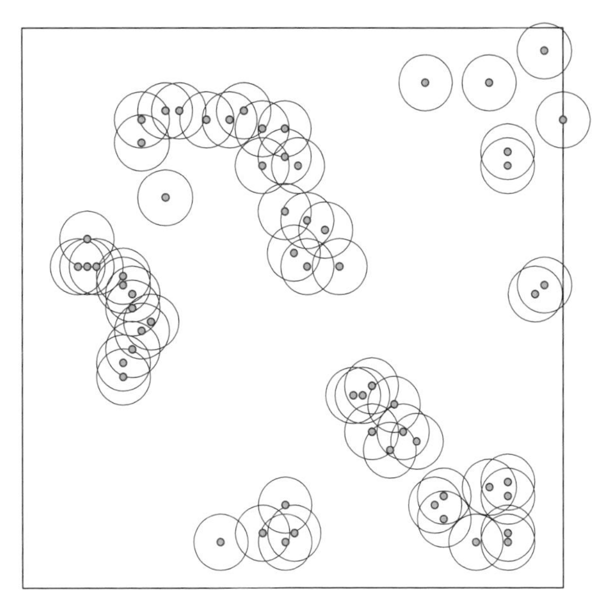

Figure 6.11 Circles with 0.05-unit radii around redwoods

centre of the circle, while those in the left-hand circle are more dispersed.

Figure 6.10 gives an example of an intensity map generated using the redwood data. The intensity estimates are made to a grid with a 0.01-unit grid spacing (that is, the grid nodes are 0.01 units apart in the x and y directions). The distance decay (that is, the weight or influence given to points according to their distance from the centre of each node of the intensity grid) reaches a value of zero at 0.1 units—in other words, events more than 0.1 units from a grid node were not included in the intensity estimate for that grid node. Intensity estimation is discussed in greater depth by Lloyd (2010b).

Kernel estimation provides a way of visually assessing spatial variations in point event distributions. Other approaches exist for identifying statistically significant (in some sense) clusters of point values, and Lloyd (2011) provides a summary of some relevant methods.

Measures based on distances

Common summaries of point patterns include the mean nearest neighbour distance. The distance from each point to its closest neighbour is measured and the mean average of these distances is then computed.

K function

With the K function, the clustering of events at different spatial scales can be explored. In simple terms, the K function is a measure of the intensity of events in neighbourhoods of different sizes. The K function for a set of events which form small distinct clusters will be very different, for example, to the K function for events which form larger, overlapping clusters. The K function can be computed by (a) finding all points within radius r of an event and (b) counting these and calculating the mean count for all events, after which the mean count is divided by the overall study area event density, which gives the K function for a particular distance, (c) the radius is then increased by some fixed step and (d) steps (a), (b), and (c) are repeated to the maximum desired distance. The output from the last stage for each distance can be plotted against the distance.

Figure 6.11 shows distance bands of 0.05 units placed around each of the 62 redwood locations. In this example, the first stage of the K function entails computing the number of events within 0.05 units of each event—thus, the events within each circle (but the count for a given circle excludes the event which defines its centre). The radii are then increased in 0.05 steps and the number of events is summed for each distance band, up to 0.5 distance units.

The K function can be given by:

$$\hat{K}(d) = \frac{|A|}{n^2} \sum_{i=1}^{n} \#(C(\mathbf{x}_i, d)) \qquad 6.4$$

Table 6.1 Number of events within specified distance bands with corresponding K function, L function, and expected number

Distance	Sum	*K*	*L*	Expected
0.05	88	0.023	0.035	0.008
0.1	250	0.065	0.044	0.031
0.15	418	0.109	0.036	0.071
0.2	554	0.144	0.014	0.126
0.25	698	0.182	−0.010	0.196
0.3	888	0.231	−0.029	0.283
0.35	1128	0.293	−0.044	0.385
0.4	1452	0.378	−0.053	0.503
0.45	1774	0.461	−0.067	0.636
0.5	2008	0.522	−0.092	0.785

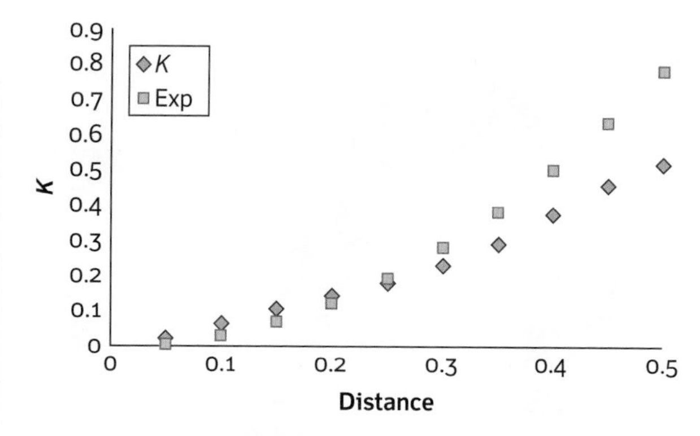

Figure 6.12 K function for the redwood data

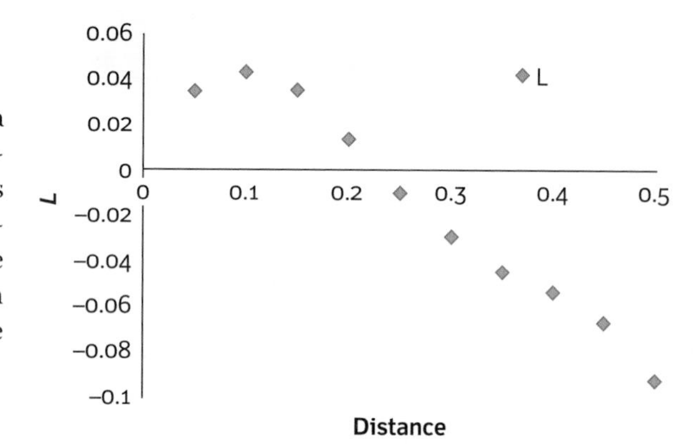

Figure 6.13 L function for the redwood data

where the hat over K indicates it is an estimate, d is a distance, $|A|$ is the area of the study region, n is the number of events, and $\#(C(\mathbf{x}_i, d))$ is the number (#) of events within a circle C, with radius d, which is centred on location $\mathbf{x}_i$. The measured value of K for a particular distance can be compared to the expected value, which is given by πd^2. Given the expected value, the K function can be transformed to the L function:

$$\hat{L}(d) = \sqrt{\frac{\hat{K}(d)}{\pi}} - d \qquad 6.5$$

For values of less than zero, the L function suggests regularity, while values of greater than zero suggest clustering. Table 6.1 gives, for the redwood data, the number of events within the specified distances of events, the K function, the L function, and the expected value. As one example, there are 88 events within 0.05 units of the events, so:

$$\hat{K}(0.05) = \frac{1}{62^2} 88 = 0.023, \hat{L}(0.05) = \sqrt{\frac{0.023}{\pi}} - 0.05 = 0.035,$$

where the expected value is $\pi 0.05^2 = 0.008$.

The K and L functions for the redwoods data are given in Figures 6.12 and 6.13, respectively. The K function indicates that the redwood point pattern is clustered at small distances (the K function values are larger than the expected values), but it is dispersed at larger distances. The L function shows this more clearly, and clustering is indicated for those distances for which values of L are greater than zero. The K and L functions provide useful summaries of the structure of a point pattern—that is, they help to understand if a point pattern clusters at all and, if so, over what spatial scales clustering occurs.

Practical applications of K functions are diverse; Bailey and Gatrell (1995) use the example of juvenile offenders. Many other applications of K functions can be found in the literature which deal with a varied range of point patterns.

In analysis of a point pattern which represents a subset of a population—for example, people with a particular disease—it is necessary to account for the population at risk. In other words, the numbers of cases of particular diseases in urban areas are likely to be greater than in rural areas, as there tend to be more people in urban areas than in rural areas. Ways of accounting for the population risk are considered by Bailey and Gatrell (1995).

6.8 Summary

This chapter has introduced a diverse range of approaches for summarizing and exploring spatial data, from simple univariate statistical descriptive statistics and regression and correlation to methods for assessing spatial structure in data values (measures of spatial autocorrelation) or in point patterns. These approaches may be used to help gain a first impression of the key characteristics of the data, or alternatively such analyses may be an end in themselves. Analysis of spatial autocorrelation lies at the heart of spatial data analysis, while tools such as GWR offer powerful approaches for assessing how geographically referenced variables are related. This chapter has introduced some key concepts and tools, and the following chapters go on to consider a variety of additional ways of exploring geographical data. Chapter 7 is concerned with assessing distances or connections between discrete entities. Chapters 8 and 9 outline some approaches for making estimates of data values at unsampled locations. Chapter 10 outlines approaches for analysing continuous fields (which are commonly represented using raster grids), while Chapter 11 considers the construction and analysis of digital elevation models.

Questions

1. What statistical measures would you employ to summarize the central tendency and dispersion (that is, variation around the 'central' value) of a set of data values?
2. How would you explore the relationship between two numerical variables such as population density and the amount of some pollutant?
3. What is meant by spatial dependence? Would you expect elevation values to be spatially dependent?
4. Why might a local (geographically weighted) regression analysis offer benefits over a conventional (global) regression analysis?
5. What would a K function of a point pattern with small isolated clusters look like?

Further reading

▼ Bailey, T. C. and Gatrell, A. C. (1995). ***Interactive Spatial Data Analysis***. Longman Scientific and Technical, Harlow.

▼ Harris, R. and Jarvis, C. (2011). ***Statistics for Geography and Environmental Science***. Prentice Hall, Harlow.

▼ Lloyd, C. D. (2010). ***Spatial Data Analysis: An Introduction for GIS Users***. Oxford University Press, Oxford.

▼ Lloyd, C. D. (2011). ***Local Models for Spatial Analysis***. 2nd edn. CRC Press, Boca Raton, FL.

▼ Rogerson, P. A. (2006). ***Statistical Methods for Geography: A Student's Guide***. 2nd edn. SAGE Publications, London.

Analysis of Discrete Entities in Space

The aim of GIS is not just to create a database of digital representations of geographical phenomena, but to provide means of selecting, retrieving, and analysing them. The previous chapter outlined some approaches to summarizing geographical data and assessing patterns. This chapter explains some of the methods available for dealing with crisp entities ('things')—how they can be selected from the database in terms of their attributes, and how new attributes can be computed ('modelled') using the rules of Boolean logic and mathematics to yield useful groups or classes, or to generalize complex map images. Many of these procedures are not really spatial because they only affect the attributes and not the size, shape, or form of the spatial entities, which can be any geographical primitive—point, line, area, or pixel. Spatial analysis often begins with the determination of spatial inclusion or exclusion, and with the intersection of lines and areas of different kinds to yield new entities. Spatial interactions are not just limited to the boundaries of existing entities, but may be extended to include neighbourhood functions such as crow's flight distances, topological proximity, and distance over networks such as roads or rivers. These procedures are illustrated by examples from several disciplines, including land evaluation and planning.

Learning objectives

By the end of this chapter, you will:

- **understand some key concepts and classes of approaches which can be used to work with discrete entities**
- **have developed a knowledge of some key ways of analysing data on discrete entities**
- **be able to apply your knowledge to reclassify data, measure distances from objects, and assess if and how features in different layers overlap.**

7.1 Spatial analysis is more than asking questions

The kinds of analytical techniques that can be used on spatial data depend greatly on the data model and the representation that have been used. It is important to realize that different data models and different kinds of representation can require different approaches to the way spatial queries can be formulated. The fundamental question is whether the basic data model refers to *entities in space* or to the *continuous variation of an attribute over space*. In the case of entities, data retrieval and analysis concern the attributes, location, and connectivity of the entities and measures of the way they are distributed in space. In the case of continuous fields, data analysis concerns the spatial properties of the fields. The matter is made more complicated by the fact that continuous fields are usually **discretized** to a set of triangles or a regular grid, and the individual triangles or grid cells (or particular sets of contiguous triangles or grid cells) can also be treated as individual entities. In Chapter 2, the vector data model was introduced and this model was shown to be appropriate for representing discrete features. Chapter 3 included a discussion of databases in GIS and made links to vector topology. Most of this chapter assumes a vector data model, although not exclusively since the raster model is often used to represent objects, with remotely sensed images (for example) containing discrete features (such as roads and buildings) as well the terrain surface.

This chapter concentrates on the methods of data analysis that are most useful for dealing with *entities in space*, either in the relational or object-oriented model; the analysis of continuous fields is covered in Chapter 10. The fundamental axioms for data modelling and analysis were presented in Chapter 2. This chapter demonstrates the applications of these axioms and how they can be translated into computer commands to solve practical problems using discrete entities. Chapter 10 extends this discussion to continuous fields.

7.2 The basic classes of operations for spatial analysis

In the entity model of objects in space, three kinds of information are important: *what is it?, where is it?*, and *what is its relation to other entities*? The nature of an entity is given by its attributes, its whereabouts by its geographical location or coordinates, and the spatial relations between different entities in terms of **proximity** and connectivity (topology). The aspects of location, proximity, and topology distinguish geographical data from many other kinds of data that are routinely handled in information systems.

We distinguish the following basic classes of data analysis options for entities:

Attribute operations

- Operations on one or more attributes of an entity
- Operations on one or more attributes of multiple entities that overlap in space
- Operations on one or more attributes that are linked by directed pointers (object orientation)
- Operations on the attributes of entities that are contained by other entities (point in polygon)

Distance/location operations

- Operations to locate entities with respect to simple Euclidean distance or location criteria
- Operations to create buffer zones around an entity

Operations using built-in spatial topology

- Operations to model spatial interactions over a connected net.

All of these operations can result in new attributes, which can be attached to the original entities, thereby increasing the size and value of the database. Certain operations also create new spatial entities, requiring the database to be expanded to include these new items. Data that have been retrieved from the database can be displayed on the screen, plotted as a paper map, or written as a digital file for future processing.

Computing areas (see Box 7.1) and perimeters are common operations on discrete entities.

7.3 Operations on the attributes of geographic entities

As explained in Chapter 2, attributes are properties of entities that define what they are. They can be divided into three types—those that refer to location (the **geographical attributes** of latitude, longitude (or easting, northing), and elevation); those that are simply attached as qualitative or quantitative descriptors of some non-spatial property; and those that are derived from the spatial properties of the entity itself. For example, the attributes of parcel number, name of the owner, and land cover describe non-spatial properties of a piece of land. The length of the fence bordering the road, the area, shape, and contiguity are attributes that are derived from the form of the piece of land.

As in conventional information systems, new attributes can be attached to entities as the result of a database operation. For example, a new attribute (or a new value of

Box 7.1 Computing polygon areas using the trapezoidal rule

A polygon may be described in terms of a series of trapeziums, as shown in the figure below.

The area under the polygon is calculated by summing the areas of the various trapeziums that make up the total shape.

The area of a trapezium = (half the sum of its sides) × (horizontal distance)

The way to derive the total area, accounting for the varying levels of the sides, is to sum all the trapeziums that make each side of the shape in one direction and then subtract the total in the opposite direction.

The area of the polygon in the figure is computed by:

Add the areas of upper trapeziums A, B, C, and subtract the areas of lower trapeziums D, E, F. Upper trapeziums: A (h, a, b, i); B (i, b, c, k); C (k, c, d, m). Lower trapeziums: D (h, a, f, j); E (j, f, e, l); F (l, e, d, m). Lloyd (2010b) illustrates another approach to area calculation.

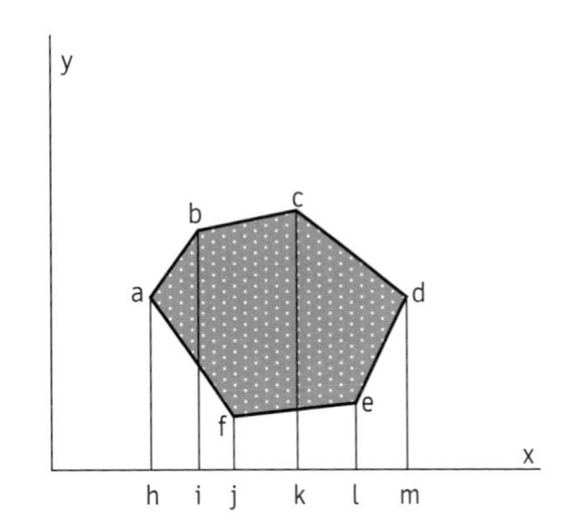

an attribute) can be computed for land parcels larger than a given size, or for those having owners that live abroad. For displaying information, the new attribute could be the colour or the symbol chosen to represent this kind of entity on the map. The new attribute can be derived by any legitimate method of logical and mathematical analysis, including operations on the proximity and topological properties of entities. Simple data retrieval can be seen as the creation of a new temporary attribute 'selected' when the set of attributes attached to an entity match the search criteria.

The process of selection or creation of new attributes can be formalized as follows. For any given location x, the value of a derived attribute U_i is given by:

$$U_i = f(A, B, C, D, ...) \qquad 7.1$$

where A, B, C, ... are the values of the attributes used to estimate U_i. The function $f()$ can be any of the following, singly or in combination:

a) logical (Boolean) operations

b) simple and complex arithmetical operations and numerical models

c) univariate statistical analysis

d) multivariate statistical methods or Bayesian statistics for classification and discrimination

e) multicriteria methods, artificial intelligence (AI)-based methods: neural networks.

Vector data in multiple separate layers can be easily combined and queries conducted using these combinations.

Data retrieval using the attributes attached to individual entities

Entities can be selectively retrieved or reclassified on their attributes by using the standard rules of Boolean algebra, which are incorporated in database languages such as SQL (Worboys 2005 provides an introduction to relational database management systems in GIS, as well as other database frameworks). **Boolean algebra** uses the logical operators AND, OR, XOR, NOT to determine whether a particular condition is true or false (Box 7.2). Each attribute is thought of as defining a *set*. The operator *AND* (symbol $\wedge$) is the *intersection* of two sets—those entities that belong to both *A* and *B*; *OR* (symbol $\vee$) is the *union* of two sets—those entities that belong either to set *A* or to set *B*; *NOT* (symbol $\neg$) is the *difference* operator, identifying those entities that belong to *A* but not to *B*, and *XOR* (symbol $\veebar$) is the *exclusive OR*, or the set of objects that belong to one set or another, but not to both. These simple set relations are often portrayed visually in the form of Venn diagrams (Figure 7.1). Note that logical operations can be applied to all data types, whether they are Boolean, nominal, ordinal, scalar, or directional.

Two simple examples illustrate the principles. Consider first a spatial database used by an estate agent. A typical retrieval query from a prospective buyer might be the following: 'Please show me the locations of all houses costing between \$200 000 and \$300 000 with four bedrooms and plots measuring at least 300 m²'. If the data set contains the attributes 'cost', 'number of bedrooms', 'area of plot', and location, a map of the desired premises

Box 7.2 Mathematical operations for transforming attribute data

a) Logical operations

Truth or falsehood (0 or 1) resulting from union (∨ logical OR), ***intersection*** (∧ logical AND), ***negation*** (¬ logical NOT), and ***exclusion*** (⊻ logical exclusive or XOR) of two or more sets.

b) Arithmetical operations

New attribute is the result of addition (+), subtraction (−), multiplication (*), division (/), raising to power (**), exponentiation (exp), logarithms (ln—natural, log—base 10), truncation, or square root.

c) Trigonometric operations

New attribute is the sine (sin), cosine (cos), tangent (tan) or their inverse (arcsin, arccos, arctan), or is converted from degrees to radians or grad representation.

d) Data type operations

New attribute is the original attribute expressed as a different data type (Boolean, nominal, ordinal, directional, integer, real, or topological data type).

e) Statistical operations

New attribute is the mean, mode, median, standard deviation, variance, minimum, maximum, range, skewness, kurtosis, etc. of a given attribute represented by *n* entities.

f) Multivariate operations

New attribute is computed by a multivariate regression model.

New attribute is computed by a numerical model of a physical process.

New attribute is computed by a ***principal component analysis***, ***factor analysis***, or ***correspondence analysis*** transformation of multivariate data.

Entity is assigned to a given ***class*** (new attribute = class name) by methods of multivariate numerical taxonomy.

Entity is assigned a ***probability*** (based on statistical chance) by discriminant analysis, maximum likelihood, or Bayesian techniques, of belonging to a given set.

Entity is assigned a ***fuzzy membership value*** for a given set.

Entity is assigned to a class using neural network methods.

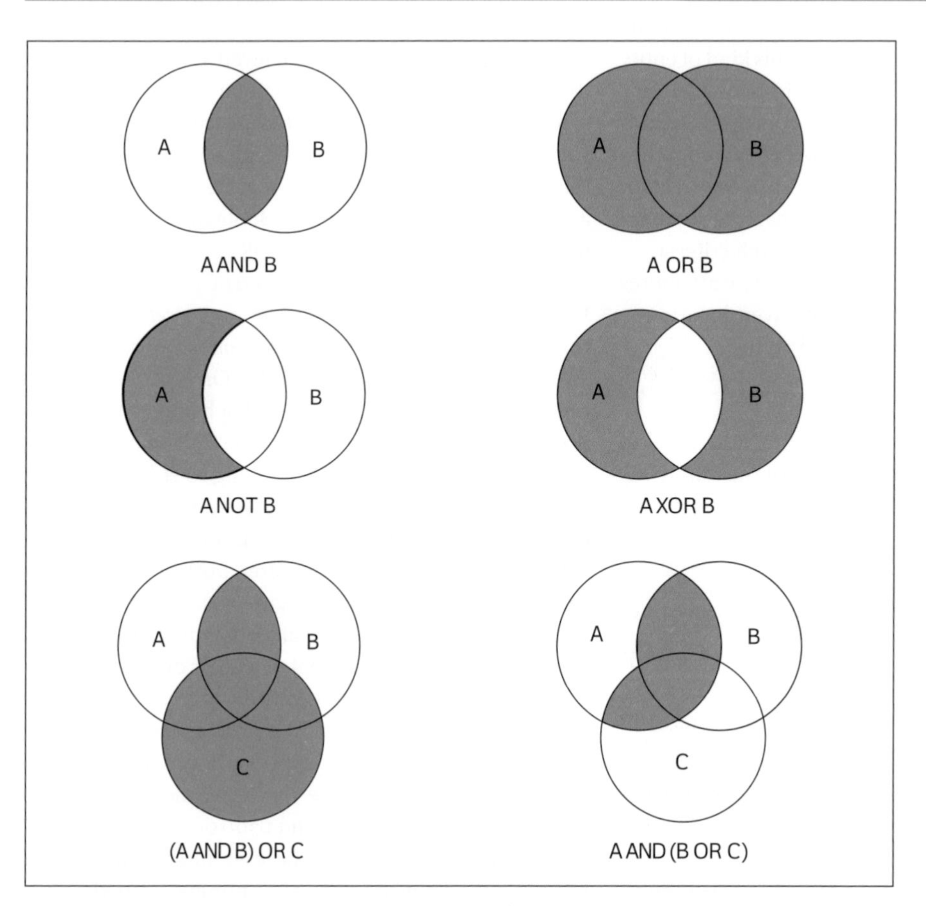

Figure 7.1 Venn diagrams showing the results of applying Boolean logic to the union and intersection of two or more sets; in each case the shaded zones are 'true'

is easily produced by a multiple AND query on the specified attributes to highlight the matching plots:

IF COST GE $200 000 AND COST LT $300 000 AND N BEDROOM = 4 AND PLOT AREA GE 300 THEN ITEM = 1 ELSE ITEM = 0

The selected entities are given a Boolean value of 1 (true) if they match the specifications, and a 0 (false) if not. Display of the results follows by assigning a new colour to entities with ITEM = 1.

Now consider a query in land suitability classification. In a database of soil mapping units, each mapping unit may have attributes describing the texture and pH of the topsoil. If set *A* is the set of mapping units called *Oregon loam* (nominal data type), and if set *B* is the set of mapping units for which the topsoil pH equals or exceeds 7.0 (scalar data type), then the data retrieval statements work as follows:

X = A AND B finds all occurrences of Oregon loam with pH ≥ 7.0

X = A OR B finds all occurrences of Oregon loam, and all mapping units with pH ≥ 7.0.

X = A XOR B finds all mapping units that are either Oregon loam, or have a pH ≥ 7.0, but not in combination.

X = A NOT B finds all mapping units that are Oregon loam where the pH is less than 7.0.

Selected entities can also be renamed and/or given a new display symbol (Figure 7.2) by statements such as: 'Give the designation "Suitable" to all mapping units with soil texture = "loam" and pH ≥ 5.5'. This is a particular instance of the logical statement 'IF condition (*C*) THEN carry out specified task'.

Note that, unlike arithmetic operations, Boolean operations are not commutative. The result of *A AND B OR C* depends on the priority of *AND* with respect to *OR*. Parentheses are usually used to indicate clearly the order of evaluation when there are more than two sets (Figure 7.1). For example, if set *C* contains mapping units of poorly drained soil, then *X = (A AND B) OR C* returns all mapping units that are either (a) Oregon loam with a pH ≥ 7.0 or (b) units of poorly drained soil. The relation *X = A OR (B AND C)* returns (a) all Oregon loam mapping units and (b) those mapping units with a combination of pH ≥ 7.0 and poor drainage.

Note also that Boolean operations may require an exact match in attributes to return data, and they take no account of errors or uncertainty unless that is specifically incorporated into the definitions of the sets.

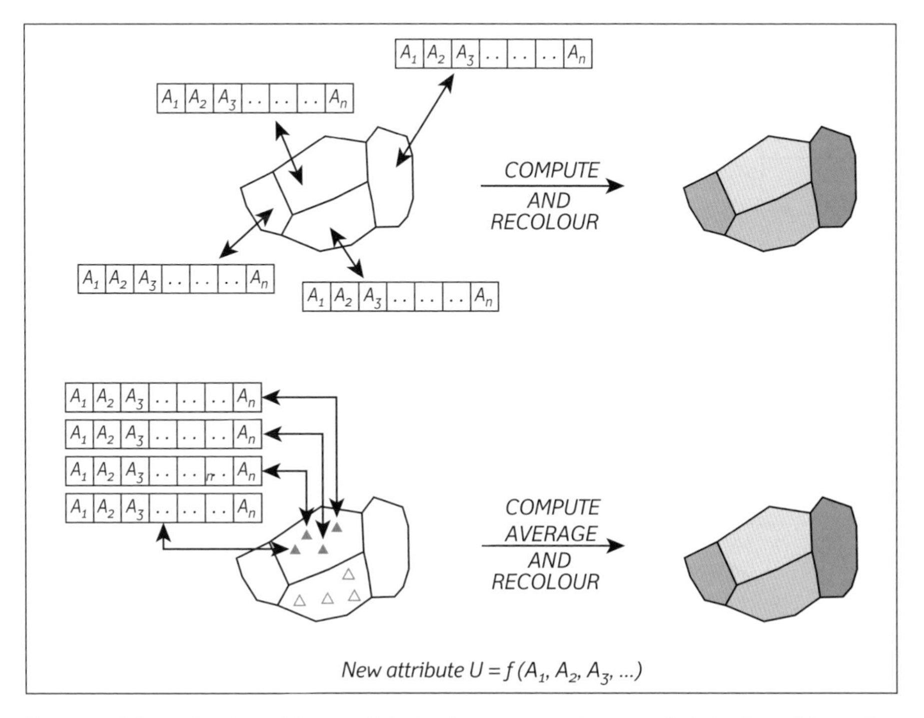

Figure 7.2 When retrieving entities on attributes alone, or computing new attributes from old ones, the spatial units of the map do not change shape, only colour or shading (top). The same occurs when a point-in-polygon search is used to find enclosed point objects that are used to compute area averages (bottom).

If the value of the attribute 'elevation' is set at 2000 m above sea level to define the class 'mountain', then hills with elevations up to 1999.999999999 ... m will be rejected. This is not a problem with ordinal and nominal data types but it can present problems when working with scalar data types that represent quantities like elevation, pH, clay content, soil depth, atmospheric pressure, salinity, population, and so on, that are subject to various sources of measurement error and uncertainty. If the error bands on these data span the boundary values of sets then strict application of Boolean rules may yield results that are counter-intuitive.

Spatial aspects of Boolean retrieval on multiple attributes of single entities

Carrying out logical retrieval and reclassification on the non-spatial attributes of spatial entities has little effect on the map image, except in terms of symbolism and boundary removal. Computing a new attribute or condition requires the preparation of a legend and a recolouring or reshading of the selected entities (see Figure 7.2). When selection leads to adjacent polygons receiving the same code it may be sensible to dissolve the boundaries between them, achieving a form of map generalization (Figure 7.3). Figure 7.4 illustrates the use of this option to simplify a complex soil map.

Boolean operations are not only applicable to the non-spatial attributes of the geographical elements—they can also be applied to the geographic location and attributes derived from the spatial or topological properties of the geographic entities. For example, one might wish to find all mapping units exceeding 5 ha in areas having soil with clay loam texture in combination with a pH > 7.0. More complicated searches may involve the shapes of areas, the properties of the boundaries of areas, or the properties of neighbouring areas such as the areas of woodland bordering urban areas. In these cases, the results of the search would have an effect on the spatial patterns.

Simple and complex arithmetical operations on attributes of single entities

New attributes can be computed using all normal arithmetical rules (+, –, /, *, logarithms, trigonometric functions, exponents, and all combinations of them, including complex mathematical models). Arithmetical and trigonometrical operations can obviously only be used on scalar data and certain kinds of ordinal data. Arithmetical operations on Boolean data types and nominal data types are nonsensical and therefore not allowed (an expression such as $X = sqrt$ (London) has no meaning). Arithmetical operations can be very simple, or very complicated, but in all cases the operation is the same—a new attribute (a result) is computed from existing data.

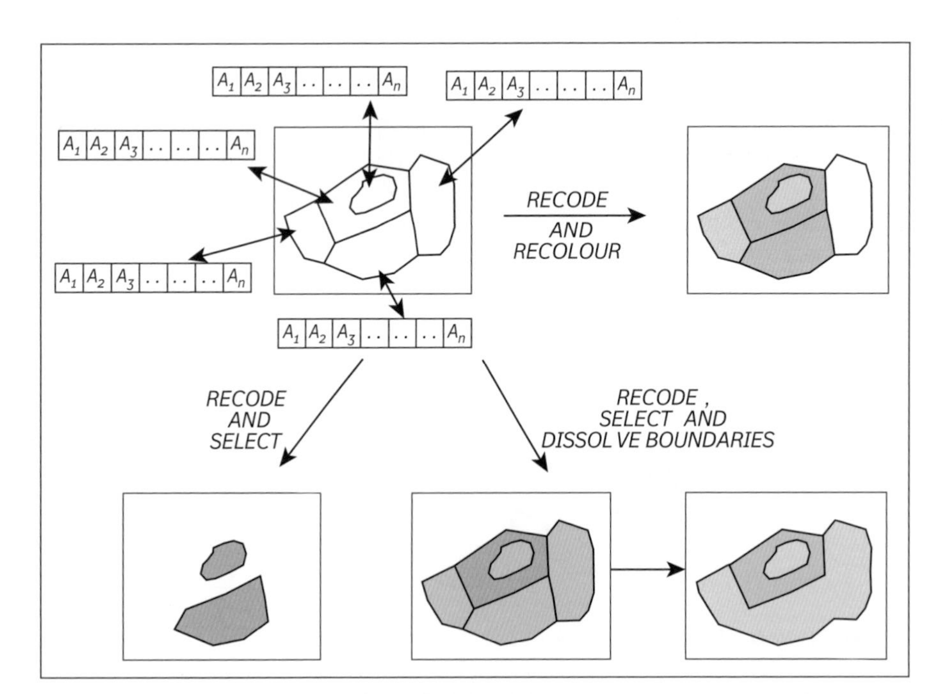

Figure 7.3 If, during a retrieval or recoding operation, two adjacent polygons receive the same new code, boundaries between them can be dissolved, leading to map generalization

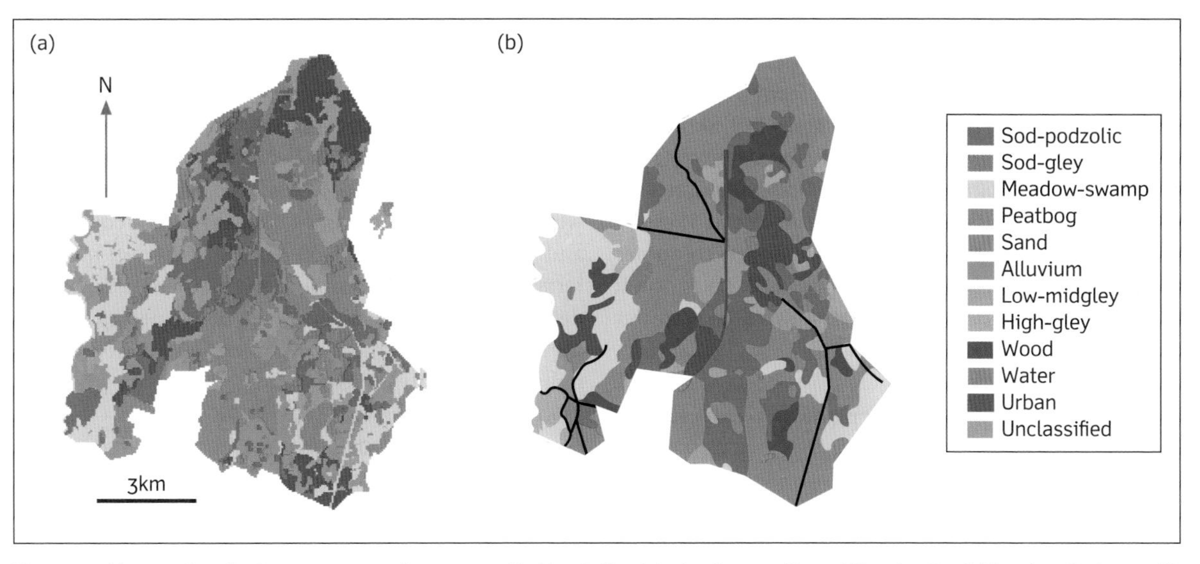

Figure 7.4 Using reclassification as a means of map generalization. Left: original soil map with 95 different units; right: reclassified map with 12 units. Note that reclassification preserves the original geometry and that the legend and shading codes only apply to the simplified map.

Some hypothetical examples of computing new attributes for a given administrative area (polygon) or point location are:

- Population increase = population 2010 − population 2000
- Total spending power = average income × number of persons
- Average wheat yield per farm = (total yield) / (number of farms)
- Predicted wheat yield = *f*(crop), where *f*(crop) is a complex mathematical model that computes wheat yield as a function of the soil, moisture, nutrients, and properties of a site (point entity)
- Class allocation = *Result* (multivariate classification) where (multivariate classification) might be any statistical or multicriteria analysis of the numerical attributes on the entity.

For a set of river catchments, the proportion of precipitation discharging through the outlet can be computed by dividing the annual precipitation for each catchment by the cumulative annual river discharge measured at the outlet.

Arithmetical operations can be easily combined with logical operators:

IF (A + B)/(C) ≥ TEST VALUE THEN CLASS = GOOD

The statistical analysis of attributes

Chapter 6 provides a brief introduction to the principles of univariate and multivariate statistical analysis. Readers requiring further details of these methods should consult a standard text such as Davis (2002) for geology and earth sciences, Rogerson (2006) for geography, Jongman et al. (1995) for ecology, and Haining (2003) for the environmental and social sciences. In this chapter we assume that statistical methods are just one set of procedures out of many for computing new attributes.

Simple statistical analysis can be used to compute means and standard deviations, and to conduct correlation and regression analysis (see Chapter 6). The operations can be applied to a set of attributes attached to single entities or to any set of entities that are retrieved by a logical search. For example, compute the mean and variance of nitrate levels for all groundwater monitoring sites included in a province P (e.g. Figure 7.2, lower example), or compute the average takings for all fast-food outlets in the postcode districts that include railway stations.

Numerical models

The range of arithmetical operations that can be applied to numerical attributes is unlimited. They are often used to compute the values of attributes that are difficult or impossible to measure, or which can be derived from cheap, readily available base data, such as data from censuses or natural resources surveys. Standard sets of mathematical operations that have been derived from empirical (regression) modelling are sometimes called **transfer functions** (see Ghermandi and Nunes 2013 for an application concerned with the values of coastal recreation) in disciplines like soil science and land evaluation. More complex sets of mathematical functions that represent a physical process such as crop growth, air quality, groundwater movement, pesticide leaching, epidemiological hazards, increase of population pressure, etc. are often referred to as **models** (described in more detail in Chapter 12). Most GIS do not provide the functionality to program these complex models; instead they are used to assemble the data and export them to the model, which might reside on another computer on the network. Development of new tools using the R programming language[1] is, however, providing greater flexibility.

Neural networks, multicriteria evaluation, and fuzzy logic

All methods of deriving new attributes so far presented are **parametric**, which is to say that they assume that the definition of a new attribute can be expressed by a logical or numerical equation in which weights or parameters can be objectively assigned. Regression analysis and the calibration of numerical models are just two examples of ways in which the 'best' parameter values are chosen for classification or calculation. It is worth adding that in most cases the numerical models are also linear—i.e. there is a direct relation between a parameter value and its effect on the output.

These assumptions derive from classical, mechanical science. But parametric methods are difficult to use in complex, non-linear situations where attribute values are not normally distributed and where causal or even statistical relations are tenuous. These difficult conditions surround many spatial data, whose interrelations may violate many of the basic tenets of parametric methods, and problems as simple as how best to classify complex spatial objects may be intractable. This problem has received particular attention in the classification of remotely sensed data into land cover classes (Lees 1996).

Neural networks are powerful ways of classifying geographic entities (entities or pixels) into sensible groups (Atkinson and Tatnall 1997; Kia et al. 2012 detail a recent application concerned with flood simulation using GIS). In many cases the analyses are not really spatial, but require an entity to be assigned to a class on the basis of a non-statistical method of computation. A neural network is a processing device, implemented as an algorithm or computer program, that is said to mimic the human brain. It comprises many simple processing elements that are connected by unidirectional communication channels carrying numeric data. The elements operate only on local data but the network as a whole organizes itself according to perceived patterns in the input data. These patterns can be created by 'self-learning'—the system determines the 'best' set of classes for the data—or 'supervised classification', in which the system is supplied with a template of the required classification.

Multicriteria evaluation and fuzzy logic. Neural networks are not the only ways of dealing with complexity and non-linearity in spatial data. Methods of **multicriteria evaluation** and multicriteria decision analysis (as defined in Section 7.7) have been developed to provide users with the means to determine new attributes that indicate alternative responses to problems involving multiple and conflicting criteria. In Chapter 13 we explore the particular use of fuzzy logic and continuous classification for spatial data analysis in GIS.

7.4 Examples of deriving new attributes for spatial entities

The results of computing new attributes or reclassification are usually displayed by reshading or recolouring the entities (Figure 7.2). As with Boolean selection, the spatial properties of the entities (location, shape, form, topology) do not change, except in the case that neighbouring entities are reclassified as being the same, when generalization can take place. Note that if the data are in raster form, these operations are carried out on each pixel separately, unless the data structure uses a map unit-based approach to raster data coding (see Chapter 3). The following sections give examples of spatial analysis based entirely on the derivation of new attributes. Section 12.1 gives an example of a regression model of surface elevation and temperature in the Swiss Alps—the regression equation can be used to derive estimates of temperature at all cell locations for an altitude matrix covering the study area.

Using multivariate clustering

Geodemographic segmentation (Harris et al. 2005) is a method used by multinational marketing companies to classify residential areas of Western countries into distinct neighbourhood types based on statistical information about the consumers who live in them. The spatial entities are provided by census districts, postal code districts, or mail order address units linked to the universe of national house addresses. We consider a case where each spatial unit is characterized by four key criteria: age (young, middle, and old), income (high, middle, and low), urbanization (metropolitan, urban, suburban, rural), and

[1] http://www.r-project.org/

family type (married couples with children, singles and childless couples, and pensioners). These attributes yield 108 possible combinations of classes, which can be recoded to ten core classes for the identification of characteristic socio-economic types. Multivariate methods of class allocation may then be used to assign a basic spatial unit to one of these ten classes, and they are also linked to other information including empirical sales data and consumer preference attributes obtained by questionnaire. Simple logical retrieval of spatial units in terms of their class and attributes provides maps at local or national level that show the spatial distribution of market opportunities and brand preferences.

Using simple Boolean logic

In many parts of the world there are insufficient data to compute crop yields with numerical models of crop growth as a function of available moisture, energy, and water, so qualitative predictions based on simple rules may be the only useful way to assess the suitability of land for agricultural development. This is the philosophy behind the now classic FAO land evaluation procedure (FAO 1976; Beek 1978).

Prescriptive land evaluation (Rossiter 1996; Castella et al. 2007) is based on the simple idea that a landscape can be divided into basic entities called *mapping units*, separated by crisp boundaries. Soil survey is often the basis for this kind of landscape division, but alternative methods use landform, ecological zones, or vegetation communities. The idea is that once basic spatial entities have been mapped and defined in terms of representative attribute values, their suitability for any given purpose may be determined by reclassification or by computing new attributes on the basis of existing information. The general procedure is called 'top-down' logic, because it starts with the presumption that global rules or physical models exist for translating primary units of information into units that can be used for a specific purpose. The procedure is exactly the same whether the data are stored and displayed as vector polygons or as grid cells; the only difference is the kind of spatial representation for the conceptual entity carrying the information.

The following is an example of this rule-based reasoning. In order to grow well, a crop needs a moist, well-drained, oxygenated, fertile soil. Growing a monocrop leads to the soil being bare, or nearly bare, for part of the year, and in this time the soil should be able to resist the effects of erosion by rain. The four attributes—available moisture, available oxygen, nutrients, and erosion susceptibility—are known as *land qualities* (LQ_i), and they may be derived from primary soil and land data using simple logical transfer functions derived by agronomists and soil experts. The overall suitability of a site (its *land quality* for the use intended) is determined by the most limiting of the land characteristics—a case of worst takes all.

Figure 7.5 illustrates the complete procedure using data from a soil series map and report and a digital elevation model of a small part of the Kisii District in Kenya (Wielemaker and Boxem 1982). The study area covers some 1406 ha (3750 × 3750 m^2) of the area mapped by the 1:12 500 detailed soil survey of the Marongo area (Boerma et al. 1974) and was chosen for its wide variety of parent material (seven major geological types), relief (altitude ranges from 4700 to 5300 feet above sea level—1420 to 1600 m), and soil (12 mapping units). Detailed soil survey information describes parent material, soil series, soil depth to weathered bedrock, stoniness and rockiness, and tabular information relating soil series to land qualities. Each attribute was digitized as a separate polygon overlay and converted to a 60 × 60 array of 62.5 m square cells. The digital elevation model was obtained by interpolation from digitized contours and spot heights and converted to local relief (minimum 40 m, maximum 560 m). Information about the climate, the chemical status of the soil, the land use, and cultural practices is also available. Slope lengths (needed for estimating erosion) were interpreted from stereo aerial photographs and digitized as a separate overlay.

In this example we consider suitability for smallholder maize, which is determined by the land qualities *nutrient supply, oxygen supply, water supply,* and *erosion susceptibility*. These land qualities can be ranked by assigning values of 1, 2, 3, respectively, to the following classes:

No limitation	assign 1
Moderate limitation	assign 2
Severe limitation	assign 3

The rules for deriving the land qualities are: water availability is a Boolean union (AND) of soil depth and soil series (*B*); oxygen availability and nutrient availability can be derived directly from the soil series information by recoding using a lookup table (*L*). Erosion susceptibility or hazard can be determined as a Boolean union (AND) of slope classes and soil series. Examples of the rules are:

If soil series is $\mathbf{S}_1$ *then assign nutrient quality* $\mathbf{W}_3$ *from lookup table* $\mathbf{L}_W$.

If soil series is $\mathbf{S}_2$ *and slope class is 'flat' assign erosion susceptibility 1.*

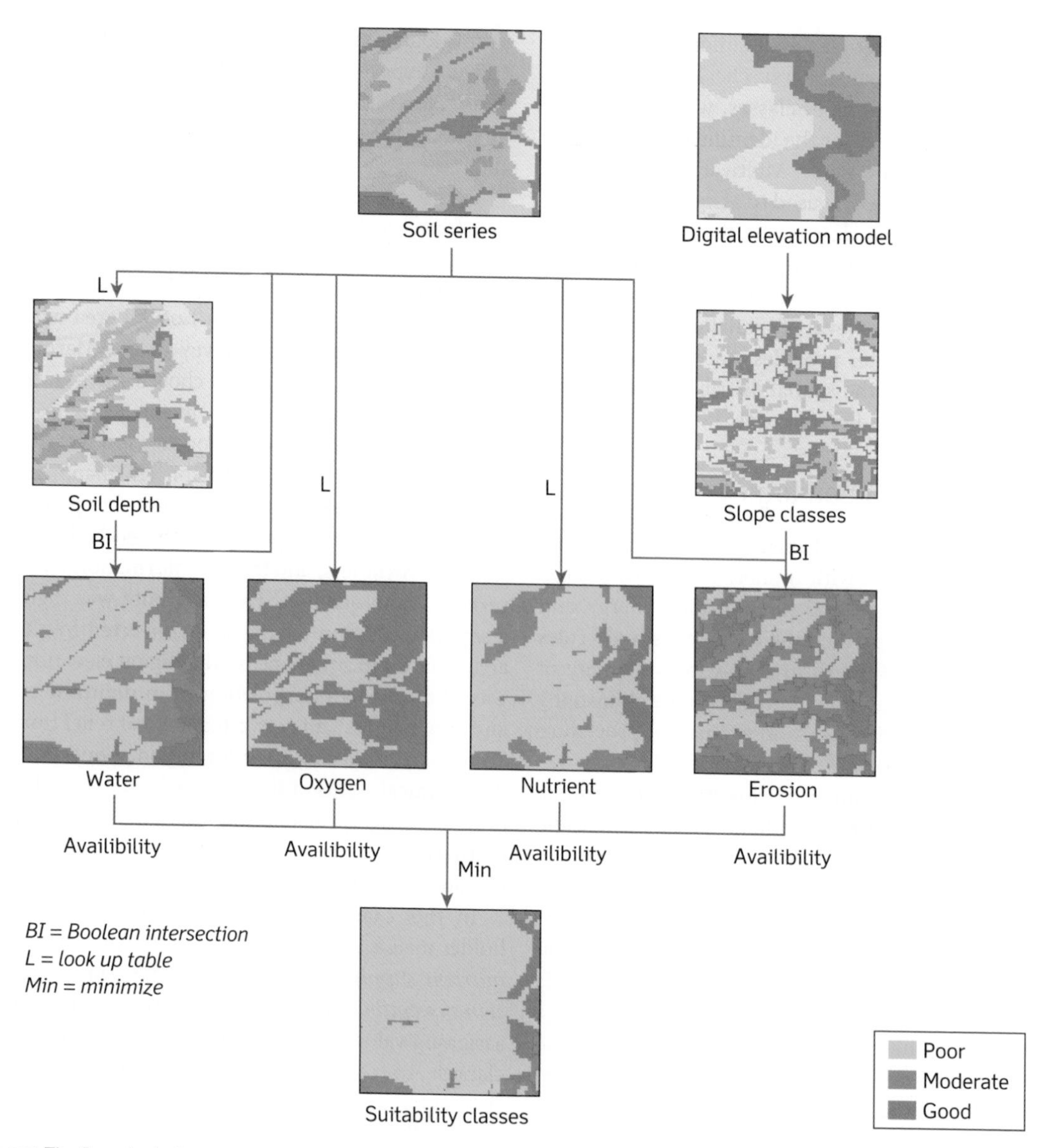

Figure 7.5 The flow chart of operations for 'top-down' land evaluation for determining the suitability of land to grow maize

If soil series is S_2 *and slope class is 'steep' assign erosion hazard value 3.*

Once the individual land qualities have been assigned, the overall suitability per polygon or pixel is determined by the land quality with the most limiting (largest) value:

$$Suitability = maximum\ (LQ_{water},\ LQ_{oxygen},\ LQ_{nutrients},\ LQ_{erosion})$$

This gives suitabilities of *poor*—at least one serious limitation; *moderate*—not severe but at least one moderate limitation; and *good*—no limitations.

It is simple to repeat the analysis for the same area using other values of the conversion factors relating soil to the land qualities, to see how the areas of suitable land may increase as limitations are dealt with by irrigation, mulching, or terracing. This *scenario exploration* can be achieved by replacing the real data with values of the land characteristics that represent a more degraded or an improved situation. In this way one can use the logical model to explore how the suitability of an area depends on the different factors. One must not forget, however, that the results are no better than the data and the insight in the land evaluation procedure allow.

7.5 Operations that depend on a simple distance between A and B: buffering

Operations of the type '*A* is within/beyond distance *D* from *B*' where *D* is a simple crow's flight distance are carried out with the help of a **buffering** command. This is used to draw a zone around the initial entity where the boundaries of the zone or buffer are all distance *D* from the coordinates of the original entity. If it is a point entity then the zone is a circle, if a straight line, the zone is a rectangle with rounded ends, or if it is an irregular line or polygon, an enlarged version of the same shape (Figure 7.6). The buffer is in effect a new polygon that may be used as a temporary aid to spatial query, or may be itself added to the database. The determination of whether an entity is inside/outside or overlaps the buffer zone is then carried out using the overlay operations, as described later in this chapter (see Section 7.7), and logical or mathematical operations on those entities proceed as before.

Typical examples of using the zoning/buffering command with other analysis options include:

- 'Determine the number of fast-food restaurants within 5 km of the White House.'
- 'Investigate the potential for water pollution in terms of the proximity of filling stations to natural waterways.'
- 'Compute the total value of the houses lying within 200 m of the proposed route for a new road.'
- 'Compute the proportion of the world population that lives within 100 km of the sea.'
- 'Compute the number of cattle grazing within 5 km of a waterhole.'
- 'Determine the potential amount of arable land within 1 hour's walk from a Neolithic village.'

Figure 7.7 shows railways in Wales, while Figure 7.8 shows a 5 km buffer computed for the railways. Buffers are very widely used in site-selection exercises where simple distance from a feature is of interest. Other approaches to dealing with distances are considered in the next section.

In such cases, distances from multiple features (e.g. roads or rivers) may be essential inputs, and so buffer polygons may be created and used to identify areas which fall within or outside of these distance bands.

7.6 Operations that depend on connectivity

These are operations in which the entities are directly linked in the database. The linkage can be spatial, as in the contiguity case where *A* is a direct neighbour of *B* or the case where *A* is connected to *B* by a topological network that models roads or other lines of communication. Entities can also be linked by an internal topology, so that

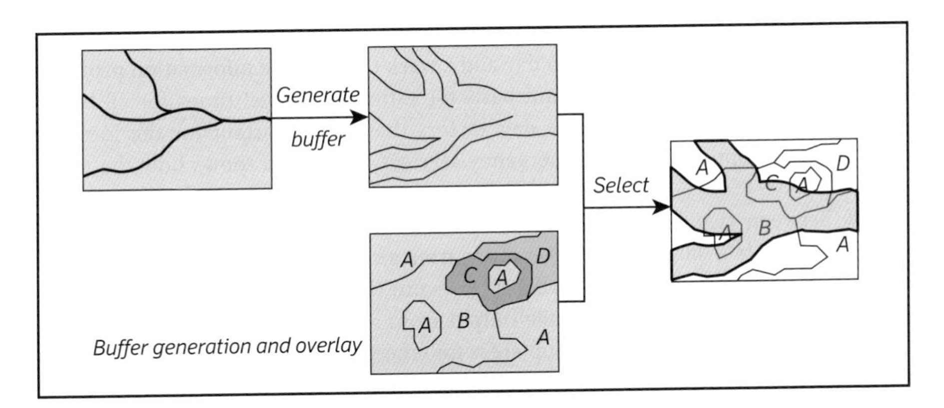

Figure 7.6 Generating buffer zones around exact entities such as points, lines, or polygons yields new polygons which can be used in polygon overlays to select defined areas of the map

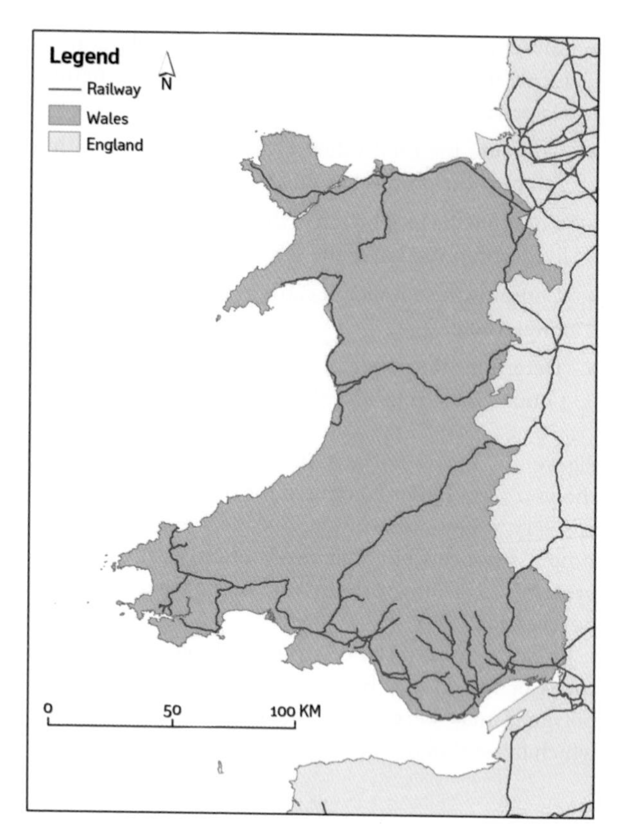

Figure 7.7 Railways in Wales (Contains Ordnance Survey data © Crown copyright and database right 2014)

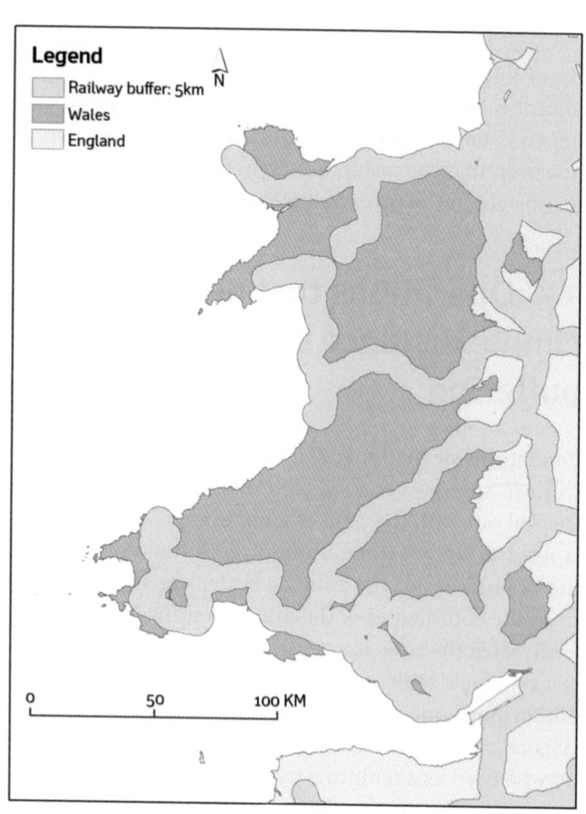

Figure 7.8 Buffer for railways in Wales: 5 km (Contains Ordnance Survey data © Crown copyright and database right 2014)

complex spatial entities are made up of sets of subentities, as is the case with object orientation.

The operations '*A* is a direct neighbour of *B*' and '*A* is connected to *B* by a topological network' are two versions of the same class of operations which, for topologically connected lines and polygons, use explicit information from the spatial database (see Chapter 3) to determine how two entities or locations are connected. Inter-entity distances over the network or other measures of connectivity such as travel times, attractiveness of a route, etc. can be used to determine indices of interaction. These operations are much used for determining the location of emergency services or for optimizing delivery routes.

For example, when a boundary between two land cover polygons is also defined as a road, it is a simple matter to select those roads/boundary lines that have particular kinds of land cover on both sides. Such an analysis would easily distinguish rural roads (agriculture on both sides) from urban roads (built-up areas on both sides) from coastal roads (sea on at least one side—to take account of sea dykes and breakwaters).

The analysis of connectivity over a topologically directed net is much used in automated route-finding (vehicle navigation systems) and for the optimum location of emergency services (see Cromley and McLafferty 2011 for a summary). Attributes attached to the line elements representing roads, rivers, or rail links can identify the character of the connector. For example a road can be identified not only by its width, surface, class, and number of lanes, but also by its visual attractiveness or otherwise (potential tourist route) or traffic densities. Linking time series data on traffic densities over a day and a week to the route information provides a sound basis for estimating travel times for all hours of the day, factors that are important for the location of emergency services. Figure 7.9 shows how the route from *A* to *B* over a network may depend on the attributes of the roads taken and may be quite different from that computed from a crow's flight path based on simple buffering. Figure 7.10 (Ritsema Van Eck 1993) shows an example output from an analysis of the accessibility of built-up areas for emergency services.

A common operation which uses data on transport networks is the **shortest path algorithm**, which is used to find the shortest (or perhaps cheapest) route between two or more points on a network (see Lloyd 2010b for more details).

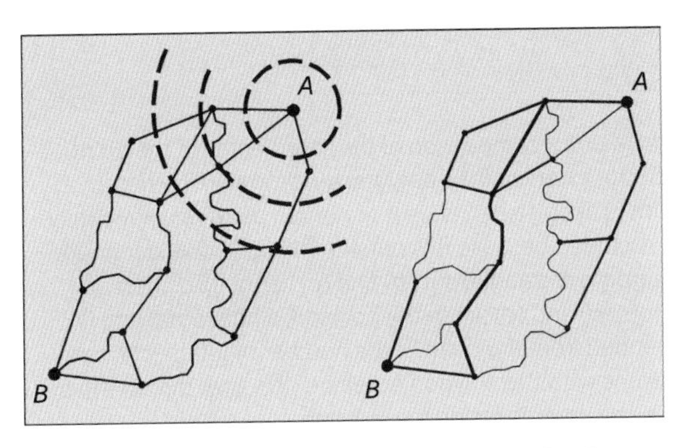

Figure 7.9 The analysis of transport times from A to B in terms of (a) crow's flight distance and (b) times along different routes in a network to determine expected travel times for different road conditions

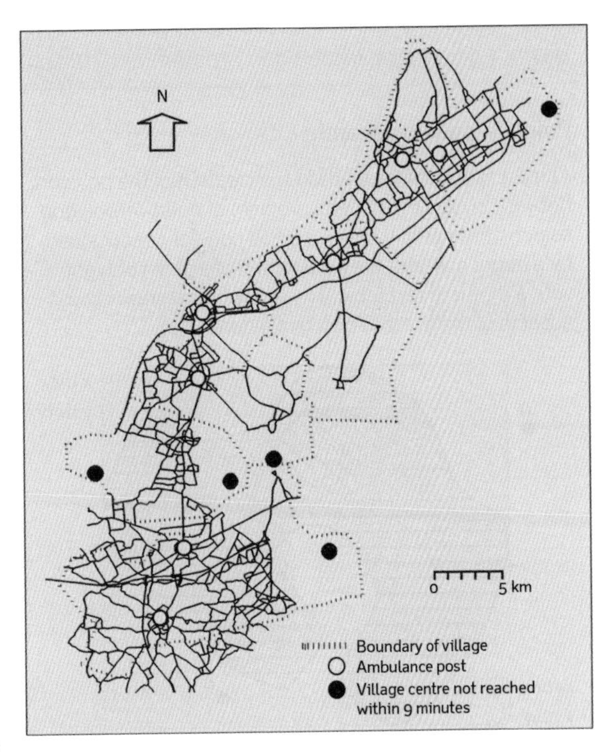

Figure 7.10 The results of a transport time analysis to see which suburban areas can be reached by ambulance within 9 minutes from the ambulance posts. Black lines show roads that can be reached within 9 minutes from the ambulance posts (open circles). Black circles show local centres that cannot be reached in that time. Pecked lines show outlines of urban areas.

7.7 Operations on attributes of multiple entities that overlap in space

Here we extend the discussion of operations on attributes to include attributes from two or more entities that completely or partially occupy or cover the same space. In other words, we consider the **inclusion problem**:

A contains *B*, or

A is contained by *B*,

and the overlap and intersection problem:

A crosses *B*

A overlaps with *B*

where *A* and *B* are two different spatial entities.

Inclusion

The cases '*A* contains *B*' and '*A* is contained by *B*' are solved by extending the rules of Boolean algebra from attributes of entities to measures of how entities occupy space. The problem is the well-known 'point-in-polygon' issue, which is outlined in Box 7.3 . The first step in the analysis is to determine which entities are included or excluded in the location sense—e.g. 'Which restaurants are located in Soho?', 'Which groundwater observation wells have been drilled in formation X?' Once the entities have been selected and tagged, the procedures for attribute analysis can be applied, either per entity, or collectively. For example, the minimum and maximum water levels could be extracted for a given year for each groundwater well, or the average water level of all wells could be computed. The result of these computations can be used to tag the enclosing polygon, which can be displayed with a new colour, shading, or label (Figure 7.2, lower example).

Other examples of applications of this kind of analysis are: from archaeology, 'determine the number of late Iron Age burial sites in parish A', or 'retrieve all passage graves and determine the kinds of soil and landscape position where they occur'; or from soil science: 'find all soil profiles located in unit *S1*, and compute the mean value and standard deviation of the clay content of the topsoil'.

Entity overlap and intersection

In certain cases of '*A* contains *B* and *A* is contained by *B*' and with '*A* crosses *B* and *A* overlaps with *B*', where *A* and *B* are lines or polygons of different forms, the first steps in logical retrieval are to define new areas or line segments. With polygons the process is known as **polygon overlay and intersection**, and it leads to the creation of new spatial entities. Operations which transfer attributes from both input layers are overlay operators, while those

Box 7.3 The point-in-polygon problem and its solution

Point-in-polygon search

At least two separate steps in the creation of the polygon network involve the general problem of ***point-in-polygon*** search: the checks to see if a small polygon is contained by a larger one, and the association of a given polygon with a digitized text label. The figure below shows two aspects of the point-in-polygon algorithms.

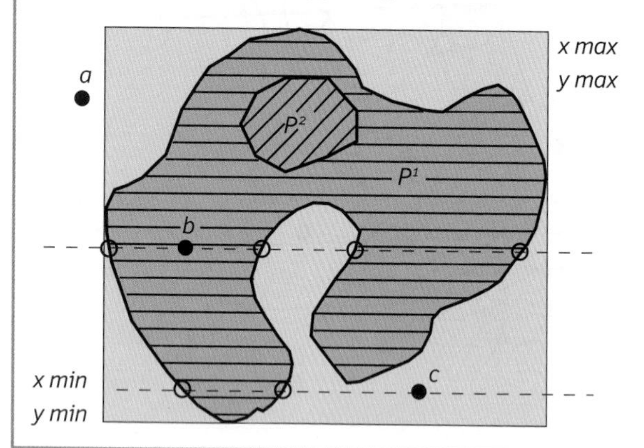

First, a quick comparison of the coordinates of the point with the extents of the polygon quickly reveals whether a point is likely to be inside it or not. So point (*a*) may easily be excluded because it is outside the polygon's minimum bounding rectangle, but (*b*) and (*c*) cannot. To check if points (*b*) and (*c*) are in the polygon, a horizontal line is extended from the point. If the number of intersections of this line with the polygon envelope (in either direction) is odd, the point is ***inside*** the polygon.

To check if an island polygon, P^2, is inside the larger polygon, P^1, first check the extents; P^1 is then divided into a number of horizontal bands and the first and last point of each band is treated in the same way as the point (*b*) above. If the number of intersections for each line is odd, then the polygon P^2 is completely enclosed.

Problems may occur if any segment of a boundary is exactly horizontal and has exactly the same Y coordinate as the point X, but these may be easily filtered out.

which simply cut out part of one layer using boundaries contained in another layer are *cookie cutters*.

Figure 7.11 shows three different results, depending on whether the operation is spatial Boolean **union** (Figure 7.11a), **covering** (Figure 7.11b), or the **clip** cookie cutter (Figure 7.11c); further examples of overlay and cookie cutter operators are given below. Polygon overlay is used to answer questions such as 'Find the area of the City of Westminster that is covered by parks', in which the first overlay may show the boundaries of the administrative areas of urban London and the second is an overlay delineating different kinds of land cover. An example of the clip cookie cutter operation (Figure 7.11c) in a physical application is that if map A is a map of soil types, and

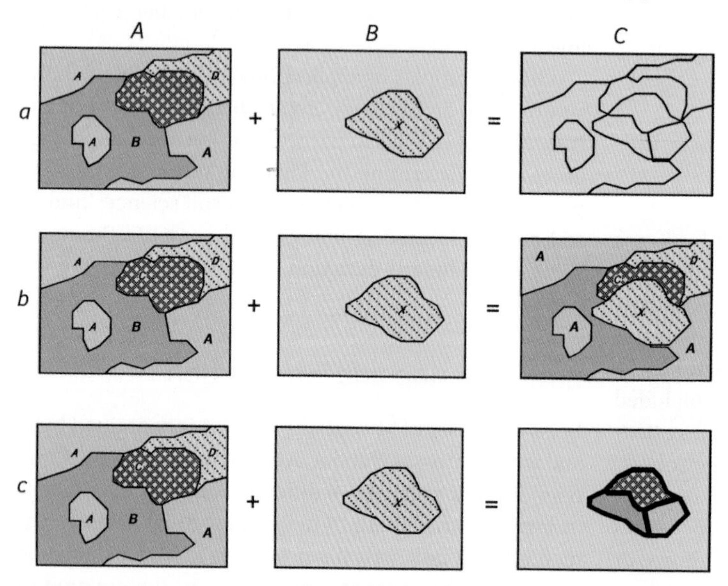

Figure 7.11 Polygon overlay leads to an increase in the number of entities in the database: (a) union overlay—all boundaries are retained; (b) second map covers the first and changes the map detail locally; (c) the covering map is used to cut out a small part of the first map (clip operator)

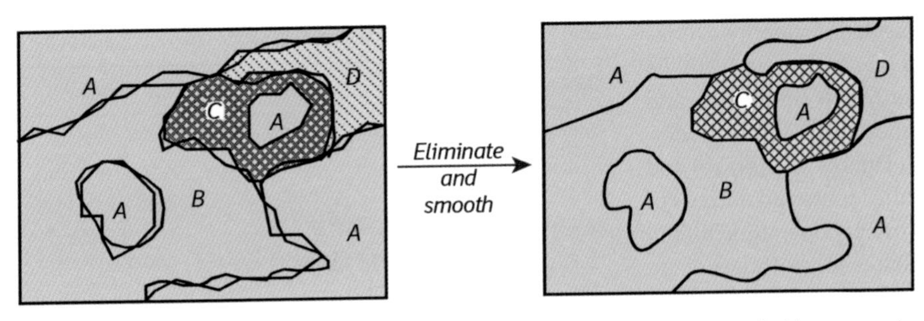

Figure 7.12 Polygon overlay can lead to a large number of spurious small polygons that have no real meaning and must be removed

map B shows the boundary of a catchment, the result is a soil map of that catchment alone.

In some situations, polygon overlap leads to the creation of so-called **spurious polygons** (Figure 7.12). This is because of small errors in the creation of boundaries that are supposed to lie in the same place. The errors can result from digitization errors or from errors made during surveying. There are several solutions. The first is to designate the boundaries on one feature layer as the dominant boundaries to which all others must defer. The second is to examine all the spurious polygons and eliminate all those which have an area smaller than some critical threshold; here it must be decided to which of the larger polygons the area covered by the spurious polygons should be added. The third is to pass a smoothing window over all the coordinates of spurious polygons along the conjugate boundary zones and to compute a new, average boundary. This is frequently over-defined and can be simplified using the Douglas–Peucker algorithm (Douglas and Peucker 1973) or other means, and smoothed for computing the boundaries of the new polygon entities and for display (Figure 7.12).

The application of overlay and **cookie cutter** operations is illustrated below using a single example. Figure 7.7 showed railways in Wales, while Figure 7.13 shows areas of woodland. A two-band buffer was computed around the railways to distances of 2.5 km and 5 km (Figure 7.14).

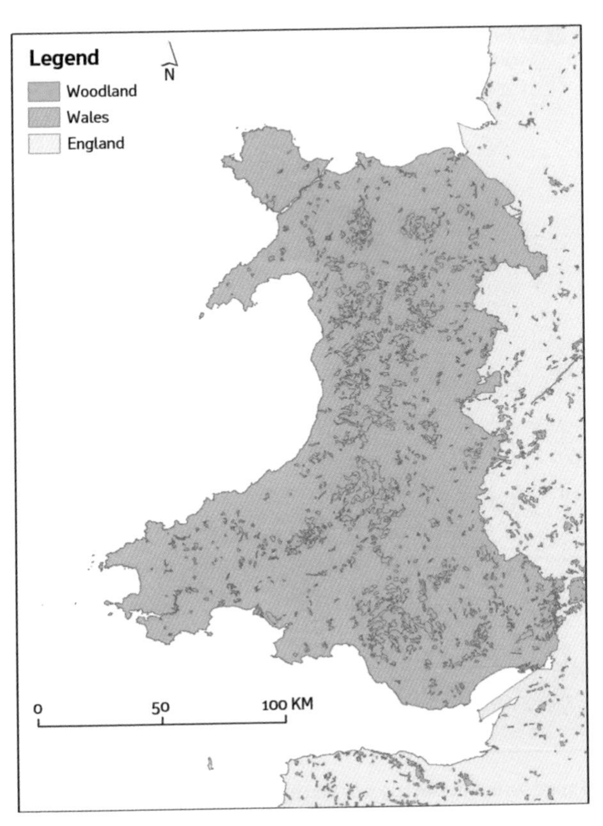

Figure 7.13 Woods in Wales (Contains Ordnance Survey data © Crown copyright and database right 2014)

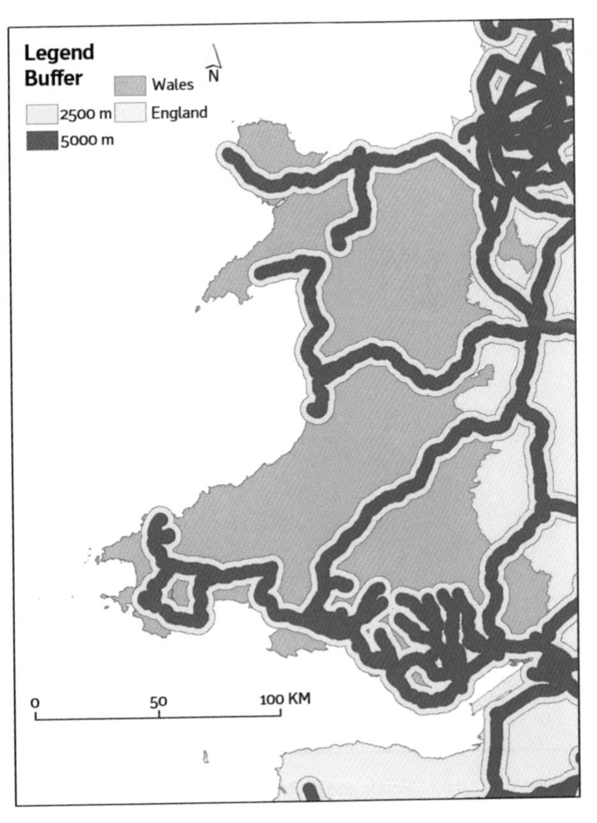

Figure 7.14 Buffer for railways in Wales: 2.5 km and 5 km (Contains Ordnance Survey data © Crown copyright and database right 2014)

The outcome of the union, **intersect**, and **identity** overlay operators with these two inputs (i.e. buffers and woods) are shown in Figures 7.15 (union), 7.16 (intersect), and 7.17 (identity). The outputs of the cookie cutter operations erase and clip are shown in Figures 7.18 (**erase**) and 7.19 (clip). Table 7.1 summarizes the inputs and outputs of each of the operations. With the overlay operators (union, intersect, and identity), all attributes for both layers are retained for the output areas. For the cookie cutter operations (erase and clip), attributes *only* for the input layers are retained—the erase and clip layers are used simply to cut out a part of the input layer. It is not always straightforward to select which overlay or cookie cutter operation is most appropriate, and often more than one operator may be suitable, in principle, for a particular task.

Overlay and cookie cutter operators are key tools in GIS multicriteria decision analysis (GIS-MCDA) (Carver 1991; Malczewski 1999; Bell et al. 2007). GIS-MCDA relates to GIS-based processes for decision-making using multiple data sources. Thus, overlay of multiple layers is needed to identify areas of overlap and extract subareas which meet criteria determined using combinations of these layers. The examples above relate to woodlands and road buffers, which can be combined in several different ways, allowing the selection of areas which are woodland or otherwise, and which fall within or outside the buffer zones. Wise (2002) shows how many key GIS algorithms work 'behind the scenes', and this includes discussion of methods for conducting overlay procedures.

Table 7.1 Overlay and cookie cutter operations for woods and buffers

Operator	Layer labels	Layers	Attributes
Union	Layers the same	All for all areas	All for all areas
Intersect	Layers the same	All for common areas	All for common areas
Identity	Input: buffers Identify: woods	All buffers and all woods inside buffers	For all buffers and all woods inside buffers
Erase	Input: woods Erase: buffers	All woods in buffer area	For woods inside buffer areas
Clip	Input: woods Clip: buffers	All woods in buffer area	For woods outside buffer areas

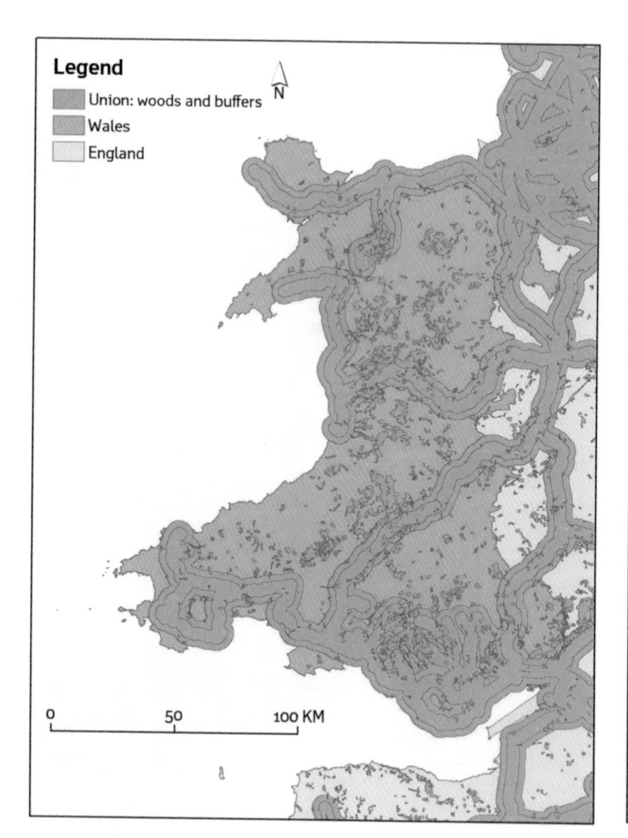

Figure 7.15 Union: woods and buffers—retains all features in both input layers (Contains Ordnance Survey data © Crown copyright and database right 2014)

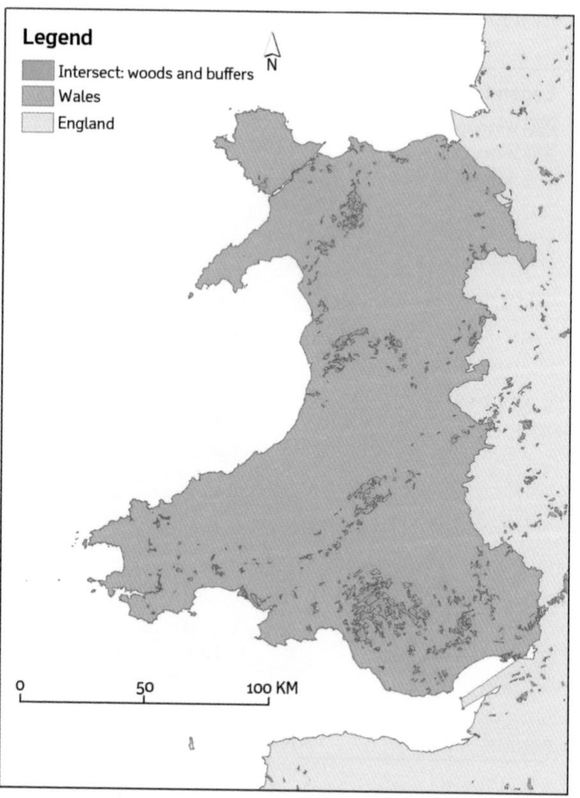

Figure 7.16 Intersect: woods and buffers—retains all features in overlapping areas (Contains Ordnance Survey data © Crown copyright and database right 2014)

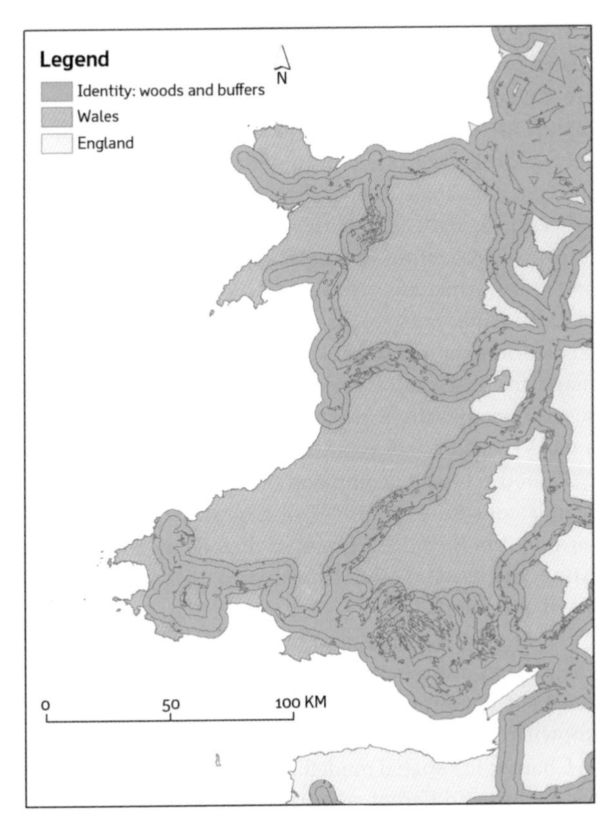

Figure 7.17 Identity: woods and buffers—retains all buffers and woods within buffers (Contains Ordnance Survey data © Crown copyright and database right 2014)

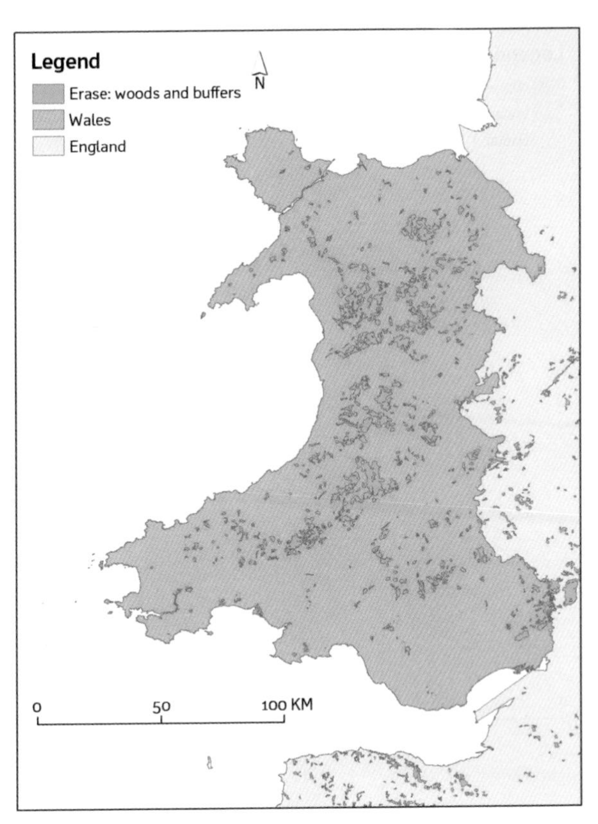

Figure 7.18 Erase: woods and buffers—retains all woods outside buffers (Contains Ordnance Survey data © Crown copyright and database right 2014)

Operations on one or more entities that are linked by directed pointers (object orientation)

Logical operations on lines and polygons that result in entities being split or removed from the database impose heavy computational costs on a spatial database because the number of entities may depend on the operations being carried out. In practical terms, if two simple polygons intersect to create a third, then the third polygon and the attributes it inherits from the two original polygons must be added to the database. If reclassifying two adjacent polygons results in the removal of a common, unnecessary boundary, then two polygons must be removed and one added to the database. In practice, the number of polygons added or removed may be large and indeterminate, so it is difficult to say just how great this overhead is. To save modifying the original database, the changes may be computed on a subset of the original data, and the results stored in a separate file or folder.

In hybrid-relational GIS, adding and removing polygons means modifying both the spatial data and the attribute data separately. Modifying the spatial data is more than just adding or deleting an entry in a table, because all the topological connections need to be recomputed. The advantage of network-relational hybrid databases is that, in principle, there is no limit to the kind and number of analysis queries that can be defined. Object-oriented GIS attempt to get around these computational problems by incorporating a large amount of information to structure the data in such a way that data volumes do not change greatly as queries are carried out. This means that the most common data retrieval and analysis options need to be thought out beforehand, which is why constructing an object-oriented database can take a great deal of time.

7.8 General aspects of data retrieval and modelling using entities

Spatial entities can be retrieved and new attributes can be computed by a wide range of logical and numerical methods. The numerical procedures can also be applied to inclusion and intersection problems and for proximity

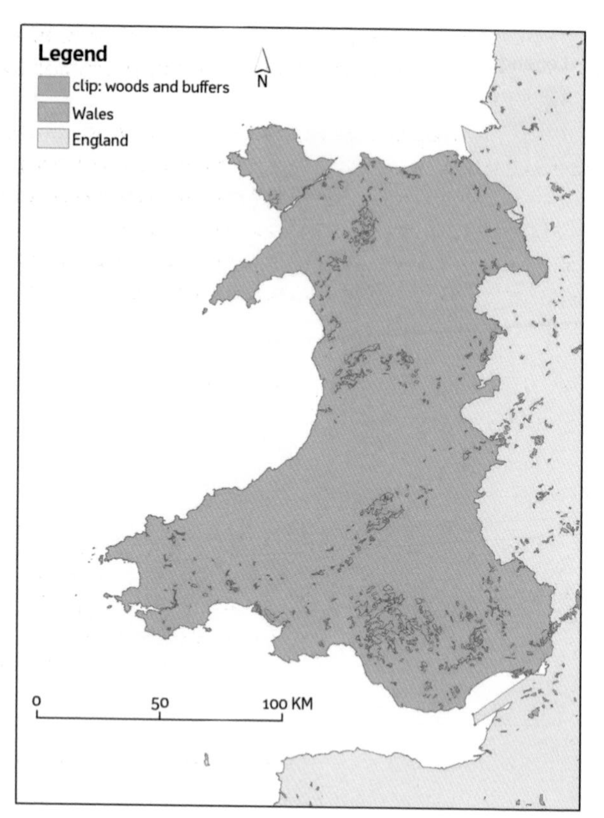

Figure 7.19 Clip: woods and buffers—retains all woods within buffers (Contains Ordnance Survey data © Crown copyright and database right 2014)

analysis, and for analysis of relations over topological connections. The methods can be combined to create complex models for addressing many different kinds of spatial problem. Note that many data-analysis operations are not commutative so the sequence in which the commands are executed is very important. While it can be very informative to sit in front of the computer browsing through a database to see what is there or how different procedures work (e.g. with exploratory data analysis), informal procedures are best for simple data retrieval and transformations. However, when a complex series of commands must be used frequently to retrieve and transform data it is sensible to create a structured command file that can be reviewed, modified, and used by several people (the R programming environment offers a wide range of packages which implement spatial analysis methods using command lines). Such a set of commands constitutes a 'model' or a 'procedure' which can be stored in the GIS, referenced directly by an icon or a name, and used on other databases to carry out the same set of operations.

None of the methods of analysis presented in this chapter pay any attention to data quality or errors; there is a tacit assumption that all data and all relations are known exactly. In spite of this (we will return to this topic in Chapter 12), spatial modelling with GIS has great value for exploring different scenarios. The tools described in this chapter have been widely used for site selection and assessment of alternative scenarios in planning contexts. GIS enables users to assess and present different possible outcomes which can then be visualized in flexible ways (see Chapter 5) to allow a wide range of interested parties to compare and contrast these alternatives.

7.9 Summary

In this chapter, the analysis of discrete features (points, lines, and areas) has been the focus. Attribute operations, such as reclassification of areas, were outlined along with explicitly spatial operations such as measurement of distances from objects, overlay operators, and cookie cutter operations. The kinds of methods detailed in this chapter allow users of GIS packages to answer a wide range of questions, yet there is much more to spatial analysis, as the following chapters show.

? Questions

1. Develop a simple entity-based model to analyse the effects of land use change annually.

2. Work out a GIS-based system for the optimum location of (a) fire stations, (b) banks, and (c) health care services in cities.
3. Explore the advantages and shortcomings of using entity-based models for ecological modelling.
4. Develop a GIS system for helping to manage the demand for building materials required for constructing a new suburb.
5. Explore the advantages of entity-based GIS in (a) real estate management, (b) hydrological modelling, and (c) archaeological site investigations.

Further reading

- Atkinson, P. M. and Tatnall, A. R. (1997). Introduction: neural networks in remote sensing. *International Journal of Remote Sensing*, 18: 699–709.
- Davis, J. C. (2002). *Statistics and Data Analysis in Geology*. 3rd edn. John Wiley & Sons, New York.
- Haining, R. (2003). *Spatial Data Analysis: Theory and Practice*. Cambridge University Press, Cambridge.
- Lloyd, C. D. (2010). *Spatial Data Analysis: An Introduction for GIS Users*. Oxford University Press, Oxford.
- Malczewski, J. (1999). *GIS and Multicriteria Decision Analysis*. Wiley, New York.
- Wise, S. (2002). *GIS Basics*. Taylor & Francis, London.

Interpolation 1: Deterministic and Spline-based Approaches

This chapter describes some approaches for the creation of surfaces from point or area data. Many properties are sampled at point locations, whereas we often require values on a grid—that is, we want to know the values of the property at all locations in our study area. Elevation, precipitation, and soil characteristics are examples of properties which are often measured at point locations. The term *interpolation* refers to the prediction of data values at locations where we have no measurements—the usual objective is to generate a set of predictions on a regular grid. The resulting grid can then be displayed as colour or greyscale maps or by contour lines. The visualization of continuous surfaces is discussed in Chapter 5.

As well as interpolating from points to grids, we sometimes need to transfer data values between one set of zones and another. For example, we may have population counts for two time periods which were reported using different zones, and to assess population change locally we would need to transfer one set of counts into the same set of zones as the other—then it would be possible to compare population counts for each zone. The transfer of values between different sets of zones is referred to as *areal interpolation*. Both point interpolation and areal interpolation are discussed in this chapter. The chapter describes spatial sampling strategies and methods of spatial prediction, including global methods of classification and regression and local deterministic interpolation methods such as **Thiessen polygons, inverse distance weighting** (IDW), and **thin-plate splines** (TPS); each method is illustrated by examples.

Learning objectives

By the end of this chapter, you will:

- **understand why and when spatial interpolation is important**
- **be able to apply your knowledge to create surfaces from point samples**
- **have developed knowledge of some widely used methods for spatial interpolation and some key pros and cons of these methods.**

8.1 Interpolation: what it is and why it is necessary

Point interpolation, as defined above, is the process of predicting the value of attributes at unsampled sites from measurements made at point locations in the same area or region. Predicting the value of an attribute at sites outside the area covered by existing observations is called **extrapolation**.

Point interpolation is used to convert data from point observations to continuous fields so that the spatial patterns sampled by these measurements can be compared with the spatial patterns of other spatial entities. Areal interpolation is used to transfer data values from one set of zones or areas to another—areal weighted overlay is often used to combine two different sets of zones and transfer counts from one set of zones (the source) to another (the target). If 25 per cent of a source zone falls in a target zone, then the target zone is assigned 25 per cent of the population from that source zone. Interpolation is necessary:

a) when the discretized surface has a different level of resolution, cell size, or orientation from that required, or
b) when a continuous surface is represented by a data model that is different from that required, or
c) when the data we have do not cover the domain of interest completely (i.e. they are samples).

An example of a) is the conversion of images (e.g. remotely sensed images) from one gridded tessellation, with a given size and/or orientation, to another. This procedure is generally known as **convolution**.

An example of b) is the transformation of a continuous surface from one kind of tessellation to another (e.g. TIN to raster, raster to TIN, or vector polygon to raster).

An example of c) is the conversion of data from sets of sample points to a discretized, continuous surface. We must distinguish situations with *dense* sampling networks from those with *sparse* sampling networks, or data collected along widely spaced transects. Dense sampling networks are common when creating hypsometric surfaces to represent variations in the elevation of the land surface (digital elevation models (DEM), also known as digital terrain models (DTM)—see Chapters 10 and 11) from aerial photographs or imagery collected by satellites and airborne platforms, where the source data are cheap and the attribute can be observed directly. Sparse sampling networks are often imposed by the costs of borings, laboratory analyses, and field surveys—most often, the spatial variation of the attribute of interest cannot be seen and must be derived indirectly. Some uses of continuous surfaces in spatial modelling are discussed in Chapters 10 and 11.

The continuous surfaces obtained from interpolation can be used as map overlays in a GIS or displayed in their own right. The surfaces can be represented by the data models of contour lines (**isopleths**), discrete regular grids, or by irregular tiles. The data structures used are regular grids (raster), contour lines, or triangular irregular networks (TINs). Because the interpolated surfaces vary continuously over space, regular gridded interpolations must be represented by a data structure in which each grid cell can take a different value: therefore space-saving raster data structures such as run-length codes and quadtrees cannot easily be used. The original data may be collected as point samples distributed regularly or irregularly in space (and/or time), or they can be taken from already gridded surfaces such as those obtained by remote sensing. Attributes predicted by interpolation are usually expressed by the same data type as those measured, but some interpolation methods provide ways of estimating indicator functions that show a **probability** that a given value is exceeded, or that a given class may occur.

8.2 The rationale behind interpolation

The rationale behind spatial interpolation and extrapolation is the very common observation that, on average, values at points close together in space are more likely to be similar than points further apart. In general, two observation points a few metres apart are more likely to have the same altitude than points on two hills some kilometres apart: this is termed spatial dependence (see Chapter 6). The concept is represented by the 'first law of geography' (Tobler 1970). However, **classification** is also a popular method for predicting values at unsampled locations from estimates of map unit means or the central concepts of taxonomic classes, irrespective of any spatial association between the measured values within the classes. When mean attribute values for 'homogeneous' classes or pieces of land have been computed, for example for a given map unit on a geological, soil, land cover, or vegetation map, all information on the short-range variation of the attributes has been lost. Most methods of interpolation attempt to use this local information to provide a more complete description of the way an attribute varies within that area. If this variation can be captured successfully, we may expect that our estimates of the value of any given attribute at unvisited sites will be better than those obtained from class averages alone.

The maps constructed in this way should result in smaller errors when used for subsequent overlay analyses and quantitative modelling in the GIS.

8.3 Data sources for interpolation

Sources of data for continuous surfaces include remote sensing (both airborne and spaceborne) and point samples of attributes measured directly or indirectly in the field on random, structured, or linear sampling patterns, such as regular transects or digitized contours. In many cases, data for interpolation come from *sampling* a complex pattern of variation at relatively few points. These measurements are often known as 'hard data'. When data are sparse, it is very useful to have information on the physical processes or phenomena that may have caused the pattern—known as 'soft information'—which can assist interpolation. In many cases, however, the physical process is unknown and we must make do with various kinds of assumptions about the nature of the spatial variation of the attribute in question. These include assumptions about the smoothness of the variation sampled, and statistical assumptions concerning the probability distribution and statistical stationarity (see Chapter 9).

Spatial sampling

The location of sample points can be critical for subsequent analysis. Ideally, for mapping, samples should be located evenly over the area. A completely regular sampling network can be biased, however, if it coincides in frequency with a regular pattern in the landscape, such as regularly spaced drains or trees, and for this reason statisticians have preferred to have some kind of random sampling for computing unbiased means and variances. Completely random location of sample points has several drawbacks too. First, every point has to be located separately, whereas a regular grid needs only the location of the origin, the orientation, and spacing to fix the position of every point. This is arguably much easier in wooded or difficult terrain, even with GPS. Second, complete randomization can lead to an uneven distribution of points unless very many points can be measured, which is usually prohibited by costs.

Figure 8.1 presents the main options available. A good compromise between random and regular sampling is

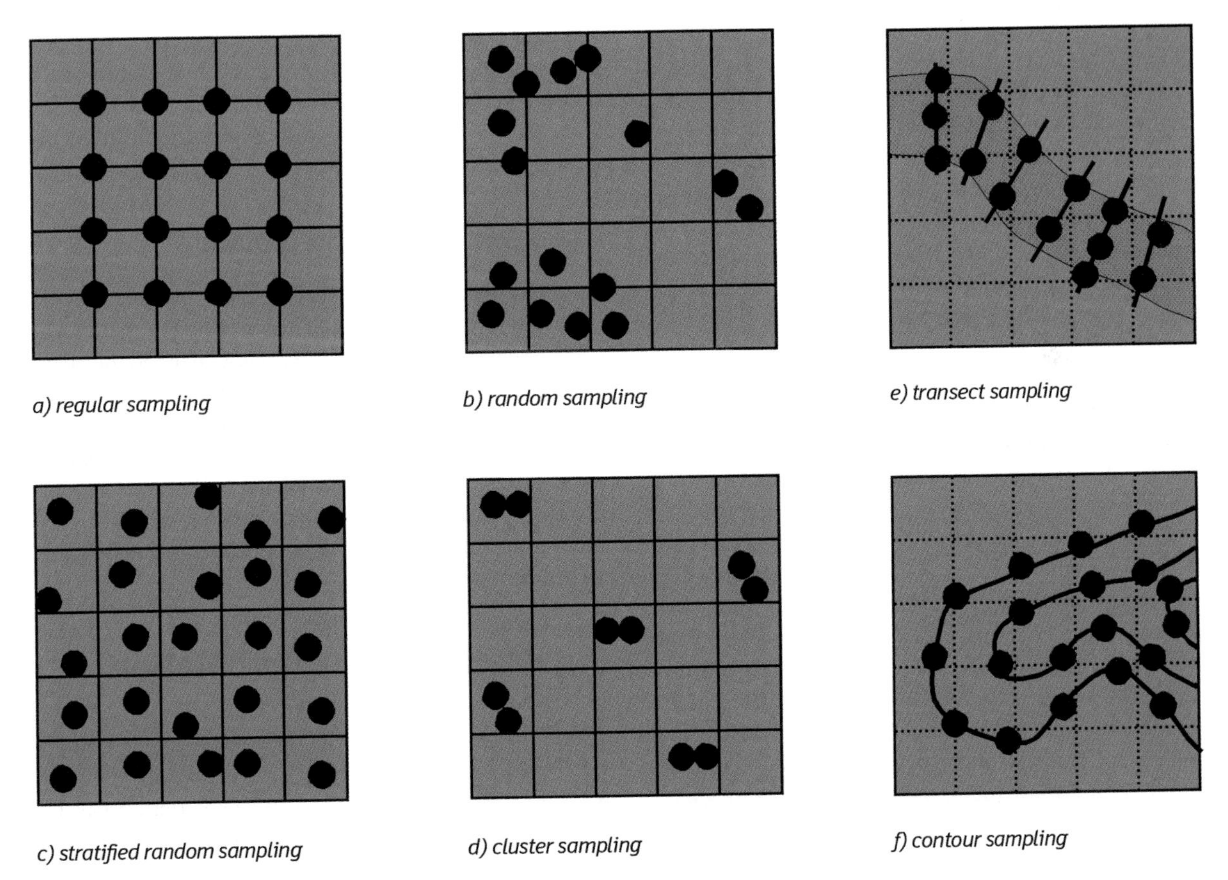

Figure 8.1 Different kinds of sampling net used to collect spatial data from point locations

given by stratified random sampling, where individual points are located at random within regularly laid out blocks or strata. Cluster (or nested) sampling can be used to examine spatial variation at several different scales. Regular transect sampling is often used to survey profiles of rivers, beaches, and hillsides. Digitizing contour lines has been a widely used method of sampling printed maps to make digital elevation models, although most applications of DEMs now tend to be based on data collected on a grid (e.g. through remote sensing), rather than on surfaces interpolated from contours.

The support

The **support** is the technical name used in geostatistics for the area or volume of the physical sample on which the measurement is made. If the sample is a kilogram of soil taken from a soil profile pit, then the support would be approximately 10×10 cm^2 in area and about 5 cm thick. If the sample is a litre of groundwater extracted from a sampling tube, then the support has the dimensions of the column. Because analytical laboratory methods usually homogenize the samples by grinding or mixing, all internal structure or variation is lost. Therefore the measurement refers to the area or volume of material sampled and not to a larger or smaller area or volume.

When data collected on a given support are used to predict values of the same attributes at unsampled locations then the predictions refer to locations that also have that support, unless procedures such as bulking or spatial averaging are used to relate the observations to larger areas or volumes. These are known collectively as '**upscaling**' procedures. The simplest upscaling procedure is to collect a bulk sample consisting of several small samples taken within a defined area around the geometrically located sampling 'point' and to homogenize this bulked sample before laboratory analysis. For example, if ten subsamples were collected and mixed within an area of 10×10 m^2, the support would then be a square of the same dimensions.

Enlarging the support by bulked sampling is sensible when the short-range spatial variation of the attribute in question is so large that long-range variations cannot be clearly distinguished. It is also useful when data collected by different methods need to be combined, such as soil or water information and information collected by remote sensors. Most remote sensing platforms collect data which are recorded as single grid values (pixel values)—pixels can vary in size, referring to on-ground areas of a few centimetres squared to several kilometres. The numbers recorded are area-weighted averages of the radiation received, so the pixel area defines the size of the support. Ground measurements, however, are usually made at much smaller locations and will detect variations within the pixels, unless suitably bulked (see Kerry et al. 2010). If the support sizes of both sets of observations are not matched, it may be difficult to combine the data sets for modelling or spatial analysis.

In population studies, the support is often an irregularly shaped area determined by a census district or postcode area. These vary in size and shape (and sometimes over time) so they provide a difficult basis for interpolation. Frequently the data collected are not measurements of a single attribute such as elevation or barometric pressure, but *counts* of people or socio-economic indicators expressed on a nominal or ordinal scale. For these kinds of data, interpolation may serve to bring data from different kinds of support (postcodes and census districts) to a common base. In such cases the term areal interpolation is used.

Terminology

Throughout this book we use the following terminology for data sampled at point locations. The attribute value at a data point is denoted by $z(\mathbf{x}_i)$, where the subscript i indicates one of a number n of possible measurements that are geographically referenced to the coordinates $\mathbf{x}$ of any convenient Cartesian grid. A predicted value at an unsampled location is indicated by $\hat{z}(\mathbf{x}_0)$.

Exact and inexact interpolators

An interpolation method that predicts a value of an attribute at a sample point which is identical to that measured is called an **exact interpolator**. This is the ideal situation, because it is only at the data points that we have direct knowledge of the attribute in question. All other methods are **inexact interpolators**. The statistics of the differences (absolute and squared) between measured and predicted values at data points $\hat{z}(\mathbf{x}_i) - z(\mathbf{x}_i)$ are often used as an indicator of the quality of an inexact interpolation method.

8.4 Methods for interpolation

The methods of interpolation discussed in this chapter include:

- *Global methods*
  Classification using external information, such as trend surfaces on geometric coordinates, regression models on surrogate attributes, methods of spectral analysis

- *Local deterministic methods*
  Thiessen polygons and pycnophylactic methods, linear and inverse distance weighting, thin-plate splines.

All these methods are relatively straightforward, requiring only an understanding of deterministic or simple statistical methods. These methods are frequently included in commercial GIS.

Geostatistical methods using measures of spatial autocorrelation, known as 'kriging', require an understanding of the principles of statistical spatial autocorrelation. These methods are used when the variation of an attribute is so irregular, and the density of samples is such, that simple methods of interpolation may give unreliable predictions. Geostatistical methods provide probabilistic estimates of the quality of the interpolation. They also enable predictions to be made for blocks of land greater than the support. In addition, geostatistical methods permit the interpolation of indicator functions and can incorporate soft data to guide interpolation, which increases the precision of the results. Some GIS include geostatistical methods, but there is a greater flexibility in specialized geostatistical packages. Because of the theoretical background, geostatistical methods of interpolation are described separately in Chapter 9.

8.5 The example data sets

In order to provide a fair comparison of all the different interpolation methods used in this chapter and Chapter 9 to interpolate from sparsely sampled populations, we use two data sets (see Figures 8.2a and 5.2). Here and in Chapter 9 we use 98 observations taken from a larger set of 155 soil samples from the top 0–20 cm of alluvial soils in a 5 × 2 km part of a larger study area on the floodplain of the River Meuse (Maas) near the village of Stein in the south of the Netherlands. All 'point' data refer to a support of 10 × 10 m, the area within which bulked samples were collected using a stratified random sampling scheme. The elevation of sample sites and their distance from the river were also recorded. Samples were chemically analysed for concentration of heavy metals (cadmium, zinc, lead, and mercury); here we use the reported zinc content in ppm for comparative illustration of all methods of interpolation. Since their use in the second edition of this book, these data have been used in many published case studies, so the results presented here can be compared with those obtained using a wide range of methods. Our second data set is that illustrated in Chapter 5 (Figure 5.2) and used in Chapter 6—precipitation amounts in Wales for January 2006. These data are used in a single comparative section at the end of the chapter.

8.6 Global interpolation

Methods of interpolation can be divided into two groups, called **global** and **local interpolators**. Global interpolators use all available data to provide predictions for the whole area of interest. Local interpolators operate within a small zone around the point being interpolated to

(a) Data points

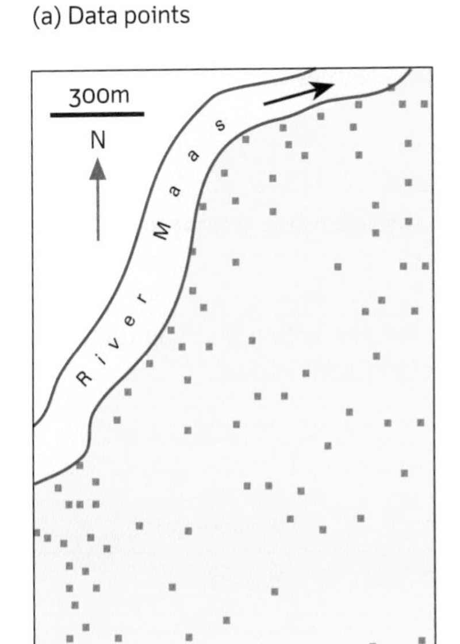

(b) Flood frequency

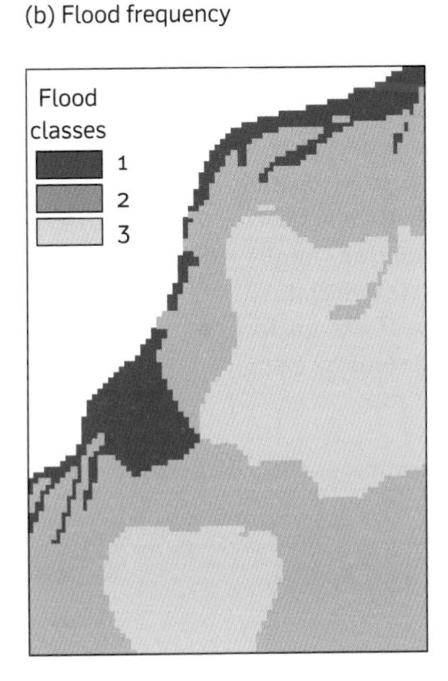

(c) Zinc levels predicted by flood frequency

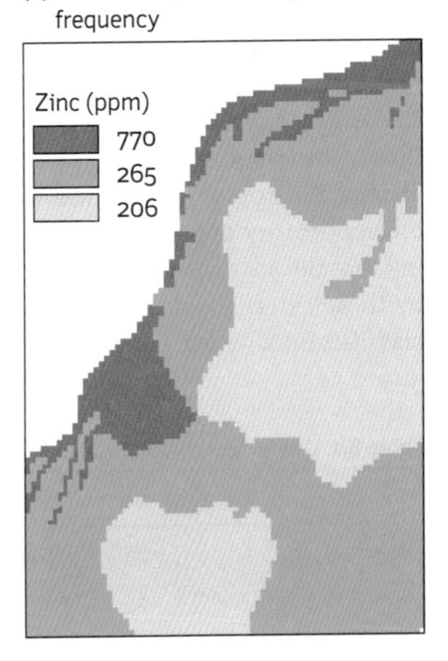

Figure 8.2 (a) Location of sites of example data set, (b) flood frequency classes, and (c) zinc levels per flood frequency class

ensure that estimates are made only with data from locations in the immediate neighbourhood.

Global interpolators are mostly used not for direct interpolation but for examining, and possibly removing, the effects of global variations caused by major trends, or the presence of various classes of land that may indicate areas that have different average values. Once the global effects have been taken care of, the **residuals** from the global variations can be interpolated locally.

Global methods are usually simple to compute and are often based on standard statistical ideas of variance analysis and regression. *Classification methods* use easily available soft information (such as soil types or administrative areas) to divide the area into regions that can be characterized by the statistical *moments* (mean, variance) of the attributes measured at locations within those regions.

Regression methods (see Chapter 6) explore a possible functional relation between attributes that are easy to measure and the attribute to be predicted. These methods can be based just on the geographical coordinates of the sample points (in which case the method is called **trend surface analysis**), or on some relationship with one or more spatially variable attributes (in which case the empirical regression model is often called a **transfer function**).

8.7 Global prediction using classification models

When spatial data are sparse it is sometimes convenient to assume that the observations are taken from a statistically stationary population (i.e. the mean and variance of the data are independent of both the location and the size of the support). Alternatively, we can decide that these observations sample some coordinated spatial change. If we opt for the former, we automatically select a classificatory approach to spatial prediction, implying that the spatial structure of variation is determined by these externally defined spatial units. If we opt for classification, we can compute our predictions by using well-known, standard analysis of variance (**ANOVA**) methods.

Classification by homogeneous polygons assumes that within-unit variation is smaller than that between units; the most important changes take place at boundaries. This conceptual model is commonly used in soil and landscape mapping to define 'homogeneous' soil units, landscape units, ecotopes, etc. where 'objects' like soil mapping units, river terraces, catchment areas, hillsides, breaks of slope, etc. have been recognized as useful features for carrying information about other aspects of the landscape.

The simplest statistical model is the ANOVA model:

$$z(\mathbf{x}_0) = \mu + \alpha_k + \varepsilon \qquad 8.1$$

where z is the value of the attribute at location $\mathbf{x}_0$, μ the general mean of z over the domain of interest, α_k is the deviation between μ and the mean of unit k, and ε is residual (pooled within-unit) error, sometimes known as **noise**.

This model assumes that for each class k the attribute values are normally distributed; ideally, each contains a distinct mode. The mean attribute value per class k is $\mu + \alpha_k$, which is estimated from a set of independent samples that are assumed to be spatially independent. The mean within-class variance is given by ε and is assumed to be the same for all classes.

The relative variance (σ_W^2 / σ_T^2), where σ_W^2 is the pooled within-class variance and σ_T^2 the total sample variance, is a measure of the goodness of the classification. Both can be estimated when all map units contain more than one point sample. The smaller the relative variance, the better the classification. The analysis of variance is set out in Table 8.1. The significance of the success of the classification can be tested with the usual statistical F-test on the variance ratio with degrees of freedom m, $n - k$.

Assumptions

This approach makes the following assumptions about the spatial variation:

- Variations in the value of z *within* the map units are random and not spatially contiguous.

Table 8.1 One-way analysis of variance of n observations spread over k classes

Source	Sums of squares	Degrees of freedom	Mean square	Variance ratio
Between	SS_B	$k-i$	MS_B	MS_B/MS_W
Within	SS_W	$n-k$	MS_W	
Total	SS_T	$n-i$	MS_r	

- All mapping units have the same within-class variance (noise), which is uniform within the polygons.
- All attributes are normally distributed.
- All spatial change takes place at boundaries, which are sharp, not gradual.

Note that these assumptions need not necessarily hold. Some map units or individual occurrences of a given map unit might be internally more variable than others. The assumption of within-class variation being random implies that differences cannot be mapped out at larger mapping scales, which is usually not true: soil maps are made over a wide range of nested scales, with differences being seen at all scales. Data are not necessarily normally distributed—they may be distributed log-normally, rectangularly, hyperbolically, or in other ways. In some cases, normalization by computing natural logarithms, logit transformations, or other suitable transformations may be necessary (Box 8.1, and see Section 6.1). When transformation is necessary, it is more sensible to treat each map unit, or indeed each delineation of each map unit, as a separate entity and to compute individual means and standard deviations if there are sufficient data.

Spatial prediction of zinc levels in the topsoil by the ANOVA method will be illustrated using soft information from a flooding frequency map with three classes: 1 frequent flooding (annual), 2 flooding every two to five years, and 3 flooding less than every five years. As the zinc is carried to the site by polluted river sediment, a reasonable hypothesis is that frequently flooded sites will have greater concentrations of zinc.

Figure 8.3 shows that, as a group, the zinc measurements do not appear to be normally distributed with a single mode. They could be log-normally distributed or they could have a multimodal distribution with different means and variances. To see the effect of non-normality, the analysis of variance is carried out both for untransformed data and data transformed to natural logarithms.

Untransformed zinc levels

The class means and standard deviations are (units ppm zinc):

	Mean	Standard deviation
1	769.76	423.17
2	264.97	176.62
3	205.77	105.32

Figures 8.2b and 8.2c show the flood frequency map and the derived map of the zinc content of the soil obtained by assigning the mean value of each class, as computed from the data points located in the respective flood frequency classes. As the maps show only average values

Box 8.1 Common data transformations

Transformations commonly used to bring data to a normal distribution

1. Logarithms. When data are very strongly positively skewed, with a small number of values that are much larger than the mean or mode, the data can be normalized by computing logarithms to base 10 or base e. There must be no zero or negative values in the data, so a small constant can be added to make all data greater than zero, i.e. the natural logarithm is given by:

$$U = \ln(A + c)$$

where U is the transformed variable, A is the original variable, and c is a small constant to ensure all values exceed zero.

2. Logit. The logit transformation is used to spread out distributions of data on proportions (range 0–1 or 0–100 as per cent) so that concentrations at the ends of the range are avoided. The logit transformation is:

$$U = \ln(p / q)$$

where U is the transformed variable, p is an observed proportion, and $q = 1 - p$. Zero and unity values of p should be avoided by adding or subtracting a small constant, as before.

3. Square root transformation. Moderate skewness can be removed by computing:

$$U = (A)^{0.5}$$

4. Angular transformations. The arcsine transformation is used for proportional counts to spread the distribution near the ends of the range. If the proportion is p then:

$$U = \sin^{-1}(p)^{0.5}$$

i.e. U is the angle whose sine is $p^{0.5}$.

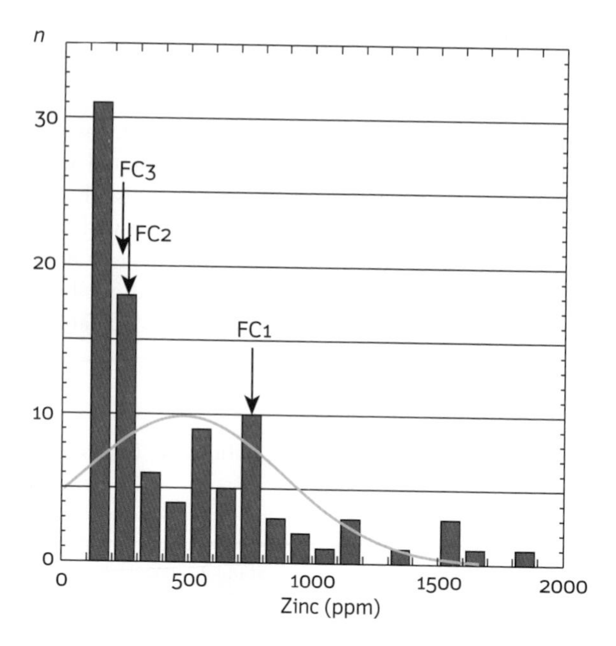

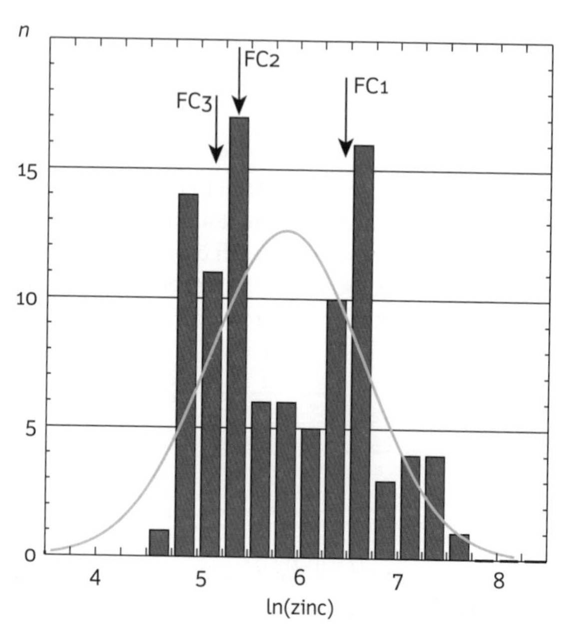

Figure 8.3 (Top) histogram of zinc levels; (bottom) histogram of ln(zinc)

per class, the maps of untransformed and transformed zinc levels look the same. The conclusion to be drawn from this analysis is that only those areas that are annually flooded have elevated levels of zinc in the soil. Clearly, the flood frequency map distinguishes the zinc content of the soil of the most frequently flooded areas from those less frequently inundated, but, as Figure 8.3 shows, there is a large overlap between classes 2 and 3.

The standard deviations per class are large because of the non-normal distribution and so the 95 per cent confidence intervals span zero, which is not sensible (-76.6–1616.0 ppm zinc).

Log-normal transformation

The class means, standard deviations, and back-transformed means are:

	Mean (ln)	Standard deviation	Mean (exponent)
1	6.484	0.609	654.58
2	5.421	0.531	226.11
3	5.239	0.415	188.48

The analysis of variance is given in Table 8.2, and the equivalent for log transformed data in Table 8.3.

These results show that the log transformation has greatly increased the variance ratio, and hence the quality of the classification. The 95 per cent confidence intervals on the log data do not span zero, so they are sensible but large (193.6–2212.8 ppm zinc). A student's *t* test on the logarithmic map unit means demonstrates that no statistical significance can be attributed to the differences between map units 2 and 3. Together with the histogram views of the data (Figure 8.3), the results confirm the large difference between flood frequency class 1 and the other two classes, which overlap considerably. The results suggest that we could usefully pool the data from flood classes 2 and 3, thereby modifying our soft information to two classes (frequently flooded versus infrequently flooded) with different means *and* variances. Simplifying the classification implies that it is the

Table 8.2 Analysis of variance of zinc by flood classes

Source	Degrees of freedom	Mean square	Variance ratio	Relative variance
Between	2	3206979	33.8	0.60
Within	95	94880		
Total	97	159048		

Table 8.3 Analysis of variance of zinc transformed to natural logarithms

Source	Degrees of freedom	Mean square	Variance ratio	Relative error
Between	2	14.55	46.60	0.52
Within	95	0.312		
Total	97	0.6056		

frequent (annual) floods that are most responsible for depositing polluted sediments (cf. Middelkoop 2000).

8.8 Global interpolation using trend surfaces

When variation in an attribute occurs continuously over a landscape, it may be possible to model it by a smooth mathematical surface. There are several ways of doing this: all of them fit some form of polynomial equation to the observations at the data points so that values at unsampled locations can be computed from their coordinates.

The simplest way to model long-range spatial variation is by a multiple regression of attribute values versus geographical location. The idea is to fit a polynomial line or surface, depending on whether our data are in one or two dimensions, by least squares through the data points, thereby minimizing the sum of squares for $\hat{z}(\mathbf{x}_i) - z(\mathbf{x}_i)$. It is assumed that the spatial coordinates (x, y) are the independent variables, and that z, the attribute of interest and the dependent variable, is normally distributed. Also, it is assumed that the regression errors are independent of location, which is often not the case.

As a simple example, consider the value of an environmental attribute z that has been measured along a transect at points $\mathbf{x}_1, \mathbf{x}_2, \ldots, \mathbf{x}_n$. If, apart from minor variation, the value of z increases linearly with location, $\mathbf{x}$, its long-range variation can be approximated by the regression model

$$z(\mathbf{x}) = b_0 + b_1 x + \varepsilon \qquad 8.2$$

where b_0 and b_1 are the polynomial coefficients known respectively as the intercept and the slope in simple regression. The residual ε (the noise) is assumed to be normally distributed and independent of the $\mathbf{x}$ values. In many circumstances z is not a linear function of $\mathbf{x}$, but may vary in a more complicated way. Quadratic or still higher order polynomial models such as

$$z(\mathbf{x}) = b_0 + b_1 x + b_2 x^2 + \varepsilon \qquad 8.3$$

can be used. By increasing the number of terms, it is possible to fit any set of points by a complicated curve exactly, thereby reducing ε to zero.

In two dimensions, the polynomials derived by *multiple regression* on x and y coordinates are surfaces of the form

$$f\{(x, y)\} = \sum_{r+s \le p} \left(b_{rs} \cdot x^r \cdot y^s\right) \qquad 8.4$$

of which the first three are:

$$b_0 \qquad \text{flat} \qquad 8.5$$

$$b_0 + b_1 \cdot x + b_2 \cdot y \qquad \text{linear} \qquad 8.6$$

$$b_0 + b_1 \cdot x + b_2 \cdot y + b_3 \cdot x^2 + b_4 \cdot xy + b_5 \cdot y^2 \qquad \text{quadratic} \qquad 8.7$$

The integer p is the order of the trend surface. There are $p = (p+1)(p+2)/2$ coefficients that are normally chosen to minimize

$$\sum_{i=1}^{n} \{z(\mathbf{x}_i) - f(\mathbf{x}_i)\}^2 \qquad 8.8$$

where $\mathbf{x}$ is the vector notation for (x, y). So a horizontal plane is zero order, an inclined plane is first order, a quadratic surface is second order, and a cubic surface with 10 parameters is third order.

Finding the b_i coefficients is a standard problem in multiple regression, so the computations are easy with standard statistical packages. Trend surfaces can be displayed by estimating the value of $z(\mathbf{x})$ at all points on a regular grid. Examples of trend surfaces computed for the untransformed zinc data are given in Figure 8.4.

The advantage of trend surface analysis is that it is a technique that is superficially easy to understand, at least with respect to the way the surfaces are calculated. Broad features of the data can be modelled by low-order trend surfaces, but it becomes increasingly difficult to ascribe a physical meaning to complex higher polynomials. The surfaces are highly susceptible to edge effects, waving the edges to fit the points in the centre of the area, with the result that second order and higher surfaces may reach ridiculously large or small values just outside the area covered by the data. Because it is a general interpolator, the trend surfaces are very susceptible to outliers in the data. Trend surfaces are smoothing functions, rarely passing exactly through the original data points unless these are few and the order of the surface is high. It is implicit in multiple regression that the residuals from a regression line or surface are normally distributed independent errors. The deviations from a trend surface are almost always to some degree spatially dependent; in fact, one of the most fruitful uses of trend surface analysis has been to reveal parts of a study area that show the greatest deviation from a general trend (Davis 2002). The main use of trend surface analysis, then, is not as an interpolator within a region, but as a way of removing broad features of the data before using some other local interpolator.

The significance of a trend surface The statistical significance of a trend surface can be tested by using the technique of analysis of variance to partition the variance between the trend and the residuals from the trend. Let n be the number of observations, so there are $(n - 1)$ degrees of variation associated with the total variation. The degrees of freedom for regression, m, are determined by the number of terms in the polynomial regression equation, excluding the b_0 coefficient.

For a linear regression, $z(x, y) = b_0 + b_1 x + b_2 y$, the degrees of freedom for regression $m = 2$. The degrees of freedom for the deviations from the regression are given by $(n - 1) - m$. Table 8.4 presents the terms in the analysis of variance; readers will note that this table is very similar to Table 8.1 for classification.

The regression line or surface can be considered to be analogous to the classes in the usual analysis of variance methods. The variance ratio, or 'F-test', estimates whether the amount of variance taken up by the regression differs significantly from that expected for an equivalent number of sites with the same degrees of freedom drawn from a random population. Just as 'relative variance' estimated how much of the variance remained after classification, so a regression goodness of fit (R^2) may also be calculated as:

$$R^2 = SS_d \,/\, SS_t \tag{8.9}$$

Using trend surfaces to model the concentration of zinc over the floodplain as a continuous surface yields different maps depending on the order of the regression surface chosen (Figure 8.4). The **goodness of fit** (R^2) values show that even the higher-order surfaces with 21 (fifth order) or 28 (sixth order) coefficients do not fully represent all the variation in the data. Indeed,

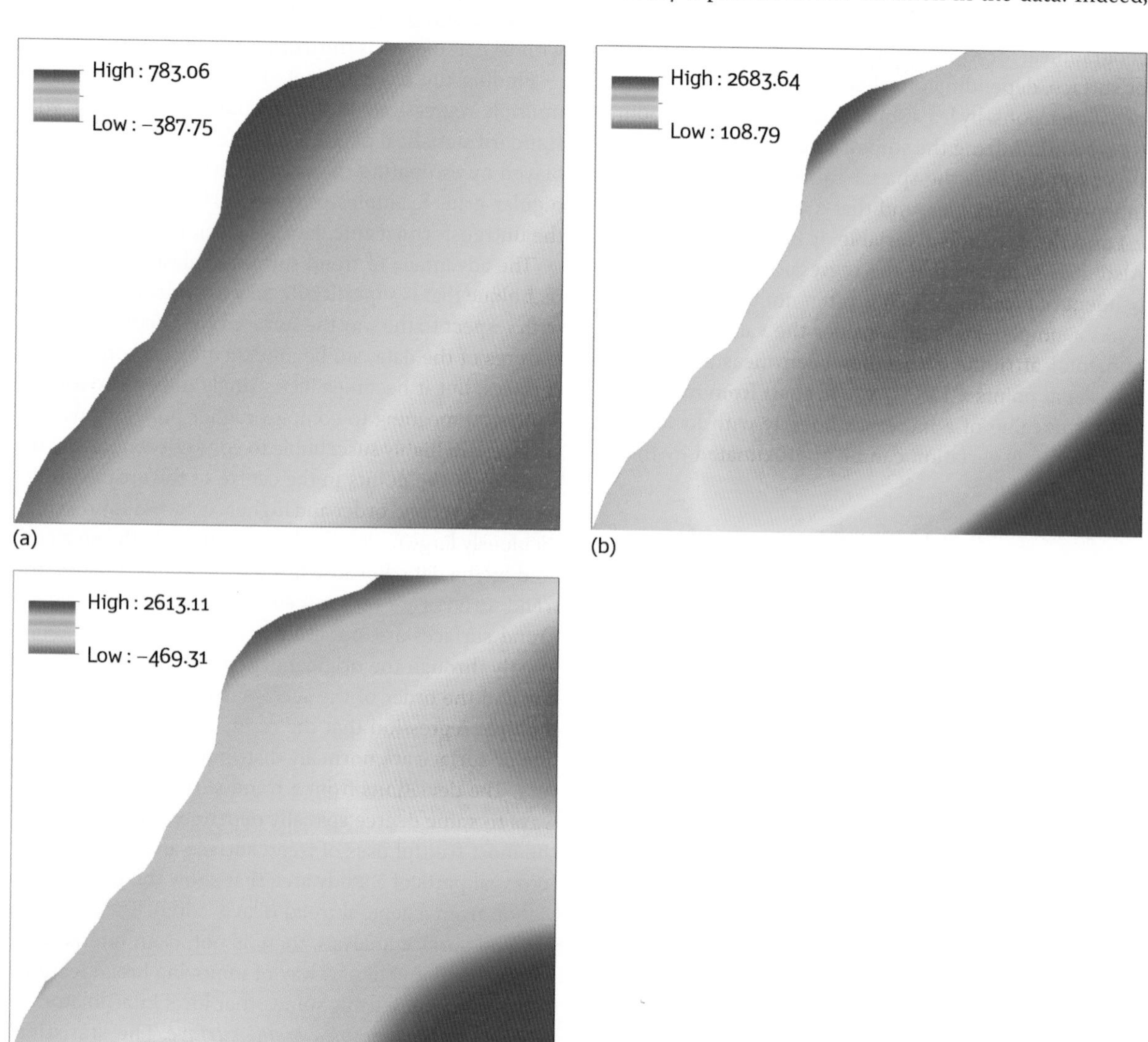

Figure 8.4 Simple global trend surface for untransformed zinc data: (a) first order, (b) second order, (c) third order

some surfaces predict *negative* values of zinc in the south-eastern corner of the area, which is clearly a serious distortion of reality.

Order:	1	2	3	4	5	6
R^2	.183	.475	.560	.687	.767	.802

Although the R^2 values increase with the order of the regression, at first sight we have no way of judging whether increasing the order of the polynomial significantly increases the fit to the data. The significance of improvement can also be estimated using an analysis of variance, as shown in Box 8.2. Even if significantly better fits can be obtained with higher-order polynomials, it is not physically sensible to choose a trend surface that has no physical explanation.

Table 8.4 Analysis of variance terms for linear regression

Source	Sums of squares	Degrees of freedom	Mean square	Variance ratio
Regression	SS_r	m	MS_r	MS_r/MS_d
Deviation	SS_d	$n-m-1$	MS_d,	
Total	SS_t	$n-1$	MS_t	

Box 8.2 Estimating the significance of using a higher-degree polynomial in trend surfaces

Significance testing for increasing the degree of the polynomial

The general ANOVA table for comparing the significance of improved fits is:

Source	Sums of squares	Degrees of freedom	Mean square	Variance ratio
Regression degree $p+1$	SS_{rp+1}	m	MS_{rp+1}	MS_{rp+1}/MS_{dp+1}
Deviation from $p+1$	SS_{dp+1}	$n-m-1$	MS_{dp+1}	
Regression degree p	SS_{rp}	k	MS_{rp}	MS_{rp}/MS_{dp}
Deviation from p	SS_{dp+1}	$n-k-1$	MS_{dp}	
Improved regression due to increase in order from p to $p+1$	$SS_{ri}\, SS_{rp+r}\, SS_{rp}$	$m-k$	MS_{ri}	MS_{ri}/MS_{dp+1}
Total	SS_t	$n-1$	MS_t	

As an example, to test for significant improvement in fit from order 2 to order 3 we compute:

Order 2 source	Sums of squares	Degrees of freedom	Mean square	Variance ratio
Regression	7320729	5	1464145	16.62
Deviation	8106885	92	88118	
Total	15427614	97		

Order 3 source	Sums of squares	Degrees of freedom	Mean square	Variance ratio
Regression	8639688	9	959965	12.45
Deviation	6787926	88	77135	
Total	15427614	97		

Both surfaces are significant according to an F-test on the variance ratio. To see if there is a significant improvement when increasing the order of the regression surface from two to three, we compute:

(*continued*)

Box 8.2 *(continued)*

Source	Sums of squares	Degrees of freedom	Mean square	Variance ratio
Regression deg $p+1$	8639688	9	959965	12.45
Deviation from $p+1$	6787926	88	77135	
Regression deg p	7320729	5	1464145	16.62
Deviation from p	8106885	92	88118	
Improved regression due to increase in order	1318959	4	329739	4.27

The variance ratio of 4.27 (degrees of freedom 4, 88) is larger than the 1% tabulated value of 3.56, so we conclude that the third order surface is significantly better.

8.9 Spatial prediction using global regression on cheap-to-measure attributes

Consideration of Figures 8.2 and 8.4 shows that for the example data set there is a clear geographical relation between the zinc content of the soil and the distance from the river. From other studies (e.g. Middelkoop 2000), it is known that heavy metal pollutants in floodplain soils depend on several factors, the most important of which is the distance from the source (i.e. the river) and the elevation of the floodplain. In this area, coarse polluted sediments are deposited with sand in the river levees, and fine polluted sediments settle out in low-lying areas where flooding is frequent and inundation persists for longer periods. Areas less frequently flooded receive a smaller pollutant load, as the example of mapping using flood frequency classes showed. Because the attributes *distance to the river* and *floodplain elevation* are cheap to map, it is possible that we could improve the spatial predictions of zinc content if we could derive an empirical model of the relation between zinc and these independent variables.

The regression model is of the form

$$z(\mathbf{x}) = b_0 + b_1 P_1 + b_2 P_2 + \varepsilon \qquad 8.10$$

where $b_0 \ldots b_n$ are regression coefficients and $P_1 \ldots P_n$ are independent properties. In this case, P_1 is *distance to river* and P_2 is *elevation.*

Figure 8.5 presents histograms for the independent variables of *distance to the river* (D_{river}—metres), *floodplain elevation* (elevation—metres), and the dependent zinc concentration (ppm); the histograms of logarithmically transformed D_{river} and zinc are also included. Inspection of Figure 8.5 indicates that it is best to use the logarithmically transformed variables in the multiple regression. Figure 8.6 shows estimates of zinc levels derived through multiple regression. Methods of multiple regression are available in standard statistical packages, as well as in some GIS packages.

For the 98 data points of the Meuse zinc example, multiple regression of ln(zinc) against elevation and ln(distance to river) gives the following relation:

$$\ln(\text{zinc}) = 10.000 - 0.292(\text{elevation}) - 0.333(\ln D_{\text{river}}) + 0.394(\text{resid error}) \qquad 8.11$$

$$R^2 = .749$$

This type of regression model is often called a *transfer function* (for an example, see Ghermandi and Nunes 2013); it can be computed easily in many GIS environments (see Chapter 6). The source maps can be obtained in several ways and may be in vector or raster format. Figure 8.6 shows the results.

The same procedure may be used with many other sets of independent and dependent variables, such as temperature and altitude, rain with distance from the sea, vegetation composition as a function of moisture surplus, number of clients and income levels. Geographical coordinates and associated attributes can be combined in one regression to use as much information from the data as possible. The most important point to consider is that the regression model makes physical sense. Note that all regression transfer models are inexact interpolators.

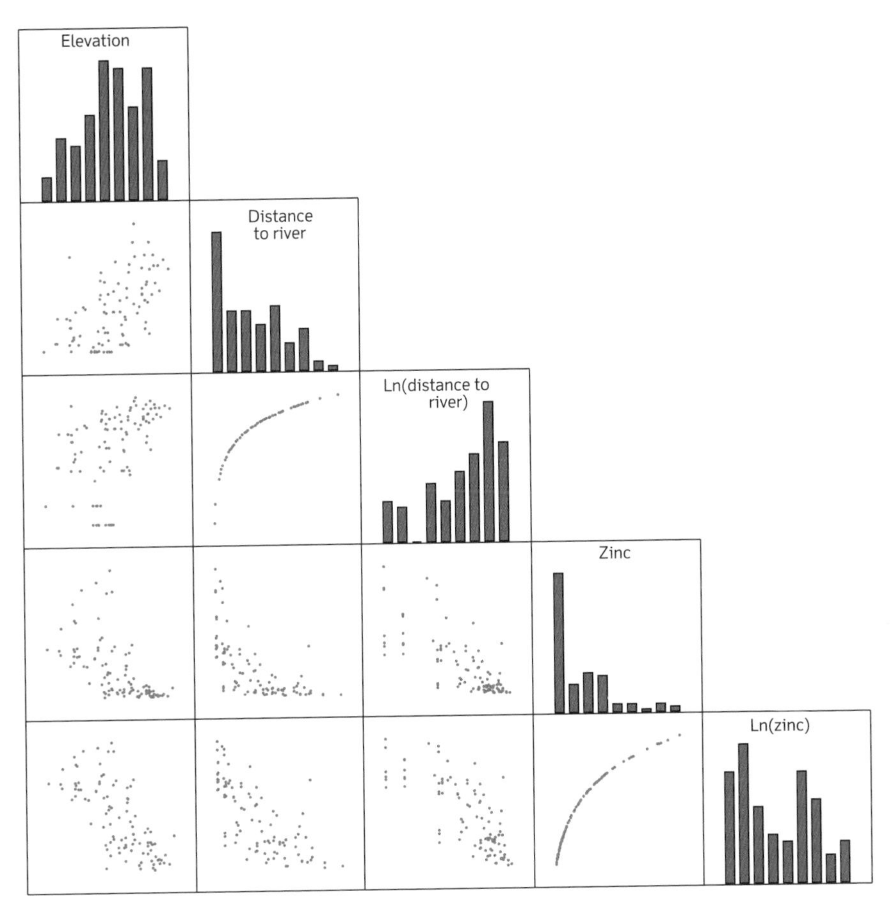

Figure 8.5 Scattergrams of independent and dependent terms in regression of zinc on distance to river and floodplain elevation

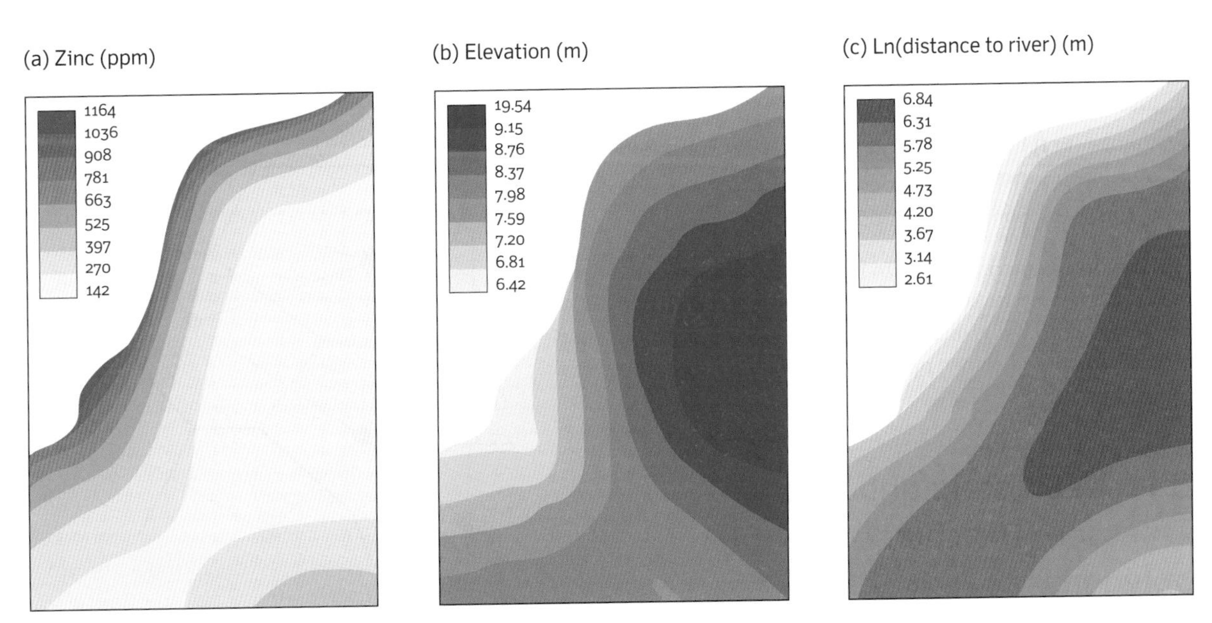

Zinc = exp(10.000 – 0.292 elevation – 0.333 ln(distance to river)

Figure 8.6 Results of computing zinc levels by regression from maps of floodplain elevation and distance to river

8.10 Local, deterministic methods for interpolation

All the methods presented so far have imposed an external, global spatial structure on the interpolation. In all cases, short-range local variations have been dismissed as random, unstructured noise. Intuitively, this is not sensible as one expects the value at an unvisited point to be similar to values measured close by. Therefore, local methods of interpolation have been sought that use the information from the nearest data points directly. For this approach, interpolation involves (a) defining a search area or neighbourhood around the point to be predicted, (b) finding the data points in this neighbourhood, (c) choosing a mathematical function to represent the variation over this limited number of points, and (d) evaluating it for the point on a regular grid. The procedure is repeated until all the points on the grid have been computed. The following issues need to be addressed:

- the kind of interpolation function to use
- the size, shape, and orientation of the neighbourhood
- the number of data points
- the distribution of the data points: regular grid or irregularly distributed?
- the possible incorporation of external information on trends or different domains.

We examine all these points in terms of different interpolation functions, specifically:

- nearest neighbours
- inverse distance weighting
- splines (exact and thin-plate smoothing) and other non-linear functions (e.g. Laplacian)
- optimal functions using spatial covariation.

All these local functions smooth the data to some degree, in that they compute some kind of average value within a *window* or search distance. The first three kinds of function are examined below; optimal functions are the subject of Chapter 9.

8.11 Nearest neighbours: Thiessen (Dirichlet/Voronoi) polygons

Thiessen (otherwise known as Dirichlet or Voronoi) polygons take the classification model of spatial prediction to the extreme in that the predictions of attributes at unsampled locations are provided by the nearest single data point. Thiessen polygons divide up a region in a way that is totally determined by the configuration of the data points, with one observation per cell. If the data lie on a regular square grid, then the Thiessen polygons are all equal, regular cells with sides equal to the grid spacing; if the data are irregularly spaced, then the result is an irregular lattice of polygons (Figure 8.7). The lines joining the data points show the Delaunay triangulation, which is the same topology as a TIN (see Chapter 3).

Thiessen polygons are often used in GIS and geographical analysis as a quick method for relating point data to space; a common, but implicit, use of Thiessen polygons is the assumption that the meteorological data for any given site can be taken from the nearest climate station. Unless there are many observations (which usually there are not), this assumption is not really appropriate for gradually varying phenomena like rainfall, temperature, and air pressure because (a) the form of the final map is determined by the distribution of the observations, and (b) the method maintains the choropleth map fiction of homogeneity within borders and all change at borders. As there is only one observation per tile, there is no way to estimate within-tile variability, short of taking replicate observations.

An advantage of Thiessen polygons is that they can be easily used with qualitative data like vegetation classes or land use if all you need is a choropleth map and you do not mind the strange geometrical pattern of the

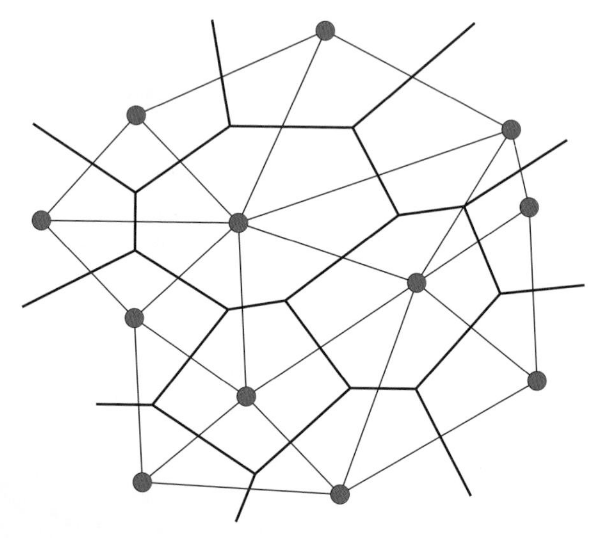

Thiessen polygons (thick lines)

Delaunay triangulation (thin lines)

Figure 8.7 An example of a Thiessen polygon net and the equivalent Delaunay triangulation

(a) Data points

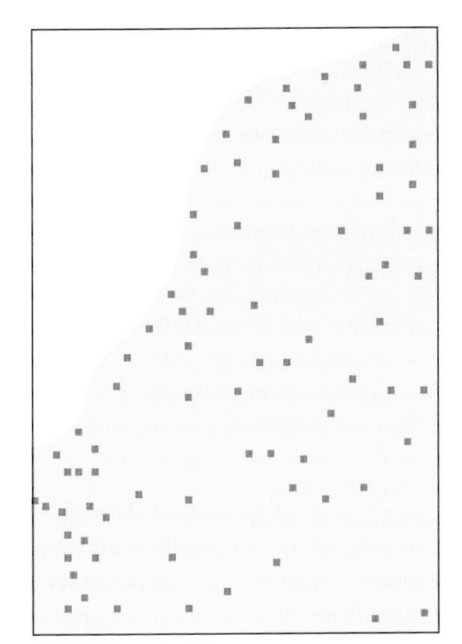

(b) Thiessen polygons

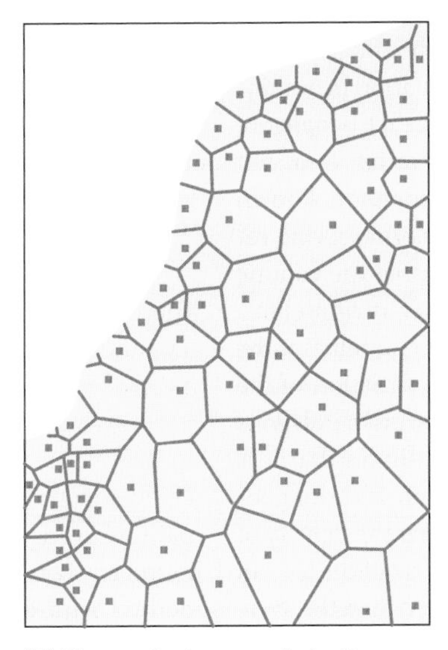

(c) Zinc content: Thiessen polygons

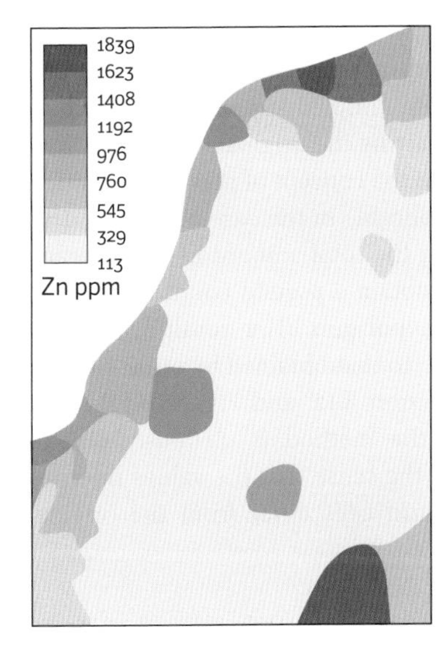

(d) Zinc content: pycnophylactic interpolation

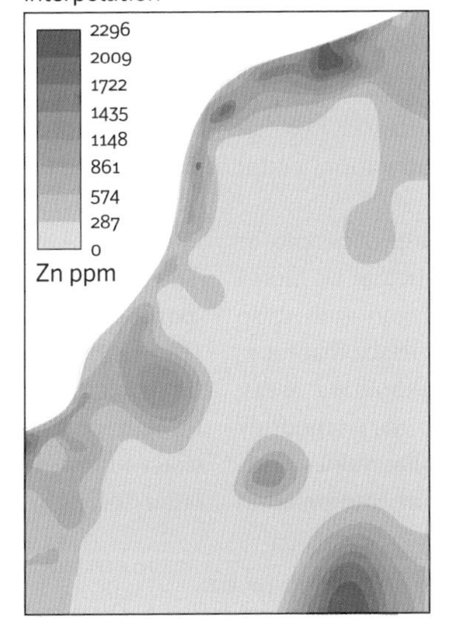

Figure 8.8 (a) Data locations, (b) Thiessen polygons, (c) zinc levels per Thiessen polygon for the zinc data, and (d) pycnophylactic interpolation of zinc from the Thiessen polygon data

boundaries. Because all predictions equal the values at the data points, Thiessen polygons are exact predictors.

Figure 8.8a, b, and c demonstrate the use of Thiessen polygons for mapping the zinc content of the floodplain soil. Instead of computing averages for externally defined units, the spatial variation of zinc over the floodplain is mapped simply by assigning the measured values to their closest neighbouring cells (Figure 8.8b and c). The Thiessen polygon map suggests that the zinc map obtained from information on flooding frequency polygons masks considerable spatial variation of zinc levels, particularly within the flood frequency classes 2 and 3 (cf. Figure 8.2c).

Tobler's pycnophylactic method

The Thiessen polygon interpolation yields the extreme case of the discrete polygon map as used for administrative units or the delineation of supposedly uniform areas of land use. A major difficulty with both approaches is that the quantities measured are assumed to be homogeneous within units and to change value only at the boundaries. This is often a gross approximation, particularly for attributes that exhibit spatial contiguity, such as population densities (people tend to congregate) or rainfall (why should rainfall amounts change abruptly at the midpoint separating the shortest distance between two measuring stations?) Tobler (1979) addressed this problem by devising a continuous, smooth interpolator that removes the abrupt changes due to inappropriate boundaries. His method was designed for counts of population in administrative areas, to deal with the problem that people are not spread uniformly over an administrative area, but tend to congregate in certain parts. The method, known as **pycnophylactic interpolation**, is a mass-preserving reallocation from primary data. It ensures that the volume of the attribute (number of people or other attribute) in a spatial entity (polygon or administrative area) remains the same, irrespective of whether the global variation of the attribute is represented by homogeneous, crisp polygons or a continuous surface. The method assumes that it is more realistic to represent the distribution of those people by a smooth surface, which implies that the density of people (or amount of rainfall) in some parts of the area in question is higher, and in other parts lower, than the average. The transfer of counts from areas to points in this way is, as noted previously, termed areal interpolation.

The primary condition for mass preservation is the invertibility condition:

$$\int_{R_i} \int f(x,y)\,dx\,dy = V_i \qquad 8.12$$

for all i, where V_i denotes the value (population, count attribute) in region R_i.

Equation 8.12 means that the total volume of people (or other attribute) per polygon is invariate, whether we model the population count by a uniform polygon with crisp boundaries or by a smooth, continuous surface that takes account of population densities in neighbouring areas. The constraining surface is assumed to vary smoothly so that neighbouring locations have similar values. This assumption is common in most kinds of interpolation because it links intuitive understanding of spatial variation to mathematical means of description. Unless there are physical barriers, the densities in neighbouring areas tend to resemble each other and so a joint, smooth surface is fitted to contiguous regions. The simplest way of doing this is to use the Laplacian condition, i.e. by minimizing:

$$\int_R \int \left(\frac{\partial f^2}{\partial x} + \frac{\partial f^2}{\partial y} \right) dx\,dy \qquad 8.13$$

where R is the set of all regions. The most general boundary condition is:

$$\frac{\partial f}{\partial \eta} = 0 \qquad 8.14$$

which constrains the gradient of the fitted surface normal to the edge of the region to be flat (η indicates the boundary between regions). Figure 8.9 illustrates the method with a simple example. Martin (1996) reviews the application of another method for surface generation from areal data, and the accuracy of estimates made using this approach is assessed by Martin et al. (2011).

The pycnophylactic method can be applied to the zinc data, or other data sampled at points such as rainfall amounts, by converting the original data into a density function, although the concept of mass preservation is more obviously applicable in the case of counts. This is done by computing the total amount of zinc in each Thiessen polygon to obtain a pseudo count of the zinc 'population' for each polygon. The resulting pattern (Figure 8.8d) is similar to that obtained by other smooth interpolators (see Section 8.12, and Figure 8.10)—but, though continuous, the method is obviously not an exact interpolator, in the sense that the values interpolated for individual grid cells differ from the original

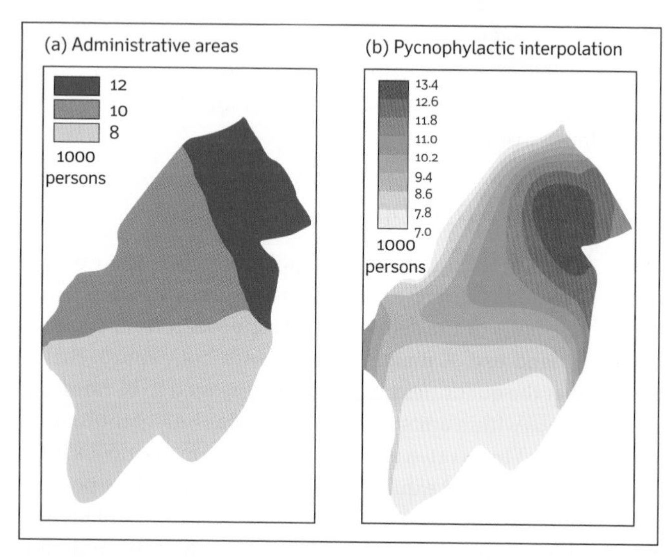

Figure 8.9 Population counts displayed by (a) uniform for administrative areas and (b) by pycnophylactic interpolation

measurements at the same cell. The largest and smallest values obtained by pycnophylactic interpolation are respectively much greater, and much smaller, than those actually measured, which for sampled data like zinc could bring problems of interpretation.

8.12 Linear interpolators: inverse distance interpolation

Inverse distance methods of interpolation combine the ideas of proximity espoused by the Thiessen polygons with the gradual change of the trend surface. The assumption is that the value of an *attribute* z at some unvisited point is a distance-weighted average of data points occurring within a neighbourhood or window surrounding the unvisited point. Typically the original data points are located on a regular grid or are distributed irregularly over an area, and estimates are made to locations on a denser regular grid in order to make a map.

Weighted moving average methods compute

$$\hat{z}(\mathbf{x}_0)=\sum_{i=1}^{n}\lambda_i\cdot z(\mathbf{x}_i) \text{ with } \sum_{i=1}^{n}\lambda_i=1 \quad 8.15$$

where the weights λ_i are given by $\Phi(d(\mathbf{x},\mathbf{x}_i))$. A requirement is that $\Phi(d) \rightarrow$ the measured value as $d \rightarrow 0$, which is given by the commonly used negative exponential functions $e^{-(d)}$ and $e^{-(d2)}$. The most common form of $\Phi(d)$ is the inverse distance weighting predictor, whose form is:

$$\hat{z}(\mathbf{x}_0)=\sum_{i=1}^{n}z(\mathbf{x}_i).d_{ij}^{-r}\Big/\sum_{i=1}^{n}d_{ij}^{-r} \quad 8.16$$

where the $\mathbf{x}_j$ are the points where the surface is to be interpolated and the $\mathbf{x}_i$ are the data points. Because in equation 8.16, $\Phi(d) \rightarrow \infty$ as $d \rightarrow 0$, the value for an interpolation point that coincides with a data point must be simply copied over. The simplest form of this is called the *linear interpolator*, in which the weights are computed from a linear function of distance between sets of data points and the point to be predicted.

Inverse distance interpolation is commonly used in GIS to create raster surfaces from point data. Once the data are on a regular grid, contour lines can be threaded through the interpolated values and the map can be drawn as either a vector contour map or as a raster shaded map.

Figure 8.10a, b, and c show the zinc data interpolated with inverse distance weighting with values of r of 1, 2, and 4. Note that the form of the resulting map is strongly dependent on the value of r, and on the number of neighbouring observations used in interpolation. Conclusions drawn about the levels of zinc pollution when $r = 1$ could be quite different from those when $r = 4$.

Inverse distance interpolation is forced to be an exact interpolator because it produces infinities when $(d_{ij}) = 0$ (i.e. at the data points), so if the output grid coordinates equal those of a sampling point the observed, unsmoothed values must be copied over.

The form of the map also depends on the clustering in the data and on the presence of outliers. Inverse distance interpolations commonly have a 'duck-egg' pattern around solitary data points with values that differ greatly from their surroundings, though this can be modified to a certain extent by altering the search criteria for the data points to account for anisotropy. The method has no inbuilt method of testing for the quality of predictions so the map quality can only be assessed by taking extra observations. Note that these must be for the same support as the original observations, though it is arguable that since inverse distance interpolation smoothes within a zone proportional to the value of r, this is the true resolution of the interpolation and not the sample support. Examples of the application of IDW are given by Lloyd (2010b) and O'Sullivan and Unwin (2010).

8.13 Splines

Before computers were used to fit curves to sets of data points, draughtsmen used flexible rulers to achieve the best locally fitting smooth curves by eye—these flexible rulers were called splines. The spline rulers were held in place by weights on pegs at the data points while the line was drawn (Pavlidis 1982). The modern equivalent is the plastic-coated flexible ruler sold in office equipment shops. It can be shown that the line drawn along a spline ruler is approximately a piece-wise cubic polynomial that is continuous and has continuous first and second derivatives.

Spline functions are mathematical equivalents of the flexible ruler. They are piece-wise functions, which is to say that they are fitted exactly to a small number of data points, while at the same time ensuring that the joins between one part of the curve and another are continuous. This means that with splines it is possible to modify one part of the curve without having to recompute the whole curve, which is not possible with trend surfaces (Figure 8.11).

The general definition of a piece-wise polynomial function $p(x)$ is:

$$p(x)=p_i(x) \qquad x_i < x < x_{i+1} \qquad i=0,1,\ldots,k-1 \quad 8.17$$

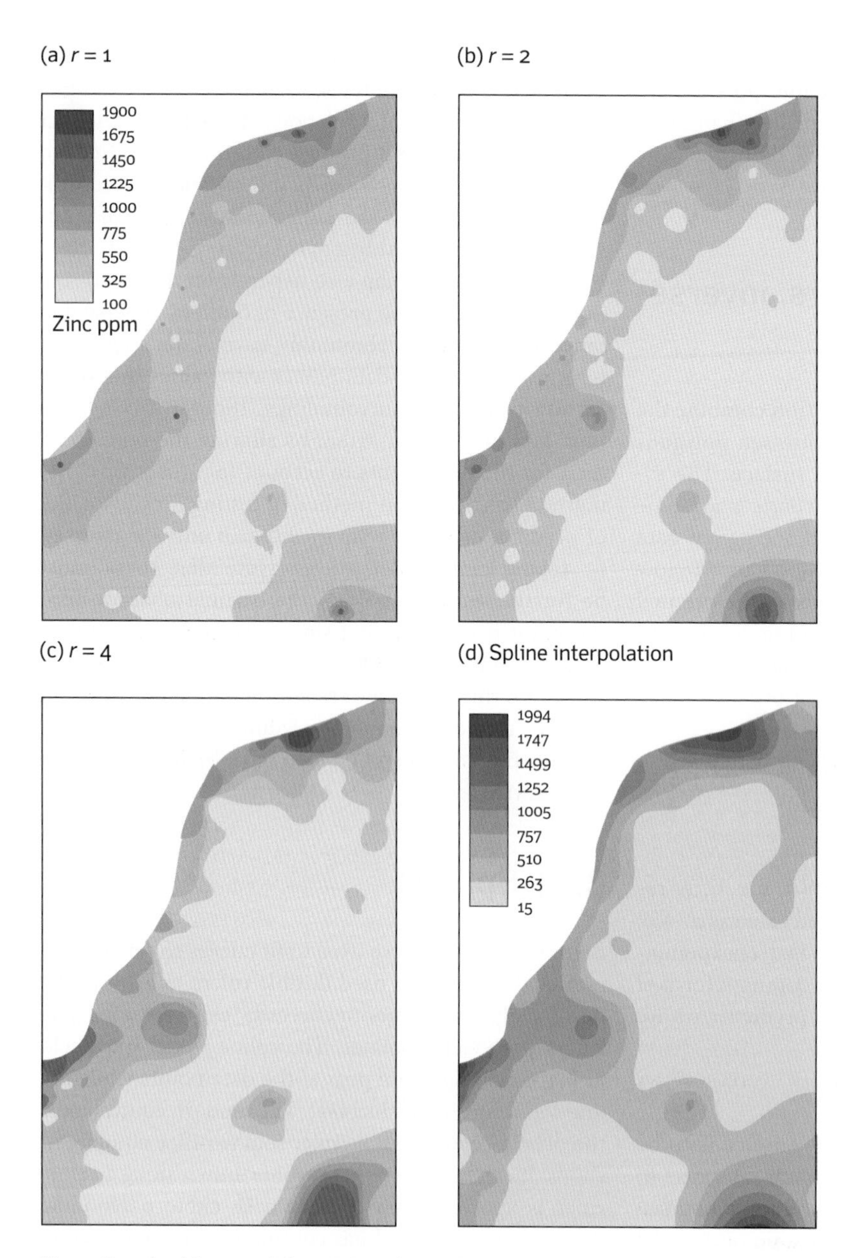

Figure 8.10 (a–c) Inverse distance interpolation showing the effect of the weighting parameter on the results; (d) thin-plate spline mapping of zinc

$$p^j(x_i) = p^j_{i+1}(x_i) \qquad j = 0, 1, \ldots, r-1; \quad i = 1, 2, \ldots, k-1 \tag{8.18}$$

The points $x_i, \ldots x_{k-1}$ that divide an interval x_0, x_k into k sub-intervals are called *break points* and the points of the curve at these values of x are commonly called *knots* (Pavlidis 1982). The functions $p_i(x)$ are polynomials of degree m or less. The term r is used to denote the constraints on the spline. When $r = 0$, there are no constraints on the function; when $r = 1$ the function is continuous without any constraints on its derivatives. If $r = m + 1$, the interval x_0, x_k can be represented by a single polynomial, so $r = m$ is the maximum number of constraints that leads to a piece-wise solution. For $m = 1$, 2, or 3, a spline is called linear, quadratic, or cubic. The derivatives are of order 1, 2, $m - 1$, so a quadratic spline must have one continuous derivative at each knot, and a cubic spline must have two continuous derivatives at each knot. For a simple spline where $r = m$ there are only $k + m$ degrees of freedom. The case of $r = m = 3$

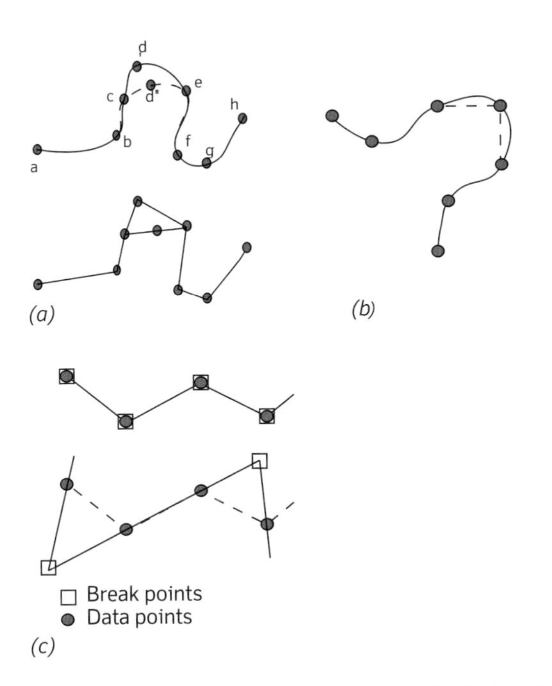

Figure 8.11 Some properties of splines: (a) local adjustments mean local changes; (b) exact splines round off sharp corners; (c) choosing to locate break points at or between data points has a large effect on the resulting spline

has particular significance because the term spline was first used for cubic piece-wise polynomial functions. The term *bicubic spline* is used for the three-dimensional situation where surfaces rather than lines need to be interpolated. Equation 8.18 means that adjacent polynomials have the same value at a break point for a given $j = r$.

Because of difficulties of calculating simple splines over a wide range of separate sub-intervals—as might be the case with a digitized line—most practical applications use a special kind of spline called B-splines. B-splines are themselves the sums of other splines that by definition have the value of zero outside the interval of interest (Pavlidis 1982). B-splines, therefore, allow local fitting from low-order polynomials in a simple way.

B-splines can be used for smoothing digitized lines for display, for instance for the boundaries on soil and geological maps where cartographic conventions expect smooth, flowing lines. Pavlidis notes that complex shapes such as those in text fonts can be more economically defined in terms of B-splines than in sets of data points. The use of B-splines for smoothing polygon boundaries can, however, lead to certain complications, particularly when computing areas and perimeters. If the area of a polygon is calculated from digitized data points using the trapezoidal rule (Box 7.1), it will be different from the area that results when the boundaries are smoothed with B-splines. Another problem may arise when high-order B-splines are used to smooth sinuous boundaries that also include sharp, rectangular corners (Figure 8.11b).

A problem with using splines for interpolation is whether one should choose the break points to coincide with the data points or to be interleaved with them. Different results for the interpolated spline may result from the two approaches (Figure 8.11c). Note that with splines the maxima and minima do not necessarily occur at the data points.

Thin-plate splines as surface interpolators

Exact splines are used in GIS and drawing packages for smoothing curves and contour lines to improve appearance where it is assumed that exact interpolation is required. When data have been sampled in two or three dimensions, effects of natural variation and measurement errors may mean that an exact spline produces local artefacts of excessively high or low values. These artefacts can be removed by using thin-plate splines, in which the exact spline surface is replaced by a locally smoothed average. As with line interpolation, spline methods of surface interpolation assume that an approximate function should pass as close as possible to the data points while also being as smooth as possible. When the data contain a source of random error, i.e.

$$y(\mathbf{x}_i) = z(\mathbf{x}_i) + \varepsilon(\mathbf{x}_i) \qquad 8.19$$

where z is the measured value of an attribute at point $\mathbf{x}_i$ and ε is the associated random error, the spline function should pass 'not too far' from the data values and so the smoothing spline is the function f that minimizes:

$$\sum_{i=1}^{n}(z(\mathbf{x}_i) - g(\mathbf{x}_i))^2 + \rho J_m(g) \qquad 8.20$$

The term $J_m(g)$ represents the 'smoothness' of the function g, and the second term represents its 'proximity' or 'fidelity' to the data; $m - 1$ is the degree of the polynomial used in the model, and ρ is called the smoothing parameter. When the smoothing parameter is set to zero, the spline passes through the data and it is an exact interpolator. When the value is greater than zero, the spline is not forced to fit to the data and it is an approximate interpolator. When the data are 'noisy'—there is clear evidence of measurement error—then a positive value for the smoothing parameter may be considered sensible. Large values of the smoothing parameter result in smoother maps. The thin-plate spline function g is made up of two elements:

$$g(\mathbf{x}) = a_0 + a_1 x + a_2 y + \sum_{i=1}^{n} \lambda_i R(\mathbf{x} - \mathbf{x}_i) \quad 8.21$$

The left-hand side of the equation $(a_0 + a_1 x + a_2 y)$ indicates the local trend in the data, and we must estimate values for a_0, a_1, a_2. The term $R(\mathbf{x} - \mathbf{x}_i)$ is called a basis function. For a thin-plate spline it is:

$$d_i^2 \log d_i \quad 8.22$$

where d_i is the distance between the prediction location $\mathbf{x}$ and the data location $\mathbf{x}_i$. In this case, $R(\mathbf{x} - \mathbf{x}_i)$ is the distance between those two locations inputted into equation 8.22. Given a distance of 7.34 m, this gives: $d_i^2 \log d_i = 7.34^2 \log 7.34 = 107.39$ (the log is the natural log).

The coefficients a_k and λ_i (where k = 0, 1, 2 and i is the index for the n local data used in interpolation) can be obtained using

$$\mathbf{R}\lambda = \mathbf{z} \quad 8.23$$

Lloyd (2010b) gives a worked example of how the coefficients are obtained.

Smoothing thin-plate splines have frequently been used for interpolating elevation to create digital elevation models, where it is necessary to interpolate large areas quickly and efficiently—see Hutchinson (1995) and Mitasova et al. (1995). Both texts demonstrate their use for multivariate interpolation of attributes such as annual mean rainfall from a trivariate spline function of longitude, latitude, and elevation.

Figure 8.10d shows the variation of zinc as mapped by splines. Note that the spline surface tends to 'draw up' the areas around sample sites with large zinc values.

Because splines are piece-wise functions using few points at a time, the interpolating values can be quickly calculated. Test data for smooth surfaces show that predictions are very close to the values being interpolated, as long as the measurement errors associated with the data are small (Mitasova et al. 1995). In contrast to trend surfaces and weighted averages, splines retain small-scale features. Both linear and surficial splines are aesthetically pleasing and quickly produce a clear overview of the data. The smoothness of splines means that mathematical derivatives can easily be calculated for direct analysis of surface geometry and topology (see Section 10.7). The incorporation of linear parametric submodels (regression models) makes the interpolation of dependent variables on point supports easy.

Some disadvantages of splines have already been mentioned. Others are that there are no direct estimates of the errors associated with spline interpolation, though these may be obtained by a recursive technique known as 'jack-knifing'. The most critical disadvantage may be that thin-plate splines provide a view of reality that is unrealistically smooth; in some situations, such as estimating attribute values for numerical models, this property could generate misleading results. A review of different forms of spline interpolators is provided by Lloyd (2011).

8.14 A comparison of simple global and local methods

We have explained several different kinds of interpolation technique, but how do they compare? Which produces the most consistent results, and when should one method be preferred to another? To give a first insight, Table 8.5 compares all the methods in terms of the minimum and maximum interpolated values, and the proportions of the area estimated to be above the thresholds of 500, 1000, and 1500 ppm zinc.

Comparing these results, we see that global trend surfaces of orders higher than 3 are unreliable. Trend surfaces are therefore really only useful for describing broad geographical trends (cf. Davis 2002). Pycnophylactic methods and thin-plate splines stretch the maximum and minimum values above and below the recorded values, but inverse distance with r = 2 gives results closest to the original data. The range of predictions of the areas exceeding the three arbitrary limits produced by all these methods is greatest for the values exceeding 1000 ppm; without independent data and a review of the performance of the methods on other data sets it is impossible to say which method is generally the best. The inverse distance and thin-plate splines seem to produce the most 'natural'-looking surfaces, but this may just be a reflection of our cultural preferences, as these surfaces are always much smoother than the underlying reality.

8.15 A comparison of IDW and TPS using cross-validation and grids

The previous section compared a range of interpolation methods. Here, a more detailed comparison of the widely used IDW and TPS approaches is presented. The

Table 8.5 Summary of results of deterministic interpolation

Method	Minimum value (ppm)	Maximum value (ppm)	Per cent area >500 ppm	Per cent area >1000 ppm	Per cent area >1500 ppm
Flood frequency (untransformed)	206	770	11.96	0.00	0.00
Trend surface—order 1	-98	791	38.53	0.00	0.00
Trend surface—order 2	109	1350	31.88	3.37	0.00
Trend surface—order 3	-13	1256	33.26	3.66	0.00
Trend surface—order 4	-100	1500	28.30	5.88	0.69
Regression on distance + elevation	142	1164	17.82	0.64	0.00
Thiessen polygons	113	1839	28.11	9.65	5.31
Pycnophylactic method	0	2296	28.35	10.15	3.87
Inverse distance $r = 1$	136	1541	28.92	1.16	0.03
Inverse distance $r = 2$	114	1827	28.82	3.99	0.61
Inverse distance $r = 4$	113	1839	28.47	7.05	2.49
Thin-plate splines	15	1994	30.44	6.74	1.87

data used are precipitation amounts in Wales for the month of January 2006. Figure 8.12 shows a map of precipitation generated using IDW with 16 nearest neighbours used for interpolation, while Figure 8.13 shows a map generated using thin-plate splines with tension (TPST), again using 16 nearest neighbours. This latter approach is a variant of thin-plate splines which allows a tension parameter to be adjusted—the tension can be

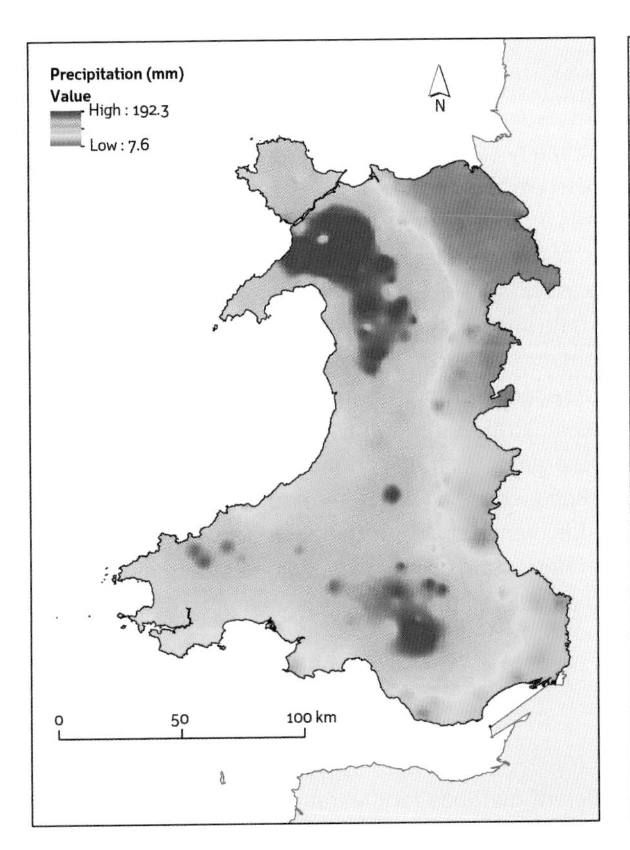

Figure 8.12 Precipitation in Wales in January 2006: IDW with 16 nearest neighbours

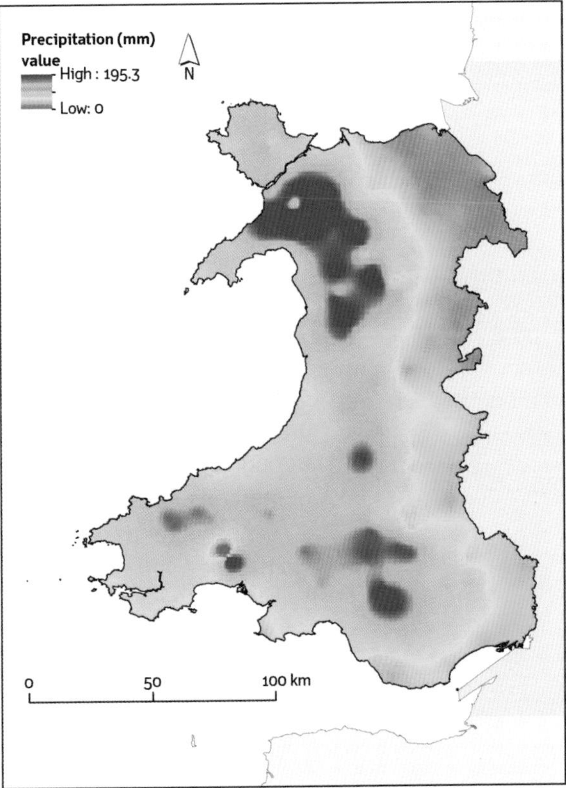

Figure 8.13 Precipitation in Wales in January 2006: TPST with 16 nearest neighbours

changed so that the surface may resemble a stiff plate or an elastic membrane. With TPST, the predicted minima were smaller than zero, and these negative predictions were removed. With IDW, all predictions must be within the original data range. The capacity of TPS (and TPST) to predict values smaller than the data minimum and greater than the data maximum is both an advantage (if the true minima and maxima are not sampled then this property is useful) and a weakness (if impossible values like precipitation amounts of less than zero are predicted). The notable presence of clusters of similar values around the data points in the IDW map (Figure 8.12) is, as noted in Section 8.12, a common feature of IDW outputs, and this undesirable feature is much less apparent in the TPST map (Figure 8.13).

Cross-validation provides one means of assessing the performance of interpolation methods. The process entails (i) temporary removal of an observation, (ii) estimation of the removed value using neighbouring data, and (iii) return of the removed value, then (iv) temporary removal of another value and back to step (i), until all values have been estimated in this way. The differences between the 'true' and estimated values can then be summarized. In Table 8.6, cross-validation prediction errors are summarized for IDW, TPS, and TPST. In this case, different numbers of nearest neighbours were specified and the number which corresponded with the smallest root mean square error (RMSE, a measure of the magnitude of errors) were retained. The mean errors, which indicate bias (over or underestimation) are also given. Judging by these results, TPST with 64 nearest neighbours provides the smallest (cross-validation) prediction errors. Cross-validation is a useful tool, but it should be used with caution.

Table 8.6 Summary of cross-validation prediction errors

Method	Mean error (mm)	RMSE (mm)
IDW 16	0.845	19.367
TPS 64	−0.083	21.825
TPST 64	0.073	18.127

8.16 Summary

Spatial interpolation is important in any case where we have a sample of spatial data but we wish to know about characteristics of the variable at locations where there are no samples. This chapter discusses a range of global and local approaches to spatial interpolation and considers some of the main issues which must be considered when implementing these methods. It was shown that trend surface analysis provides a summary of global trends in a spatial data set, but that such approaches are unsuitable for making local estimates of the property of interest. In contrast, the widely used inverse distance weighting and thin-plate spline interpolators are well suited to making rapid estimates of spatial variations, as illustrated through the mapping of soil properties and precipitation amounts. The assessment of the performance of alternative methods through cross-validation was outlined. The following chapter discusses another class of methods for spatial interpolation, which build on the approaches outlined in the present chapter by making use of information on how variables are structured spatially so that the interpolation approach is tailored to the particular characteristics of a data set.

? Questions

1. Compare and contrast the different methods for creating spatial classifications in physical sciences such as hydrology, soil science, and geology with those encountered in the human sciences, such as demography, epidemiology, and politics.

2. Explore the assumptions inherent in the ANOVA method and how these affect the integrity of the classification method of spatial prediction.
3. Explain the assumptions behind trend surface analysis and show how these may seriously affect the quality of the results. Consider residual errors, outliers, assumptions of stationarity (homogeneity of variation), and independence. Does a large value of R^2 necessarily mean that unsampled points will be predicted correctly? How would you check the quality of the predictions independently?
4. Discuss the kinds of spatial process encountered in natural and social sciences (physical and human geography, ecology, hydrology and meteorology, demography, etc.) that lead to spatial variation that can be modelled by a global trend or regression surface.
5. Explain how you would set up a method of objective testing that would compare the predictive quality of different interpolation techniques, such as inverse distance weighting versus thin-plate splines or regression surfaces.
6. Explain why the TIN may be unsuitable for modelling the continuous variation of physical attributes other than elevation measured at point locations.
7. Explore ways of generalizing to three dimensions the interpolation methods presented in this chapter.

Further reading

▼ Davis, J. C. (2002). ***Statistics and Data Analysis in Geology***. 3rd edn. John Wiley & Sons, New York.

▼ Hutchinson, M. F (1995). Interpolating mean rainfall using thin plate smoothing splines. ***International Journal of Geographical Information Systems***, 9: 385–404.

▼ Lloyd, C. D. (2010). ***Spatial Data Analysis: An Introduction for GIS Users***. Oxford University Press, Oxford.

▼ O'Sullivan, D. and Unwin, D. J. (2010). ***Geographic Information Analysis***. 2nd edn. Wiley, Hoboken, NJ.

Interpolation 2: Geostatistical Approaches

When data are abundant, most interpolation techniques give similar results. When data are sparse, however, the assumptions made about the underlying variation that has been sampled, and the choice of method and its parameters, can be critical if one is to avoid misleading results. Geostatistical methods of interpolation, popularly known as **kriging**, attempt to optimize interpolation by dividing spatial variation into three components—(a) deterministic variation (different levels or trends) that can be treated as useful, soft information, (b) variations that are **spatially autocorrelated** but physically difficult to explain, and finally (c) uncorrelated noise. The character of the spatially correlated variation is encapsulated in functions such as the **autocovariogram** and (semi) **variogram**, and these provide the information for optimizing interpolation weights and search radii. Experimental variograms are computed from sample data in one, two, or three spatial dimensions. These experimental data are fitted by one of a limited number of variogram models, which serve to provide data for computing interpolation weights.

Geostatistical methods provide great flexibility for interpolation, providing ways to interpolate to areas or volumes larger than the support (**block kriging**), methods for interpolating binary data (**indicator kriging**), methods for incorporating information about trends (universal kriging; also known as kriging with a trend model) or stratification (**stratified kriging**), and methods which make use of secondary variables (for example, **cokriging**). All these methods of interpolation yield smoothly varying surfaces accompanied by an estimation variance surface. In contrast to smooth interpolators which give a single, local average value, the methods of conditional simulation, given the variogram and the original observations, provide sets of realizations which give an indication of the range of values possible. Combining soft information and conditional simulation is useful for computing inputs for raster-based environmental models (as described in Chapter 12).

Finally, the information in the variogram can be used to help optimize sampling schemes for mapping from point data.

Learning objectives

By the end of this chapter, you will:

- **understand the purpose of, and some of the key principles behind, geostatistical analysis**

➤ **be able to apply your knowledge to create maps from point samples using information on the spatial structure of the variable you want to map**

➤ **have developed knowledge of how geostatistical approaches may offer benefits over the interpolation methods described in Chapter 8.**

9.1 A brief introduction to regionalized variable theory and kriging

When comparing all the maps made in Chapter 8 of the zinc concentration in the floodplain soils we see that the different methods return large differences in the global patterns, the amount of local detail, and in the minimum and maximum values predicted. These differences were summed up in Table 8.5 in terms of the estimates given by the various methods of the percentages of the area thought to be above 500, 1000, and 1500 ppm zinc, respectively. Clearly, if we were faced with all these different results and had to make a decision on the costs of cleaning up soil with more than a certain level of zinc, it would be very difficult to choose the best method. In addition, cross-validation error statistics for the Welsh precipitation data were provided in Table 8.6. None of the methods of interpolation discussed so far can provide direct estimates of the quality of the predictions made in terms of an estimation variance for the predicted value at unsampled locations. In all cases, the most appropriate way to determine the goodness of the predictions would be to compute estimates for a set of extra **validation points** that had not been used in the original interpolation. Apart from research studies to judge the quality of performance of a technique, this is rarely done because it costs money. An alternative is to use cross-validation, as defined in Section 8.15.

A further objection to all the methods discussed so far is that there is no a priori method of knowing whether the best values have been chosen for the weighting parameters or if the size of the search neighbourhood is appropriate. As we have seen, the control parameters of trend surfaces and inverse distance weighting can be varied to produce quite different maps, which in turn provide different estimates of the distribution of the variable being mapped. Moreover, no method studied so far provides sensible information on:

- the number of points needed to compute the local average
- the size, orientation, and shape of the neighbourhood from which those points are drawn
- whether there are better ways to estimate the interpolation weights than as a simple function of distance
- the errors (uncertainties) associated with the interpolated values.

These questions led the French geomathematician Georges Matheron and the South African mining engineer D. G. Krige to develop optimal methods of interpolation for use in the mining industry. The methods are widely used in groundwater modelling, soil mapping, and related fields, and packages for geostatistical interpolation have become important modules of commercial GIS.

Geostatistical methods for interpolation start with the recognition that the spatial variation of any continuous attribute is often too irregular to be modelled by a simple, smooth mathematical function. Instead, the variation can be better described by a stochastic surface. The attribute is then known as a **regionalized variable**; the term applies equally to the variation of atmospheric pressure, elevation above sea level, or the distribution of continuous demographic indicators. Interpolation with geostatistics is known as kriging, after D. G. Krige.

Geostatistical methods provide ways to deal with the limitations of the deterministic interpolation methods listed above, and ensure that the prediction of attribute values at unvisited points is optimal in terms of the assumptions made. The methods developed by Matheron, Krige, and their co-workers are optimal in the sense that the interpolation weights are chosen so as to optimize the interpolation function—i.e. to provide a **best linear unbiased estimate (BLUE)** of the value of a variable at a given point. The same theory can be used for optimizing sample networks.

Regionalized variable theory assumes that the spatial variation of any variable can be expressed as the sum of three major components (Figure 9.1). These are (a) a structural component, having a constant mean or trend; (b) a random but spatially correlated component, known as the variation of the regionalized variable; and (c) a spatially uncorrelated random noise or residual error term. Let $\mathbf{x}$ be a position in one, two, or three dimensions. Then the value of a random variable Z at $\mathbf{x}$ is given by

$$Z(\mathbf{x}) = m(\mathbf{x}) + \varepsilon'(\mathbf{x}) + \varepsilon'' \qquad 9.1$$

where $m(\mathbf{x})$ is a deterministic function describing the 'structural' component of Z at $\mathbf{x}$, $\varepsilon'(\mathbf{x})$ is the term denoting the stochastic, locally varying, but spatially dependent residuals from $m(\mathbf{x})$—the regionalized variable—and ε'' is a residual, spatially independent Gaussian noise term having zero mean and variance σ^2. Note the use of the capital letter to indicate that Z is a random function and not a measured attribute z. The first step is to decide on a suitable function for $m(\mathbf{x})$. In the simplest case, where no trend or *drift* is present, $m(\mathbf{x})$ equals the mean value in the sampling area, and the average or expected difference between any two places $\mathbf{x}$ and $\mathbf{x}+\mathbf{h}$, separated by a distance vector $\mathbf{h}$, will be zero:

$$E[Z(\mathbf{x})-Z(\mathbf{x}+\mathbf{h})]=0 \tag{9.2}$$

where $Z(\mathbf{x})$, $Z(\mathbf{x}+\mathbf{h})$ are the values of random variable Z at locations $\mathbf{x}$, $\mathbf{x}+\mathbf{h}$. Also, it is assumed that the variance of differences depends only on the distance between sites, $\mathbf{h}$, so that

$$E\left[\left\{Z(\mathbf{x})-Z(\mathbf{x}+\mathbf{h})\right\}^2\right] = E\left[\left\{\varepsilon'(\mathbf{x})-\varepsilon'(\mathbf{x}+\mathbf{h})\right\}^2\right]=2\gamma(\mathbf{h}) \tag{9.3}$$

where $\gamma(\mathbf{h})$ is known as the semivariance. The two conditions, stationarity of difference and variance of differences, define the requirements for the **intrinsic hypothesis** of regionalized variable theory. This means that once structural effects have been accounted for, the remaining variation is homogeneous in its variation so that differences between sites are merely a function of the distance between them. So, for a given distance $\mathbf{h}$, the variance of the random component of $Z(\mathbf{x})$ is described by the semivariance:

$$\operatorname{var}[\varepsilon'(\mathbf{x})-\varepsilon'(\mathbf{x}+\mathbf{h})]=2\gamma(\mathbf{h}) \tag{9.4}$$

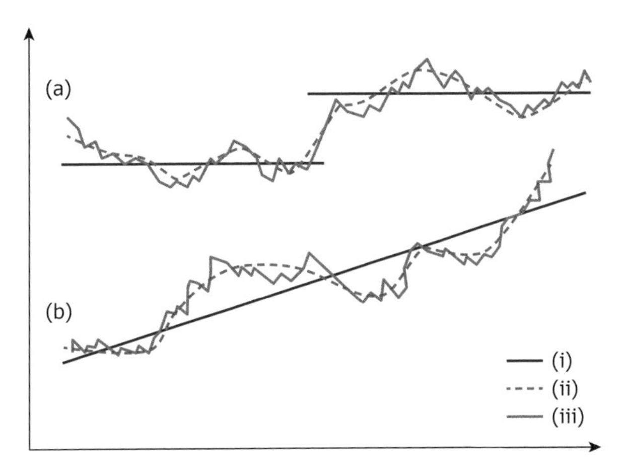

Figure 9.1 Regionalized variable theory divides complex spatial variation into (i) average behaviour such as differences in mean levels (upper) or a trend (lower), (ii) spatially correlated, but irregular ('random') variation, and (iii) random, uncorrelated local variation caused by measurement error and short-range spatial variation

If the conditions specified by the intrinsic hypothesis are fulfilled, the semivariance can be estimated from sample data:

$$\hat{\gamma}(\mathbf{h})=\frac{1}{2n}\sum_{i=1}^{n}\left\{z(\mathbf{x}_i)-z(\mathbf{x}_i+\mathbf{h})\right\}^2 \tag{9.5}$$

where n is the number of pairs of sample points of observations of the values of attribute z separated by distance $\mathbf{h}$ (see Worked example 9.1). A plot of $\hat{\gamma}(\mathbf{h})$ against $\mathbf{h}$ is

Worked example 9.1 Computing the first and second moments of a simple series

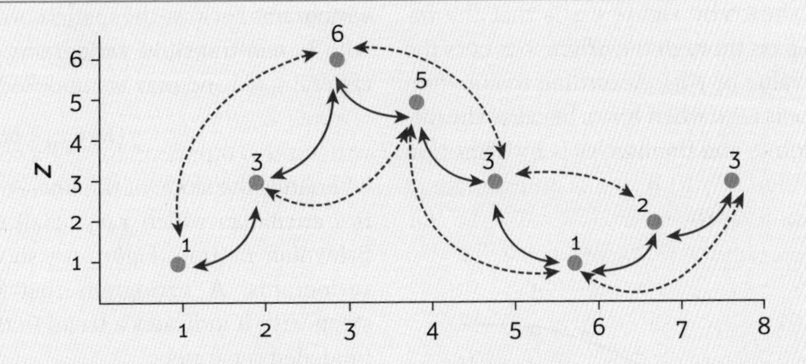

Mean = 24/8 = 3.0

Variance = [(1 − 3)² + (3 − 3)² + (6 − 3)² + (5 − 3)² + (3 − 3)² + (1 − 3)² + (2 − 3)² + (3 − 3)²]/8 = (4 + 0 + 9 + 4 + 0 + 4 + 1 + 0)/8 = 22/8 = 2.75

Covariance(1) = [(1 − 3)(3 − 3) + (3 − 3)*(6 − 3) + (6 − 3)*(5 − 3) + (5 − 3)*(3 − 3) + (3 − 3)*(1 − 3) + (1 − 3)*(2 − 3) + (2 − 3)*(3 − 3)]/7 = [0 + 0 + 6 + 0 + 0 + 2 + 0]/7 = 8/7 = 1.14*

*Semivariance(1) = [(1 − 3)² + (3 − 6)² + (6 − 5)² + (5 − 3)² + (3 − 1)² + (1 − 2)² + (2 − 3)²]/2*7 = [4 + 9 + 1 + 4 + 4 + 1 + 1]/2*7 = 24/14 = 1.715*

known as the **experimental variogram**. The experimental variogram is the first step towards a quantitative description of the regionalized variation. The variogram provides useful information for interpolation, optimizing sampling, and determining spatial patterns. To do this, however, we must first fit a theoretical model to the experimental variogram.

9.2 Fitting variogram models

Figure 9.2 shows a typical experimental variogram of data from a not too smoothly varying attribute, such as a soil property. The curve that has been fitted through the experimentally derived data points displays several important features. First, at large values of the lag, **h**, it levels off. This horizontal part is known as the **sill**; it implies that at these values of the lag there is no spatial dependence between the data points because all estimates of variances of differences will be invariant with sample separation distance. Second, the curve rises from a low value of $\gamma(\mathbf{h})$ to the sill, reaching it at a value of **h** known as the **range**. This is the critically important part of the variogram because it describes how inter-site differences are spatially dependent. Within the range, the closer together sites are the more similar they are likely to be. The range gives us an answer to the question posed in weighted moving average interpolation about the maximum size the window should be. Clearly, if the distance separating an unvisited site from a data point is greater than the range, then that data point can make no useful contribution to the interpolation; it is too far away.

The third point shown by Figure 9.2 is that the fitted model does not pass through the origin, but cuts the γ-axis at a positive value of $\gamma(\mathbf{h})$. According to equation 9.5, the semivariance is zero when $\mathbf{h} = 0$, because the differences between points and themselves is by definition zero. The positive value of $\gamma(\mathbf{h})$ $\mathbf{h} \rightarrow 0$ is an estimate of

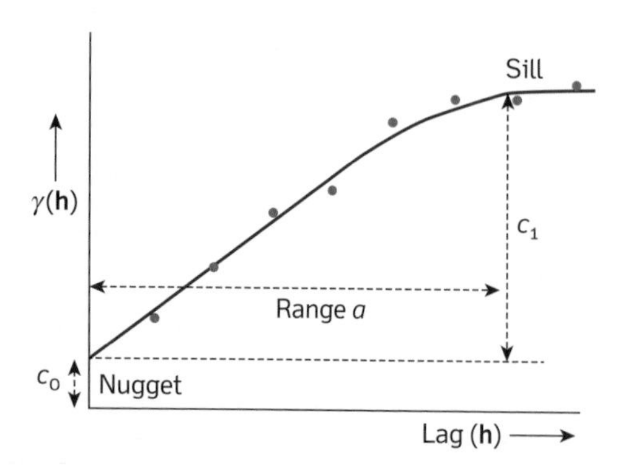

Figure 9.2 An example of a simple transitional variogram with range, nugget, and sill

ε'', the residual, spatially uncorrelated noise. ε'' is known as the **nugget** variance—the variance of measurement errors combined with that from spatial variation at distances much shorter than the sample spacing, which cannot be resolved.

The form of the variogram can be quite revealing about the kind of spatial variation present in an area, and can help to decide how to proceed further. When the nugget variance is important but not too large, and there is a clear range and sill, a curve known as the spherical model,

$$\gamma(\mathbf{h}) = c_0 + c_1\left\{\frac{3\mathbf{h}}{2a} - 1/2(\mathbf{h}/a)^3\right\} \quad \text{for } 0 < \mathbf{h} < a$$
$$= c_0 + c_1 \quad \text{for } \mathbf{h} \geq a \qquad 9.6$$
$$\gamma(0) = 0$$

where a is the range, **h** is the lag, c_0 is the nugget variance, and $c_0 + c_1$ equals the sill, often fits observed variograms well.

If there is a clear nugget and sill, but only a gradual approach to the range, the exponential model is often a good choice:

$$\gamma(\mathbf{h}) = c_0 + c_1\{1 - \exp(-\mathbf{h}/a)\} \qquad 9.7$$

If the variation is very smooth and the nugget variance ε'' is very small compared to the spatially dependent random variation $\varepsilon'(\mathbf{x})$, then the variogram can often best be fitted by a curve having an inflection, such as the Gaussian model:

$$\gamma(\mathbf{h}) = c_0 + c_1\{1 - \exp(-\mathbf{h}^2/a^2)\} \qquad 9.8$$

All the models defined above are known as **transitive variograms** because the spatial correlation structure varies with **h**; **non-transitive variograms** have no sill within the area sampled and may be modelled by the linear model:

$$\gamma(\mathbf{h}) = c_0 + bh \qquad 9.9$$

where b is the slope of the line. A linear variogram typifies attributes which vary at all scales, such as simple Brownian motion. Figure 9.3 shows examples of these variograms. A variogram that becomes increasingly steep with **h** indicates a trend in the data that should be modelled separately.

Variogram estimation and modelling is extremely important for structural analysis and for interpolation. The variogram models cannot be any haphazardly chosen function, as they must obey certain mathematical constraints (see, for example, Webster and Oliver 2007). Also, when fitting models it is sensible to take into account the numbers of pairs of points at each lag (so the model fit is influenced most by semivariances computed from large numbers of pairs). Because the data are used repeatedly when estimating variograms, the effective degrees of

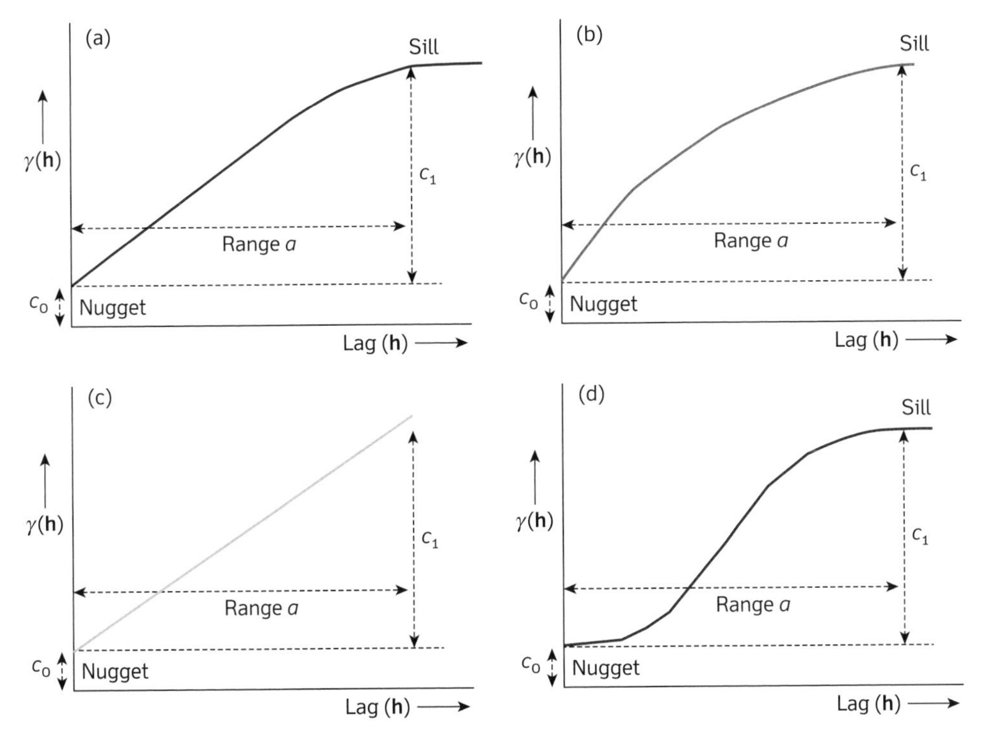

Figure 9.3 Examples of the most commonly used variogram models: (a) spherical; (b) exponential; (c) linear; and (d) Gaussian

freedom are greatest for the shortest lag and then decrease in a complex way with the lag. Consequently, the fitting of variogram models usually proceeds using a weighted least squares method, where the weights are computed from the numbers of pairs. Even so, variogram fitting is an interactive process requiring considerable judgement and skill. Geostatistics software usually includes interactive routines for variogram fitting (e.g. Pebesma and Wesseling 1998; Bivand et al. 2008).

9.3 Using the variogram for spatial analysis

The variogram is an essential step on the way to determining optimal weights for interpolation. When the nugget variance c_0 so dominates the local variation that the experimental variogram shows no tendency to diminish as $\mathbf{h} \rightarrow 0$, the interpretation is that the data are so noisy that interpolation is not sensible. In this situation, the best estimate of $z(\mathbf{x})$ is the overall mean computed from all sample points in the region of interest without taking spatial dependence into account: interpolation is meaningless and a waste of time and money.

A noisy variogram, in which the experimentally derived semivariances are scattered, suggests that too few samples have been used to compute $\hat{\gamma}(\mathbf{h})$. A rule of thumb suggests that possibly at least 50–100 data points are necessary to achieve a stable variogram, depending on the kind of spatial variation encountered, though smooth surfaces require fewer points than those with irregular variation—smoother variograms can also be obtained by increasing the size of the lag.

The range of the variogram provides clear information about the size of the search window that should be used. If the distance from a data point to an unsampled point exceeds the range, then it is too far away to make any contribution; if all data points are further away than the range, the best estimate is the general mean. These distances can be modified by anisotropy, which modifies the shape of the search neighbourhood from a circle to an ellipse.

The presence of a **hole effect** in the experimental variogram (a dip in semivariance at distances greater than the range) may indicate a pseudo-periodic pattern caused by long-range variation over a study area that is too small to encompass the total range of variation. True periodicity will give a variogram with a periodic variation in the sill that matches the wavelength of the pattern, as long as the original field sampling is in step with the periodicity.

If the range is large, then long-range variation dominates: if it is small, then the major variation occurs over short distances. Anisotropy in the experimental variogram suggests a directional effect in pattern, but directional differences can occur if there are insufficient samples to get robust estimates in all directions.

A variogram that can be fitted by a Gaussian variogram model indicates a smoothly varying pattern, such as often occurs with elevation data. A variogram modelled by a spherical variogram model has a clear transition point which implies that one pattern is dominant. The choice of an exponential variogram model may suggest that the pattern of variation shows a gradual transition over a spread of ranges, or that several patterns are interfering.

9.4 Isotropic and anisotropic variation

In the foregoing, we have implied that the source data for the variogram are collected on regularly spaced transects, or possibly a regular grid. In many cases we only have irregularly spaced measurements, so it is useful to be able to compute the experimental variogram from such data. To do this we use a circular search radius, rather like the tyre of a bicycle, to define a zone whose midpoint is **h** from its centre. This wheel is placed over a data point and all data points falling inside the tyre are used to estimate the contribution of $(z_i - z_j)^2$ from all pairs (Figure 9.4). As a general rule of thumb, to avoid edge effects it is not usually sensible to compute a variogram for more lags than a total separation distance of half the dimensions of the study area. If we ignore directional effects, the resulting variogram is known as **isotropic**; it averages the variogram over all directions. However, as Figure 9.4 shows, it is easy to compute the variogram for specific directions β. These are known as **anisotropic** variograms, and if they are different in range or sill for different values of β they may indicate that the spatial variation varies with direction. This could happen in sediments perpendicular or parallel to a river, for example.

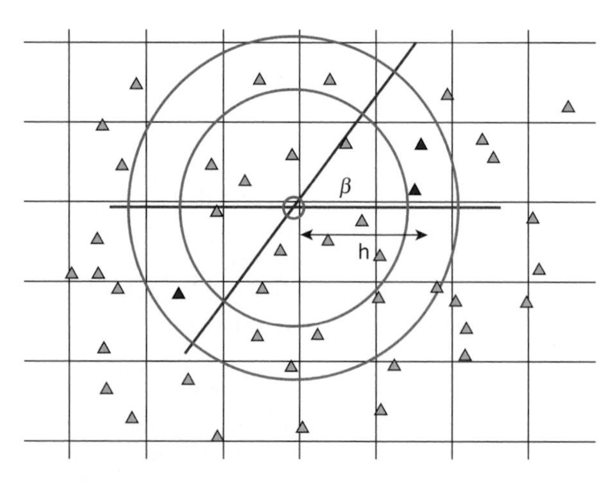

Figure 9.4 Circular search window for estimating semivariance in any direction β

By computing $(z_i - z_j)^2$ for all possible directions one creates a variogram surface, which can be plotted as an ellipsoid map with centre point 0,0. The **major axis** of the variogram is the longer axis of the ellipse (that is, the direction with the most continuous spatial variation), and its orientation is given by the angle of that axis with respect to north.

9.5 Variograms showing spatial variation at several scales

The original model of covariance simply assumes that the spatial variation of the attribute of interest can be divided into the three components modelled by (i) a general mean or trend, (ii) a variogram, and (iii) residual nugget. In some situations, particularly if there is sufficient data, it may be possible and useful to distinguish more than one variogram component, for example where two random patterns that have two or more widely separated ranges interfere with each other. In this case, the complex variogram $\gamma_T(\mathbf{h})$ can be split into several components according to the linear model of coregionalization (Isaaks and Srivastava 1989):

$$\gamma_T(\mathbf{h}) = \gamma_1(\mathbf{h}) + \gamma_2(\mathbf{h}) + \gamma_3(\mathbf{h}) + \ldots \qquad 9.10$$

so for two levels of nesting we replace equation 9.4 by:

$$\text{var}[\varepsilon'(\mathbf{x}) - \varepsilon'(\mathbf{x}+\mathbf{h})] = \gamma_1(\mathbf{h}) + \gamma_2(\mathbf{h}) + \ldots \qquad 9.11$$

Each of the subvariograms is defined by its own set of parameters.

9.6 Local variograms

The straightforward approach to computing the variogram is to assume that all data are located in the same cover class or domain. In this case the computed variogram is a global model that determines the interpolation weights over the whole area. But this need not necessarily be so: different parts of an area may have important differences in land cover, rock type, soil type, flooding frequency, or land tenure pattern, each of which has its own unique spatial correlation structure or pattern. For example, the pattern of distribution of heavy metals in river floodplains is likely to be quite different from that on the hillsides bordering the river, because the sources of heavy metals and the transport processes will be quite different.

When faced with the problem of important differences in domain it may be sensible to see if the study area can be better modelled by a set of domain-specific variograms rather than a single global model. The main problem is usually insufficient data: should one risk using an inappropriate, but well-defined global variogram, or several poorly defined local models? The pragmatic approach is to attempt to use local models whenever possible, because then for each spatial unit (polygon or mapping unit) the variogram for any given variable becomes a new, unique attribute of that polygon. Another approach is to estimate and automatically model the variogram for a moving window, so that there is a unique variogram for each window location (see Lloyd 2011 for a summary of such approaches).

9.7 Using the variogram for interpolation: ordinary kriging

Given that the spatially dependent random variations are not swamped by uncorrelated noise, the fitted variogram can be used to determine the weights λ_i needed for local interpolation. The procedure is similar to that used in weighted moving average interpolation except that now the weights are derived from a geostatistical analysis of the data rather than from a general, and possibly inappropriate, model. The 'true' value $z(\mathbf{x}_0)$ is given by:

$$\hat{z}(\mathbf{x}_0) = \sum_{i=1}^{n} \lambda_i \cdot z(\mathbf{x}_i) \qquad 9.12$$

with $\sum_{i=1}^{n} \lambda_i = 1$. The weights λ_i are chosen so that the estimate $\hat{z}(\mathbf{x}_0)$ is unbiased, and that the estimation variance σ_e^2 is less than for any other linear combination of the observed values.

The minimum variance of $[\hat{z}(\mathbf{x}_0) - z(\mathbf{x}_0)]$, the prediction error, or 'kriging variance' is given by:

$$\hat{\sigma}_e^2 = \sum_{i=1}^{n} \lambda_i \gamma(\mathbf{x}_i, \mathbf{x}_0) + \phi \qquad 9.13$$

and is obtained when

$$\sum_{i=1}^{n} \lambda_i \gamma(\mathbf{x}_i, \mathbf{x}_j) + \phi = \gamma(\mathbf{x}_j, \mathbf{x}_0) \text{ for all } j \qquad 9.14$$

The quantity $\gamma(\mathbf{x}_i,\mathbf{x}_j)$ is the semivariance of z between the sampling points $\mathbf{x}_i$ and $\mathbf{x}_j$; $\gamma(\mathbf{x}_j, \mathbf{x}_0)$ is the semivariance between the value at the sampling point $\mathbf{x}_i$ and the unvisited point $\mathbf{x}_0$. Both these quantities are obtained from the fitted variogram. The quantity ϕ is a Lagrange multiplier required for the minimalization. The method is known as **ordinary kriging** (OK): it is an exact interpolator in the sense that when the equations 9.12 to 9.14 are used, the interpolated values, or best local average, will coincide with the values at the data points. In mapping, values will be interpolated for points on a regular grid that is finer than the spacing used for sampling, if that was itself a grid. The interpolated values can then be converted to a contour map using the techniques already described. Similarly, the estimation error σ_e^2, known as the **kriging variance**, can also be mapped to give valuable information about the reliability of the interpolated values over the area of interest. Often the kriging variance is mapped as the **kriging standard deviation** (or kriging error), because this has the same units as the predictions; this convention is followed in this chapter. Worked example 9.2 and Figure 9.5 show how

Worked example 9.2 Computing kriging weights

Computing kriging weights for the unsampled point $z(x_{i=0})$ (x=5, y=5) in Figure 9.5

Let the spatial variation of the attribute sampled at the five points be modelled by a spherical variogram with parameters c_0=2.5, c_1=7.5, and range a=10.0. The data at the five sampled points are:

i	*x*	*y*	*z*
1	2	2	3
2	3	7	4
3	9	9	2
4	6	5	4
5	5	3	6

In matrix terms we have to solve the following:

$$A^{-1} \cdot b = \begin{bmatrix} \lambda \\ \phi \end{bmatrix}$$

where A is the matrix of semivariances between pairs of data points, b is the vector of semivariances between each data point and the point to be predicted, and λ is the vector of weights. ϕ is a Lagrangian for solving the equations.

Start by creating a distance matrix between the data points:

(continued)

Worked example 9.2 *(continued)*

i	*1*	*2*	*3*	*4*	*5*
1	0.0	5.099	9.899	5.0	3.162
2	5.099	0.0	6.325	3.606	4.472
3	9.899	6.325	0.0	5.0	7.211
4	5.0	3.606	5.0	0.0	2.236
5	3.162	4.472	7.211	2.236	0.0

and the vector of distances between the data points and the unknown site:

i	0
1	4.243
2	2.828
3	5.657
4	1.0
5	2.0

Substitute these numbers into the variogram to get the corresponding semivariances (matrices *A* and *b*).

A=*i*	1	2	3	4	5	6
1	0	7.739	9.999	7.656	5.939	1.000
2	7.739	0	8.667	6.381	7.196	1.000
3	9.999	8.667	0	7.656	9.206	1.000
4	7.656	6.381	7.656	0	4.936	1.000
5	5.939	7.196	9.206	4.936	0	1.000
6	1.000	1.000	1.000	1.000	1.000	0

Note the extra row and column (*i*=6) to ensure that the weights sum to 1.

b=*i*	0
1	7.151
2	5.597
3	8.815
4	3.621
5	4.720
6	1.000

$\boldsymbol{A^{-1}}$=*i*	1	2	3	4	5	6
1	−0.106	0.028	0.015	0.008	0.054	0.247
2	0.028	−0.108	0.024	0.038	0.018	0.201
3	0.015	0.024	−0.081	0.034	0.008	0.312
4	0.008	0.038	0.034	−0.148	0.068	0.097
5	0.054	0.018	0.008	0.068	−0.148	0.143
6	0.247	0.201	0.312	0.097	0.143	-6.267

with λ *as: i*	*weights*		*distances*
1	.0662		4.243
2	.2321		2.828
3	.0019	Σ=1	5.657
4	.4533		1.000
5	.2465		2.000
6	.4017	ϕ	

Worked example 9.2 *(continued)*

Therefore the value at $z(x_{i=0}) = .0662 * 3 + .2321 * 4 + .0019 * 2 + .4533 * 4 + .2465 * 6$

$$= \mathbf{4.423}$$

with estimation variance $\sigma_e^2 = \begin{bmatrix} .0662 * 7.151 + .2321 * 5.597 + .0019 * 8.815 \\ +.4533 * 3.621 + .2465 * 4.720 \end{bmatrix} + \varphi$

$$= 4.5942 + 0.4017$$

$$= \mathbf{4.9959}$$

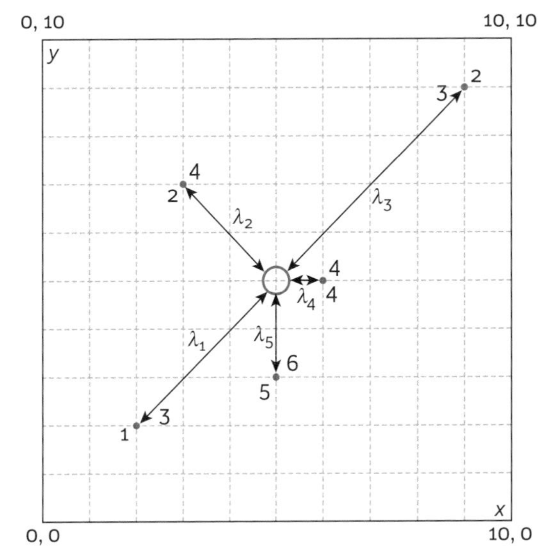

Figure 9.5 Simple example to illustrate the prediction of z at unsampled sites by ordinary kriging

the equations for ordinary kriging are set up and solved using the semivariances. Note that if $\gamma(\mathbf{h}) = 0$, A cannot be inverted, so most geostatisticians solve the equations using the covariances instead (Isaaks and Srivastava 1989: 292).

Example: Figure 9.6 presents the results of using ordinary kriging for the untransformed zinc data. The map of prediction variances is also shown. Compare the prediction map with the inverse distance maps and spline map (Figure 8.10): it is smoother than these, but also avoids the local large values that are linked to isolated data points. The standard deviation map clearly shows how prediction variance is linked to the density of the data points. Though comparisons cannot strictly be made because variance analysis and kriging use different statistical models, note that the maximum kriging standard deviation of 458 ppm is similar to the maximum within-class standard deviation of class 1 of the flooding frequency map (Figure 8.2), but that this maximum only occurs in the kriging map in areas where there are no data points. In most of the areas covered by class 1 (standard deviation 423 ppm) the kriging prediction standard error is about 140 ppm. These results suggest that kriging prediction that takes account of gradual spatial change appears to be much better than that given by global, crisp, choropleth mapping.

9.8 Using kriging to validate the variogram model

Cross-validation is the practice of using the kriging equations retrospectively to assess the appropriateness of the variogram model and estimation neighbourhood. As noted in Section 8.15, it involves computing the moments of the distribution of residuals $(\hat{z}(\mathbf{x}_i) - z(\mathbf{x}_i))$ for all data points, when each data point is successively left out and predicted from (a subset of) the rest of the data (Figure 9.7). The procedure can be used to test the variogram for self-consistency and lack of bias—indicated by a mean difference of residuals of zero and a variance of 1. Cross-validation is used in Section 9.15 in an assessment of alternative kriging approaches.

9.9 Block kriging

Clearly, kriging fulfils the aims of finding better ways to determine interpolation weights and of providing information about errors. The resulting map of interpolated values may not be exactly what is desired, however, because the point kriging, or punctual kriging, equations (9.13 and 9.14) imply that all interpolated values relate to the support, which is the area or volume of an original sample. Very often, as in sampling for soil or water quality, this sample is only a few

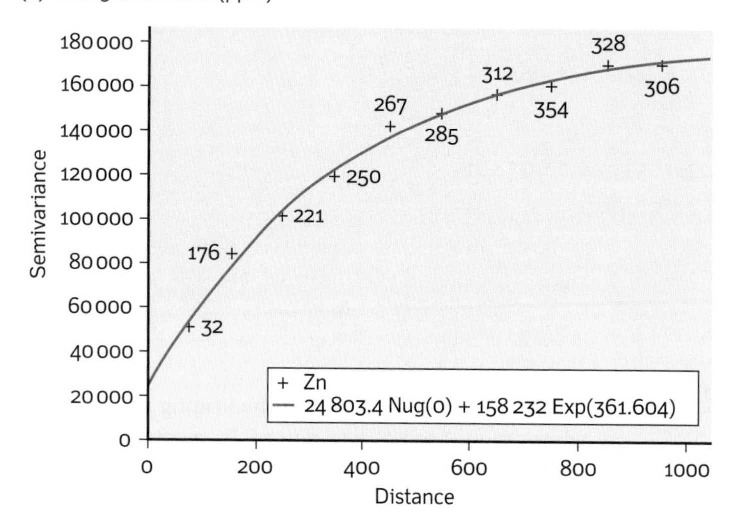

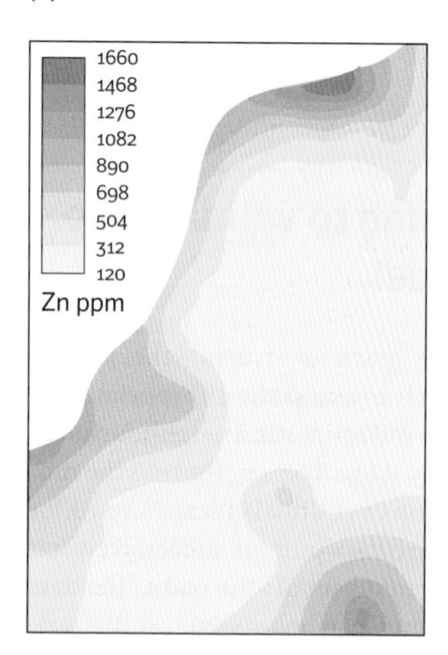

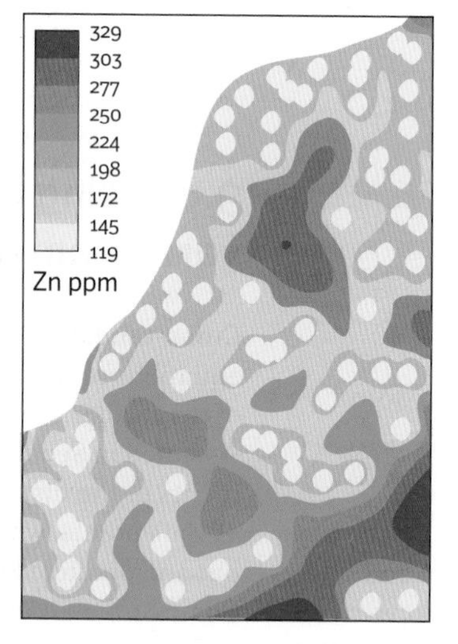

Figure 9.6 Results of interpolating the Maas data set using ordinary kriging: (a) variogram with fitted exponential model; (b) kriging predictions of zinc levels; (c) kriging standard deviation

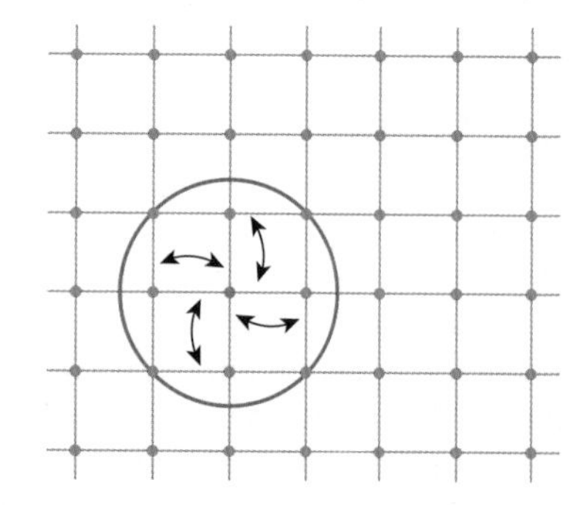

Figure 9.7 Cross-validation is the process of assessing the variogram and estimation neighbourhood using the original data

centimetres across. Given the often large amplitude, short-range variation of many natural phenomena like soil or water quality, ordinary point kriging may result in maps that have many sharp spikes or pits at the data points. This can be overcome by modifying the kriging equations to estimate an average value $z(B)$ of the variable z over a block of land B. This is useful when one wishes to estimate average values of z for experimental plots of a given area, or to interpolate values for grid cells of a specific size for quantitative modelling (Figure 9.8).

The average value of z over a block B, given by

$$z(B) = \int_b \frac{z(\mathbf{x})d\mathbf{x}}{\text{area}B} \qquad 9.15$$

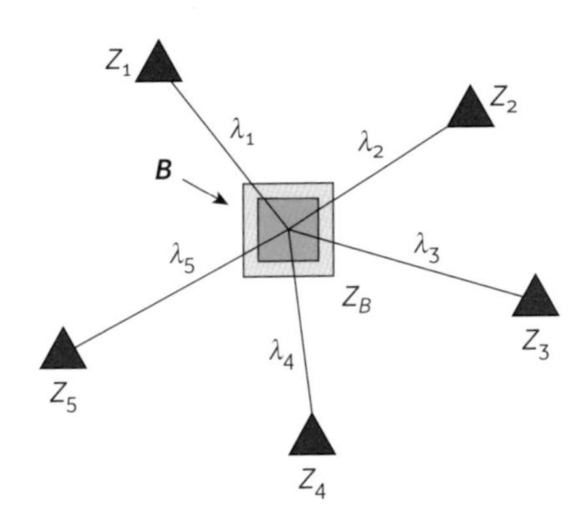

Figure 9.8 Predicting z for blocks of different size

is estimated by

$$\hat{z}(B)=\sum_{i=1}^{n}\lambda_i\cdot z(\mathbf{x}_i) \quad 9.16$$

with $\sum_{i=1}^{n}\lambda_i=1$, as before.

The minimum variance is now

$$\hat{\sigma}^2(\mathbf{B})=\sum_{i=1}^{n}\lambda_i\bar{\gamma}(\mathbf{x}_i,\mathbf{B})+\varphi-\bar{\gamma}(\mathbf{B},\mathbf{B}) \quad 9.17$$

and is obtained when

$$\sum_{i=1}^{n}\lambda_i\gamma(\mathbf{x}_i,\mathbf{x}_j)+\varphi=\bar{\gamma}(\mathbf{x}_j,\mathbf{B}) \quad \text{for all } j \quad 9.18$$

where $\bar{\gamma}(\mathbf{x}_i,\mathbf{B})$ is the point-to-block semivariance and $\bar{\gamma}(\mathbf{B},\mathbf{B})$ is the within-block semivariance. Estimation variances obtained for block kriging are usually substantially less than for point kriging. When these equations are used, the resulting smoothed interpolated surface is free from the pits and spikes that result from point kriging. Lloyd (2014) introduces methods for kriging from blocks to points.

9.10 Other forms of kriging

Simple kriging

Simple kriging is prediction by generalized linear regression under the assumption of second order stationarity with a known mean (Webster and Oliver 2007; Chilès and Delfiner 2012). Because the assumption of second order stationarity is often too restrictive for most data from the physical environment, ordinary kriging (no a priori mean) is most often used, though.

Non-linear kriging

Log-normal kriging is the interpolation of log-normally distributed, rather than normally distributed, data (Webster and Oliver 2007). The data are first transformed to natural logarithms or base-10 logarithms so that variogram modelling and interpolation proceeds with the transform $y(\mathbf{x})=\ln(z(\mathbf{x}))$ to give an estimate $\hat{y}(\mathbf{x})$ for $\ln(z(\mathbf{x}))$. The predicted values can be transformed back after interpolation, but care must be exercised because the antilog back-transform $e^{y*(u)}$ is a biased estimator of $Z(\mathbf{x})$. Deutsch and Journel (1998) recommend using an unbiased back-transform

$$z^*(\mathbf{x})=\exp\left[y^*(\mathbf{x})+\frac{\sigma_{sk}^2(\mathbf{x})}{2}\right] \quad 9.19$$

where $\sigma_{sk}^2(\mathbf{x})$ is the simple log-normal kriging variance. Deutsch and Journel (1998) point out that the extreme sensitivity of the errors for antilog back-transformation make log-normal kriging difficult to use. They recommend multi-Gaussian kriging (MG) or indicator kriging (see Section 9.12) as alternatives. Nevertheless, the log-normal transform is useful for many physical data with positive skew distributions, such as soil chemical and physical attributes (Webster and Oliver 2007).

Figure 9.9a presents the experimental variogram of the zinc data using the log-transformed data, and Figure 9.10a and b show the maps of predictions and standard deviations of the log-transforms. The general patterns are similar to those in Figure 9.6.

Ordinary kriging with anisotropy or nested variograms

Incorporating anisotropy into the ordinary kriging procedure is simply a matter of modifying the conversion of the distance matrix into the matrix of semivariances A (see Worked example 9.2), taking into account the variation of semivariance with direction. Figure 9.11 presents the variograms computed in a NW–SE direction (perpendicular to the river) and in a NE–SW direction (parallel to the river). Note the double variogram fitted to the major axis (NE–SW), which is an attempt to force both variograms to have the same nugget. Figure 9.12

shows the effect of incorporating anisotropy into the interpolation, producing long, linear streaks parallel to the river. The results are plausible, but in spite of the clear variograms they look rather contrived compared to the other maps.

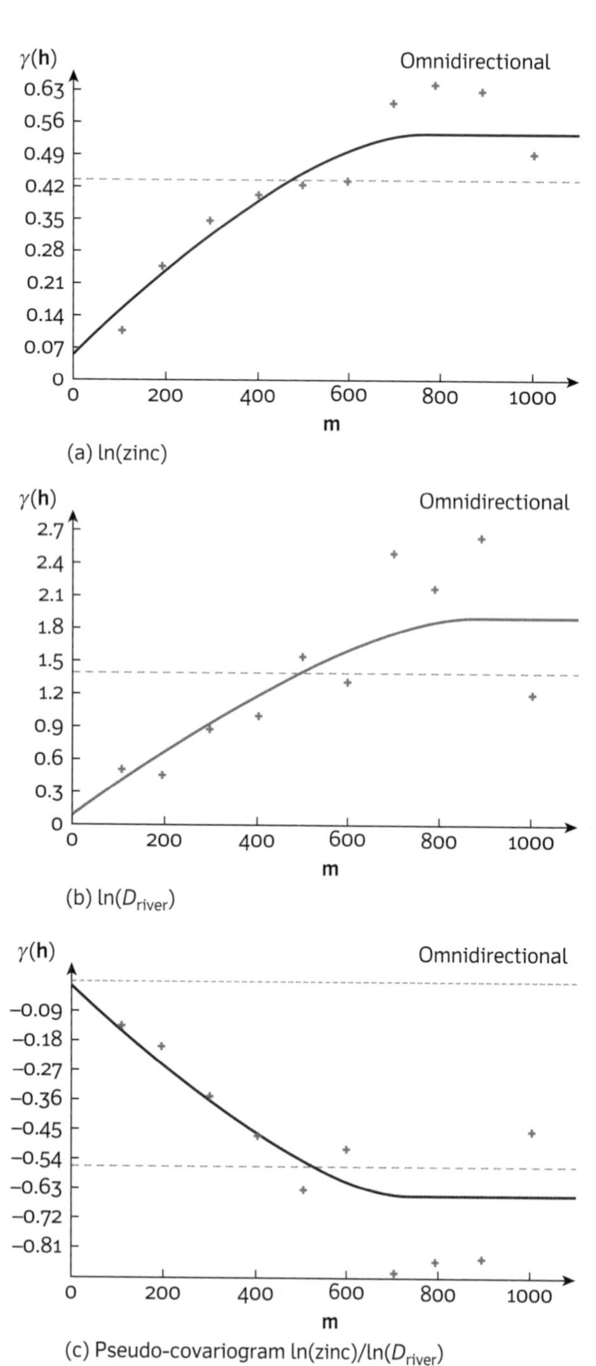

Figure 9.9 Variograms for (a) ln(zinc), (b) ln(distance to river), and (c) pseudo cross-variogram

9.11 Kriging using extra information

Frequently, the data points are not the only source of information about the distribution of z, and we may be able to draw on other knowledge that can help with interpolation. The main sources are: (a) an appropriate stratification into clearly different domains, (b) data from a cheap-to-measure co-variable that has been sampled at many more data points, and (c) a physical or empirical spatial model of a driving process.

Stratified kriging

When there is enough soft information to classify the area into meaningful subareas and there are enough data to compute variograms for each different domain, the interpolation can be carried out on points or blocks for each area separately, adjusting the kriging equations to avoid discontinuities at class boundaries. The analysis of variance of zinc concentration according to flood frequency classes (Chapter 8; Figure 8.2) suggested that there were two classes of importance: flood class 1 (annual flooding) and the rest. Figure 9.13 presents the variograms computed for these two classes. It shows that both units have strongly correlated spatial variation, but that the variance in flood class 1 is almost ten times as great as in the less frequently flooded areas. Figure 9.14 shows the interpolations which show how the zinc concentrations vary within the classes. Comparing the stratified kriging standard deviation map with Figure 9.6a and b (ordinary point kriging) suggests that the stratification reduces the interpolation uncertainty. In this case, stratification also preserves the fine spatial structure of the narrow flooded channels, which is smaller than the average spacing between data points and is therefore lost by interpolating from the 'hard' point data alone.

Cokriging

Often data may be available for more than one attribute per sampled location. One set (U) may be expensive to measure and is therefore sampled infrequently, while another (V) may be cheap to measure and has more observations. If U and V are spatially correlated, then it may be possible to use the information about the spatial variation of V to help map U. This can be done by using an extension of the normal kriging technique known as cokriging. Cokriging can produce predictions for both points and

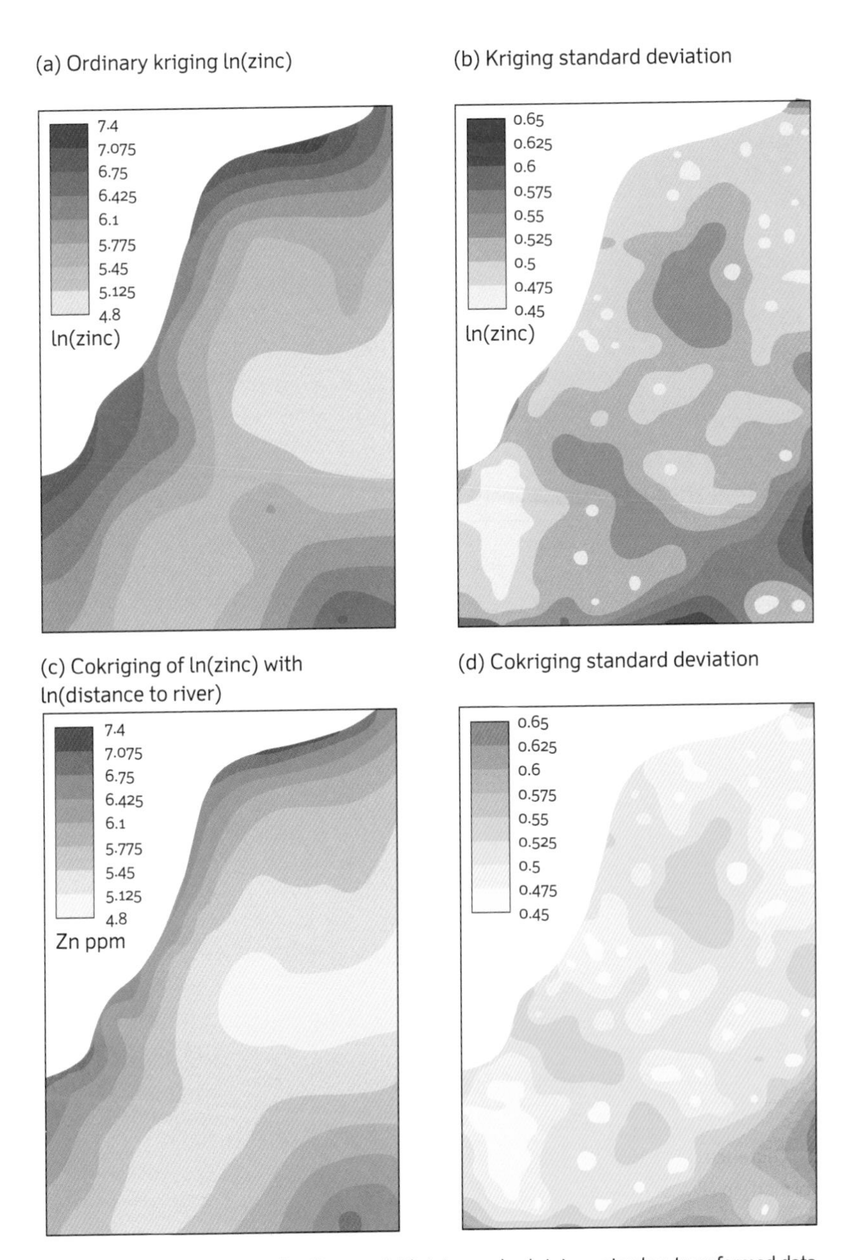

Figure 9.10 A comparison of ordinary point kriging and cokriging using log-transformed data

blocks in analogy to ordinary kriging (Goovaerts 1997; Webster and Oliver 2007).

Besides the variograms describing the non-structural variation of both U and V, cokriging needs information on the *joint* spatial covariation of the two variables. For any pair of variables U and V, the cross-semivariance $\gamma_{UV}(\mathbf{h})$ at lag $\mathbf{h}$ is defined as:

$$2\gamma_{UV}(\mathbf{h}) = E[\{Z_U(\mathbf{x}) - Z_U(\mathbf{x}+\mathbf{h})\}\{Z_V(\mathbf{x}) - Z_V(\mathbf{x}+\mathbf{h})\}] \quad 9.20$$

where Z_U, Z_V are the values of U, V at places $\mathbf{x}, \mathbf{x}+\mathbf{h}$.

The cross-variogram is estimated directly from the sample data using:

$$\hat{\gamma}_{UV}(\mathbf{h}) = 1/2n(\mathbf{h}) \sum_{i=1}^{n(h)} \{Z_U(\mathbf{x}_i) - Z_U(\mathbf{x}_i+\mathbf{h})\}\{Z_V(\mathbf{x}_i) - Z_V(\mathbf{x}_i+\mathbf{h})\} \quad 9.21$$

where $n(\mathbf{h})$ is the number of data pairs of observations of Z_U, Z_V at locations $\mathbf{x}_i$, $\mathbf{x}_i + \mathbf{h}$ for the distance vector $\mathbf{h}$. Cross-variograms can increase or decrease with $\boldsymbol{h}$ depending on the correlation between U and V. When

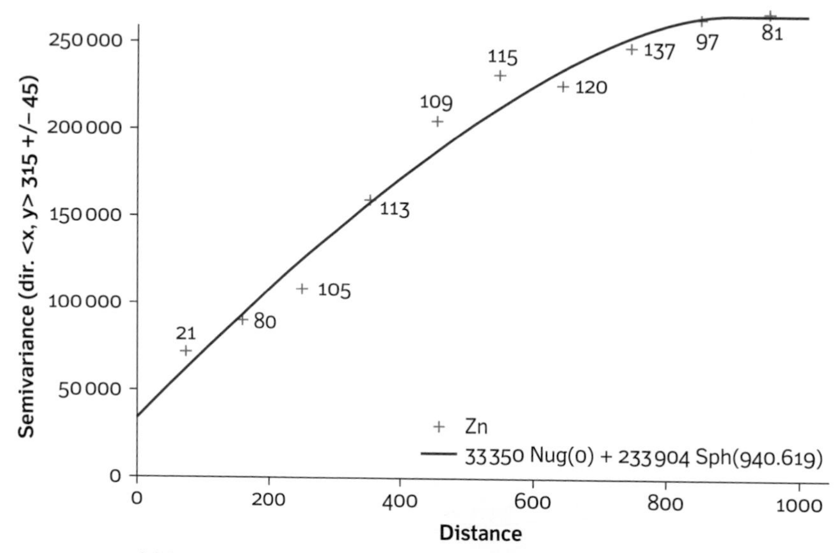

(a) Variogram of zinc (untransformed) in NW–SE direction

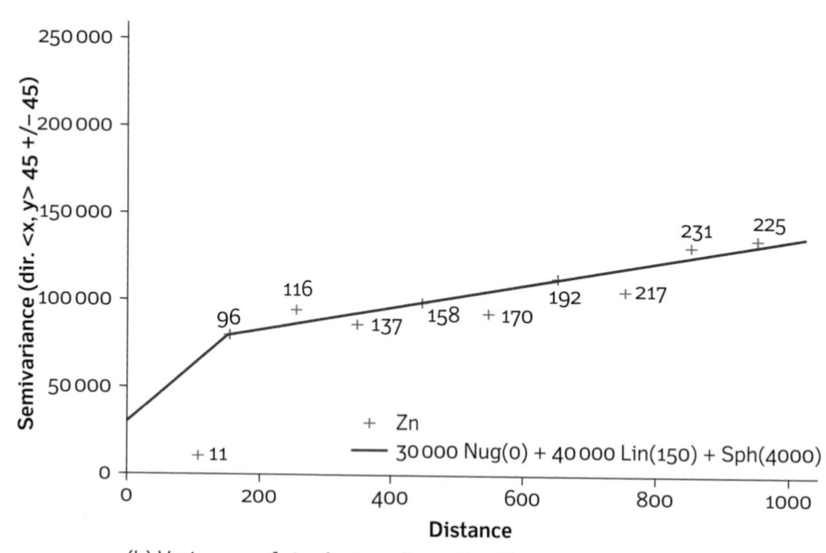

(b) Variogram of zinc (untransformed) in NW–SE direction

Figure 9.11 Computing variograms in different directions: (a) NW–SE, (b) NE–SW
Parameters of the fitted spherical model: $c_0=33\,350$, $c_1=233\,909$, $a=940.6$
Parameters of the fitted double model: $c_0=30\,000$
Linear $c_2=40\,000$, $a_1=150$
Spherical $c_2=177\,250$, $a_2=4000$

cross-variograms are fitted, the Cauchy–Schwarz relation:

$$|\gamma_{UV}(\mathbf{h})| \le \sqrt{(\gamma_U(\mathbf{h})^* \gamma_V(\mathbf{h}))} \ \textit{for all}\ \mathbf{h} > 0 \qquad 9.22$$

must be checked to guarantee a positive cokriging estimation variance in all circumstances (Chilès and Delfiner 2012).

A cokriged estimate is a weighted average in which the value of U at location $\mathbf{x}_o$ is estimated as a linear weighted sum of co-variables V_k. If there are k variables, $k=1, 2, 3, \ldots$ V, and each is measured at n_V places, $\mathbf{x}_{ik}=1, 2, 3, \ldots N_k$, then the value of one variable U at $\mathbf{x}_o$ is predicted by:

$$\hat{z}_U(\mathbf{x}_0) = \sum_{k=1}^{V}\sum_{i=1}^{n_v} \lambda_{ik} z(\mathbf{x}_{ik}) \ \textit{for all}\ V_k \qquad 9.23$$

To avoid bias—i.e. to ensure that $E[z_U(\mathbf{x}_0) - \hat{z}_U(\mathbf{x}_0)] = 0$ —the weights λ_{ik} must sum as follows:

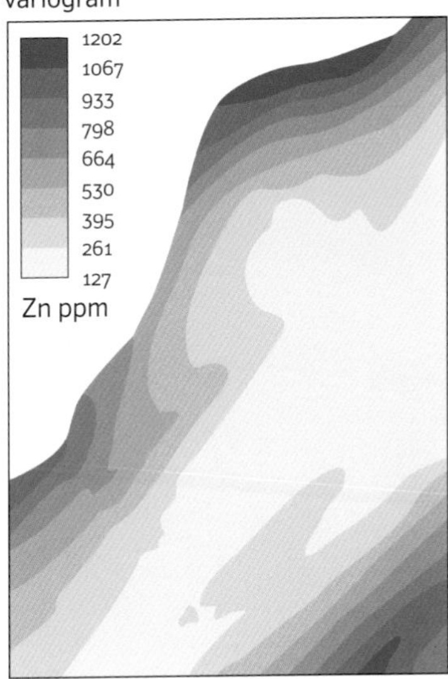

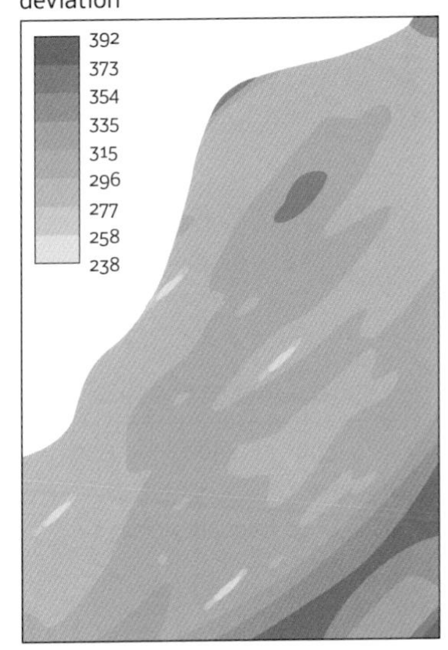

Figure 9.12 Results of ordinary point kriging of untransformed zinc data using anisotropic variograms

$$\sum_{i=1}^{n_v} \lambda_{ik} = 1 \text{ for } V = U \text{ and}$$
$$\sum_{i=1}^{n_v} \lambda_{ik} = 0 \text{ for all } V_k \neq U \qquad 9.24$$

The first condition implies that there must be at least one observation of U for cokriging to be possible. The interpolation weights are chosen to minimize the variance:

$$\sigma_{\hat{U}}^2(\mathbf{x}_0) = E[\{z_U(\mathbf{x}_0) - \hat{z}_U(\mathbf{x}_0)\}^2] \qquad 9.25$$

There is one equation for each combination of sampling site and attribute, so to estimate the value of variable j at site $\mathbf{x}_0$ the equation for the g-th observation site of the k-th variable is:

$$\sum_{j=1}^{V} \sum_{i=1}^{n_v} \lambda_{ij} \gamma_{ij}(\mathbf{x}_{ij}, \mathbf{x}_{gk}) + \varphi_k = \gamma_{UV}(\mathbf{x}_0, \mathbf{x}_{gk}) \qquad 9.26$$

for all $g=1$ to n_v and all $k=1$ to V, where ϕ_k is a Lagrange multiplier. These equations together make up the cokriging system.

Cokriging only has an advantage if the cheap-to-measure attributes and the expensive attribute are measured at different numbers of data points. If both (or more) attributes are all measured at the same points then the cokriging will not give estimates that are different from ordinary kriging.

If the cross-variogram is computed between variables measured at the same set of observation points it is called a true cross-variogram. If it is computed for pairs of attributes measured on observations at different locations then it is known as the **pseudo cross-variogram** because an extra assumption has been made: that both sets of sample points belong to comparable realizations of the regionalized variables.

Example. In Chapter 8, equation 8.9 and Figure 8.5 we noted a strong linear correlation between ln(zinc) levels and ln(distance to river). For illustration, assume that we have 98 measurements of 'distance to river' but only 49 of zinc. Figure 9.9 shows the variograms for ln(zinc) and ln(distance to river) and the cross-variogram between both variables. Figure 9.10c shows the point cokriging of ln(zinc) based on the 49 observations for zinc, aided by the 98 of ln(distance to river). Comparison of Figure 9.10d (cokriging standard deviation) with Figure 9.10b (standard deviation for ln(zinc) based on 98 data points) shows that the interpolation error of the cokriged map (on the logarithmic scale) is approximately the same as that of the point kriging of ln(zinc) alone. This illustrates the potential of cokriging to save on expensive laboratory analyses if there is strong spatial correlation with a cheap-to-measure co-variable.

Goovaerts (1997), Webster and Oliver (2007), and Chilès and Delfiner (2012) provide reviews and examples of cokriging applications. Cokriging is most successful when the patterns of the variables used are related by a common physical process. Dungan (1998) provides an example of the application of cokriging to mapping vegetation quantities using ground and image data.

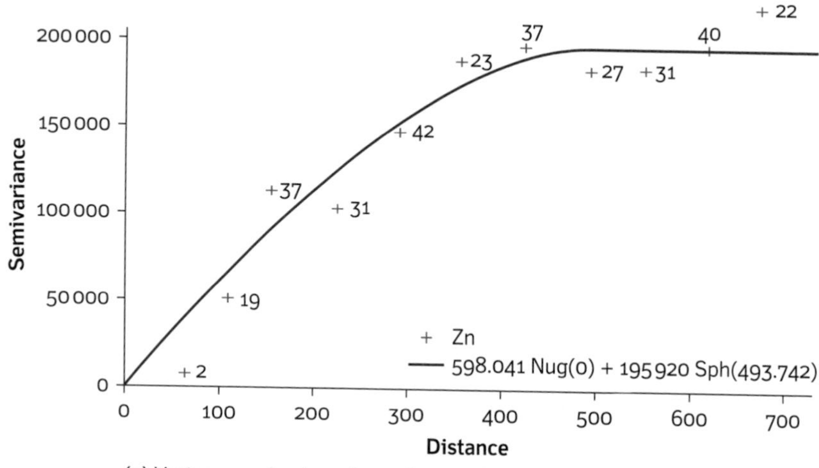

(a) Variogram of untransformed zinc in flood frequency class 1

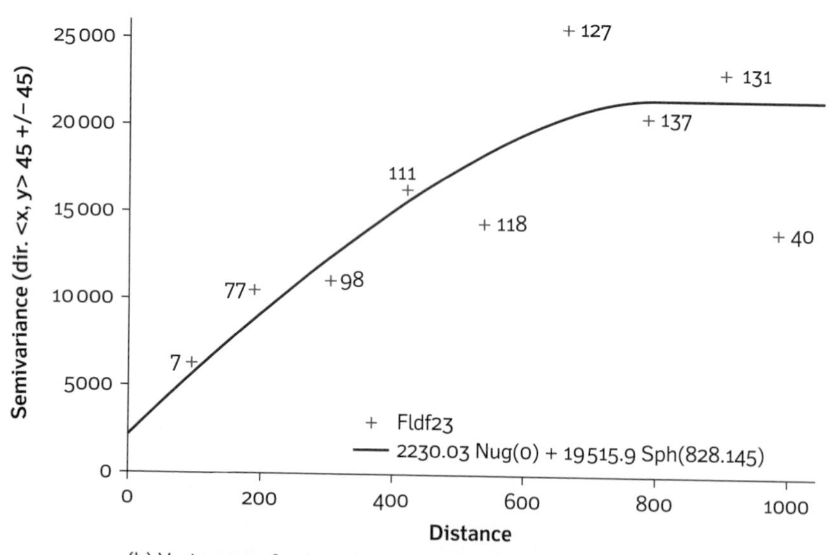

(b) Variogram of untransformed zinc in flood frequency class 2 and 3

Figure 9.13 Computing the variograms of zinc for flood frequency zones separately: (a) variogram for flood frequency class 1; (b) variogram for flood frequency classes 2 and 3—note that the sill for class 1 is nearly ten times higher than for classes 2 and 3

Universal kriging—interpolation with a built-in trend

Cokriging is a difficult technique to carry out, requiring considerable insight into the theory of geostatistics, and some authorities feel that the effort in fitting cross-variograms that obey the limitations of the intrinsic hypothesis and the linear model of co-regionalization is not worthwhile. Another approach is to use information on local trends in the data as a part of the kriging process. Isaaks and Srivastava (1989) point out that the term $m(\mathbf{x})$ in equation 9.1, where regionalized variable Z at $\mathbf{x}$ given by

$$z(\mathbf{x}) = m(\mathbf{x}) + \varepsilon'(\mathbf{x}) + \varepsilon'' \qquad 9.27$$

can be modelled by a deterministic trend, such as a regression equation that is incorporated in the kriging equations, so that

$$z(\mathbf{x}) = \sum_{j=1}^{p} f_j(\mathbf{x})\beta_j + \varepsilon'(\mathbf{x}) + \varepsilon'' \qquad 9.28$$

where $\gamma'(\mathbf{h})$ is the variogram of the residuals from the trend. This model is referred to as **universal kriging**, or kriging with a trend model—here, the local mean is a function of the coordinates (see Webster and Oliver 2007).

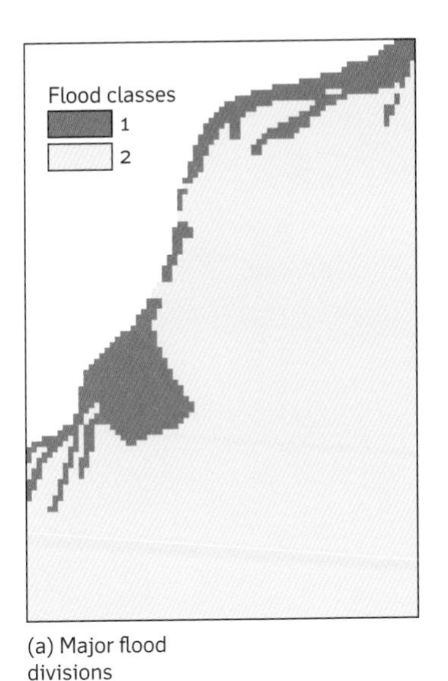

(a) Major flood divisions

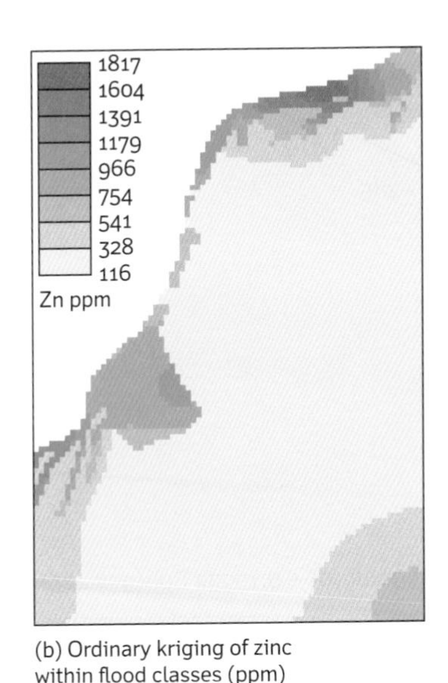

(b) Ordinary kriging of zinc within flood classes (ppm)

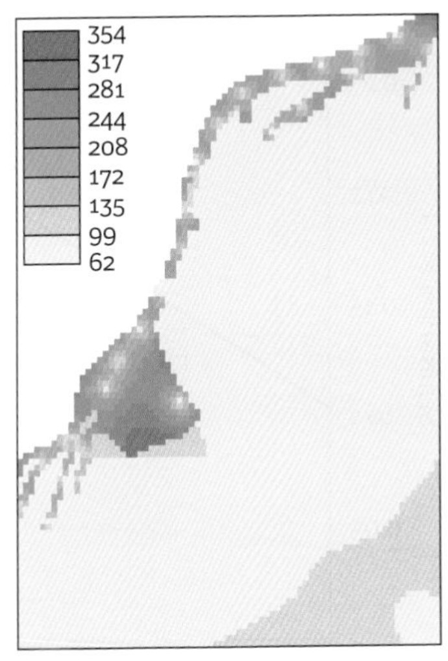

(c) Kriging standard deviation within flood classes

Figure 9.14 Prediction of zinc levels using the stratified variograms in Figure 9.13

Deriving the mean with secondary data

With universal kriging the mean is estimated using a local trend model which is derived as a part of the kriging system. The term *regression kriging* (see, for example, Hengl et al. 2007; Meng et al. 2013) is sometimes used to refer to kriging in which the deterministic component is modelled using a separate regression model and the residuals are used to derive the variograms and for kriging, with the trend model being added back to the estimate afterwards. Simple kriging with a locally varying mean (SKlm) refers to simple kriging with the local mean determined using secondary data. Kriging with an external drift (KED) is equivalent to universal kriging (kriging with a trend model), but the local trend model is a function of secondary data, rather than the data coordinates (Goovaerts 1997; Deutsch and Journel 1998; Lloyd 2011).

Multivariate kriging

Multivariate kriging is the application of geostatistics to multivariate transformations, such as the results of regression models, principal component transformations, reciprocal averaging, or fuzzy *k*-means. Sometimes constraints must be applied—e.g. the sum of all fuzzy *k*-means values must equal 1—so the kriging equations need to be modified.

9.12 Probabilistic kriging

Given the wide variations in results among the different kriging and deterministic interpolation methods, some users may decide that for their purposes they are not really interested in the best estimates of $z(\mathbf{x}_0)$, but only in the **probability** that the value of the attribute in question exceeds a certain threshold. For them, the probability of a threshold being exceeded may be sufficient to support a decision to mine, to commission operations to clean up polluted soil, or to set up a marketing base. **Indicator kriging** is a non-linear form of ordinary kriging in which the original data are transformed from a continuous scale to a binary scale (1 if $z \leq T_j$, and 0 otherwise, or vice versa). Variograms are computed for the binary data in the usual way, and ordinary kriging proceeds with the transformed data. The resulting maps display continuous data in the range 0–1, indicating the probability that T_j has been exceeded or not exceeded, as the case may be. Computing variograms and interpolating for other thresholds provides insight into the way the probabilities of threshold exceedance vary with threshold levels.

Figure 9.15 presents indicator variograms for cut-offs in zinc levels at 500 ppm and 1000 ppm measured on all 98 points. The resulting maps (Figure 9.16a and b) show the *probability that the zinc level exceeds the given threshold*. The results are clear and easy to understand, and could easily be incorporated in a GIS database for decision-making where the managers do not want to have to understand the intricacies of kriging.

Indicator kriging can easily be combined with soft information, such as the flood frequency classes: Figures 9.16c and d show the probabilities of zinc levels exceeding 500 ppm within the areas defined by the flood frequency classes, and these figures provide some interesting insights into the process of flooding and contamination around frequently flooded creeks.

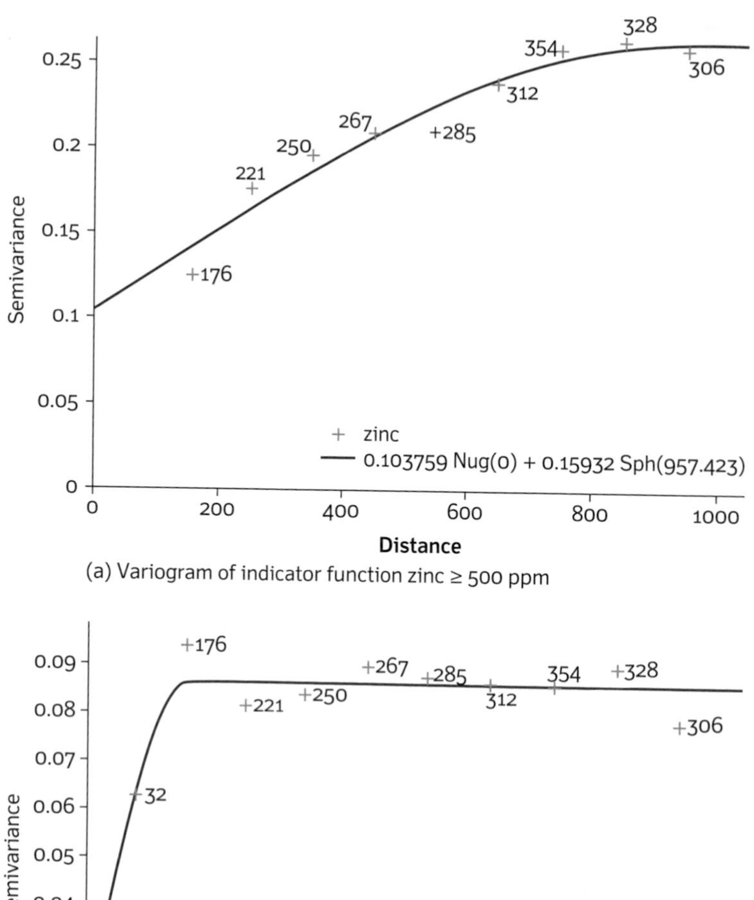

(a) Variogram of indicator function zinc ≥ 500 ppm

(b) Variogram of indicator function of zinc ≥ 1000 ppm

Figure 9.15 Indicator variograms of zinc: (a) for > 500 ppm level, (b) for > 1000 ppm level

There are several published examples of indicator kriging and indicator cokriging in soil science, soil pollution, and geology (see, for example, Goovaerts 1997). If the original data do not follow any simple distribution (multimodal Gaussian or log-normal), **disjunctive kriging** provides a non-linear, distribution-dependent estimator (Rivoirard 1994; Jaber et al. 2013). When the data are not normally distributed, they are transformed to normality by a linear combination of Hermite polynomials. The method is computationally demanding, but may be useful for attributes with unusual distributions, such as can be found in pollution studies (e.g. Atkinson and Lloyd 2001).

9.13 Simulation

Unfortunately, very few earth science processes are understood well enough to permit the application of deterministic models. Though we know the physics and chemistry of many fundamental processes, the variables of interest are the end result of a vast number of processes

(a) Indicator kriging: probability of zinc >500 ppm

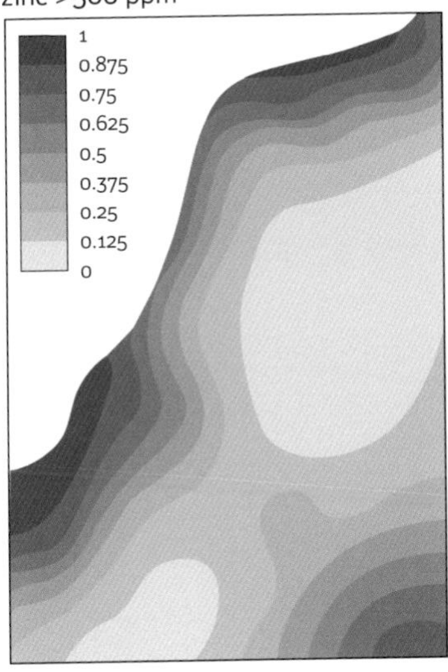

(b) Indicator kriging: probability of zinc >1000 ppm

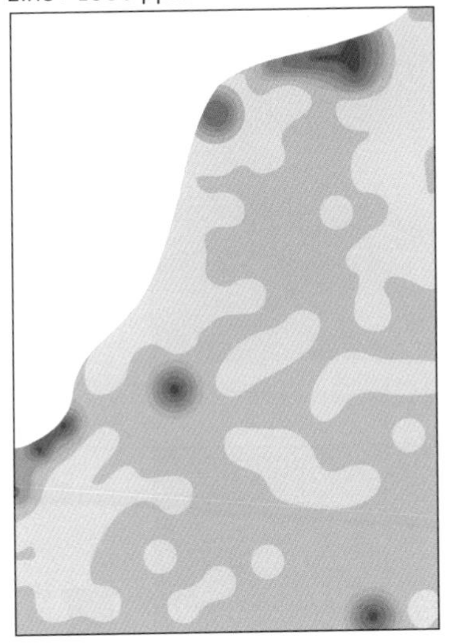

(c) Flood frequency classes

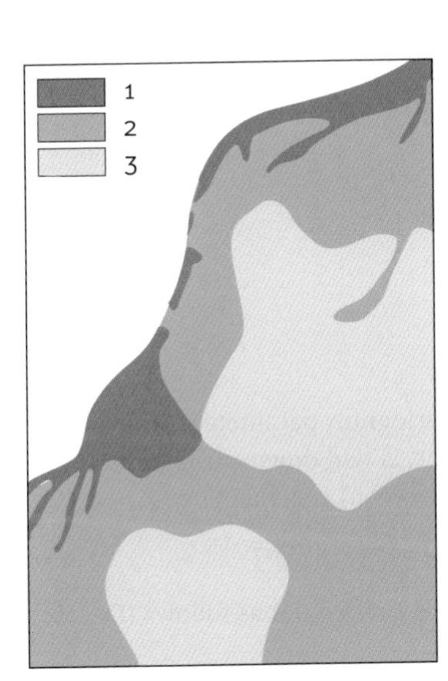

(d) Probability of zinc >500 ppm using indicator kriging with stratification according to flood frequency

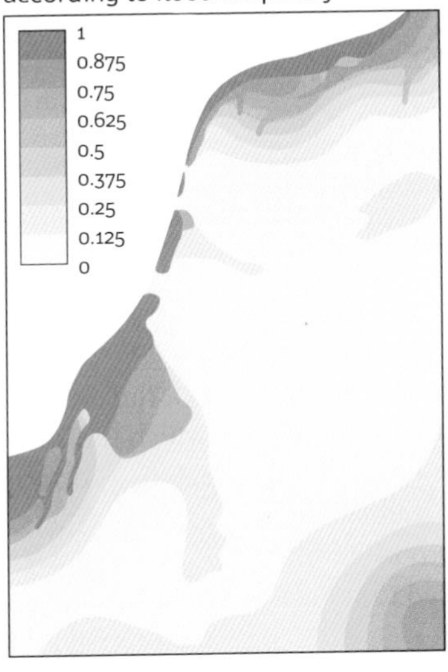

Figure 9.16 (a) Indicator kriging 500 ppm level; (b) indicator kriging 1000 ppm level; (c) flood frequency classes; (d) indicator kriging with stratification in flood frequency classes 1 and 2+3 (500 ppm)

which we cannot describe quantitatively (Isaaks and Srivastava 1989). For many GIS applications, the user only needs to interpolate point data so that they can be displayed or combined simply with other data. However, GIS are often used to provide data for quantitative models of environmental processes, such as climate change, groundwater pollution, surface run-off, and the diffusion of plants, animals, or people. For some models only the best prediction per cell is needed; for others we need to have an idea of how the model reacts to the whole range of variation that is possible in every cell. As we cannot measure data values for every cell we must resort to **stochastic simulation**. Smooth interpolation may not be what we want—we may want to generate a rough, realistic surface or volume!

Kriging interpolation methods provide each cell with a local, optimal prediction and a standard deviation that depends on the variogram and the spatial configuration of the data. To examine how a given numerical model might react to input values larger or smaller than the local mean, we might compute upper and lower bounding surfaces from the original interpolation and its standard deviations. These surfaces, however, only

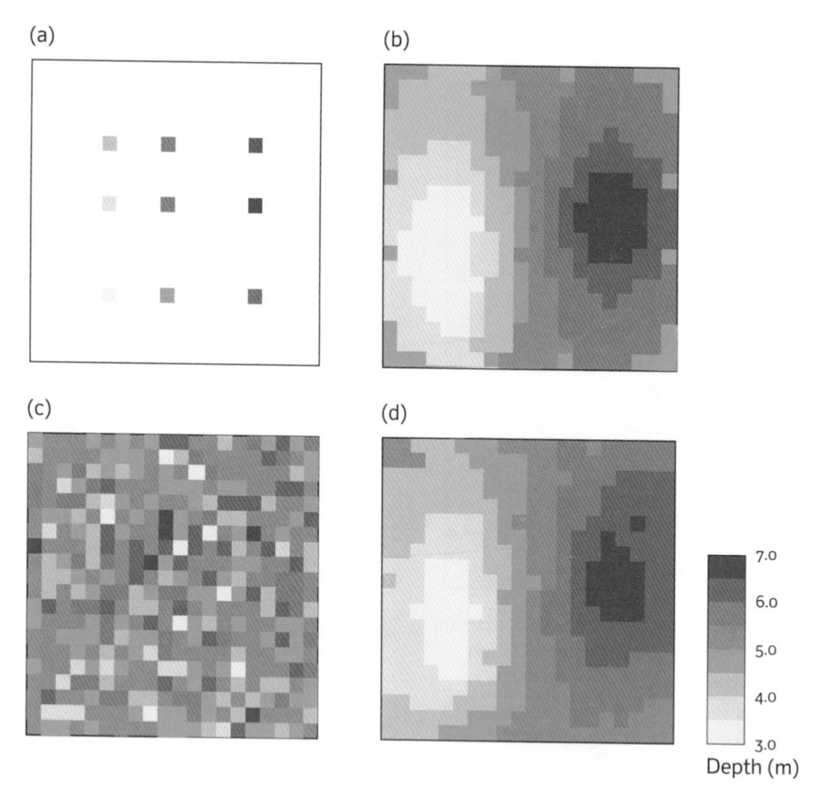

Figure 9.17 Monte Carlo simulation: (a) original data; (b) ordinary kriging map; (c) mean of 100 realizations of a stationary noise field having the same mean and variance as the data; (d) mean of 100 realizations of a simulation conditioned to pass through the data points and with the same variogram as the original data

provide smooth estimates of possible values; for evaluating the sensitivity of models in which spatial contiguity is important (groundwater models, erosion, run-off, diffusion models), it is much better to have error data in which each cell has a unique and an equiprobable value.

Simple Monte Carlo methods can be used to simulate values for each cell individually:

$$z(\mathbf{x}) = \Pr(Z) \qquad 9.29$$

where Z is a spatially independent, normally distributed probability distribution function (PDF) with mean μ and variance σ^2, but this treats every cell as spatially independent so the result is a stationary noise surface (Figure 9.17c).

If the variogram is known, then several methods such as the nearest neighbour method or the turning bands method (Deutsch and Journel 1998) can be used to simulate a surface that has similar spatial characteristics (nugget, sill, range, mean) to the original surface, but which does not match the patterns of the sampled area. **Conditional simulation**, otherwise known as stochastic imaging (Chilès and Delfiner 2012), combines the data at the observation points with the information from the variogram to compute the most likely outcomes per cell as a function of the variogram parameters. The surface (which is discontinuous, is tied down at the observation points, and varies in between) is defined by:

$$z(\mathbf{x}) = \Pr(Z), \gamma(\mathbf{h}) \qquad 9.30$$

Stochastic imaging is carried out as follows (Deutsch and Journel 1998):

1. Set up the usual equations for simple kriging with an overall mean.
2. Select an unsampled data point at random. Compute Kriging prediction and standard deviation using data from the data set in the locality of the cell.
3. Draw a random value from the probability distribution defined by the prediction and standard deviation. Add this to the list of data points.
4. Repeat steps 2 and 3 until all cells have been visited and the simulation of one realization is complete.
5. Repeat steps 1 to 4 until sufficient realizations have been created.
6. Compute surfaces for the mean value and the standard deviation from all realizations if required.

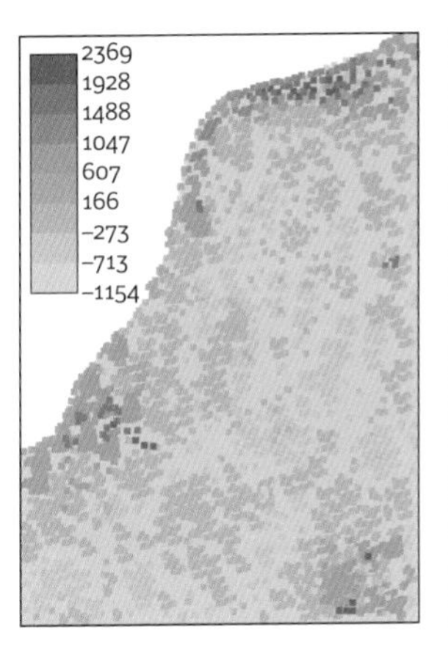

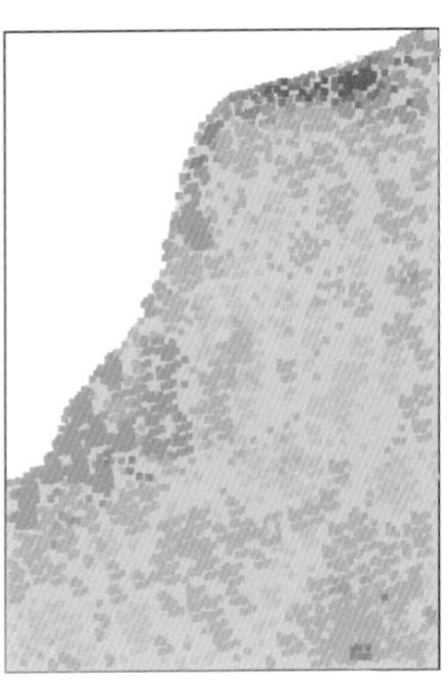
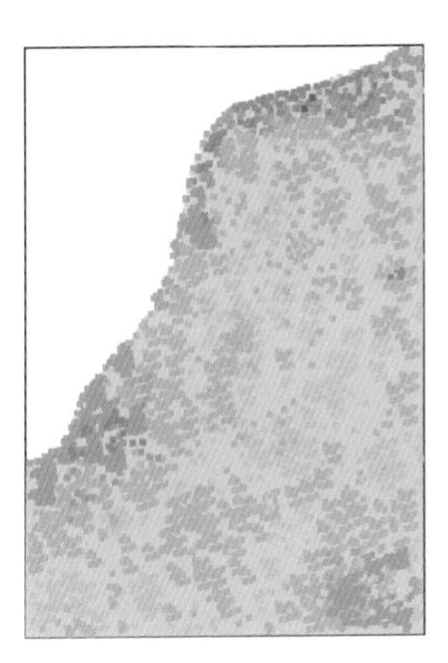

Figure 9.18 Four of 100 realizations of zinc levels produced by conditional simulation

7. Run the environmental model with each realization to see how results vary with the different inputs.

It is sensible to use this method when possible realizations, rather than smoothed averages, are needed to fill data cells for modelling. The differences between ordinary kriging, simple simulation, and conditional simulation are shown in Figure 9.17. Usually one simulates at least 500 *realizations* of the random surfaces, which can be used as different inputs for a Monte Carlo analysis of models (see Chapter 12). Figure 9.18 shows four realizations from among 100 realizations of zinc content, and Figure 9.19 shows the mean and standard deviation surfaces for all 100 realizations. Note that the standard deviation surface produced by conditional simulation does not reflect the distribution of data points as directly as kriging. Note also that the mean surface of the conditional simulations should be very similar to the ordinary kriging interpolation (if enough replicates are used it should be the same). With fewer than 500 realizations it is possible that the standard deviation of the simulated surfaces is substantially different from ordinary kriging, merely due to computing too few realizations.

Conditional simulation can also be combined with soft information in an attempt to obtain still better precision. Figure 9.19c and d shows the results of computing an average surface and standard deviation surface for 100 realizations using the variograms and stratification for the two flood frequency classes of Figure 9.14a.

9.14 The relative merits of different interpolation methods

Geostatistical methods are not one, but a wide range of techniques that rely on an understanding of the underlying spatial correlation structure of the data and use that to guide interpolation. Theoretical assumptions include ideas of stationarity, the intrinsic hypothesis, and normality, and these are sometimes difficult to meet with real data. The validity of the results depends not only on the method but also on assumptions about the uniformity of your area. Geostatistical methods have the advantage of providing estimates for points or blocks; information on spatial anisotropy, nested scales of variation, and external information can be combined to get the most out of expensive data (Box 9.1).

Box 9.1 The steps in kriging

Kriging requires the following steps:

1. Examine the data for normality and spatial trends, and carry out appropriate transformations. If using indicator kriging transform to binary values (0/1).
2. Compute the experimental variogram and fit a suitable model if the variable is spatially dependent. If the data are not spatially dependent (100 per cent nugget variance) then interpolation is not sensible.
3. Check the model by cross-validation.
4. Choose either kriging or conditional simulation.
5. If kriging, use the variogram model to interpolate sites on a regular grid where the sites are either equal in size to the original samples (point kriging) or are larger blocks of land (block kriging). If conditional simulation, compute at least 100 realizations on the regular grid. From these, compute average and standard deviation surfaces.
6. Display results as grid cell maps or by threading contours (not with conditional simulation), either singly or draped over other data layers (e.g. a DEM).
7. Use the results in conjunction with the other data.

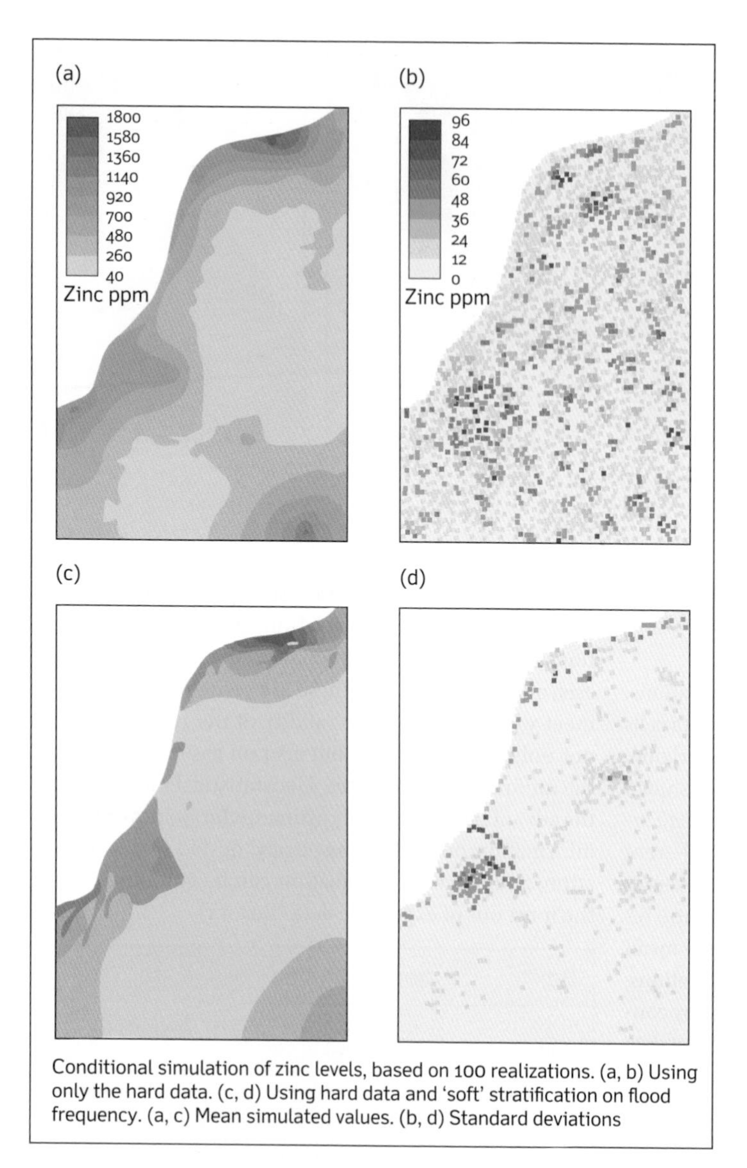

Figure 9.19 Conditional simulation can be carried out for whole areas or can be stratified according to flood frequency classes, which yields much more precise results

There are many published comparisons of geostatistical interpolation with other methods. The presence of large outliers can seriously affect the performance of all prediction methods, largely by distorting the estimated variogram, and several authors recommend the removal of outliers by methods of exploratory data analysis (EDA) or the use of robust methods (see Chilès and Delfiner 2012). Ordinary kriging is least successful when abrupt boundaries are present, though stratification into clearly different areas may bring considerable improvement, as shown in this chapter. Conventional choropleth mapping is generally poor at predicting site-specific values when spatial variation is gradual and splines can behave unpredictably.

Comparing the results of the different interpolation methods in Chapters 8 and 9

In Table 8.5 we compared the results of the deterministic methods of interpolation; Table 9.1 does the same for the geostatistical methods. Although there are no absolute standards for comparison, such as an independent data set for validation, we note that both deterministic methods and geostatistical methods show a wide range in results. Generally all methods (deterministic and geostatistical) that use only the data from the point observations estimate larger proportions of the area to be above the critical threshold values given in the table than those that use external information such as flooding classes and distance/elevation regression models.

Comparison of the kriging standard deviation maps suggests that stratified kriging has produced by far the best results (Tables 9.1 and 9.2). In general, the greater the information from data points and external sources (flood frequency classes), the smaller the prediction standard deviations. Geostatistical prediction is effective at reducing interpolation errors where spatial variation is continuous; in combination with the stratification, even smaller prediction errors can be achieved. However, the kriging standard deviation is not usually an accurate guide to interpolation errors. So, whether these results are generally true requires testing with an independent data set—although an alternative approach is described in Section 9.15. The evidence presented so far strongly suggests that using geostatistics and soft data together greatly improves the predictive power of GIS (Case study 9.1).

Table 9.3 gives a comparative overview of all the methods of interpolation discussed in both Chapters 8 and 9.

Table 9.1 Summary of results of geostatistical interpolation

Method	Minimum value (ppm)	Maximum value (ppm)	Per cent area >500 ppm	Per cent area >1000 ppm	Per cent area >1500 ppm
Ordinary point kriging with isotropic variogram	119	1661	29.28	4.91	0.36
The same, but with transformation to natural logarithms	122	1348	20.24	1.45	0.00
Ordinary point kriging with anisotropic variogram	127	1202	31.43	3.92	0.00
Ordinary point kriging in major flood categories	116	1817	15.56	3.53	0.85
Cokriging on ln(distance to river)	144	1300	16.20	0.97	0.00
Universal point kriging with a regression model on ln(distance) and elevation	114	1483	18.89	2.33	0.00
Conditional simulation with one general variogram	140	1606	28.89	3.96	0.21
Conditional simulation with stratification according to flood frequency and 2 variograms	105	1800	15.11	3.53	0.86

Table 9.2 Comparison of prediction standard deviations for different methods

Method	Minimum standard deviation	Maximum standard deviation
Flood frequency map		
class 1	423	423
class 2	177	177
class 3	105	105
Ordinary point kriging with isotropic variogram for whole area	119	329
Ordinary point kriging with anisotropic variogram for whole area	238	392
Ordinary point kriging in major flood categories (2 variograms)	62	354

Table 9.3 A comparison of methods of interpolation

Method	Deterministic/ stochastic	Local/ global	Transitions abrupt/ gradual	Exact interpolator	Limitations of the procedure	Best for	Computing load	Output data structure	Assumptions of interpolation model
Classification	Deterministic 'soft' information	Global	Abrupt if used alone	No	Delineation of areas and classes may be subjective Error assessment limited to within-class standard deviations	Quick assessments when data are sparse Removing systematic differences before continuous interpolation from data points	Small	Classified polygons	Homogeneity within boundaries
Trend surfaces	Essentially deterministic (empirical)	Global	Gradual	No	Physical meaning of trend may be unclear Outliers and edge effects may distort surface Error assessment limited to goodness of fit	Quick assessment and removal of spatial trends	Small	Continuous, gridded surface	Phenomenological explanation of trend, normally distributed data
Regression models	Essentially deterministic (empirical statistical)	Global with local refinements	Gradual if inputs have gradual variation	No	Result depends on the fit of the regression model and the quality and detail of the input data surfaces Error assessment possible if input errors are known	Simple numerical modelling of expensive data when better methods are not available or budgets are limited	Small	Polygons or continuous, gridded surface	Phenomenological explanation of regression model
Thiessen polygons (proximal mapping)	Deterministic	Local	Abrupt	Yes	No error assessment, only one data point per polygon Tessellation pattern depends on distribution of data	Nominal data from point observations	Small	Polygons or gridded surface	Best local predictor is nearest data point
Pycnophylactic interpolation	Deterministic	Local	Gradual	No, but conserves volumes	Data inputs are counts or densities	Transforming step-wise patterns of population counts to continuous surfaces	Small moderate	Gridded surface or contours	Continuous, smooth variation is better than ad hoc areas
Linear interpolation	Deterministic	Local	Gradual	Yes	No error assessment	Interpolating from point data when data densities are high as in converting gridded data from one projection to another	Small	Gridded surface	Data densities are so large that linear approximation is no problem
Moving averages and inverse distance weighting	Deterministic	Local	Gradual	Not with regular smoothing window, but can be forced	No error assessment Results depend on size of search window and choice of weighting parameter Poor choice of window can give artefacts when used with high data densities such as digitized contours	Quick interpolation from sparse data on regular grid or irregularly spaced samples	Small	Gridded surface, contour lines	Underlying surface is smooth

Table 9.3 A comparison of methods of interpolation (*continued*)

Method	Deterministic/ stochastic	Local/ global	Transitions abrupt/ gradual	Exact interpolator	Limitations of the procedure	Best for	Computing load	Output data structure	Assumptions of interpolation model
Thin-plate splines	Deterministic with local stochastic component	Local	Gradual	Yes, within smoothing limits	Goodness of fit possible, but within the assumption that the fitted surface is perfectly smooth	Quick interpolation (univariate or multivariate) of digital elevation data and related attributes to create DEMs from moderately detailed data	Small	Gridded surface, contour lines	Underlying surface is smooth everywhere
Kriging	Stochastic	Local with global variograms Local with local variograms when stratified Local with global trends	Gradual	Yes	Error assessment depends on variogram and distribution of data points and size of interpolated blocks Requires care when modelling spatial correlation structures	When data are sufficient to compute variograms, kriging provides a good interpolator for sparse data Binary and nominal data can be interpolated with indicator kriging Soft information can also be incorporated as trends or stratification Multivariate data can be interpolated with cokriging	Moderate	Gridded surface	Interpolated surface is smooth; statistical stationarity and the intrinsic hypothesis
Conditional simulation	Stochastic	Local with global variograms Local with local variograms when stratified Local with global trends	Irregular	No	Understanding of underlying stochastic process and models is necessary	Provides an excellent estimate of the range of possible values of an attribute at unsampled locations that are necessary for Monte Carlo analysis of numerical models, also for error assessments that do not depend on distribution of the data but on local values	Moderate to heavy	Gridded surfaces	Statistical stationarity and the intrinsic hypothesis

Case study 9.1
January 2006 precipitation in Wales

Selected methods are applied to the interpolation of precipitation amounts in Wales in January 2006. In Chapter 8, inverse distance weighting (IDW) was used for this purpose. Here, OK is applied to the same data. The variogram estimated from the data is shown in Figure 9.20. A map generated using IDW was shown in Figure 8.12, while the thin-plate splines with tension-derived map was given in Figure 8.13. The OK-derived map is given in Figure 9.21.

Another variant of kriging, **kriging with an external drift** (KED) model (see Section 9.11), was also used in this analysis. KED allows the use of secondary data to inform the prediction. Specifically, if there is a secondary variable which describes well the general trend in the data, then using this secondary variable may increase the accuracy of estimates. Elevation tends to be linearly related to precipitation amount, and thus elevation describes well, in at least some parts of Wales, the general trend in precipitation amount. With KED, the variograms used should be free of trend—that is, they should represent localized variations in precipitation amount rather than regional trends. One way of deriving trend-free variograms is to estimate the variograms for different directions and retain the variograms which show least evidence of trend. In this case, the variogram with the smallest variance was selected and a model fitted. The model had a nugget effect of 165.55, a structured component of 1133.79, and a range of 36 025.2.

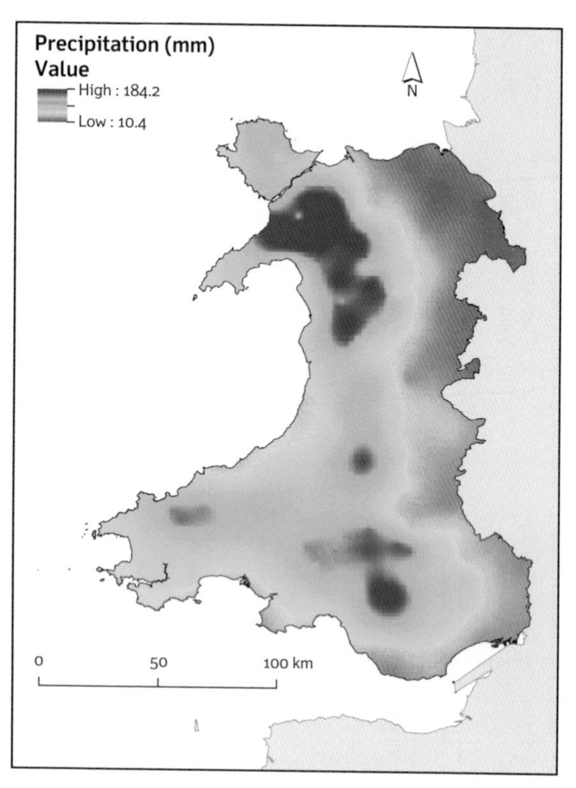

Figure 9.21 Precipitation in January 2006: OK with 16 nearest neighbours

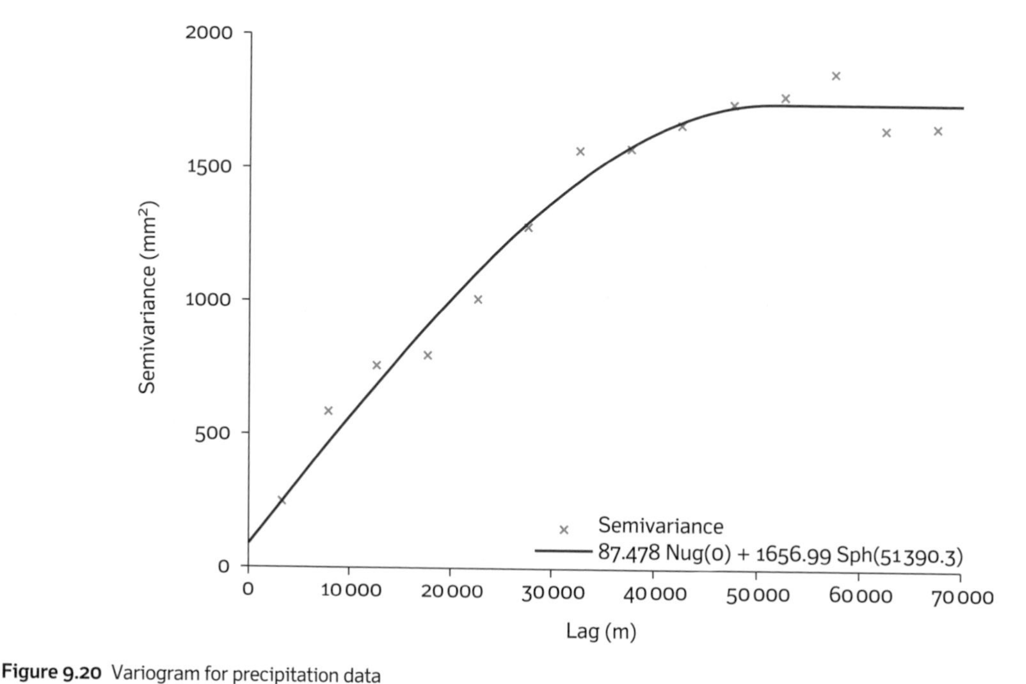

Figure 9.20 Variogram for precipitation data

Case study 9.1 (continued)

Table 9.4 summarizes cross-validation prediction errors for OK and KED, using a range of different numbers of nearest neighbours for prediction. An independent validation set would be preferable, but cross-validation provides a provisional guide. Judging by the root mean square errors (RMSE), KED with 128 nearest neighbours provides the smallest errors, and therefore making use of elevation data as a part of the prediction process appears to be beneficial in this case. Comparison of these results with those in Table 8.1 suggests that, for this example, KED provides more accurate predictions than IDW or the variants of thin-plate spline interpolators assessed. A detailed comparison of kriging approaches for mapping precipitation amounts is provided by Lloyd (2010a).

Table 9.4 Summary of cross-validation prediction errors

Method	Mean error	RMSE
OK 16	0.026	18.039
OK 32	0.174	17.987
OK 64	0.368	18.088
OK 128	0.153	18.129
KED 16	0.740	18.732
KED 32	0.666	18.128
KED 64	0.516	17.798
KED 128	0.233	17.347

9.15 Using variograms to optimize sampling

This chapter has been about methods of interpolation, their advantages and disadvantages, and their strengths and weaknesses. Clearly, even with the best methods one is severely limited by the data, and though many GIS users may not collect their own data, for those that do it is worth considering whether different numbers of data or arrangements of data points might not yield better information. Here we show how the variogram can be used to relate sample spacing (and layout) to the size of the cells used for block kriging.

Note that in ordinary block kriging (equations 9.17 and 9.18) the prediction errors of z_B are controlled only by the variogram and the sampling configuration. Therefore, once the variogram is known it is possible to design sampling strategies that will result in any required minimum interpolation error. In particular, the prediction error σ_B^2 for a block of land B depends on:

a) the form of the variogram (linear, spherical, or other function), the presence of anisotropy or non-normality, and the amount of nugget variance or noise

b) the number of neighbouring data points that is used to compute the point or block estimate

c) sampling configuration—which is most efficient: irregular sampling, a regular square grid, or a triangular grid?

d) the size of the block of soil for which the estimate is made—is it an area equivalent to the support or a larger block of land?

e) how the sample points are arranged with respect to the block.

These points are reported in Webster and Oliver (1990); here we only consider points (b) and (e)—the number of neighbouring data points on a regular grid and the sample spacing needed to yield a maximum value of σ_B^2 for a given block.

Once the variogram is known, one can calculate the combination of block size and sample spacing on a regular triangular or rectangular grid to yield predictions of block averages with a given prediction error. The prediction variance of a block B is the expected squared difference between the kriging prediction of the block mean and the true value. The standard error of the block mean is the square root of the prediction variance of the block B. The prediction variance is given by:

$$\sigma_B^2 = E\{[z_B - \hat{z}_B]^2\}$$
$$= 2\sum_{i=1}^{n} \lambda_i \gamma(\mathbf{x}_i, B) - \sum_{i=1}^{n}\sum_{j=i}^{n} \lambda_i \lambda_j \gamma(\mathbf{x}_i, \mathbf{x}_j) - \gamma(B, B). \quad 9.31$$

where $\gamma(\mathbf{x}_i, \mathbf{x}_j)$ is the semivariance of the attribute between points $\mathbf{x}_i, \mathbf{x}_j$, taking account of the distance $\mathbf{x}_i - \mathbf{x}_j$ between them (and the angle in cases of anisotropy), $\gamma(\mathbf{x}_i, B)$ is the average semivariance between $\mathbf{x}_i$ and all points within the block, and $\gamma(B, B)$ is the average semivariance within the block (i.e. the within-block variance).

Equation 9.31 shows that the prediction variances are not constant, but depend on the size of block B, the form of the variogram, and the distance between the data points (i.e. the configuration of sampling points in relation to the block to be estimated). Note that these variances do not depend on the observed values themselves (except through the variogram). We can compute values of σ_B^2 for different block sizes and square grid spacing and plot them as a map: see Figure 9.22. These curves can help us decide what intensity of sampling and sample spacing is needed (and how much the sampling campaign will cost) to ensure that σ_B^2 is less than a given maximum value. This could be useful when trying to link data from point observations to data collected as grid cell averages (e.g. remotely sensed data) or where point data are to be used to drive a quantitative process model that uses a predetermined spatial resolution. Atkinson and Lloyd (2007) use per-zone (local) variograms for determining optimal sample spacings.

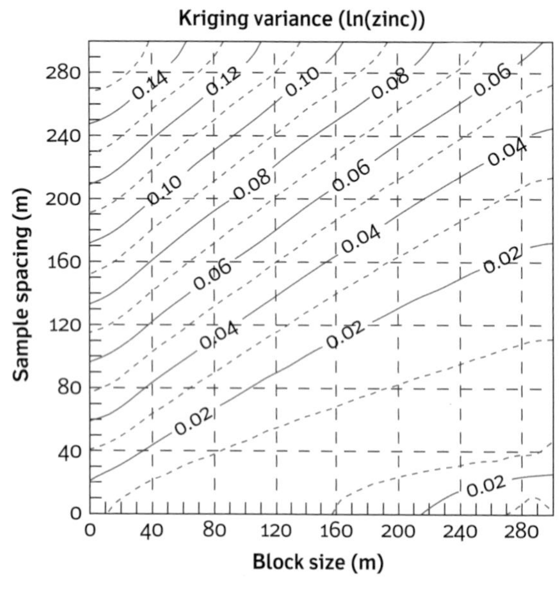

Figure 9.22 Isoline plot of equal prediction variances for ln(zinc) for different combinations of sample spacing and block size

Optimizing sampling when the variogram is unknown

If the variogram is not known it can be estimated approximately by a reconnaissance method known as **nested sampling** (Webster and Oliver 2007). Clusters of samples are laid out over the area of interest in such a way that pairs of samples are located at short distances from each other: these pairs are at a larger, known distance apart, and these groups are still further apart, and so on. Nested analysis of variance is used to estimate how the cumulative variance varies with mean sample spacing, which was shown by Miesch (1975) to be equivalent to the variogram. Nested sampling is not sufficient to provide sound estimates of the variogram for interpolation, but experience shows that the results of nested sampling can indicate near-optimal sample spacing for regular mapping by interpolation. Consequently, when variograms are unknown it can be cost-effective to perform a nested sampling first, to determine the best sample spacing before regular sampling. Nested sampling can be made more efficient by the unbalanced approach, which omits some replicates at the closer sample spacings (Webster and Oliver 2007).

9.16 Summary

This chapter has introduced and illustrated a powerful set of methods for exploring, mapping, and sampling spatially referenced variables. By estimating and modelling the variograms, we can use the characteristics of the data to make the most of them. There are potential benefits over methods such as inverse distance weighting (as explored in Section 8.12), particularly where the samples are sparse. The integration of secondary data, using methods such as cokriging and kriging with an external drift, may increase the accuracy of estimates further. The subject of geostatistics is a complex one, but one which may offer real benefits over simpler deterministic methods for spatial interpolation. There are many published introductions to geostatistics for general users as well as specific disciplinary areas. Some introductions for beginners include Isaaks and Srivastava (1989), Armstrong (1998), Burrough (2001), and Atkinson and Lloyd (2009). More advanced texts include the books by Goovaerts (1997), Webster and Oliver (2007), and Chilès and Delfiner (2012).

? Questions

1. Explain why it is so important in kriging to have a good model of the variogram.
2. Compare ordinary point kriging and thin-plate splines as methods for interpolating elevation data to make a DEM.
3. Examine the costs and benefits of the different ways in which kriging interpolation can be assisted by using extra hard and soft information.
4. Discuss ways of using indicator kriging to interpolate presence/absence data derived from social science surveys or vegetation studies.
5. Explain how you might use block kriging to ensure that point data are interpolated to a grid that has the same spacing and level of spatial generalization as a remotely sensed image.
6. Explain to a decision-maker why it is worth having interpolated data with a known level of uncertainty.

→ Further reading

▼ Atkinson, P. M. and Lloyd, C. D. (2009). Geostatistics and spatial interpolation. In A. S. Fotheringham and P. A. Rogerson (eds.) *The SAGE Handbook of Spatial Analysis*. SAGE Publications, London, pp. 159–181.

▼ Chilès, J. P. and Delfiner, P. (2012). *Geostatistics: Modelling Spatial Uncertainty*. 2nd edn. Wiley, New York.

▼ Goovaerts, P. (1997). *Geostatistics for Natural Resources Evaluation*. Oxford University Press, New York.

▼ Isaaks, E. H. and Srivastava, R. M. (1989). *An Introduction to Applied Geostatistics*. Oxford University Press, New York.

▼ Lloyd, C. D. (2010). *Spatial Data Analysis: An Introduction for GIS Users*. Oxford University Press, Oxford.

▼ Webster, R. and Oliver, M. A. (2007). *Geostatistics for Environmental Scientists*. 2nd edn. John Wiley & Sons, Chichester.

Analysis of Continuous Fields

The paradigm of continuous fields provides a rich foundation for spatial modelling, particularly when data are held in regular, square rasters (grids). Methods of map algebra allow mathematical operations to be carried out on whole raster overlays just as easily as if each overlay were only a single number and this facilitates the writing of numerical models. Mathematical operations on continuous fields can be divided into point operations and spatial operations. Point operations are the same as those discussed for attributes in Chapter 7; spatial operations include **spatial filtering**, the computation of **surface derivatives** (**slope**, **aspect**, **convexity**), **surface topology** and **drainage nets**, **spatial contiguity**, **linear** and **non-linear proximity determination**, and properties of whole surfaces such as **viewsheds**, **shaded relief**, and **irradiance** calculations. In conjunction with Chapter 11, this chapter explains each of these operations; the methods are illustrated by applications in hydrology, erosion, and surface run-off.

Learning objectives

By the end of this chapter, you will:

- **understand some key concepts and classes of approaches which can be used to work with continuous fields**
- **have developed a knowledge of some key ways of analysing data on continuous fields**
- **be able to apply your knowledge to analyse grids in combination, using map algebra, and separately, using spatial filters and other grid operators**
- **be able to assess surface form using slope (gradient), aspect, and other derivatives.**

As we explained in Chapter 2, there are two main ways of representing continuous fields. The first is the Delaunay triangulation (the TIN of digital elevation modelling); the second is the more common altitude matrix or grid used in raster GIS and image analysis. Delaunay networks are often used outside GIS to support the finite element modelling of dynamic flow processes in groundwater movement (MODFLOW—Harbaugh 2005), discharge over floodplains (see Di Baldassarre 2012 for an extensive review), or air quality (Oliver et al. 2012). Camelli et al. (2012) discuss construction of surfaces for transport and dispersion modelling in GIS contexts. Finite element modelling (FEM) is not usually part of the standard generic toolkit of most GIS. Numerical models that use FEM are usually loosely coupled to the GIS, with the GIS being used to assemble the data and pass them to the model via an interface. The results from the model are returned to the GIS, converted to a square grid or contour lines for ease of handling, and then overlaid on digitized base maps for display.

Finite element modelling is outside the scope of this book. Here we describe the operations for spatial analysis of continuous fields that are represented by the regular square grid, where each attribute is represented by a separate overlay, and each grid cell is allowed to take a different, scalar value. Note that although in many examples we refer to an **altitude matrix** (i.e. a gridded digital elevation model), the *z* attribute can represent any other continuously varying attribute or regionalized variable, such as levels of pollutants in soil, atmospheric pressure, annual precipitation, an index of marketing potential, population density, or the costs of access to a given location. As well as introducing general methods for the analysis of continuous surfaces, some methods for the analysis of DEMs specifically are introduced as context for Chapter 11, which has DEMs as its focus.

10.1 Basic operations for spatial analysis with discretized continuous fields

Map algebra and cartographic modelling

The Kisii land evaluation examples given in Chapter 7 demonstrated that when one has to use data from several geographically overlapping entities it is much easier to maintain the database and to compute new attribute values if all the data are referenced to a uniform geometry—that of the regular square grid. The loss of information due to rasterizing smooth polygon boundaries is more than offset by the advantage of not having to create new polygons by intersection. Moreover, by choosing a grid size to match that used by remotely sensed imagery, one has the added advantage that the satellite data can also be used as input for data analysis and modelling. As computers increase in power and data sources increase in resolution, the degradation of a spatial pattern by rasterizing becomes more acceptable than the creation of large numbers of topologically linked lines and polygons.

A further major advantage with raster representation in which each attribute is recorded in a separate overlay, is that any mathematical operation performed on one or more attributes for the same cell can easily be applied to all cells in the overlay. This means that one can use exactly the same algebraic notation to operate on gridded data as on single numbers. The method is called **map algebra** (Tomlin 1983); and see Mennis 2010 for an expansion of the concept to multidimensional data) and the procedure of using algebraic techniques to build models for spatial analysis is called **cartographic modelling**.

The methods of map algebra mean that the user needs only to specify the spatial operations to be used, the names of the source overlays, and the result—the computer program then applies the operation to all the cells in the overlays. This makes it very easy to write computer models as sequences of computations, and makes the extension of formerly point models to two-dimensional space very easy. For example, the command:

$$\text{NEWMAP} = \text{MAP1} + \text{MAP2} + \text{MAP3} \qquad 10.1$$

is all that is necessary to compute the sum of the values of the attributes on the three overlays called MAPI, MAP2, and MAP3. The command:

$$\text{NEWMAP} = (\text{MAP1} + \text{MAP2} + \text{MAP3})/3 \qquad 10.2$$

computes the average value, and the command:

$$\text{NEWMAP} = (\text{SQRT(MAP1)} + \text{SQRT(MAP2)} + \text{SQRT(MAP3)}\ )^{**}2 \qquad 10.3$$

computes the squared sum of square roots.

All these examples compute new values on a cell-by-cell basis: they are known as **point operations** and are formally equivalent to the same mathematical operations that were applied to the attributes of single point, line, or polygon entities in Box 7.2. The extra advantages of the grid-based approach become clear when we see that by using concepts of surface differentiation and smoothing it is possible to compute attributes that are some **spatial function** of the area or neighbourhood surrounding a given cell. As we explain in this chapter, there are many useful neighbourhood functions which can be used for spatial analysis. Finally, because the grid-based approach is an approximation of a continuous surface, it is possible to determine new attributes such as views over the surface, or to extract the topology of the steepest downhill paths. These operations provide a rich toolkit for studying phenomena as varied as cross-country visibility and hydrological processes, or for optimizing access to particular kinds of terrain.

From this, we see that if the command language interface (CLI) allows the user to express the basic spatial functions in a mathematical language then it will be easy to write mathematical models to operate on the gridded data. Many GIS provide a programming language called a **macro language** for this purpose. In this book we assume that all operations can be expressed in simple, general, mathematical terms so that the generic model code used to explain the examples can easily be implemented on various systems. Not all authors of map algebra are happy with mathematical formulations: Tomlin (1983) provides an English-language alternative for those who cannot cope with the mathematical conventions. It is

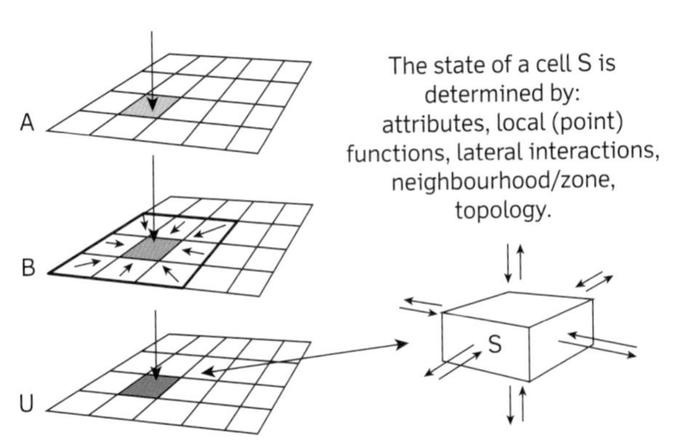

Figure 10.1 The state of a cell is a function of local operations and lateral interactions with its neighbours

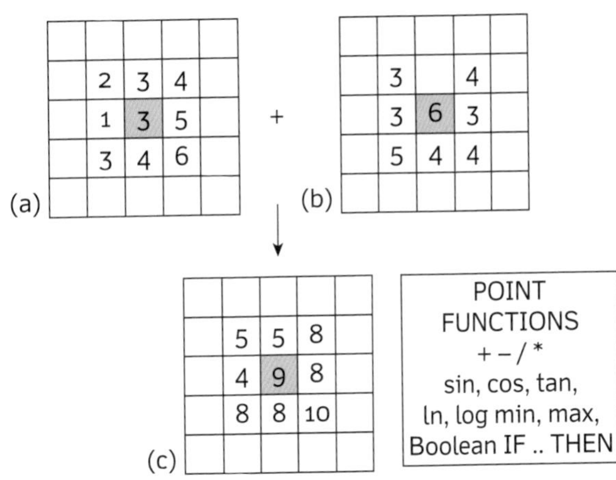

Figure 10.2 Examples of point operations on a cell

our belief, however, that the use of conventional mathematical terminology is much preferable.

In this chapter we review the main kinds of mathematical operation that can be used with gridded data, and illustrate the operations with a range of simple and more complex practical examples.

Point operations

All the logical and numerical operations presented in Chapter 7 for simple point, line, and area entities linked to a RDBMS can be used on the individual grid cells of discretized continuous fields. This means that the values for the same grid cell in different overlays can be logically selected, added, subtracted, or manipulated by any mathematical method that is permitted for the data type in question (Figures 10.1 and 10.2). Therefore one can write commands that can add up or subtract real numbers, but not numbers that are coded as a Boolean or nominal data type. Boolean or nominal data can be operated on by logical commands in the same ways as given in the previous chapter.

Spatial operations

Using gridded data has advantages and disadvantages compared with a topologically linked vector database of defined entities. The disadvantages include the problems that the exact shapes of entities are only approximated by the grid cells and that directed operations over a network cannot be carried out without first deriving the topology from the properties of the surface. The advantages are that the continuous field model provides a much richer suite of truly spatial analysis operations that have many practical uses.

The following operations compute a new attribute for a given cell as some function of the attributes of cells within a certain **spatial neighbourhood**. The neighbourhood is often (but does not have to be) isomorphic (i.e. square or circular). In most cases in GIS, the cell size is fixed and uniform over the whole domain of interest. The adoption of grids of variable density would require modifications to the algorithms but would not essentially change the character of the operations being carried out. These spatial operations include:

- interpolation
- spatial filtering
- first and higher order derivatives
- the derivation of surface topology: drainage networks and catchment delineation
- contiguity assessment (clumping)
- non-linear dilation (spreading with friction)
- viewsheds, shaded relief, and irradiance.

Discussion of these methods spans Chapters 8 and 9 (interpolation), Chapter 10 (spatial filtering, first and higher order derivatives, derivation of surface topology, clumping, non-linear dilation), and Chapter 11 (viewsheds, shaded relief, and irradiance).

10.2 Interpolation

Interpolation is the prediction of a value of an attribute $\hat{z}$ at an unsampled site (x_0) from measurements made at other sites x_i falling within a given neighbourhood. As shown in Chapters 8 and 9, interpolation is used to create discretized continuous surfaces from observations at sparsely located points or for resampling a grid to a different density or orientation as in remote sensing images. Interpolation can be seen as a particular class of spatial filtering, where the input data are not necessarily already located on a continuous grid. All other methods discussed here assume that the grid has already been created. Remember that the range of the variogram can be used to define the radius of the interpolation search radius (Chapter 9).

Interpolation is often a complicated operation, and while interpolation operations can be expressed in a mathematical command language many users will encounter specialist packages so that standard terminology cannot be used.

10.3 Spatial analysis using square windows

Spatial filtering

The simplest and perhaps most widely used method of spatial filtering a discretized, continuous surface involves passing a square window (otherwise known as a kernel or filter) over the surface and computing a new value of the central cell of the window $C_{i,j}$ as a function of the cell values covered by the window. This kind of operation is also commonly known as **convolution**. The window is frequently of size 3 × 3 cells, but any other kind of square window (5 × 5, 7 × 7 cells, or distance measurements) is possible. The general equation is:

$$C_{i,j} = f\left(\sum_{i-m}^{i+m}\sum_{j-n}^{j+n} c_{i,j} \cdot \lambda_{i,j}\right) \quad 10.4$$

where f stands for a given window operator on windows of sides $2m + 1$, $2n + 1$, and λ_{ij} is a weighting factor.

$C_{i-1,j-1}$	$C_{i,j-1}$	$C_{i+1,j-1}$
$C_{i-1,j}$	$C_{i,j}$	$C_{i+1,j}$
$C_{i-1,j+1}$	$C_{i,j+1}$	$C_{i+1,j+1}$

The most commonly used window operations (f) are low- and high-pass filters.

Smoothing (low-pass) filter The value for the cell at the centre of the window is computed as a simple arithmetic average of the values of the other cells (Figure 10.3). In remote sensing systems and image analysis, the mean values are computed by multiplying the cell values in the window by the $n \times m$ values in the filter. For example, for a 3 × 3 filter, the mean value for the window centre can be computed by multiplying each cell value by a weight of 1/9 and adding all the results. For a 5 × 5 window, the weight of each cell is 1/25. Extra weights can be given to the central cell by introducing weights that are nonlinear—i.e. those cells closest to the central cell have larger weights than those further away, much like the idea of distance weighting in ordinary interpolation.

For example, for a 3 × 3 window:

1	1/15	2/15	1/15
2	2/15	3/15	2/15
3	1/15	2/15	1/15
	1	2	3

or for a 5 × 5 window:

1	1/65	2/65	3/65	2/65	1/65
2	2/65	3/65	4/65	3/65	2/65
3	3/65	4/65	5/65	4/65	3/65
4	2/65	3/65	4/65	3/65	2/65
5	1/65	2/65	3/65	2/65	1/65
	1	2	3	4	5

The low-pass filter has the effect of removing extremes from the data, producing a smoother image (Figures 10.4 and 10.5). For nominal and ordinal data (and also integer and ratio data) the mean can be replaced by the mode, which is the most common value (**majority**). Using a modal filter on nominal data (e.g. soil units) can be a useful way of simplifying a complex map (Figure 10.6), but note that smoothing a gridded image with a modal filter is a different kind of operation than the procedure of generalizing a map by reclassifying the attributes and merging the soil polygons given in Chapter 7.

Generic commands for filtering

To compute low-pass and high-pass filters:

Low-pass = *windowaverage* (continuous_surface, *n*)

High-pass = continuous_surface – low-pass

where *n* is the side of the square window in cells or distance units.

To compute a modal filter:

Modalmap = *windowmajority* (continuous_surface, *n*)

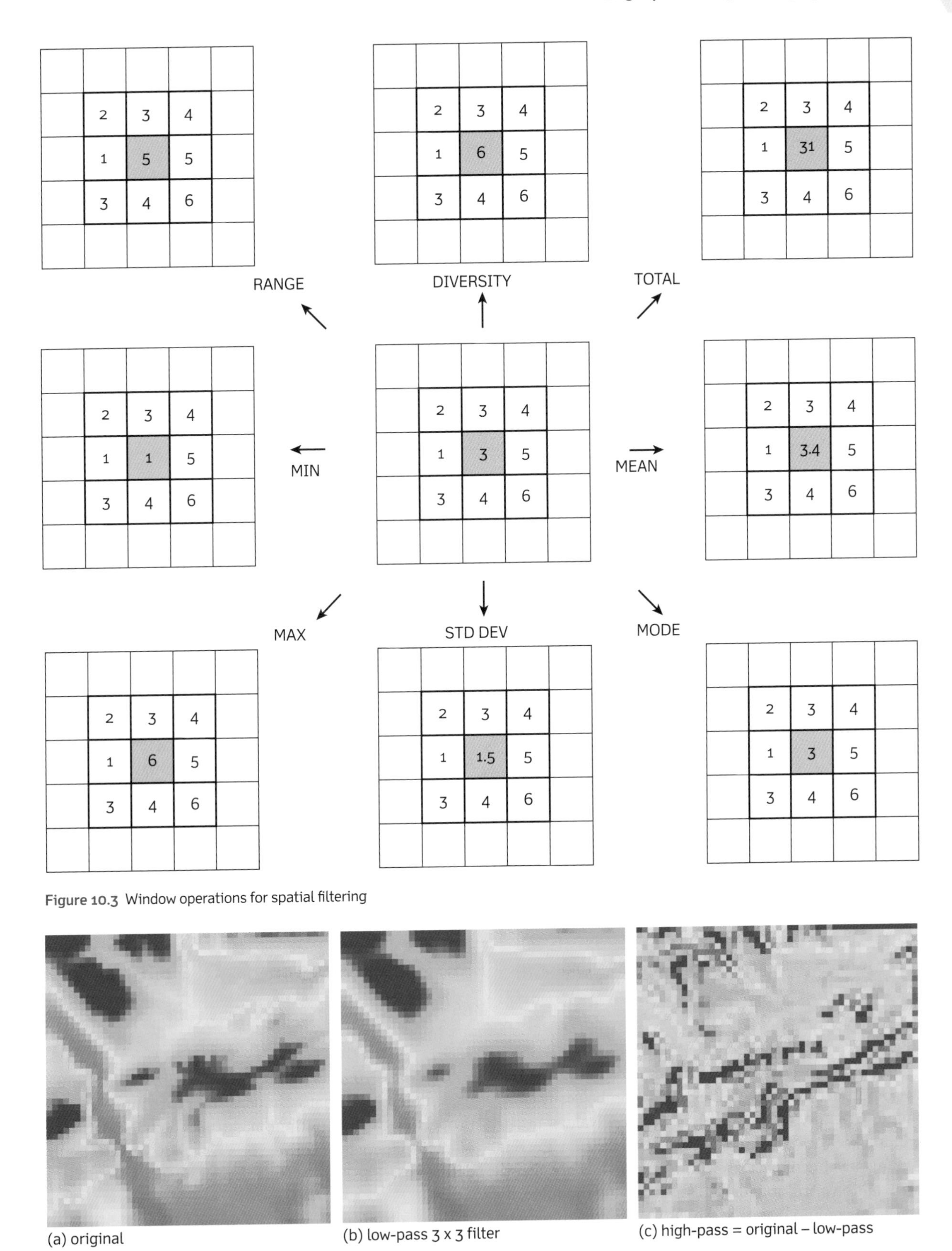

Figure 10.3 Window operations for spatial filtering

(a) original (b) low-pass 3 x 3 filter (c) high-pass = original – low-pass

Figure 10.4 Smoothing a surface with a low-pass filter

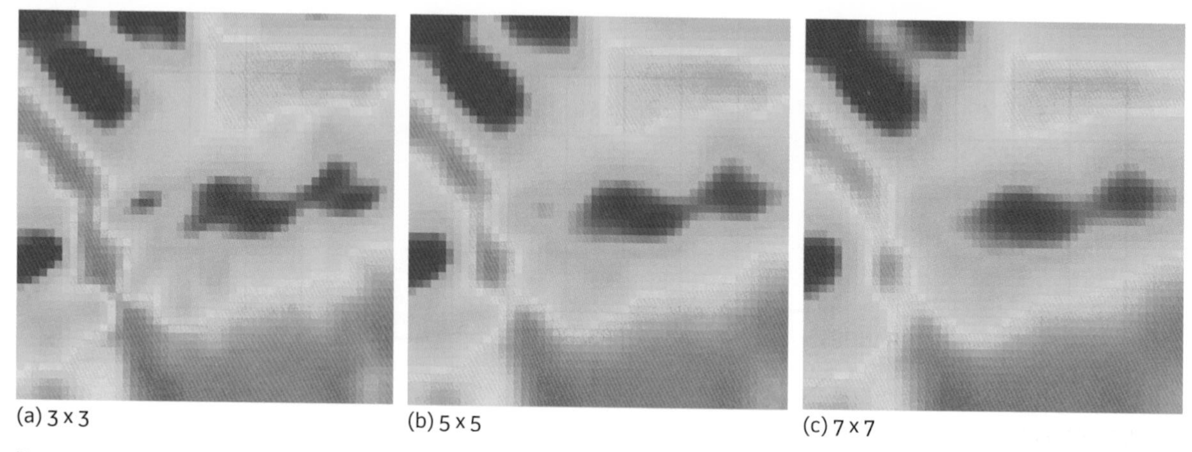

Figure 10.5 The effect of increasing window size on smoothing

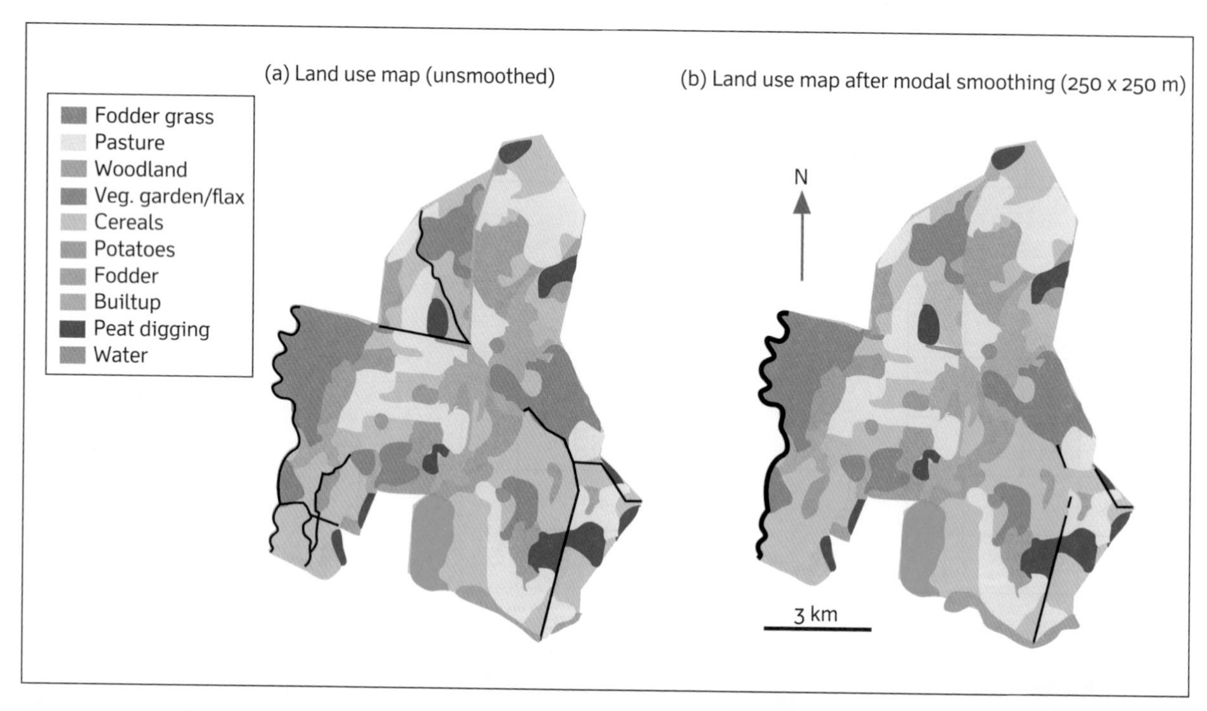

Figure 10.6 Smoothing a complex polygon map with median smoothing aggregates areas but does not reduce the number of classes (cf. Figure 7.4)

In a similar way, the local maximum or minimum values, and their difference—the range—can be easily computed. **Diversity** (the number of different values in the window) or the differences between two cells on any one of the four directional axes within the window are alternative options. For nominal and ordinal data, the minority (the least common) and the diversity are useful operations for indicating the local complexity of the spatial pattern.

High-pass and edge filters The inverse of the low-pass filter is one that enhances the short-range spatial properties of the continuous surface, enhancing areas of rapid change or complexity. The high-pass filter is defined as:

$$\textbf{Original surface} - \textbf{low-pass image} = \textbf{high-pass image} \quad (10.5)$$

(a) Original surface

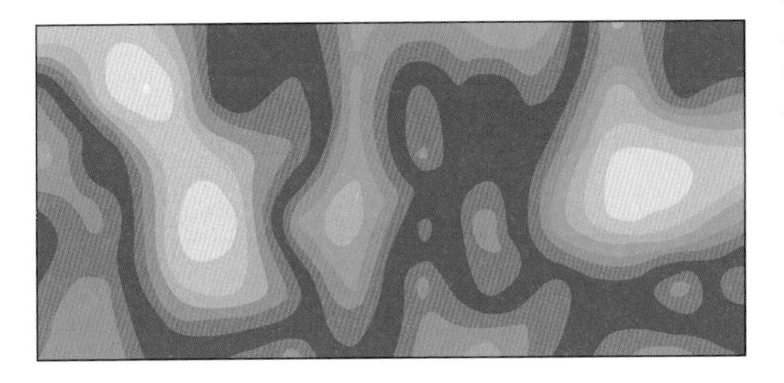

(b) High-pass filter

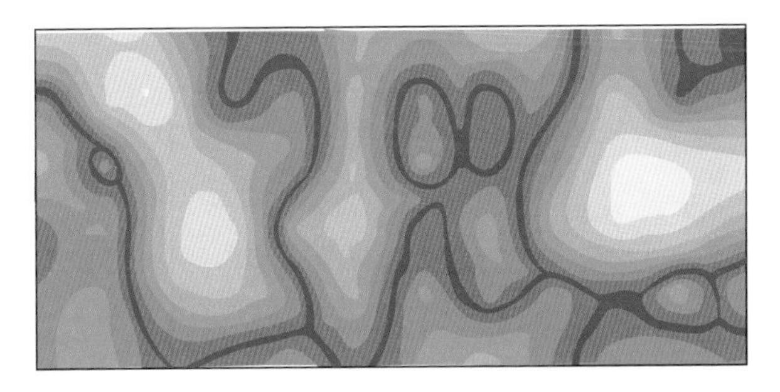

(c) Maximum rate of change yields boundaries

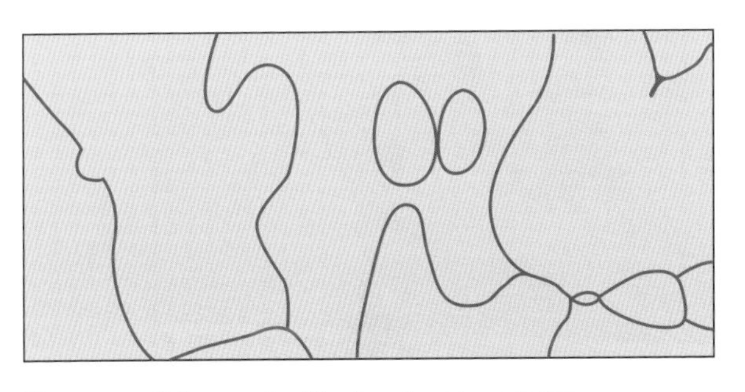

Figure 10.7 Using an edge filter to extract boundaries

The qualities of the high-pass filter can therefore depend on how the low-pass filter is defined. Alternatively, a set of weights can be defined for the window (Pavlidis 1982). A commonly used set of weights is the **Laplacian filter**:

0	1	0
1	−4	1
0	1	0

Figure 10.7 shows an example of applying an edge filter to determine the locations of maximum rates of change in a continuous surface. Edge filters are also used to enhance relatively uniform areas in the continuum provided by a remotely sensed image. The derivation of sharp edges and sets of boundary pixels is often used for inferring the presence of discrete spatial entities in the image, which ultimately could be extracted and vectorized as required.

10.4 Filtering case studies

This section builds on the discussion above by introducing two case studies which are intended to show how filters can be used to modify the characteristics of raster data. Case study 10.1 deals with a continuous property—specifically, elevation in Wales. Case study 10.2 demonstrates filtering of an orthoimage of part of Washington, DC.

These examples illustrate how filters can be used to smooth images (mean and median filters) or enhance contrast in some way (by generating the high-pass image or by applying a local measure of variability, such as the standard deviation).

Case study 10.1
Filtering elevation data for Wales

Figure 10.8 shows a DEM of Wales, with a spatial resolution of 661 m. In Figure 10.9, the output of a 3 x 3 mean filter is shown. The range of data values is smaller in the filtered image—in particular, the maximum elevation is much smaller than in the original image (Figure 10.8)—so Figure 10.9 presents a smoothed version of the original DEM.

Figure 10.10 shows the output of a 3 x 3 pixel median filter; again, the maximum elevation is clearly smaller than in the original image although, as with the mean filter output, the general pattern of elevation values is the same. The original grid (Figure 10.8) minus the 3 x 3 pixel mean filter output (Figure 10.9) is given in Figure 10.11. This is termed a high-pass image, as defined

Case study 10.1 (continued)

in equation 10.5, and notable breaks of slope are particularly clear in this map. Figure 10.12 shows the output of the standard deviation filter, which highlights areas of local contrast in elevation values.

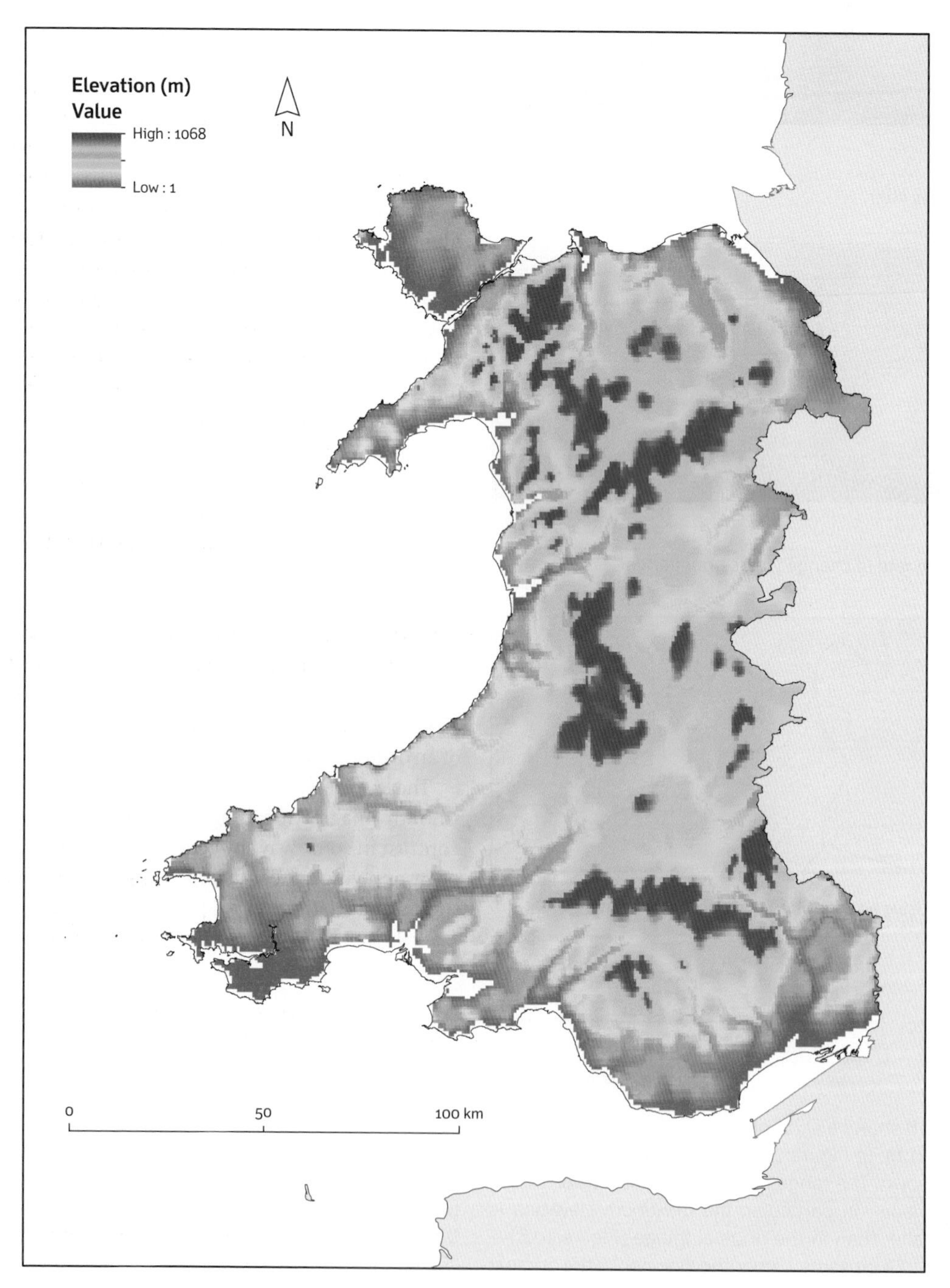

Figure 10.8 Digital elevation model of Wales; spatial resolution of 661 m (image courtesy of the US Geological Survey)

Case study 10.1 (continued)

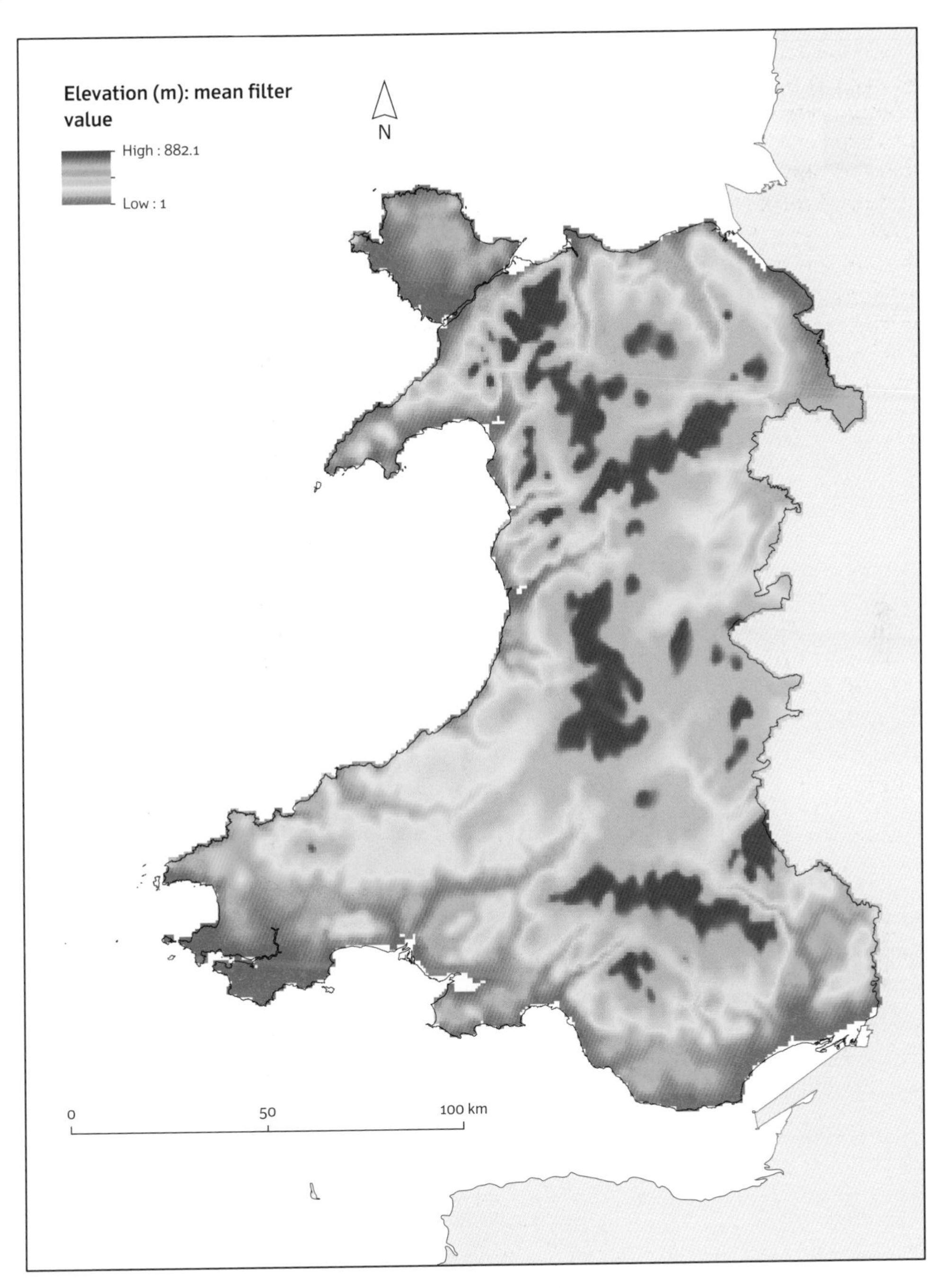

Figure 10.9 3 x 3 pixel mean filter of digital elevation model of Wales

Case study 10.1 (*continued*)

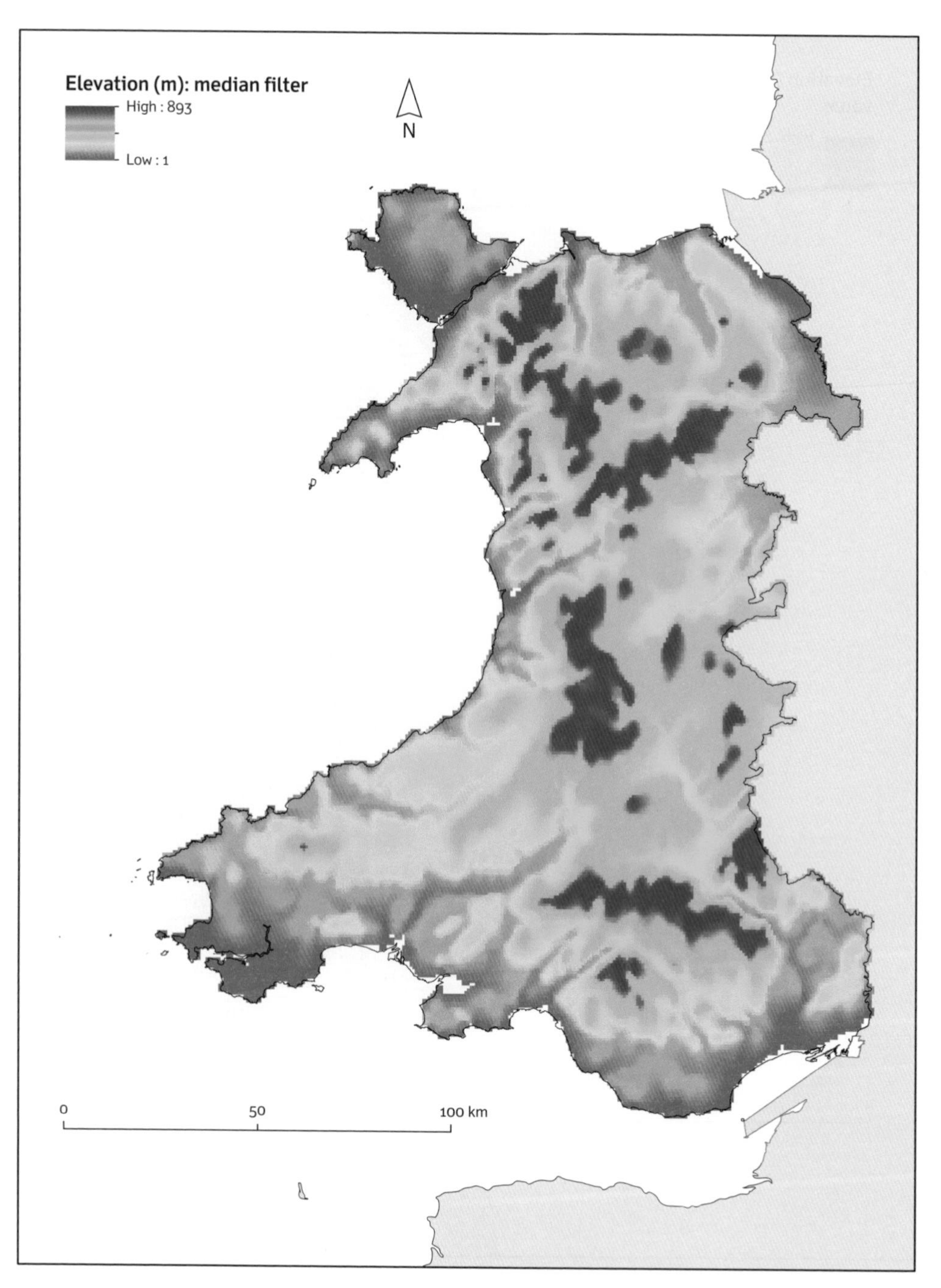

Figure 10.10 3 x 3 pixel median filter of digital elevation model of Wales

Case study 10.1 (*continued*)

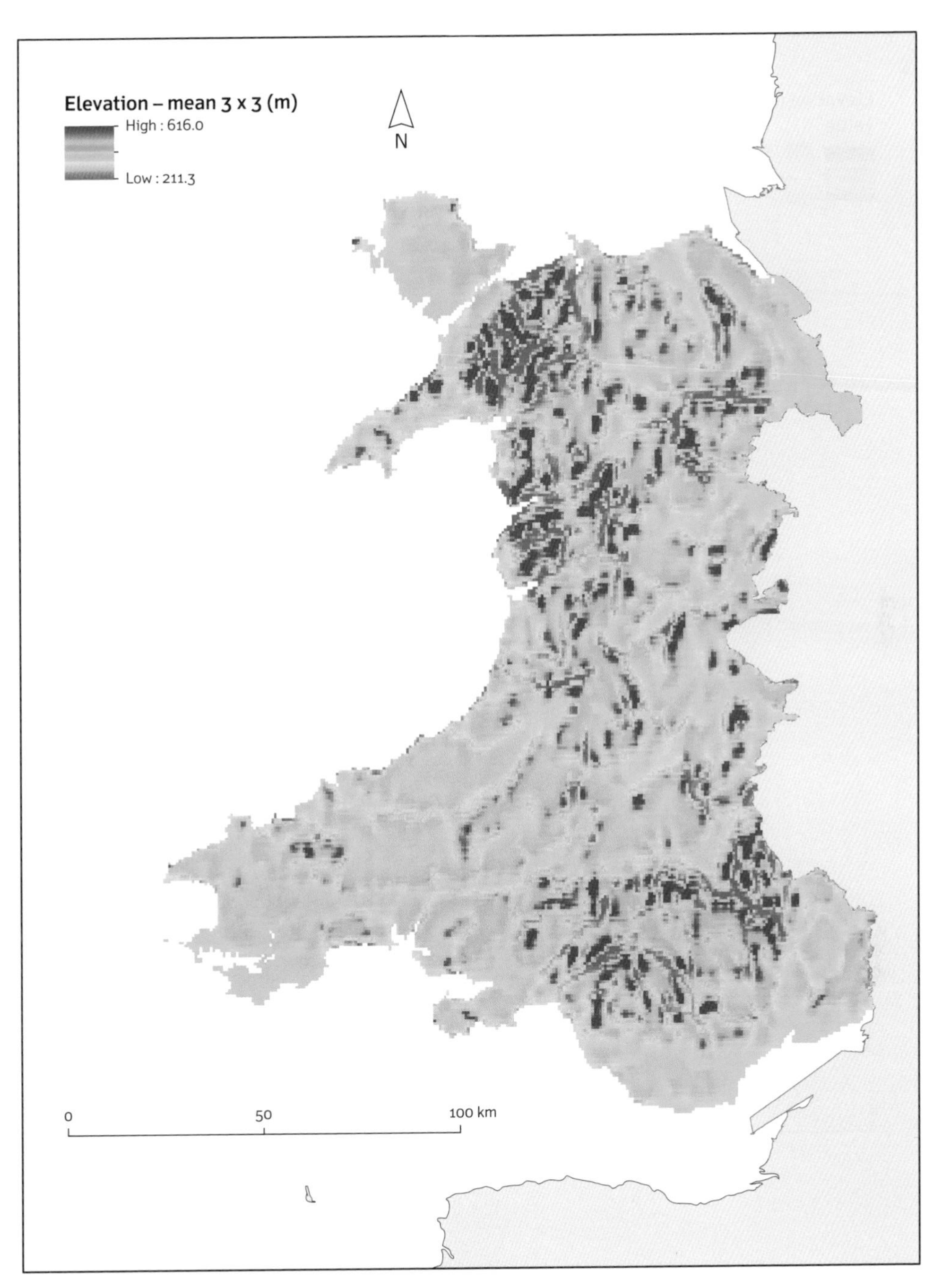

Figure 10.11 Elevation minus 3 x 3 mean filter

Case study 10.1 (continued)

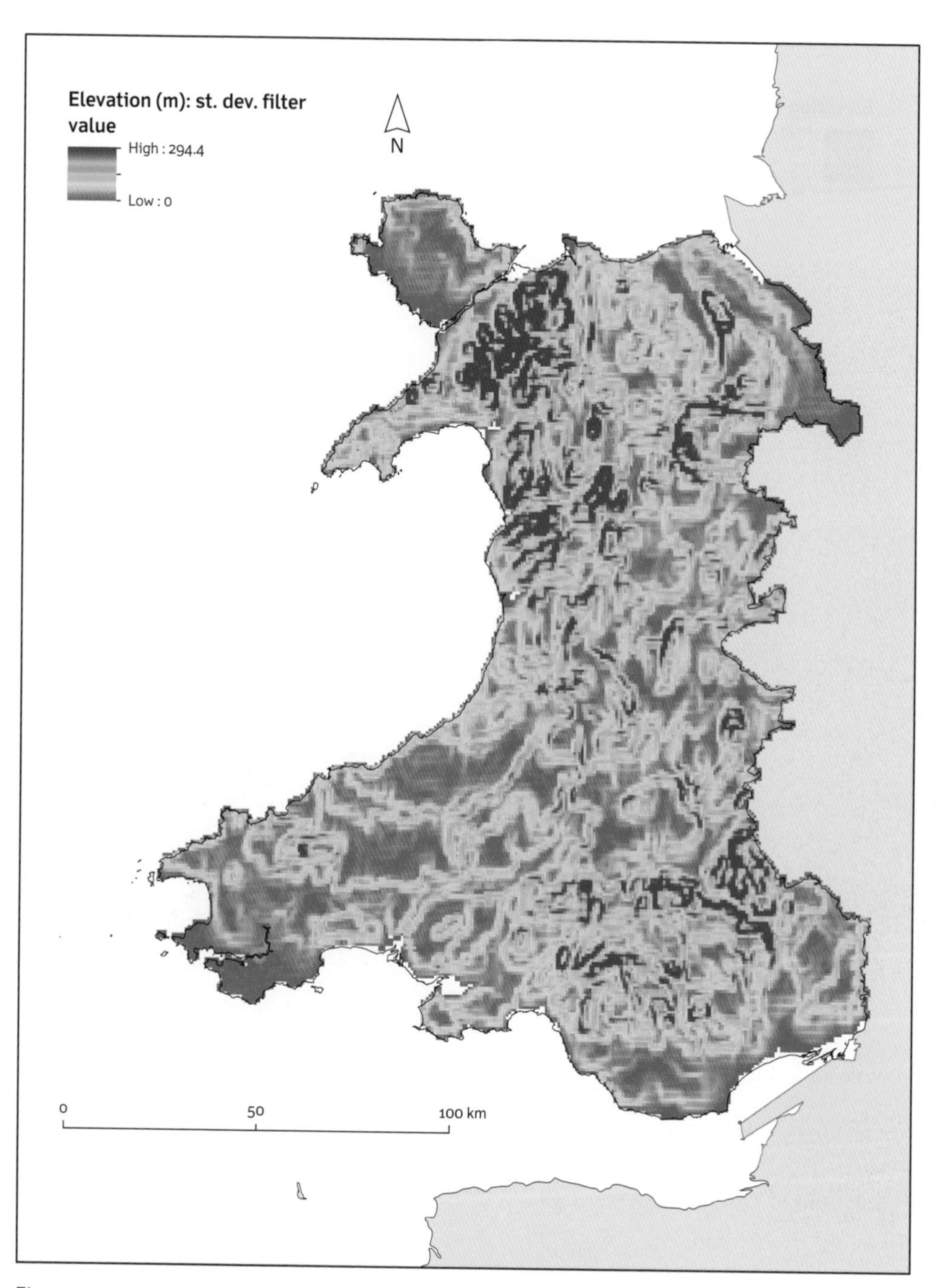

Figure 10.12 3 x 3 pixel standard deviation filter of digital elevation model of Wales

Case study 10.2
Enhancing contrast with filters in Washington, DC

A set of filters were also applied to an orthoimage (geometrically corrected aerial image) of a part of Washington, DC. The image is shown in Figure 10.13. Moving windows of 3 x 3 (Figure 10.14), 21 x 21 (Figure 10.15), and 51 x 51 (Figure 10.16) pixels were applied for deriving the local means. Note that the White House has been 'blocked out' by the data suppliers as a security measure, and thus appears uniform. Figure 10.17 gives the 3 x 3 pixel standard deviation. The White House buildings show clearly, as do the street network and other buildings in this part of the city.

Figure 10.13 Orthoimage of part of Washington, DC; spatial resolution: 0.3 m (image courtesy of the US Geological Survey)

Case study 10.2 (continued)

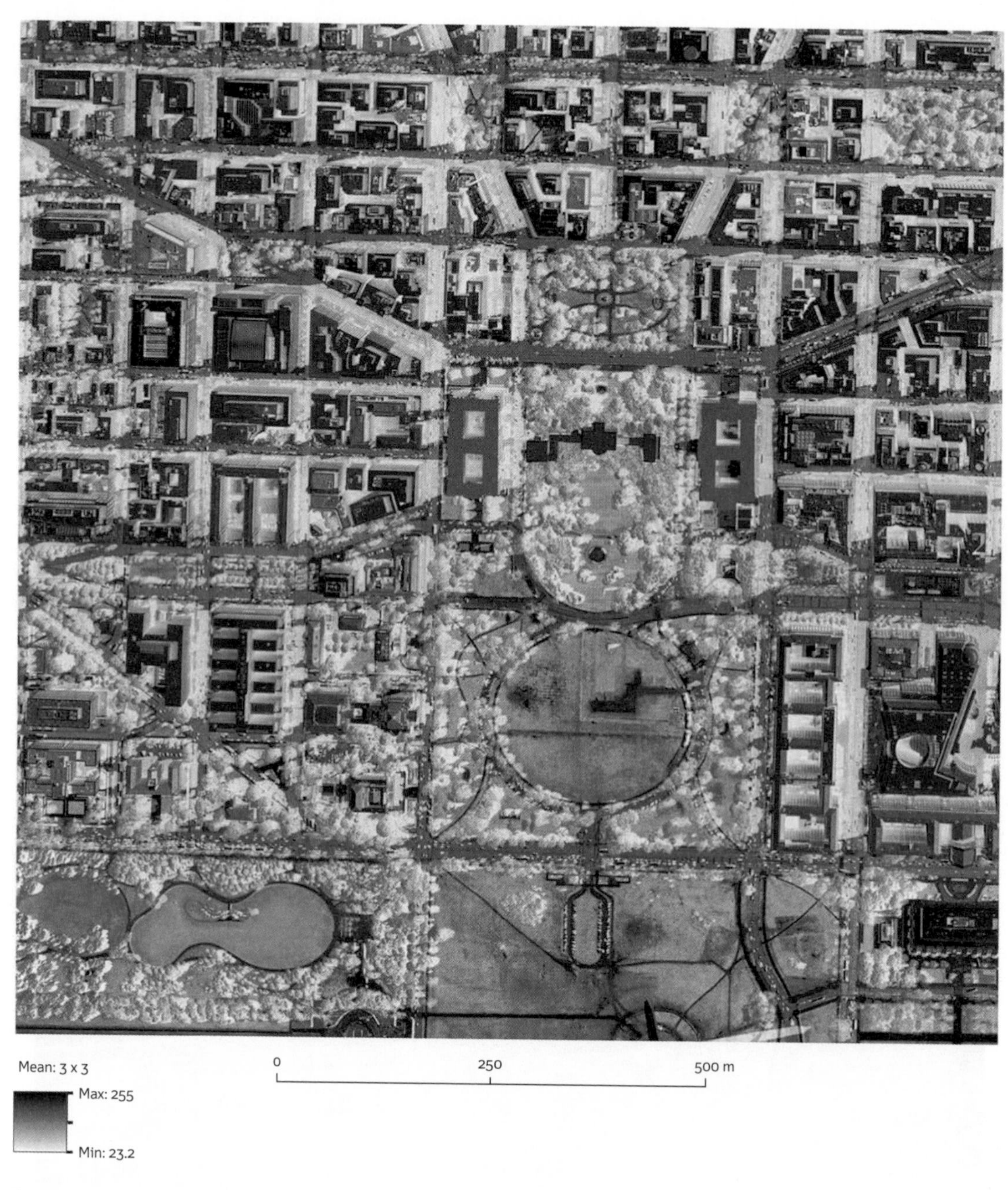

Figure 10.14 3 x 3 pixel mean filter of orthoimage of part of Washington, DC

Case study 10.2 (continued)

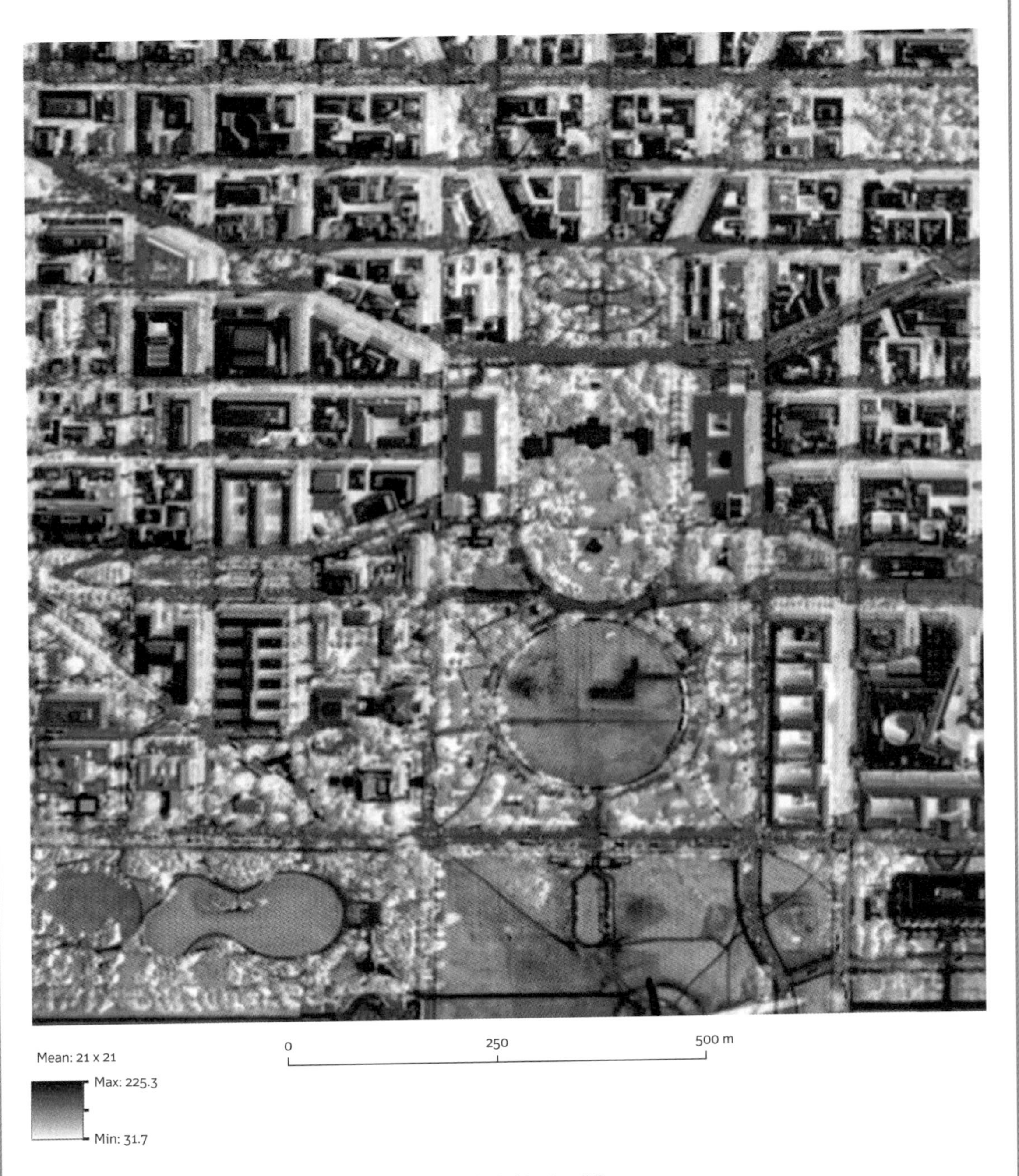

Figure 10.15 21 x 21 pixel mean filter of orthoimage of part of Washington, DC

Case study 10.2 (continued)

Figure 10.16 51 x 51 pixel mean filter of orthoimage of part of Washington, DC

Case study 10.2 (continued)

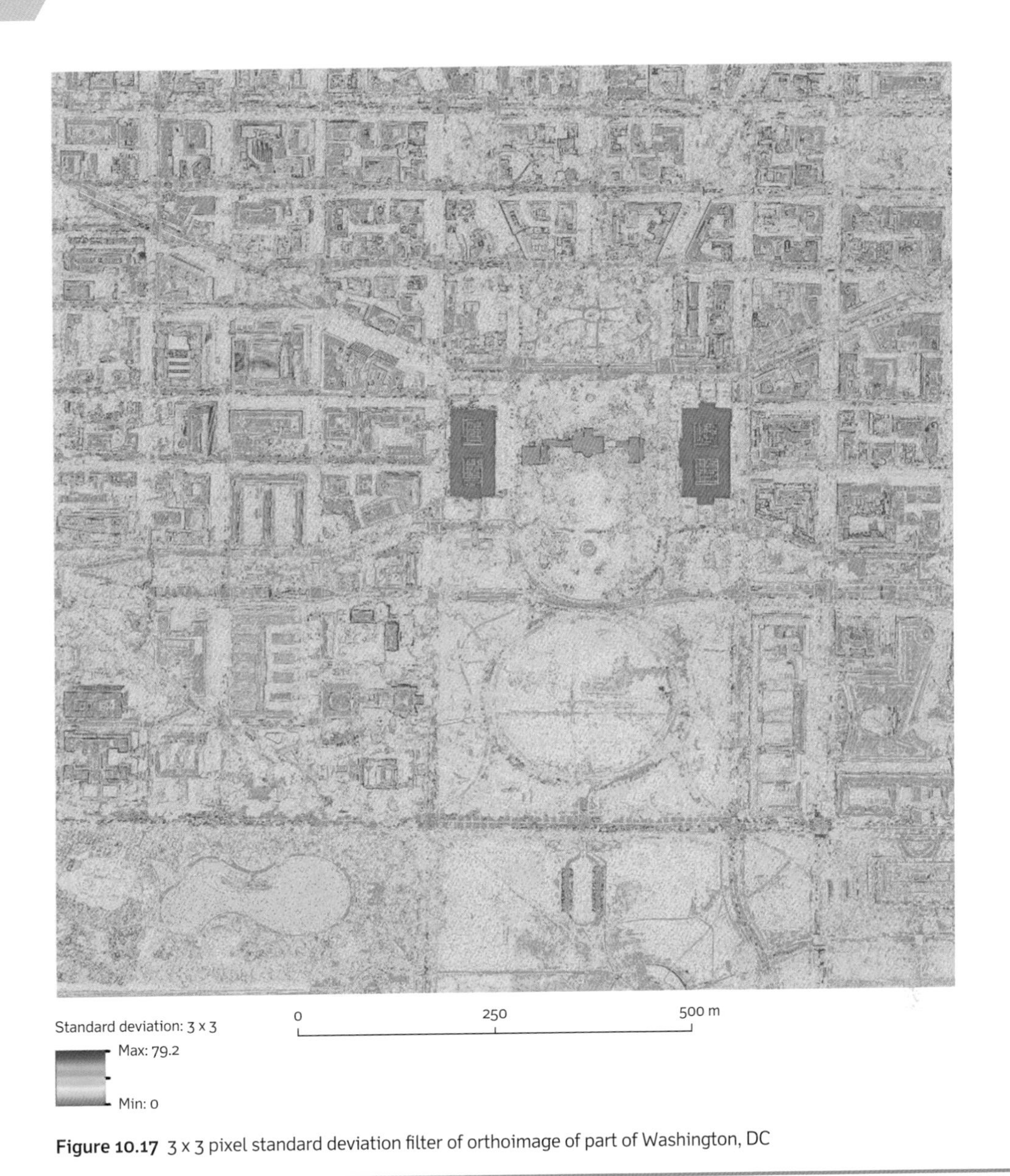

Figure 10.17 3 x 3 pixel standard deviation filter of orthoimage of part of Washington, DC

10.5 Other grid operators

The chapter has so far focused primarily on focal operators such as the smoothing filter. This section describes some other key classes of grid operators. Measurement of cell distances from prespecified source cells is a common objective—Figure 10.18 shows distances of all cells in a grid from two source cells (the shaded cells). In this case, the distances are in cell units and refer to distance from the nearest source cell to each cell in the grid. It is easy to think of potential applications for this approach, given that distances are central to many spatial analysis operations.

Chapter 7 dealt with, among other topics, overlay between discrete features. Similarly, with raster data, it is sometimes necessary to ascertain which data values fall within particular zones or areas and to summarize these data values in some way. Operators which allow for this kind of analysis are known as **zonal operators**.

4.00	3.00	2.00	1.00	0
3.16	3.00	2.24	1.41	1.00
2.24	2.00	2.24	2.24	2.00
1.41	1.00	1.41	2.24	3.00
1.00	0	1.00	2.00	3.00

Figure 10.18 Euclidean distances from source cells (shaded)—the shortest distance from either of the two source cells is given

Figure 10.19 gives an example of zonal sum and zonal mean. In words, the cell values (top left) which overlap with each zone (top right) are identified and summarized. For example, in the case of the zonal sum, the values which fall within zone 1 are summed: 5 + 3 + 3 + 7 + 5 + 4 + 6 = 33. The value 33 is then written to all of the zone 1 cell locations in the output—so the value 33 appears in the output in each of the seven cells which correspond to zone 1. For the zonal mean, the summed value is divided by the number of cells in the zone, giving 33/7 = 4.7. So, given a set of zones (note that cells in zones do not need to be contiguous), the overlapping data values can be summarized using any standard statistical summary including the minimum, maximum, mean, and sum. Sonwalkar et al. (2010) apply the zonal function to summarize NDVI (Normalized Difference Vegetation Index) values in particular soil types.

10.6 Other cell-based analysis operations

The cell-based operations for continuous fields given here are merely the most common subset found in GIS and image analysis. Other operations of similar type have been developed in the areas of mathematical morphology (see Shih 2009) and **cellular automata** (see Liu 2008).

Temporal change With gridded data it is very easy to adapt the operations given in this chapter and also the attribute calculations of Chapter 7 in such a way that they can be carried out many times, thereby providing a means to model dynamic processes. Most operations with cellular automata involve temporal change.

5	4	11	7	8
3	2	9	10	11
3	2	4	5	8
7	4	6	8	9
5	6	4	7	12

Values

Zones

Zonal sum

Zonal mean

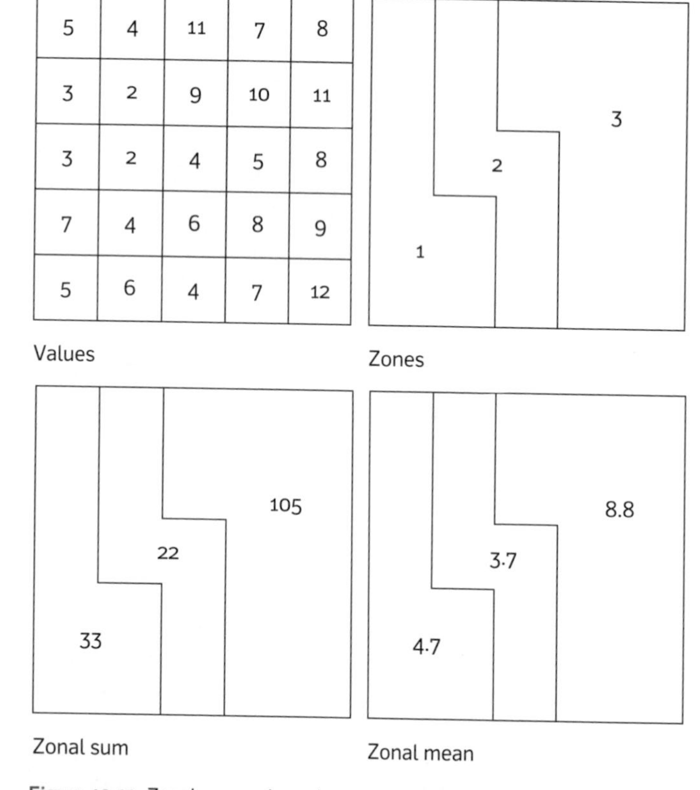

Figure 10.19 Zonal sum and zonal mean operators, given the values in the top left and the zones on the top right

10.7 First and higher order derivatives of a continuous surface

Chapter 11 focuses on one form of continuous surface—the digital elevation model (DEM). The remainder of this chapter introduces some key products which can be derived from DEMs, providing some context for Chapter 11. For the purposes of this chapter, a DEM is defined as a model of a physical surface—in the case of a raster-based model, the cells of the grid are elevations about some arbitrary datum. Detailed discussions about DEM types and sources, as well as further methods for their analysis, are provided in Chapter 11.

Because the gridded surface is supposed to be mathematically continuous, it is in principle possible to calculate the mathematical derivatives at any location. In practice, because the surface has been discretized, the derivatives are approximated either by computing differences within a square filter or by fitting a polynomial to the data within the filter. The two first order derivatives are the slope and the aspect of the surface; the two second order derivatives are the **profile convexity** and **plan convexity** (Evans 1980). Slope is defined by a plane tangent to the surface as modelled by the DEM at any given point, and comprises two components: **gradient** (the maximum rate of change of altitude) and aspect (the compass direction of this maximum rate of change). These terms follow the terminology of Evans (1980); many users of GIS packages use 'slope' to mean 'gradient' as just defined. Gradient and aspect are sufficient for many purposes, being the first two derivatives of the altitude surface or hypsometric curve, but for geomorphological analysis, the second differentials, convexity (the rate of change of slope expressed as plan convexity and profile convexity) and **concavity** (i.e. negative convexity), are also useful. Gradient is usually measured in per cent, degrees, or radians, aspect in degrees (converted to a compass bearing), while convexity is measured in degrees per unit of distance (e.g. degrees per 100 m).

Using directional filters to estimate slope and aspect

The derivatives of the hypsometric curve are usually computed locally for each cell on the altitude matrix from data within a 3 × 3 cell kernel or 'window' that is successively moved over the map (Figure 10.20). The simplest finite difference estimate of gradient in the x direction at point i,j is the *maximum downward gradient*:

$$[\delta z / \delta x]_{ij} = \max[(z_{i+lj} - z_{i-lj})/2]/dx \qquad 10.6$$

where δx is the distance between cell centres. (Note that for comparisons along diagonals the $\sqrt{2}$ correction to δx should be applied to reflect the greater distance between cell centres; see Section 10.8). This estimator has the disadvantage that local errors in terrain elevation contribute quite heavily to errors in slope. A better, much-used, second-order finite difference method (Zevenbergen and Thorne 1987) uses a second order finite difference algorithm fitted to the four closest neighbours in the window. This gives the slope by

$$\tan S = [(\delta z / \delta x)^2 + (\delta z / \delta y)^2]^{0.5} \qquad 10.7$$

where z is altitude and x and y are the coordinate axes.

The aspect is given by

$$\tan A = -(\delta z / \delta y)/(\delta z / \delta x) \qquad (-p < A < p) \qquad 10.8$$

Zevenbergen and Thorne (1987) show how these attributes and the convexity and concavity are computed from a six-parameter quadratic equation fitted to the data in the kernel—see Box 10.1.

A third order finite difference estimator using all eight outer points of the window, given by Horn (1981), is:

for the east–west gradient

$$[\delta z / \delta x] = [(z_{i+1,j+1} + 2z_{i+1,j} + z_{i+1,j-1}) - (z_{i-1,j+1} + 2z_{i-1,j} + z_{i-1,j-1})]/8\delta x \qquad 10.9$$

and for the south–north gradient

$$[\delta z / \delta y] = [(z_{i+1,j+1} + 2z_{i,j+1} + z_{i-1,j+1}) - (z_{i+1,j-1} + 2z_{i,j-1} + z_{i-1,j-1})]/8\delta y \qquad 10.10$$

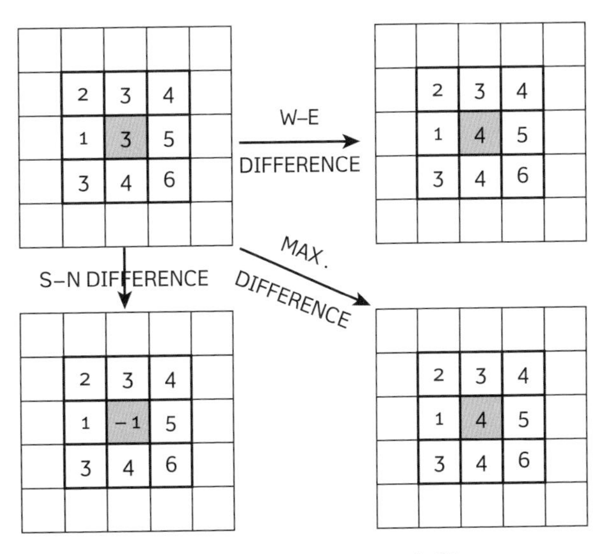

Figure 10.20 Computing derivatives with simple filters

Box 10.1 Computing slopes using Zevenbergen and Thorne's method

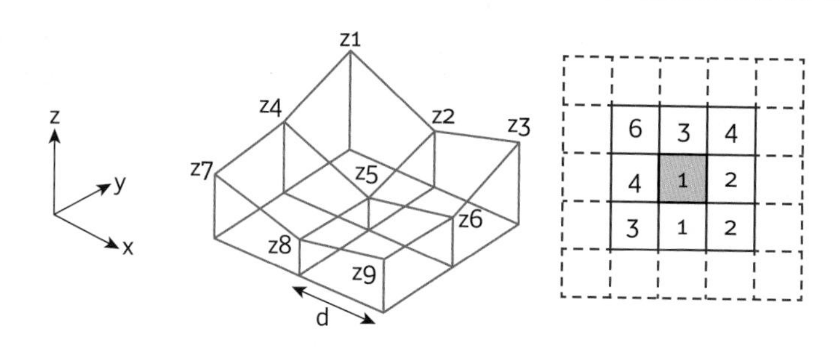

$A = [(z1 + z3 + z7 + z9)/4 - (z2 + z4 + z6 + z8)/2 + z5]/d^4$

$B = [(z1 + z3 - z7 - z9)/4 - (z2 - z8)/2]/d^3$

$C = [(-z1 + z3 - z7 + z9)/4 + (z4 - z6)/2]/d^3$

$D = [(z4 + z6)/2 - z5]/d^2$

$E = [(z2 + z8)/2 - z5]/d^2$

$F = (-z1 + z3 + z7 - z9)/4d^2$

$G = (-z4 + z6)/2d$

$H = (z2 - z8)/2d$

$I = z5$

SLOPE

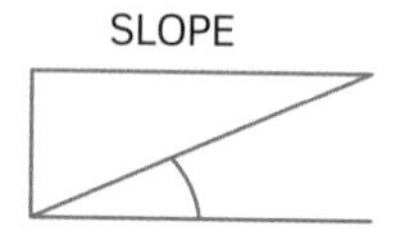

ASPECT

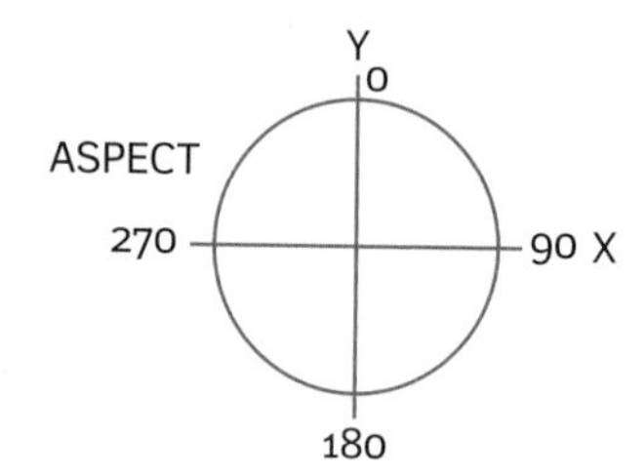

SLOPE = SQRT($G^2 + H^2$)

ASPECT = arctan(−H/−G)

Profile curvature

Plan curvature

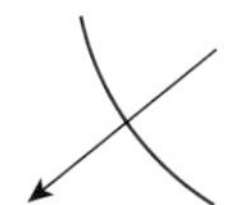

$PrC = 2(DG^2 + EH^2 + FGH)/(G^2 + H^2)$

$PlC = -2(DH^2 + EG^2 - FGH)/(G^2 + H^2)$

concave = positive

convex = negative

Alternative methods fit a multiple regression to the nine elevation points in the 3 × 3 window and derive the slope and aspect from that.

Given the variety of methods available for computing slope and aspect, it is useful to know which is best. Skidmore (1989) reviewed six methods of estimating slope and aspect, including those given in this section. He concluded that both the second and third methods given above were superior to the simple algorithm in equation 10.6, but that there was little difference in the results returned by Horn's method and the polynomial method. Li et al. (2004) provide a good summary of alternative approaches. Jones (1998) has carried out another analysis of eight algorithms for computing slope and aspect using both real and synthetic DEM surfaces. Based on analysis of slope and aspect algorithms, as tested by RMS (root mean squared) residual error values derived from the difference between the values generated by the algorithm and the true values for the test surfaces, the algorithms of Zevenbergen and Thorne (1987) and Horn (1981) perform well. These algorithms are used by several well-known GIS packages, so there is general agreement on the better algorithms. These findings are supported by Zhou and Liu (2004), whose comparison was based on deriving slope from synthetic surfaces with no data errors and with errors (random noise) added to the surfaces.

Displaying maps of slope and aspect After the appropriate derivative has been calculated for each cell in the altitude matrix, the results may need to be classified in order to display them clearly on a map. For visual appreciation, display of the thematic data (slope, aspect, etc.) draped over a digital elevation model is very effective (see Section 5.6). Figure 10.21 gives examples of the slope, profile convexity, and plan convexity for a small catchment with moderate relief having a 30 × 30 m resolution.

Aspect maps can be displayed by nine classes—one for each of the main compass directions N, NE, E, SE, S, SW, W, NW, and one for flat terrain (see Figure 10.22). An alternative is to use a continuous, circular greyscale which is chosen so that NE-facing surfaces are lightest: this gives a realistic impression of a 3D surface.

Slope often varies quite differently in different regions, and although adherents of standard classification systems usually want to apply uniform class definitions, the best maps are arguably produced by calibrating the class limits to the mean and standard deviation of the frequency distribution at hand. Six classes, with class limits at the mean, the mean ±0.6 standard deviations, and the mean ±1.2 standard deviations usually give very satisfactory results (Evans 1980; see also Mitasova et al. 1995 for original ways of displaying slope information). Figure 10.23 shows a map of slope using a continuous colour scale.

It is a general feature of maps derived from altitude matrices that the images are more noisy than the original surface—in general, roughness increases

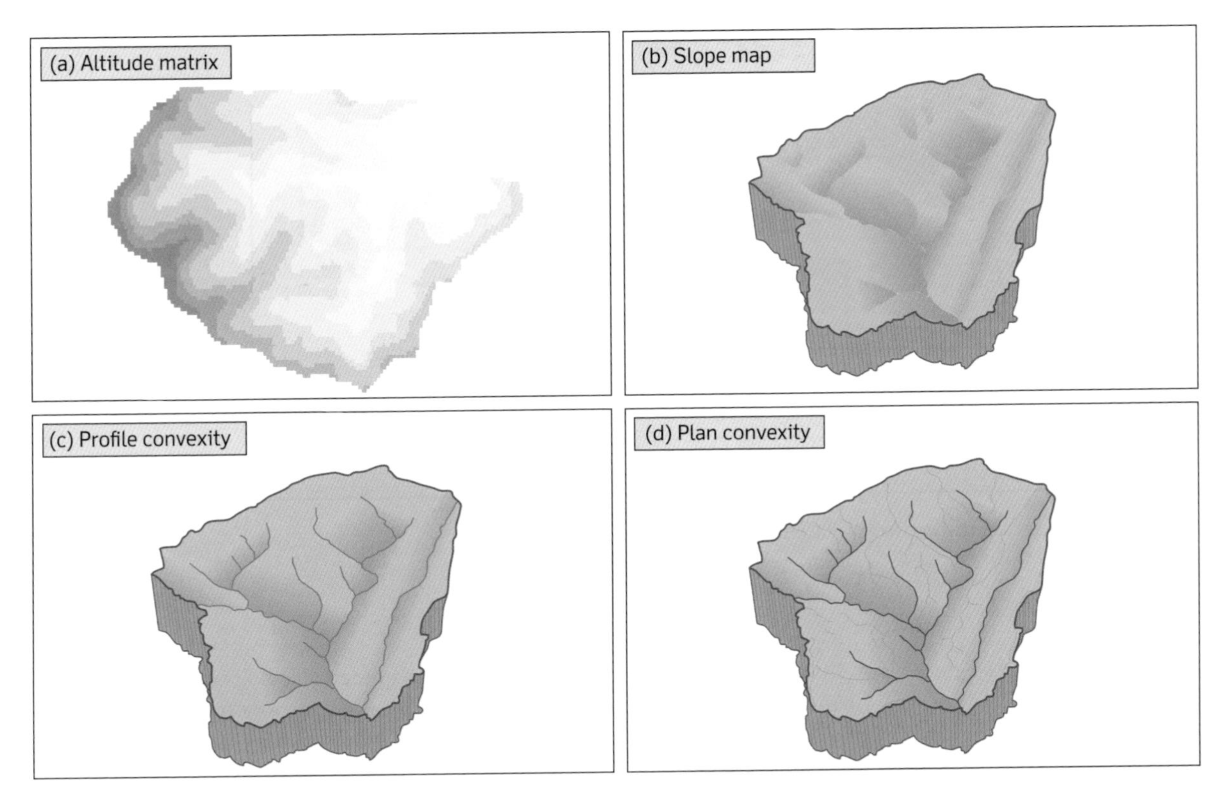

Figure 10.21 First and second order derivatives of a DEM

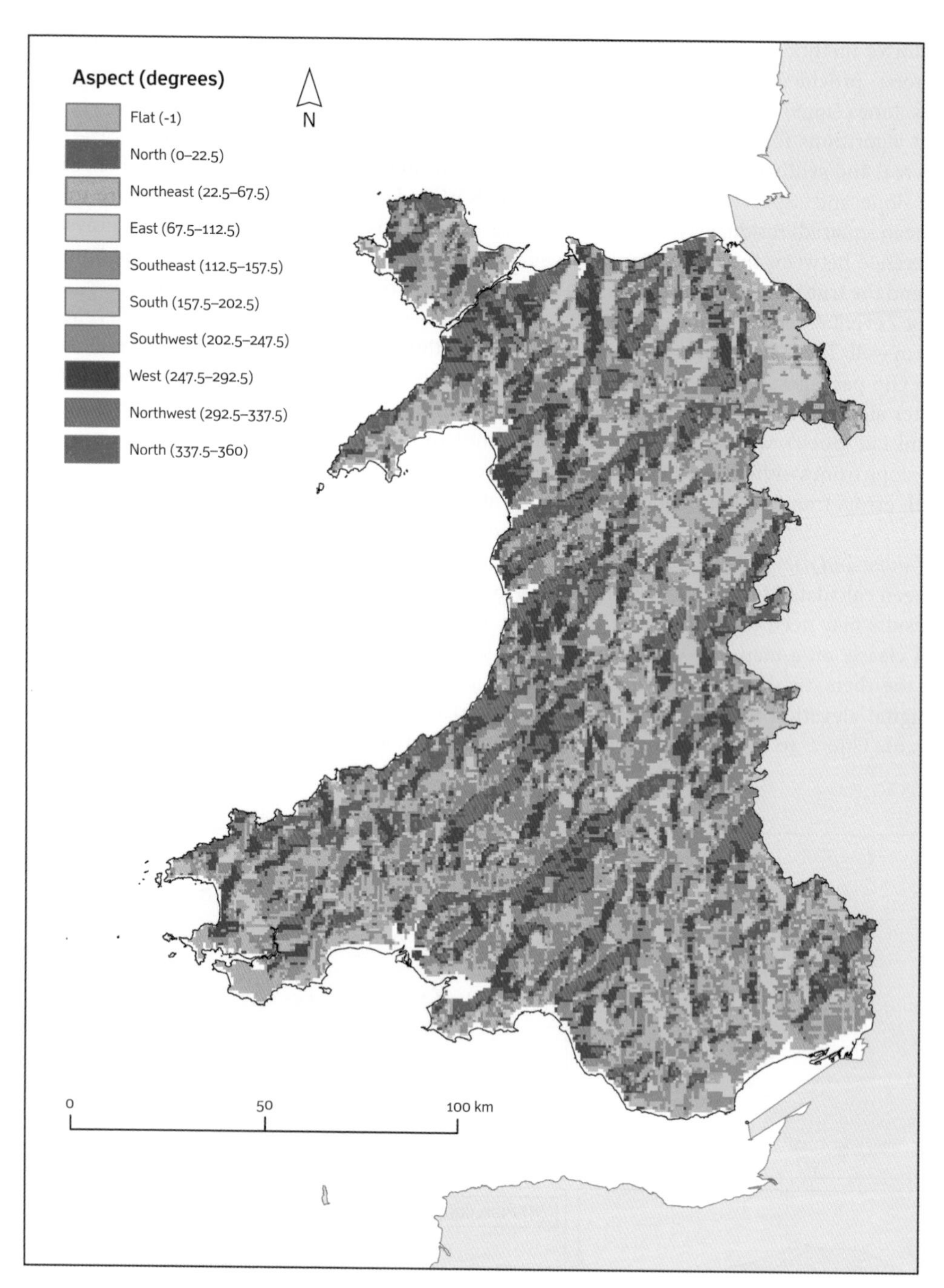

Figure 10.22 Aspect map, derived from digital elevation model of Wales

with the order of the derivative. The derivatives can be smoothed by a low-pass filter before the results are plotted. Smoothing the DEM with a low-pass filter before computing the derivatives also reduces noise, but this is at the expense of removing the extremes from the data and results in underestimates of slope angles.

Slope and aspect maps can also be prepared from TINs by computing the slope or aspect for each triangular facet separately and then shading it according to the gradient class.

The following section defines some other key products which can be derived from DEMs.

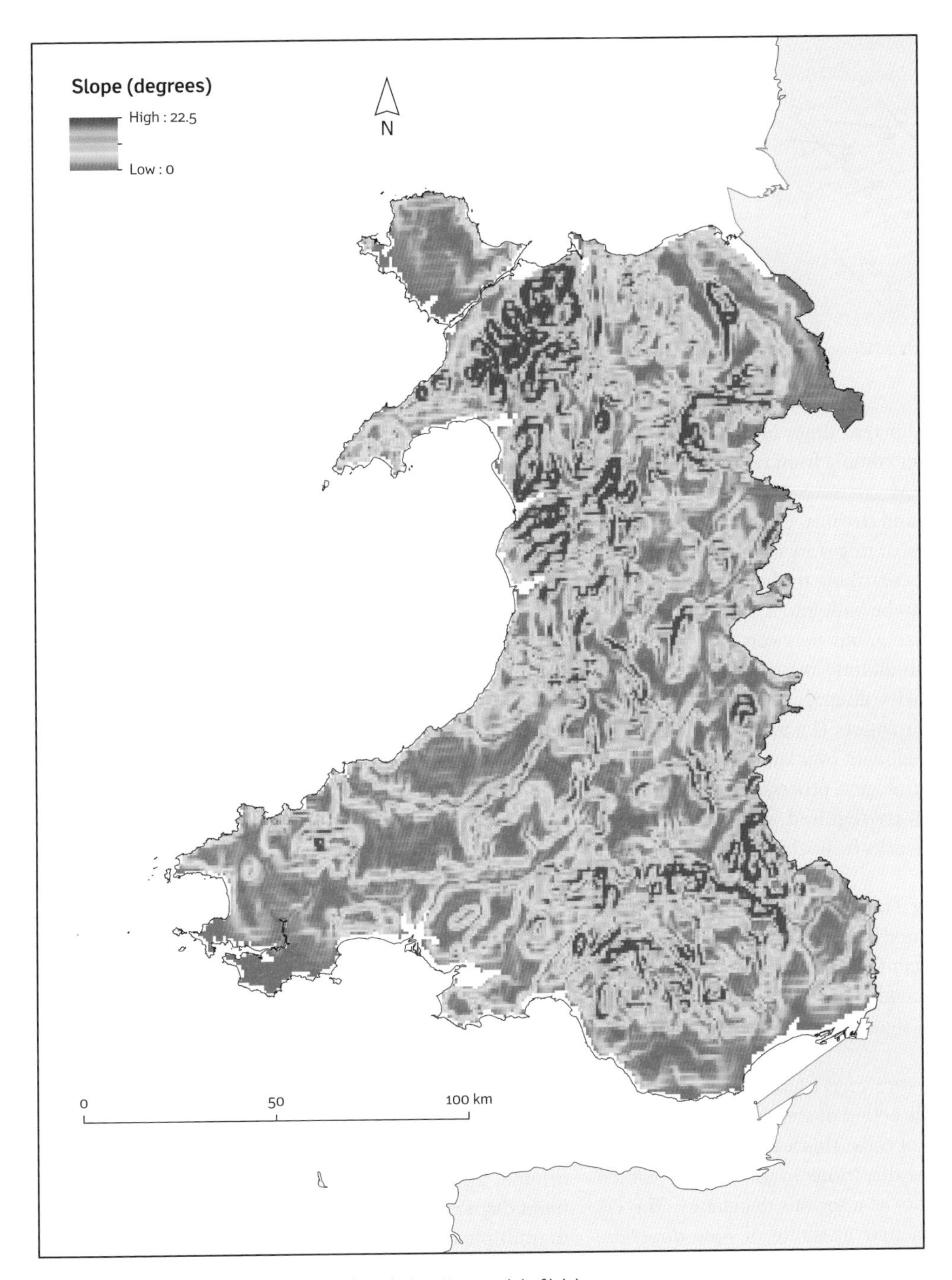

Figure 10.23 Slope map, derived from digital elevation model of Wales

10.8 Deriving surface topology and drainage networks

Before drainage basins and drainage networks could be analysed quantitatively, they had to be laboriously copied from aerial photographs or printed topographical maps. Besides being tedious, this work inevitably led to an increase in the errors in the data. In areas of gentle relief it is not always easy to judge by eye on aerial photographs where the boundary of a catchment should be, and under thick forest it may be difficult to even see the streams. Even on very detailed topographical maps the drainage network as represented by the drawn blue lines may seriously underestimate the actual pattern of all potential water courses. It could be useful, for example, to be able to separate water-carrying channels

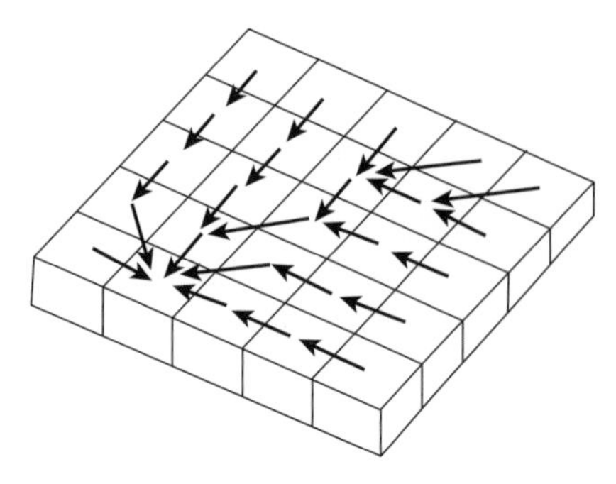

Figure 10.24 Local drain direction vectors to indicate steepest downhill path

from dry channels at different times of the year, with the information about water coming from remote sensing and the channels from a DEM.

Drainage networks and streams, catchments (or watersheds), drainage divides, or ridges are important properties of real landscapes that contribute to the understanding of material flows. They can be built into a TIN or an altitude matrix by direct digitizing, but they can also be derived automatically from the altitude matrix (Li et al. 2004). Automatic derivation of the drainage network has provided powerful tools for hydrologists (Vieux 2004) to estimate the flow of water and sediment over landscapes and to link dynamic models of hydrological processes to GIS.

The following steps are required when automatically deriving drainage networks from altitude matrices.

1. Determine routing The flow of material over a gridded surface is determined by considering the direction of steepest downhill descent. There are several algorithms for calculating this, called the D8 or 8-point pour algorithm, slope-weighted algorithms, and stream-tube algorithms (Moore 1996).

The D8 (deterministic) algorithm approximates the flow direction by the direction of steepest downhill slope within a 3 × 3 window of cells. This leads automatically to a discretization of flow directions to units of 45°, which is seen by some authors as a serious deficiency. The D8 algorithm computes a new attribute of *flow direction,* which can take eight different directional values that can be expressed as degrees or as numeric codes. One useful implementation is to indicate the directions by the numbers given on the numeric pad of a computer keyboard, so that a *sink* or *pit* is indicated by a '5', but other direction numbering conventions are also used (Moore 1996).

7	8	9
4	5	6
1	2	3

The resulting new grid overlay is called the set of **local drain directions**, or **ldd** (Figure 10.24). Each cell contains a *directional*-type integer of value *FD* (flow direction) where:

$$FD = d \text{ where } d = f \text{ for } \max_{(f=1,8)}[w_f \left| z_{ij} - z_{i,i\pm1,j,j\pm1} \right|] \qquad 10.11$$

The distance weight w_f is 1 for NSEW neighbours and $1/\sqrt{2}$ for diagonals (to reflect longer distances across, rather than along, cells).

Figure 10.25b is an example of the ldd map displayed over the background of the DEM. Because of its simplicity, the D8 algorithm has been incorporated in several commercial GIS. On uniformly sloping surfaces it produces long, linear flow lines, and uniform flow directions, and it is not uncommon to get parallel flow lines that do not converge. It cannot model flow dispersion.

The Rho8 (random) algorithm is a statistical version of the D8 algorithm which was introduced to represent better the stochastic aspects of terrain. It replaces the w_f of $1/\sqrt{2}$ for diagonals by $1/(2 - r)$, where r is a uniformly distributed random variable between 0 and 1. Moore (1996) claims that this simulates more realistic flow networks, though like the D8 algorithm it cannot model dispersion (Wilson and Gallant 2000a, 2000b). A D8-based alternative to the Rho8 which might be even more realistic can be obtained by Monte Carlo simulation. An RMS error can be added to the DEM and the D8 algorithm is used to compute a network, which is stored. This is repeated, for example 100 times, to yield a most probable network (see Chapter 12). The extra advantage of the Monte Carlo simulation is that the error on the DEM can be adjusted to realistic levels and probabilistic flow paths are generated.

FD8 and FRho8 algorithms are modifications of the original algorithms, allowing flow dispersion or catchment dispersion to be modelled. Flow can be distributed to multiple nearest-neighbour nodes in situations where there is overland flow, rather than concentration of flow in channels, where the D8/Rho8 algorithms are used. The proportion of flow to the multiple downstream nodes is computed on a slope-weighted basis (Wilson and Gallant 2000b).

Stream tube methods Costa-Cabral and Burges (1993) determine the amount of flow as a fraction of the area of the source pixel entering each pixel downstream as determined by the intersection of a line indicating the drainage direction (aspect) and the edge of the pixel.

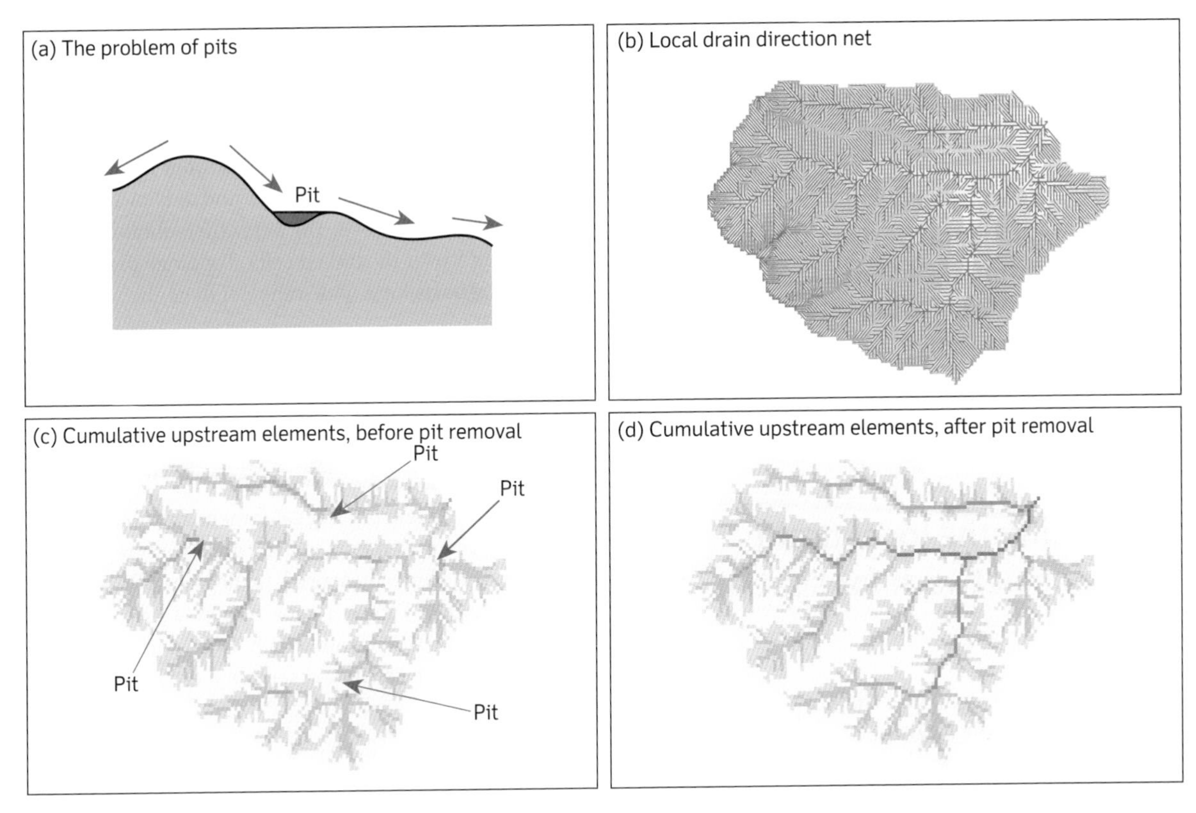

Figure 10.25 Deriving a drainage network from a gridded DEM

2. Removal of pits When a smooth continuous surface is approximated by a square grid it is inevitable that some cells will be surrounded by neighbours that all have higher elevations. These pits could be real closed depressions or merely artefacts of the gridding process. Pits that are artefacts are often generated in narrow valleys where the width of the valley bottom is smaller than the cell size, and they can occur at all levels of resolution. They can also occur in areas of gentle relief through errors in interpolation (e.g. Figure 10.25a).

The problem with artefact pits is that they disrupt the drainage topology and need to be removed to obtain a continuous ldd net. They can be removed by one of two strategies: cutting through or filling up. Cutting through one or more layers of boundary cells to find the next downstream cell requires enlarging the size of the search window to find a cell or series of cells of the same elevation or lower as the core (pit) cell. Once this path has been found, the appropriate topological links are written into the cells along the path, irrespective of their true elevation. Filling up involves increasing the elevation of the core cell until it is equal to one or more of its neighbours, and then examining whether the neighbour drains downhill to another destination. If this does not happen, the elevation is increased again until a linkage is found.

Pit removing is an interactive process regarded by some as a necessary evil. Hutchinson (1989) has developed a spline interpolator for ensuring that pits do not occur, but it is not always sensible to remove all pits automatically because closed and semi-closed depressions may be real features in some landscapes. Yamazaki et al. (2012) describe an algorithm for adjusting DEMs derived from spaceborne sensors (SRTM3 DEM; Farr et al. 2007) using drainage network information with the intention of removing pits in the DEMs.

Once pits and plateaux have been identified, then the DEM can be adjusted to remove them. For large and complex data sets with large-area pixels a practical alternative is to obtain a vector representation of the river network, convert the river vectors to grid cells, and then 'burn in' the river cells at a lower level in the drainage network.

Example of a generic command for extracting surface topology from a gridded DEM

lddmap = (dem.map, a, p1, p2, p3, p4,...)

where lddmap is the derived topology, a is the algorithm used, and p1, p2, p3,... are parameters for removing pits according to their outflow depth, core volume, core area, etc.

10.9 Using the local drain direction network for spatial analysis

Irrespective of the algorithm used to compute the flow directions, the result is to create a gridded overlay in which the surface topology has been made explicit (e.g. Figure 10.25b). This ldd network is extremely useful for computing other properties of a digital elevation model because it explicitly contains information about the connectivity of different cells. This knowledge makes it possible to address problems of directed flow and the transfer of fluids or material without recourse to ad hoc search windows.

Accumulating fluxes of material over a net

Because, in a topologically correct network, each cell is linked to a downstream neighbour, it is very easy to compute attributes such as the cumulative amount of material that passes through each cell. The **accumulation operator** computes the new state of the cell as the sum of the original cell value plus the sum of the upstream elements draining to the cell

$$S(c_i) = S(c_i) + \sum_{u}^{n} (c_u) \qquad 10.12$$

If the material value for each cell is 1, the result gives the **upstream element map**, or in other words the cumulative number of cells upstream of the current cell that discharge through that cell. The upstream element map is usually displayed on a logarithmic scale.

If the material value is supplied from another overlay, for example effective precipitation, then the accumulation operator will compute the cumulative flow over an ideal surface. For example, it is easy to compute a mass balance for each cell in terms of

$$S = P - I - F - E \qquad 10.13$$

where S is surplus water per cell, P is input precipitation, I is interception, F is infiltration, and E is evaporation. The cumulative flow over the net is then obtained by accumulating S over the linked cells. Topological networks are also the basis for a wide range of dynamic modelling tools in GIS.

The upstream element map can itself be useful for computing other indices of the terrain. For example, a **wetness index map** can be defined as:

$$\text{wetnessindexmap} = \ln\left(A_s / \tan\beta\right) \qquad 10.14$$

where A_s is the contributing catchment area in m^2 (number of upstream elements $\times$ g the area of each grid cell) and β is the slope measured in degrees (Wilson and Gallant 2000c). Figure 10.26a shows a wetness map draped over the DEM from which it was derived.

The **Stream Power Index** (see Fried et al. 2000) is defined as

$$\omega = A_s * \tan\beta \qquad 10.15$$

This is directly proportional to the stream power $P = pgq$ $\tan\beta$, where p is the density of water, g is the acceleration due to gravity, and q is the overland flow discharge per unit width, which is the rate of energy expenditure over time and is a measure of the erosive power of overland flow.

The Sediment Transport Index is defined as

$$\tau = [A_s / 22.13]^{0.6*} [\sin\beta / 0.0896]^{1.3} \qquad 10.16$$

This index characterizes the processes of erosion and deposition, in particular the effects of topography on soil loss; it resembles the length-slope factor of the Universal Soil Loss Equation (see Winchell et al. 2008) but is applicable to three-dimensional surfaces. Figure 10.26b shows that the sediment transport index can vary along the length of a stream.

Other products that can be derived from the local drain direction map

Stream channels can be defined as cells having more than N contributing upstream elements. Stream channels can be determined by using a Boolean operator such as:

$$\text{Streams} = \text{if(upstreamelements} \geq 50 \text{ then 1 else 0)} \qquad 10.17$$

This creates a binary map in which all cells with 50 or more upstream elements are defined as belonging to the set of 'streams' (Figure 10.26c).

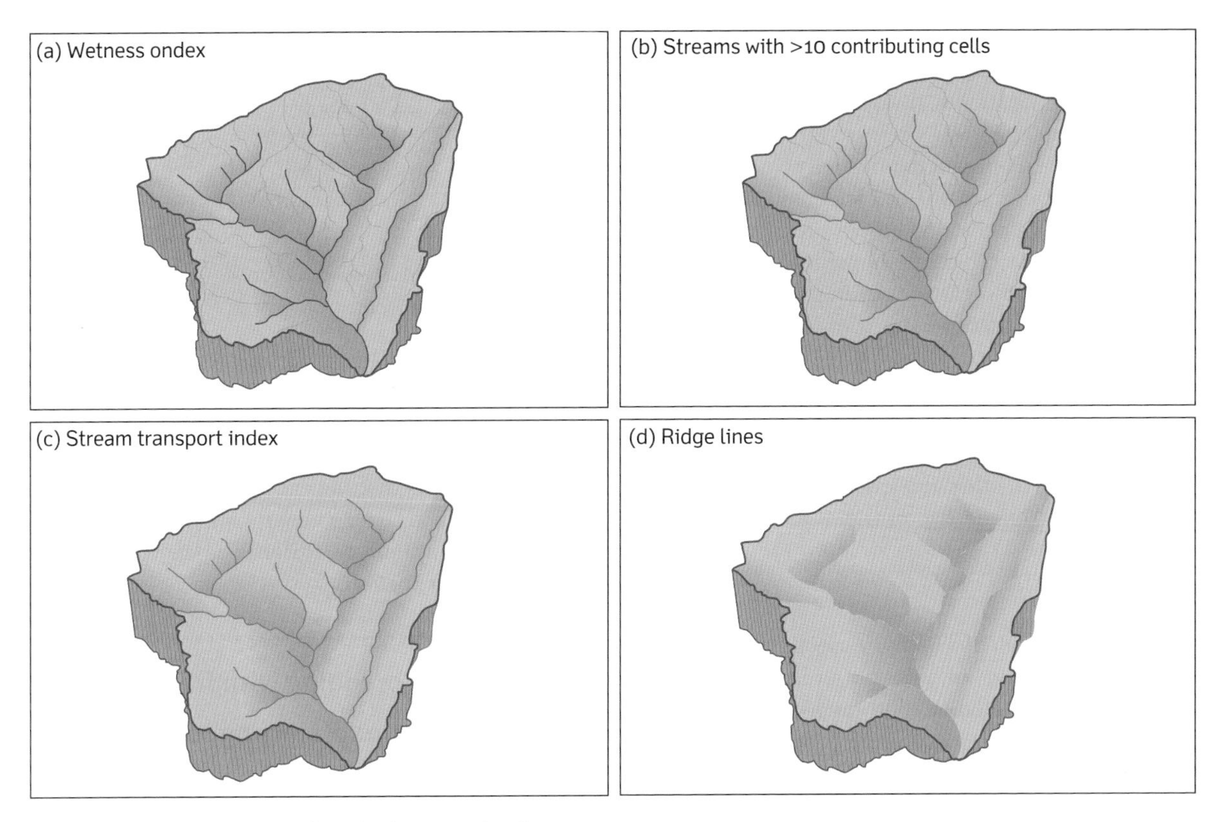

Figure 10.26 Properties derived from the drainage network

Ridges. By definition, ridges have no upstream elements, so selecting all cells with an upstream elements value of 1 provides a first estimate of ridges (Figure 10.26d).

Catchments. Because all cells that drain through a given cell are part of the catchment of that cell, counting upstream over the ldd automatically computes the area and defines the catchment of the cell. A catchment mask can be computed by assigning a 'one' to all cells in the catchment and a 'zero' to those outside. This can be used as a 'cookie cutter' to identify catchment-specific data from remotely sensed imagery or other sources at the same level of resolution. Using a high-pass or edge filter yields a linear catchment boundary which can be vectorized by converting the cell representation to a chain code (Chapter 3).

The **slopelength operator** is similar to the accumulation operator but it computes a new attribute of a cell as the sum of the original cell value and the upstream cells, multiplied by the distance travelled over the network, d_u.

$$S(c_i) = S(c_i) + \sum_u^n (c_u^* d_u) \qquad 10.18$$

Figure 10.27 shows slope lengths computed in this way. The distance travelled can be a simple Euclidean distance depending on the size of the cells (1 * unit cell size for N–S and E–W; 1.414 for diagonals), or it can include a friction term to deal with resistances within the cells on the network (see Section 10.10).

Difficulties with drainage nets derived from altitude matrices

Though there are many benefits of deriving the drainage network from the altitude matrix, there are also

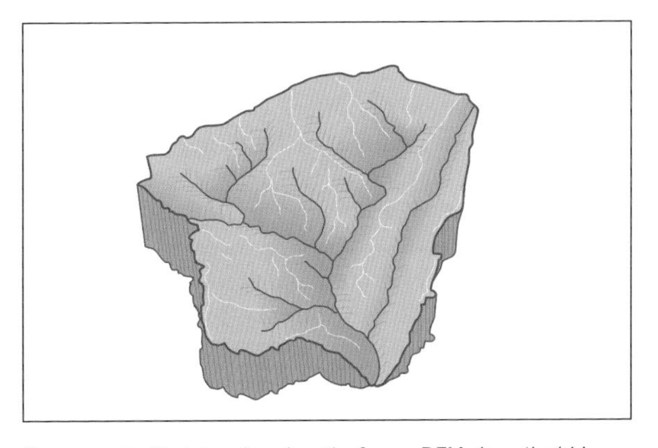

Figure 10.27 Deriving slope lengths from a DEM along the ldd

several difficulties. One is the problem of modelling dispersion and diffusion. Others involve the errors in DEM—landforms are not always smooth and differentiable. The choice of cell size may affect estimates of slope, aspect, and stream connectivity. In the altitude matrix the streams are one cell wide, but real streams vary in width over their length and may be narrower or wider than the cell dimensions. Modelling the accumulation of flows assumes simple gravity-driven processes and ignores the inertia of fast-flowing water masses, which need to be approached using kinematic wave equations. In spite of all these difficulties, altitude matrix DEMs are finding a place in the hydrologist's toolkit and the analysis of continuous surfaces can lead to useful results in other application areas, as shown by the examples given in Chapter 11.

Clumping

Very often the result of a Boolean selection or classification on the attributes of cells will result in sets of cells that are spatially contiguous but which cannot be identified as being part of a spatial 'entity'. The **clump** operator examines every cell to see if any of its immediate neighbours in a 3 × 3 window have the same class—if so, then both cells are assigned to the same clump and given a value that identifies that clump as distinct from others. The result is that each contiguous group of cells is aggregated into a larger spatial unit, which could be useful for many purposes. For example, identifying all 'ridges' via the upstream element map may create several loose aggregations of cells that belong to different ridges. Applying the clump operator will identify each cell with a specific clump.

10.10 Dilation/spreading with or without friction

This is not a window operation, but a continuous analogue of the dilation or buffering operations on exact entities. Whereas dilation (or buffering) of exact entities is usually limited to isotropic and isomorphic spreading (a buffer around a circle is just a larger circle—Figure 10.28), spreading over a continuous surface can be carried out heterogeneously to reflect the variations in resistance to the spreading process (Figure 10.29).

In non-isotropic spreading, two components contribute to the accumulation of values from the starting point. The first is distance, counted as cell steps or in

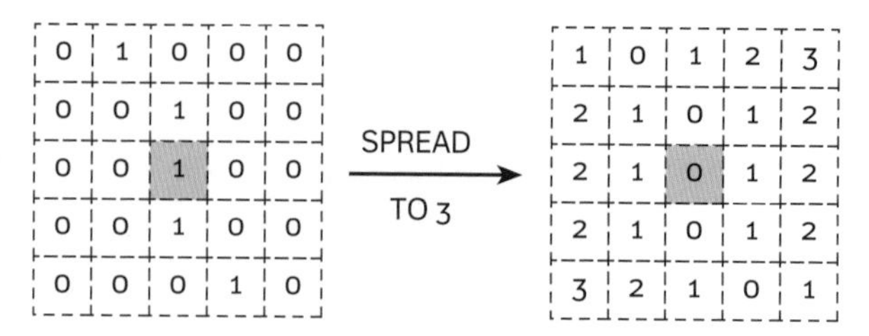

Figure 10.28 Isotropic spreading with grids

real units. The second depends on the attributes of the cells through which the distance accumulation takes place. The larger the value of the 'friction' attribute, the greater the accumulation of 'distance' when traversing a cell. The result is that the effective spreading distance accumulates much faster where resistance is greatest, so that geometrically longer paths may be 'cheaper' ways to reach a given destination.

Operators for spreading with friction

Spreadmap = spread(startingpoints, v, friction)

where startingpoints gives the locations (cells) from which to start the spreading or buffering, v is an initial value, and friction gives the internal resistance on a cell-by-cell basis

---

Both simple and frictional spreading can be used to estimate slope lengths perpendicular to the stream nets derived from the upstream element maps (or from any other linear feature such as roads or railways).

Commands to create slope length and inverse slope length

slopelength **= spread(strm, 0, slp.map)**

slmx **= mapmaximum(*slopelength*)**

report slopelength = ((1 – (*slopelength*/ *slmx*))**slmx*)**

where *strm* is the map of stream locations, *slp.map* is the map of slopes, and *slopelength* is the resulting slope length perpendicular to streams

---

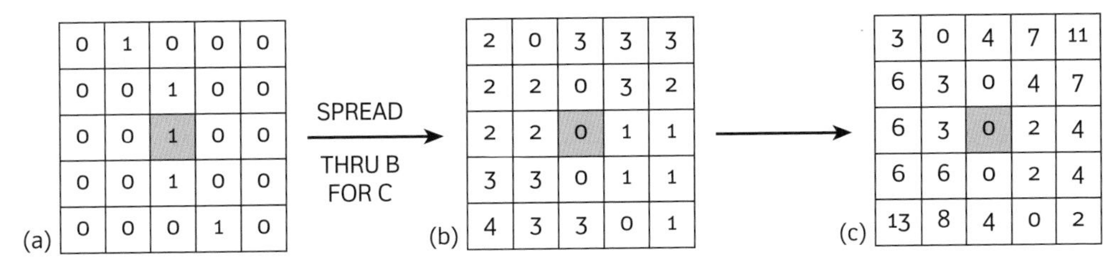

Figure 10.29 Spreading through a resistance function

Non-isotropic spreading yields a pit-free, continuous surface. The 'drainage network' over such a surface defines the set of optimal paths from each cell to the starting point. Computing the 'catchments' and upstream elements for such a surface can indicate the best routes to follow.

Instead of spreading over all cells, non-isotropic spreading can be confined to follow the routes defined by the ldd map, or by a subset of the ldd. When the resistance varies over the network this can reveal areas where flow problems might accumulate.

10.11 Summary

This chapter has demonstrated that there is a large range of products that can be derived from continuous surfaces that have been discretized as regular grids (Table 10.1). Some derived data, such as slope and aspect (and line of sight, hill shading, and irradiance, as considered in the following chapter), can also be easily obtained from TINs, but it is not usual to find systems that provide facilities for deriving the topology of the surface and for computing material flows over these derived networks. Frequently, the users of TINs will have input their networks explicitly as topologically connected lines or objects.

The most commonly encountered continuous field is the DEM, and most of the derivatives mentioned above have a direct bearing on the use and interpretation of terrain elevation. The operators presented can be used on any continuous field, however, such as remotely sensed images or the results of interpolation or spatial modelling, as will be described in the following chapter.

Table 10.1 summarizes the functional capabilities of a GIS that can deal both with exact entities and continuous fields. Frequently, but not always, the continuous surfaces represent landform, as discussed in more detail in the next chapter. These methods have applications in many fields, not just including hydrology, but also erosion and land degradation studies, forest management, and soil and water pollution. Although all the examples relate to physical problems of the landscape, these kinds of analyses can also be applied to any study in which one or more attributes can be modelled as a continuous surface, such as surfaces of market potential, exposure to disease, or economic well-being.

Table 10.1 Functional capabilities of GIS for analysis of entities and continuous fields

Geometric	Convert geographic coordinates from one correction cartographic projection to another
Interactive	Interactively update and edit geographic editing and attribute data
Sorting	Sort attribute or geographic data as required
Location	Locate entities having defined sets of attributes
Summarize	Summarize attributes per geographic entity (point/line/polygon/cell)
Compute statistics	Compute statistics (means, areas, enclosures, etc.) for points/lines/ polygons/cells
Proximity	Conduct nearest neighbour and proximity searches—create buffer zones and carry out corridor analyses
Interpolate	Interpolate from point data to regular grid or isolines (contours)
Block diagram	Compute block diagrams of three-dimensional data
Overlay analysis	Overlay and combine several maps in either vector (polygon) or raster (grid) mode using Boolean (AND/OR/NOT) logic and arithmetical functions/filters to manipulate both the geographic and the attribute data
Polygon to raster	Convert graphic representation from polygon to grid cell representation
Edge detection	Semi-automated detection of edges of images in raster representation
Network analysis	Find shortest path along a road network with weighted route parameters for traffic density Link specific entities (roads, cables) to each other and follow the network through
Digital terrain analysis	Represent landform as a 3D surface Compute slope, aspect, intervisibility, rate of change of slope, shaded relief, direction of flow, determine watershed boundaries
Models	Ability to interface with simulation models

? Questions

1. Compare and contrast the effects of (a) spatial filtering and (b) polygon reclassification and entity merging, for generalizing a soil or land use map.
2. Provide an outline of a procedure which could be used to identify edges in an image and convert the edges into vector features.
3. Outline an application where the zonal sum operator might be useful.

→ Further reading

▼ Li, Z., Zhu, Q., and Gold, C. (2004). ***Digital Terrain Modeling: Principles and Methodology***. CRC Press, Boca Raton, FL.

▼ Lloyd, C. D. (2010). ***Spatial Data Analysis: An Introduction for GIS Users***. Oxford University Press, Oxford.

▼ Moore, I. D. (1996). Hydrological modeling and GIS. In M. F. Goodchild, L. T. Steyaert, B. O. Parks, C. Johnston, D. Maidment, M. Crane, and S. Glendinning (eds.) ***GIS and Environmental Modeling: Progress and Research Issues***. GIS World Books, Fort Collins, CO, pp. 143–148.

▼ Wilson, J. P. and Gallant, J. C. (2000). Digital terrain analysis. In J. P. Wilson and J. C. Gallant (eds.) ***Terrain Analysis: Principles and Applications***. John Wiley, New York, pp. 1–27.

Digital Elevation Models

The shape of the surface of the earth is fundamentally important to humans. Terrain form impacts on the direction of the flow of water, the clarity of the signal received using a mobile telephone, the perceived degree of attractiveness of the scenery in any one place, as well as many other factors. The measurement of the shape of the surface of the earth (landform) is a concern in many disciplines, and a wide array of techniques exist for (a) measuring the height at any location above some datum and (b) producing a model of the terrain for some area. This chapter reviews some approaches for capturing data on landform and builds on the previous chapter to outline some further approaches for deriving information from digital elevation models (DEMs). In addition, some tools which make use of DEMs are outlined. The central importance of DEMs in many GIS applications provides justification for a separate chapter which describes DEMs and their applications. The previous chapter discussed surface derivatives and derivation of drainage networks and friction surfaces from DEMs; the present chapter describes some further approaches which make use of data on landform, with a particular focus on visualization and visibility.

Learning objectives

By the end of this chapter, you will:

- **understand some key concepts and classes of approaches which can be used to work with digital elevation models**
- **have developed a knowledge of some key ways of analysing digital elevation models**
- **be able to apply your knowledge to analyse the form of surfaces and derive intervisibility, shaded relief, and irradiance maps.**

Since the shape of the earth's surface affects a wide range of earth surface processes, DEMs are very widely used across the geosciences. DEMs are used extensively in geology, geomorphology, hydrology, and glaciology, among other disciplines—this chapter discusses what DEMs are, and how and why they are acquired and used. A DEM may be used directly where elevation is a defining value in a study (for example, to assess the effect of flooding on a particular area or to assess the visibility of one location from another location). Alternatively, users may be interested in deriving variables from the DEM that are a product of the elevation maps, such as slope (the rate of change of elevation over a distance) or aspect (the compass direction a part of the landscape is facing) (see Chapter 10). Additionally, there may be a desire to divide the landscape into subsets according to surface features (for example, river catchments).

Clearly, specific applications will necessitate access to DEMs with a wide range of different spatial resolutions. For example, some users require DEMs with a fine spatial resolution (5 m or finer), while for some applications such as broad-scale atmospheric modelling a DEM with a spatial resolution of 50 km may be suitable. A useful review of recent developments in digital elevation modelling is provided by Wilson (2012). More detailed accounts of DEMs and their applications are provided by Wilson and Gallant (2000a) and Li et al. (2004). Ruzickova (2012) provides an introduction to a special issue of the journal *Transactions in GIS* which presents some recent research on the development and use of DEMs.

Applications of digital terrain data can be divided into several principal classes, including:

- topographic mapping
- earth science
- civil engineering
- planning and resource management
- military.

The first application area is the focus of national mapping and other government agencies, which provide DEMs for use by others in many different applications. Methods for mapping topography are detailed below. DEMs of specific regions and at greater precision are developed and used by surveyors and civil engineers for the design of building schemes such as roads and airfields, or other activities that involve extensive modification of the landscape. A major application is the use of DEMs in volumetric analysis (cut-and-fill problems). The impact on the landscape of developments is a major concern (and is discussed below) which may be assessed through the use of DEMs. DEMs have been used for many years as a means to assess the visual impact of the landscape as viewed from a particular location, and they are now used widely in planning and resource management—in environmental and urban planning and other applications which deal with the management of natural resources (for example agriculture, forestry, and geological exploration). Risk assessment and disaster management are informed by DEMs. Also, service providers often rely on DEMs to help develop their facilities. For example, telecommunications companies use DEMs to help site mobile phone base stations (see, for example, Muralikrishnan et al.2007). In military contexts, battlefield management is a major application where movement of military personnel and equipment is a concern. Terrain intervisibility (viewshed) analysis has been widely applied in these contexts. Also, DEMs are a necessity for missile guidance systems: cruise missiles were one of the main drivers for the development of global DEMs. In addition, aircraft guidance systems and flight simulators (both military and civilian) make use of DEMs.

Some of the uses to which DEMs are put are summarized in Box 11.1.

DEMs are sometimes derived from point measurements or from contour lines, and thus there are links to Chapters 8 and 9 (however, interpolation is often not needed as DEMs may be derived directly in gridded data from remote sensing; exceptions include cases where there are gaps due to shadows and other factors, and the missing values must be estimated). There is a range of terms used to refer to ways of representing topographic form digitally, and different authors use these terms to

Box 11.1 Uses of DEMs

Some common uses of digital elevation models

- Storage of elevation data for digital topographic maps in national databases.
- Creation of digital and analogue orthophoto maps.
- Cut-and-fill problems in road design and other civil and military engineering projects.
- Three-dimensional display of landforms for military purposes (weapon guidance systems, pilot training) and for landscape design and planning (landscape architecture).
- Analysis of cross-country visibility (also for military and landscape planning purposes).
- Planning routes of roads, locations of dams, etc.
- Statistical analysis and comparison of different kinds of terrain.
- Source data for derived maps of maps, aspect, profile curvature, shaded relief insolation, and hydrological and ecological modelling.
- As a background for displaying thematic information or for combining relief data with thematic data such as soils, land use, or vegetation.
- Providing data for simulation models of landscapes and landscape processes.

mean different things. The terms digital elevation model (DEM) and digital terrain model (DTM) are often used interchangeably. The term digital surface model (DSM) is also sometimes used. Some authors regard a DTM as including 'above-surface' features such as vegetation and buildings, whereas a DEM includes information only on the ground surface.

11.1 Methods of representing DEMs

The variation of surface elevation over an area can be modelled in many ways. DEMs can be represented either by mathematically defined surfaces or by point or line images. Line data can be used to represent contours and profiles, and critical features such as streams, ridges, shorelines, and breaks in slope. In GIS, DEMs are modelled by regular grids (altitude matrices) and triangular irregular networks (TINs). The two forms are inter-convertible, and the preference for one or the other depends on the kind of data analysis that needs to be carried out.

Altitude matrices are the most common form of discretized elevation surface. The altitude matrix is a regular grid of numbers which represent elevation above an arbitrary datum (such as mean sea level). Altitude matrices may be derived directly through remote sensing or through interpolation from a grid of data points that are regularly or irregularly spaced. Altitude matrices have several advantages, including the ease with which they may be manipulated.

Because of the ease with which matrices can be handled in the computer, in particular in raster-based geographical information systems, the altitude matrix has become the most available form of DEM. DEMs with a fairly fine spatial resolution are freely available for the whole of the earth. For example, the Advanced Spaceborne Thermal Emission and Reflection Radiometer (ASTER) Global Digital Elevation Model Version 2 has a posting interval of 30 m.[1] At national and local levels, DEMs at a very fine spatial resolution may be available (i.e. less than 5 m).

Figures 11.1 and 11.2 give examples of altitude matrix DEMs. Figure 11.1 shows a DEM with a cell size of 200 × 200 m. Elevation data may be displayed in several ways—Figure 11.1 shows a colour-scale image of simple elevation and a colour image of surface aspect (Chapter 10 explains how this is derived). Figure 11.2 shows a similar image from another area, but now the aspect information is 'draped' over the DEM to yield a **block diagram** that portrays the relief naturally. Altitude matrices are the starting point for deriving much useful information about landform, such as slope, profile convexity, solar irradiance, lines of sight, and surface topology, as explained in Chapter 10, and later in this chapter.

[1] http://asterweb.jpl.nasa.gov/gdem.asp

The triangular irregular network (TIN)

Although altitude matrices are useful for calculating contours (see Figure 11.3 for an example), slope angles and aspects, hill shading, and automatic basin delineation (as discussed later in this chapter), the regular grid system is not without its disadvantages. These disadvantages include (a) the large amount of data redundancy in areas of uniform terrain, (b) the inability to adapt to areas of differing relief complexity without changing the grid size, and (c) the exaggerated emphasis along the axes of the grid for certain kinds of computation such as line-of-sight calculations.

The triangular irregular network (or TIN; see Chapter 3) was designed by Peucker and co-workers (Peucker et al. 1978) for digital elevation modelling that avoids the redundancies of the altitude matrix, and which at the same time would be more efficient for many types of computation (such as slope) than systems of that time that were based only on digitized contours. A TIN is a terrain model that uses a sheet of continuous, connected triangular facets based on a Delaunay triangulation of irregularly spaced nodes or observation points (Figure 8.7). Unlike altitude matrices, the TIN allows extra information to be gathered in areas of complex relief without the need to gather huge amounts of redundant data from areas of simple relief. Consequently, the data capture process for a TIN can specifically follow ridges, stream lines, and other important topological features that can be digitized to the accuracy required. TINs provide efficient, accurate data storage of elevation data, but at the expense of introducing a triangular discretization that may hinder some kinds of spatial analysis, such as the derivation of surface geometry and topology.

TINs are modelled with a topological vector structure similar to those used for polygon networks, and the TIN data structure was explained in Chapter 3. The main difference with vector polygons is that the TIN does not have to make provision for islands or holes. Peucker et al. (1978) demonstrated that the TIN structure can be built up from data captured by manual digitizing or from automated point selection and triangulation of dense raster data gathered by automated orthophoto machines. They have also shown that the TIN structure can be used to generate maps of slope, shaded relief, contour maps,

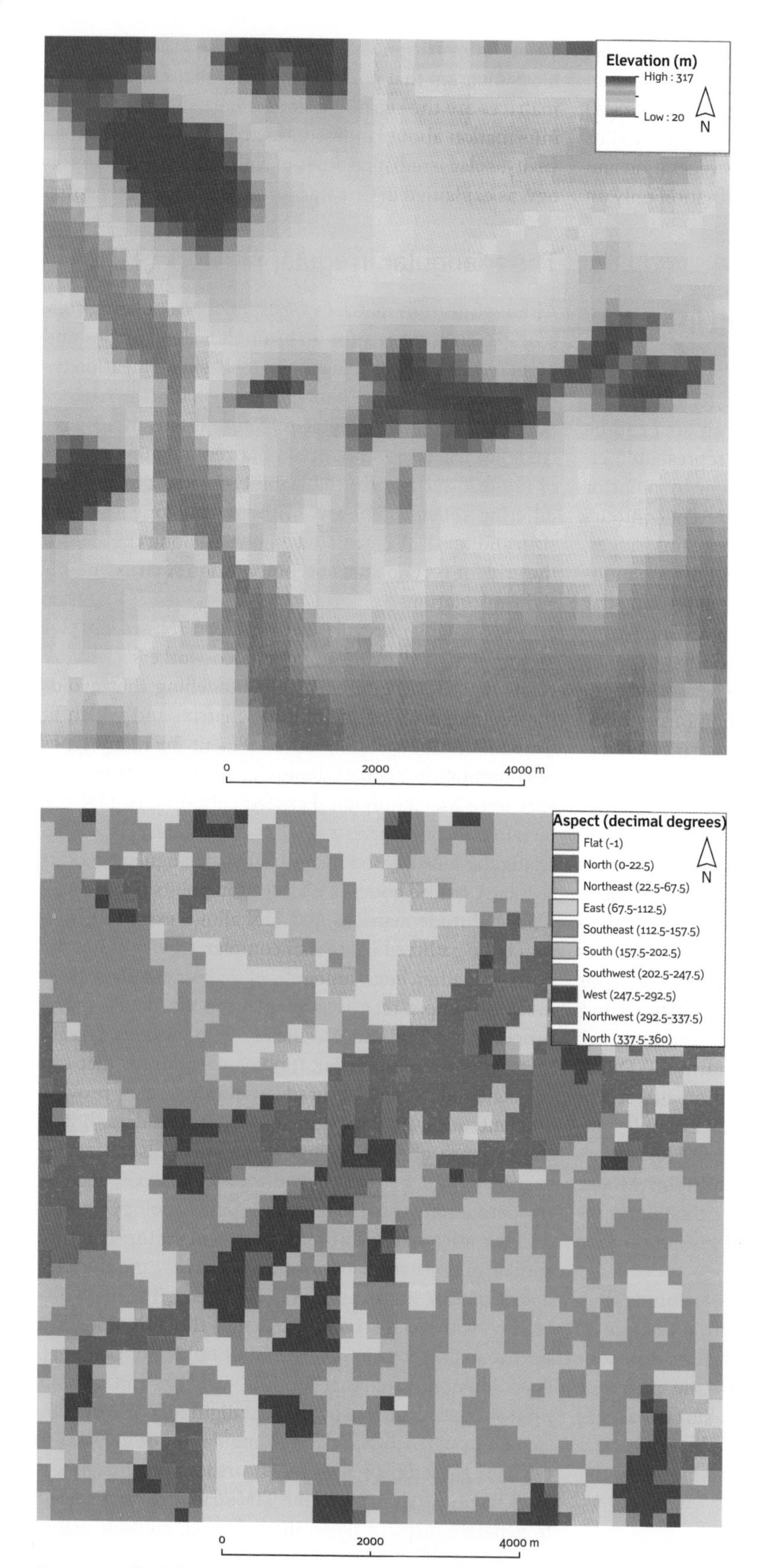

Figure 11.1 (Top) Colour-scale altitude matrix (pixel size 200 × 200 m); (bottom) aspect map

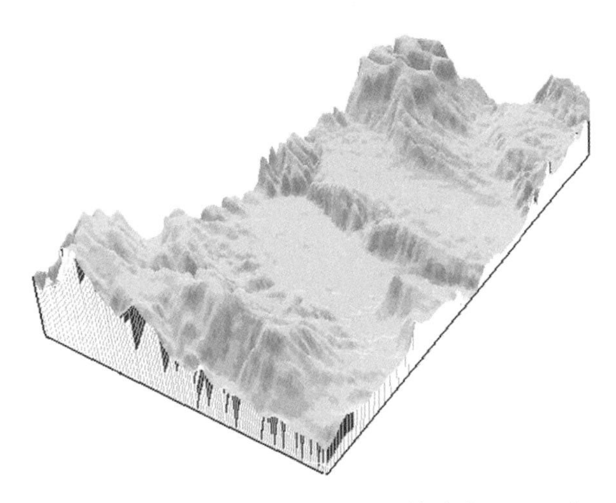

Figure 11.2 Shaded relief data draped over a block diagram creates a strong impression of the relief
(Data courtesy S. M. de Jong)

profiles, horizons, block diagrams, and line of sight maps, though the final map images retain an imprint of the Delaunay triangulation (Figure 11.4). Information about surface cover can be incorporated by overlaying and intersecting the TIN structure with the topological polygon structure used for many thematic maps of discrete variables.

Converting altitude matrices to TINs

The key objective of conversion from an altitude matrix to a TIN is to extract the smallest possible set of irregularly spaced elevation points that provides a maximum of information about topographic structures such as peaks, ridges, valley bottoms, and breaks of slope. DEM to TIN conversion can be conducted using the LandSerf software developed by Wood (2009) (and see De Smith et al. 2007 for an example). Li et al. (2004) describe the VIP—'very important points' method: selection of VIPs is based on assigning a significance value to each point so that points with the highest significance values can be retained. Relevant methods for identifying significant points and for determining an appropriate significance threshold are detailed by Li et al. (2004).

Converting TINs to altitude matrices

The transformation of elevation data from TIN to raster is achieved by laying a regular grid of the required resolution and orientation over the TIN. Each grid cell is visited in turn, checked to see in which cells the nearest TIN apices are, and a linear or bicubic average of the TIN heights is computed.

11.2 DEM data sources

Unlike many other kinds of quantitative data, elevation of the land is easy and cheap to measure. Data sources include direct measurement in the field with total stations and GPS, stereo aerial photographs, scanner systems in aeroplanes and satellites, and (increasingly rarely) the digitizing of contour lines on paper maps. For systematic mapping, elevation data are derived by the methods of photogrammetry from overlapping stereoscopic aerial photographs and satellite imagery (as highlighted in Chapter 4). For special purposes, other kinds of scanners can be used, such as airborne laser scanners for high-accuracy surface measurements. Sonar scanners mounted on boats, submarines, or hovercraft are also used for surveying the elevation patterns of sea and lake beds, or ground-penetrating radar and seismic technologies for mapping the elevation of subsurface layers. An introduction to methods for spatial data generation was provided in Chapter 4.

There is a wide range of free or low-cost DEMs available currently. As for all forms of spatial data, some countries have more enlightened attitudes to free data access than others. In the United States, in particular, several different products are available at no cost to users. The United States Geological Survey (USGS) is a major source of such data. The global 30 arc-second (GTOPO 30) DEM is available for all countries, as are the ASTER data mentioned above. Other than these freely available sources, most DEMs must be purchased through national mapping agencies or private organizations. The development of fine resolution DEMs with a global coverage using such technologies as interferometric radar will have a major impact for a wide range of users.

As there is often an abundance of data, local, simple (linear) methods of interpolation are often better than complex interpolation methods, because they do not need to make assumptions about the spatial interactions and they are quick to compute. When stereo aerial photographs and satellite images are the source of elevation data, we have complete coverage of the landscape at the level of resolution of the image. The creation of a DEM is then the extraction of data to create a proper **hypsometric surface** at a level of resolution appropriate for the application, including the geometric correction and removal of distortion with respect to the chosen spheroid, projection, and orientation. Distortions in the data caused by tilt and wobble in the viewing platform (aeroplane or satellite), variations in land elevation, and atmospheric effects must also be removed to give a good product.

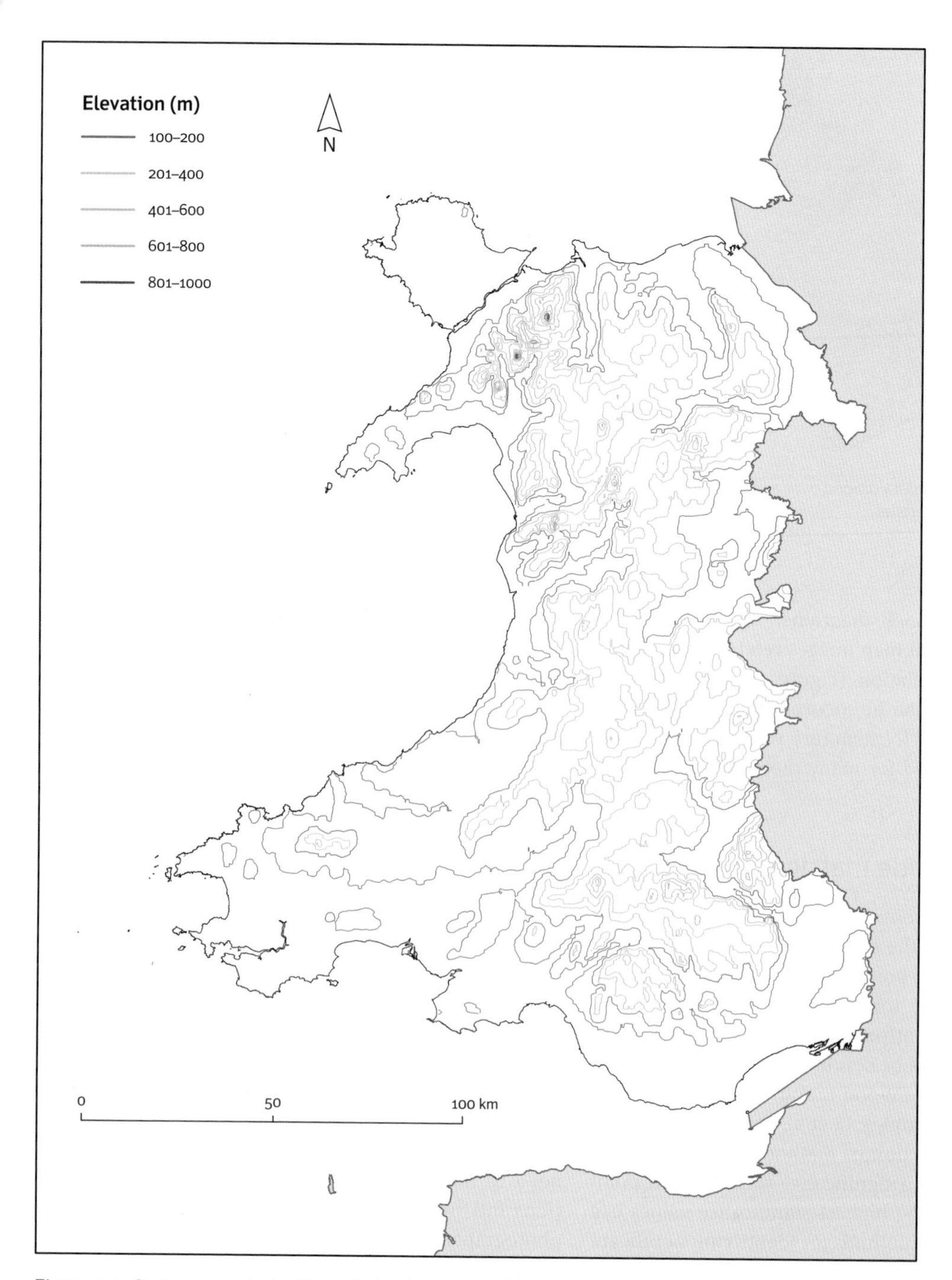

Figure 11.3 Contour map, derived from digital elevation model of Wales (Figure 10.8).

Ground survey

Ground-based survey (for a wide-ranging introduction see Bannister et al. 1998) provides the most accurate measurements of elevation. In addition, ground-based survey may be particularly useful in mapping ground elevation for areas that are inaccessible to remote sensing, such as wooded regions. Furthermore, remotely sensed images are often georeferenced using positional measurements obtained through ground survey. Ground survey may use a benchmark location or, if this is not available, a local coordinate system. Ground survey can be very time-consuming and, irrespective of the approach employed, it may be most profitable to make

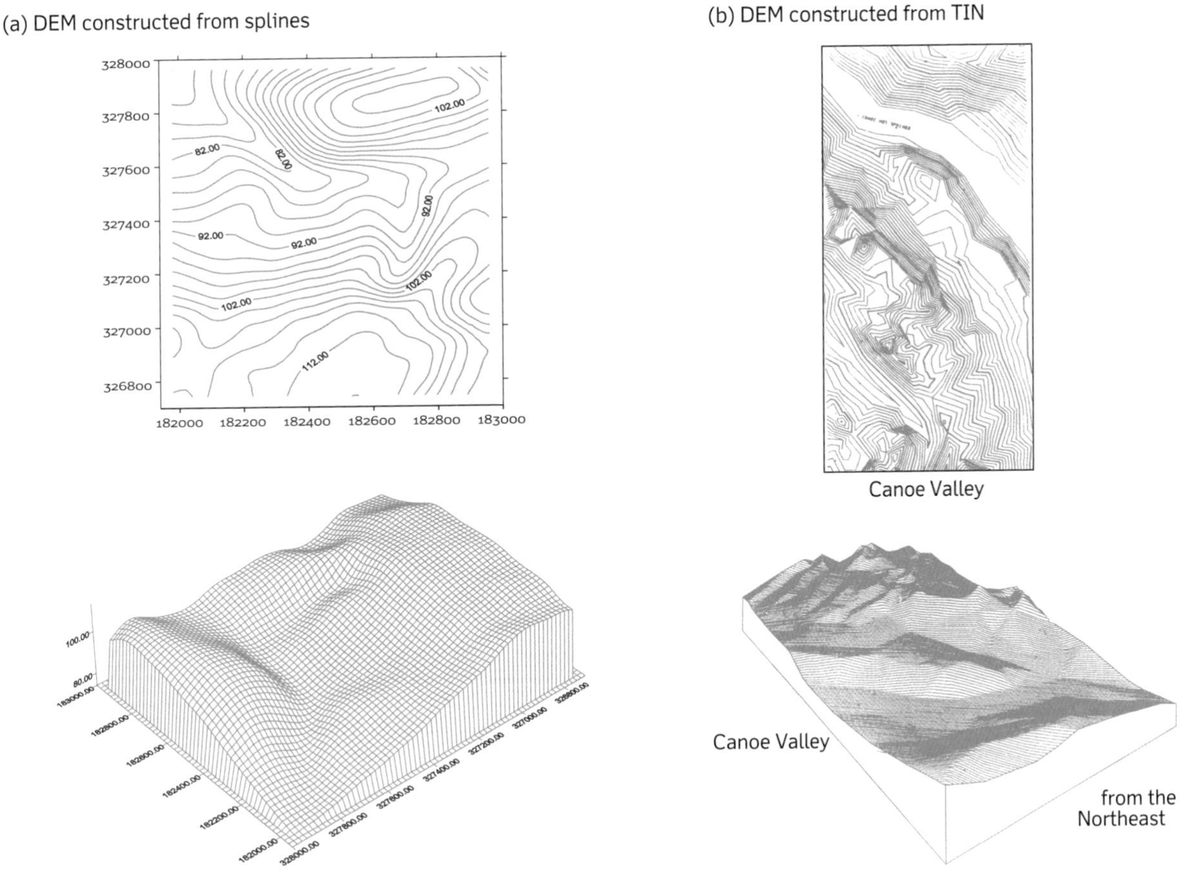

(a) DEM constructed from splines

(b) DEM constructed from TIN

Canoe Valley

Canoe Valley

from the Northeast

Figure 11.4 Digital elevation models generated (a) from splines, and (b) from a TIN
(Data courtesy of A. de Roo and T. Poiker)

measurements only at significant locations such as high and low points and breaks of slope.

The measurement of angles and distances from other positions enables measurement of the relative positions of points on a surface. Angles may be measured using tools such as **theodolites**. In basis, a theodolite comprises a telescope fitted with a spirit level that is mounted on a tripod. The telescope is mounted so that it can move on both its vertical and horizontal axes. The vertical angle is measured from the horizontal and the horizontal angle from true North. A graduated scale measures the amount of rotation about the axes. Tape measures are a simple means of measuring distances. Stand-alone theodolites and tape measures have, however, been largely superseded by more technologically advanced solutions which combine measurement of both angles and distances. The measurement of angles and distances may be combined using a high-precision system called a **total station**. With a total station, angles are measured using an electronic theodolite, and distances from the sighting device to the target can be measured through electronic (or electromagnetic) distance measurement (EDM). There are two types of EDM instrument: electronic (microwave) EDMs and electro-optical EDMs (usually working with an infrared beam). Total stations are capable of obtaining positional measurements accurate to one millimetre. For the foreseeable future, total stations are likely to remain the standard technology in micro-scale analyses.

The term global navigation satellite system (GNSS) was introduced in Chapter 2 as indicating a series of systems for determining point locations on the surface of the earth. The global positioning system (GPS) is the most well known of these systems, and a summary of some key features of the system was provided in Box 4.1. There is a vast range of GPS receivers now available. These vary from small handheld systems capable of obtaining

Figure 11.5 Differential GPS in operation

positional measurements accurate to within perhaps 10 m, to expensive differential systems capable of obtaining positional measurements to within 1 cm. Figure 11.5 shows differential GPS in use in mapping an archaeological earthwork. With **differential GPS**, a stationary base station receiver is used to refine measurements made by one or more roving receivers. Gallay et al. (2013) assess the accuracy of DEMs derived using a total station, differential GPS, terrestrial laser scanning, and airborne laser scanning (see the discussion about laser scanning in 'Laser-based systems'). In that study, DEMs derived using the different methods were assessed in relation to the total station point measurements. The performance of different methods was shown to vary according to the steepness of the terrain in parts of the study area.

Creating digital elevation models from stereo aerial photography and satellite images

Chapter 4 provided an introduction to some principles of remote sensing, and some more details are provided here with particular reference to digital elevation modelling. Most DEMs now available publicly are based in whole or in part on measurements made using some form of remote sensing technology, both airborne and spaceborne. Aguilar et al. (2005) consider the accuracy of DEMs derived using photogrammetry (see 'Photogrammetry') with interpolation methods used to fill gaps between grid nodes where, for example, there are shadows and photogrammetry cannot provide sufficiently accurate measurements.

In the following section, the main technologies that may be used to derive models of topographic form are reviewed. Measurements for a complete altitude matrix may be obtained through the use of a range of remote sensing technologies. These include photogrammetry, **radargrammetry**, **interferometric radar**, and **LiDAR** (light detection and ranging). Each of these technologies is summarized below; detailed introductions to key remote sensing technologies are provided by Lillesand et al. (2008). DEMs derived through any remote sensing technology must usually be post-processed in some way. In particular, if the desire is to derive a DEM (model of ground heights only) rather than a DTM (including above-surface features such as vegetation and buildings) then it will be necessary to identify and digitally remove these above-surface features. One of the major advantages in sampling elevation using remote sensing is the possibility of sampling over the same region on a regular basis. Thus, analysis of terrain evolution becomes possible with relatively high temporal frequency. The following sections discuss, in turn, photogrammetry, radar-based systems, and laser-based systems.

Photogrammetry

Photogrammetry may be used to determine the heights of features in the terrain. This is achieved using overlapping aerial photographs or images. Where photographs are used, the process is termed hardcopy photogrammetry. Where raster images are used, it is termed digital or softcopy photogrammetry. In the area of overlap of the paired photographs (a stereopair), there are two views of the same terrain taken from different vantage points. The relative positions of features lying closer to the camera (high elevations) will appear to change more from one photo to the other than will features lying at a lower elevation. The change in relative position is called **parallax**, and it can be measured on overlapping photographs and used to determine the heights of features on the terrain. Photogrammetry is usually used to obtain complete coverage. Alternatively, spot heights or contours may be extracted using photogrammetry, either manually or through some semi-automated process. Interpolation is then necessary to produce an altitude matrix. DEMs have been derived photogrammetrically using images derived using a range of remote sensing technologies. These include scanned aerial photographs and SPOT (Système Pour l'Observation de la Terre) imagery. Photogrammetry may be used to produce DEMs at any scale from the microscopic upwards if appropriate imagery is available.

Radar-based systems

Radar (radio detection and ranging) now plays a major role in deriving models of the earth's surface. A particular benefit of radar is its ability to penetrate cloud as well as limited surface penetration. In this section, the two main approaches to obtaining DEMs from radar data will be examined briefly. Airborne and spaceborne radar remote sensing is based on what are termed side-looking radar (SLR) systems. SLR is based on the transmission of short bursts of microwave energy from an antenna. The wider the antenna, the finer the spatial resolution in the along-track direction. A major area of research in radar remote sensing is synthetic aperture radar (SAR). With SAR, it is possible to avoid the need for long antennae by synthesizing the effect of a long antenna. SAR has been applied extensively in topographic mapping.

When radar images are obtained from two different flight lines, image parallax is caused by differential relief displacements. For radar images the process is termed radargrammetry. An alternative approach to deriving DEMs from radar data is interferometry. This is based on the phase difference (the interferogram) between the radar signals received by two antennas located at different positions in space. The antennae can be mounted on one platform (this is termed single-pass), or two acquisitions can be made at two different time periods (termed two-pass). The application of interferometric SAR (InSAR; Rosen et al. 2000), of which there are several varieties, for mapping topographic form has been a focus of much research. The errors of measurement made using InSAR will vary according to the platform used and other factors such as land cover type.

Laser-based systems

Light detection and ranging (LiDAR) is based on the laser range finders that make accurate line of sight measurements. In basis, a laser pulse is fired at a target and the time taken for the pulse to hit the target and return is measured. LiDAR consists of two parts: the airborne module and the ground-based module. The airborne module includes the airborne GPS, an inertial measuring unit, a rapidly pulsing laser, and a highly accurate clock. The ground-based module includes a surveyed ground base location with differential post-processing corrections. LiDAR has been shown to be an appropriate technology for obtaining accurate topographic data. The utility of LiDAR for obtaining digital elevation data in a variety of conditions has been recognized by many organizations in the UK (for example, the Environment Agency) and in other countries.

Since LiDAR measurements are obtained in an irregular configuration and on a point (rather than a pixel) basis, it is usually necessary to interpolate from these point data to a regular grid. Where raw LiDAR data contain information on above-surface features such as vegetation and buildings, the aim is often to remove these features and predict the surface elevations beneath them, thereby obtaining a DEM. LiDAR is particularly useful for predicting between-canopy ground heights in wooded areas. Some of the pulses will be returned from the ground through the canopy, and the differences between the top of the canopy and the ground may be used as a basis to remove the forest canopy from the derived DEM. Some researchers have noted significant increases in accuracy for DEMs derived by LiDAR over photogrammetrically-derived DEMs. The use of ground-based GPS for the post-processing of plane position can provide an accuracy of at least 5 cm in the x and y position. The TopoSys system, for example, has a measurement frequency of 80 000 measurements per second (Lohr 1998). This gives on average five measurements per square metre. A typical product from this system is an altitude matrix with a spatial resolution of 1 m. As well as airborne laser systems, terrestrial (ground-based) LiDAR systems are now used widely to obtain detailed terrain models over small areas (see Jones 2006 for an example).

Interpolating from digitized contours to an altitude matrix

An increasingly rare approach is to derive DEMs from digitized contour lines. Great efforts have been made in the past by national mapping agencies to capture them automatically using scanners (see Chapter 4). Unfortunately, digitized contours are not especially suitable for computing slopes or for making shaded relief models and so they must be converted to an altitude matrix. The reasons for the errors can be seen in Figure 11.6. Accurately digitized contour lines return sets of data points that all have the same *z* value. Estimating a *z* value for an unsampled location on a grid involves finding a certain number of data points within a given search radius and then computing a weighted average. If all data points have the same *z* value, the result will also be that very same *z* value. The net result is that narrow areas bordering each contour zone are all interpolated with the same *z* value, so each contour is transformed into a 'padi' (or rice) terrace. The problem is usually greater in areas of low relief, where contour lines are further apart and the chance is greatest that the search algorithm will only catch data from one contour line. Nowadays, there is usually no need to rely on DEMs derived from contours, as DEMs are often derived directly as altitude matrices using some form of remote sensing.

Geometry correction for altitude matrices and other raster data

All scanned imagery collected from airborne and space platforms contains geometrical distortions because of the curvature of the earth, tilt and wobble in the platform, atmospheric effects, and the variation in elevation of the land surface. Scanned paper maps contain distortions induced by paper stretch and folding. All these sources of error need to be removed before the data can be correctly registered to an accurate geometrical reference base.

The methods used to correct the geometry of a discretized surface, possibly also to produce one with a different level of resolution, cell size, or orientation from the original, are collectively known as convolution (see Chapter 10 for detail on this). The problem is solved by assuming that the original gridded data are discrete samples from a continuous statistical surface. A set of values for a new tessellation is obtained by using a local interpolation technique to convert data from the original array to the new. For large raster arrays this is a very intensive computing problem because every output pixel must be computed separately, using data from its neighbours. Consequently, warping or transforming large raster images used to take considerable amounts of computer time, but technological developments in computer processors and memories have reduced the time required to rotate a large raster array from several hours to a fraction of a second. Mather and Koch (2011) provide a detailed introduction to methods for geometry correction. Warping involves two separate processes. The first is the computation of the addresses of the output pixels relative to the input data matrix. The general equation for a first order warp (translation, rotation, scale) is given by:

$$\begin{aligned} u &= a_0 + a_1 a_2 x + a_1 a_3 y \\ v &= b_0 + b_1 b_2 x + b_1 b_3 y \end{aligned} \qquad 11.1$$

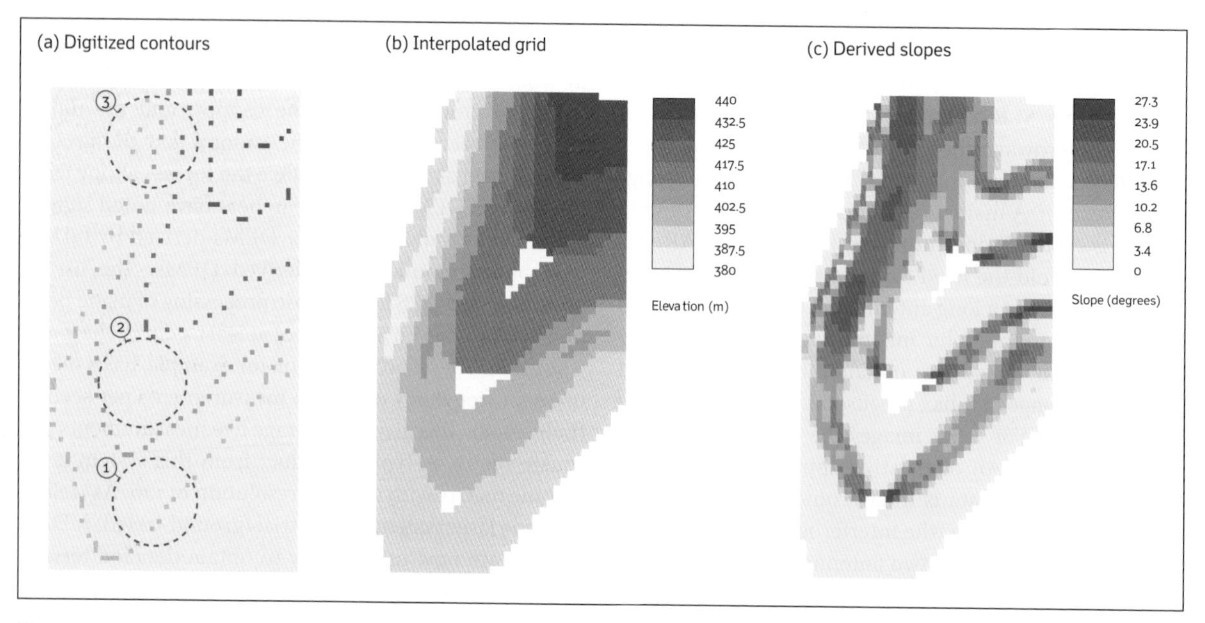

Figure 11.6 Interpolating from digitized contour lines with a simple search circle can create serious distortions and errors in the interpolated surface

where x, y are the original pixel coordinates, u, v are the new coordinates, a_0 and b_0 are translation values, a_1 and b_1 are x and y scale values, respectively, and a_2, b_2 and a_3, b_3 are dependent on the angle of rotation θ as given by

$$a_2 = \cos\theta \qquad b_2 = -\sin\theta$$
$$a_3 = \sin\theta \qquad b_3 = \cos\theta \qquad (11.2)$$

For warps that are not coplanar, such as in the problem of fitting satellite imagery to the curved surface of the earth to match a conventional map projection such as the Universal Transverse Mercator, higher-order warps must be used. Several methods are used for the local interpolation. The simplest, and most limited, is to interpolate the cell value on the warped surface from its closest neighbour on the original surface. A better alternative is a bilinear interpolator in which the new value is computed from the four input pixels surrounding the output pixel (Figure 11.7a). The best interpolator is probably the cubic convolution (Figure 11.7b), which uses a neighbourhood of 16 pixels and weighted sum approach based on a two-dimensional version of the sinc $x(\sin x/x)$ function.

Digital orthophotos

The previous chapter and the latter part of this chapter explain how many useful products can be derived from altitude matrices, but perhaps the most useful digital cartographic product that can be obtained from a DEM is the **digital orthophoto** (see Chapter 4 for an introduction to

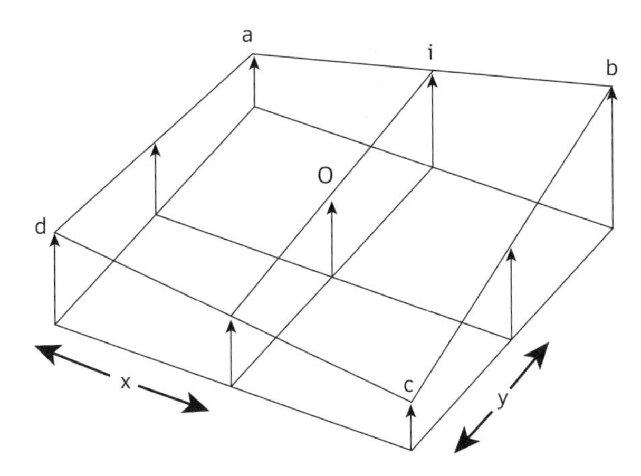

(a) Bilinear interpolation from four input pixels (a–d) to compute a new value O

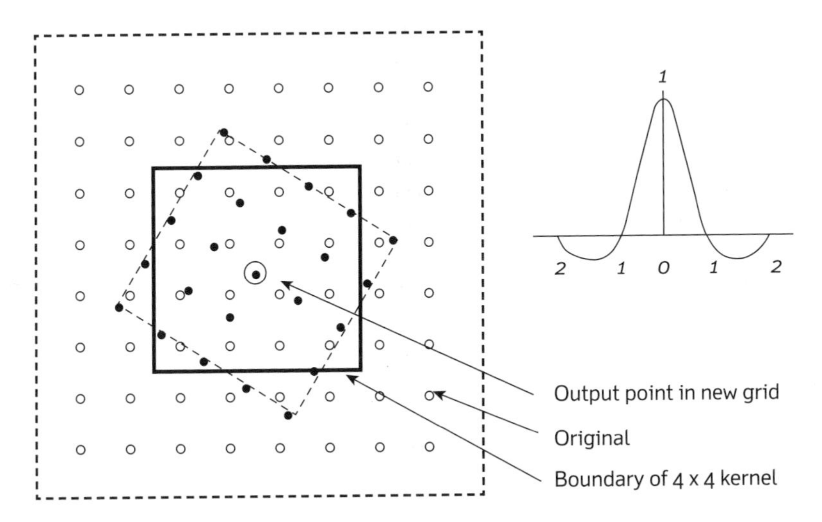

(b) Cubic convolution using the sine function (sin x/x)

Figure 11.7 (a) Bilinear interpolation from four input pixels to compute a new value at cell O; (b) cubic convolution using the sine function (sin x/x)

orthophotos). Digital orthophoto maps are used to provide geometrically correct, highly detailed photographic images for deriving data on land use and land cover and as an informative background to data on utilities, municipal administration, and environmental studies.

The orthophoto is a photo map that is geometrically correct in the same way that a topographical map is geometrically correct with respect to map scale and projection. Orthophoto maps are photogrammetric products that were designed initially to reduce the costs and speed up the production of large-scale topographic mapping in areas where full ground surveys had not been carried out, or where there were financial constraints for producing full topographic coverage at large map scales, or for updating maps at short notice. Unlike topographic maps, in which the terrain features are depicted in standard codes and symbols on a scale-correct base map using a standardized map projection and map legend, orthophotos are aerial photo mosaics that have been geometrically corrected to a standard scale and projection. Orthophoto map scales are usually larger than 1:25 000. Orthophoto maps and digital products may carry a limited amount of topographic information such as contour lines, administrative boundaries, and thematic data.

Remotely sensed satellite images can also be corrected for scale and projection and used as surrogates for topographic maps. The resulting maps are usually at smaller scales than orthophoto maps because of the coarser spatial resolution. The main difference between the orthophoto map (or satellite image map) and the topographic map is that the former contains photographic information on all aspects of the landscape that are visible at the level of resolution used, while the topographic map presents structured, classified information about selected aspects of the landscape. The orthophoto map is easier to use for orientation and object recognition; the topographic map is a rapid key to the structure, location, and connectivity of predefined objects.

Digital orthophoto maps are produced from digital aerial photographs. The images are corrected for distortion by mathematical correction methods using information from ground control points (at least six per aerial photo), as well as for camera optics, colour balance, and terrain elevation differences (which requires an accurate DEM).

Digital orthophoto maps can be produced at a wide range of scales or, perhaps better, levels of resolution. As a rule of thumb, the scale of the map is usually reckoned to be about a maximum of three times the scale of the aerial photographs. Scales of 1:1000 to 1:3500 may be used for urban areas; scales of 1:50 000 can be used for regional mapping.

Updating orthophotos

Producing the DEM for the geometric correction of aerial photographs is the most expensive part of making digital orthophotos. However, since in most cases the shape of the landscape does not change over time, the DEM does not need to be recomputed every time new aerial photography is flown, and the existing surface can be used to support the correction of new photography. When the digital data are in a GIS it is very easy to compare the original situation (e.g. land use) with the new situation, and quantitative estimates of change can easily be made. In certain situations, such as open-cast mining, the erosion of dunes on the coast, or in civil engineering cut-and-fill operations for construction, DEM updating allows volume changes to be computed quickly and accurately. If imagery is available for several years, the whole process of change can be studied, which could be of value in coastal or fluviatile areas.

Much structured topographic data can be collected directly from aerial photographs, particularly for large- to medium-scale applications. Structured data on invisible aspects of the landscape, such as buried pipelines and surveyed parcel boundaries, must be collected by field survey. GIS permit structured vector data to be viewed and analysed with raster data, such as scanned data from satellite images or digital orthophotos. Consequently, the structured topographic data, which shows an abstract view of the world, can be seen in its proper context. This can be useful for checking on data quality and correctness and for seeing the relations between structured data and other aspects of the location.

11.3 Quality of DEMs

Error in DEMs (as well as other spatial data) may be divided into two primary groups. Systematic errors (**bias**) are consistent and may be corrected through the use of a simple transformation. Random errors, in contrast, are those which occur relative to neighbouring elevations (**precision**). In addition to errors in the source data, the quality of a DEM is a function of several factors, including:

1. scale or spatial resolution of the input source
2. density of data
3. spatial variation in the topography
4. spatial resolution of the output DEM.

In the recent past, most DEMs available freely or through purchase from mapping agencies were derived through interpolation from digitized contours or spot height data. Increasingly, grids or finely spaced point measurements are generated when measuring elevations. Contour lines may ensure a good selection of elevation values; they are not, however, well suited for interpolation. Local minima and maxima must be represented by spot heights (a procedure which retains the value at sampled locations) if an exact predictor is used. Problems related to interpolation from contours are well documented. The most widely encountered problem is the clustering of values around contour line locations. That is, DEMs derived from contours often exhibit regular platforms formed around the contour lines. They are manifested as spikes, corresponding to the contour intervals, in the histogram of the derived DEM.

The relationship between factors 2 and 3 should concern all producers and users of DEMs. The spatial resolution of a DEM should be fine enough to resolve variation in the features of interest. If terrain varies over a wavelength of perhaps a metre, then a 10 m spatial resolution will effectively average out this variation. However, we require only just enough data to resolve the variation of interest. Acquisition of data which is surplus to that actually needed to resolve the variation of interest is clearly wasteful. So, where the spatial variation in the topography is of a high frequency (also termed short range) more samples (i.e. greater density) will be required than where the spatial variation is of low frequency (long range). In other words, a grid with a 5 m spacing may provide insufficient information in a mountainous area, but too much information in a river floodplain. As a consequence, many researchers have expended much effort in attempting to measure information and redundancy, and to use this information to design an optimal sampling strategy. Geostatistics is one appropriate set of tools that has been applied to the problem of sampling topographic form (see Chapter 9).

The spatial resolution of an output DEM will usually be coarser than the spatial resolution of the source data. DEMs derived through remote sensing are often subject to various sources of noise, so these DEMs may be filtered to reduce the effect of this noise. Often this will entail coarsening the spatial resolution.

11.4 Viewsheds, shaded relief, and irradiance

Three strongly related methods concern the computation of the paths of light between a light source on or above the DEM and its effect at other locations. Whereas the spatial analysis methods discussed so far concern the attribute values of cells, or the differences in attribute values between cells in the plane of the map overlay, the following operators are concerned with establishing new attributes that refer to the three-dimensional form of the continuous surface. The methods are for the computation of line of sight (determining the viewshed), for computing surfaces with shaded relief for quasi-3D display, and for computing the diurnal or annual inputs of solar energy.

Line of sight maps

This is the simplest operation; the aim is to determine those parts of the landscape that can be seen from a given point. **Intervisibility** is often coded as a binary variable—0 invisible, 1 visible. The collective distribution of all the 'true' points is called the **viewshed**. Over large distances it is necessary to take the curvature of the earth into account, and also the transparency of the atmosphere may be important.

Determining intervisibility from conventional contour maps is not easy because of the large number of profiles that must be extracted and compared. Intervisibility maps can be prepared from altitude matrices and TINs using tracking procedures that are variants of the hidden line algorithms. The site from which the viewshed needs to be calculated is identified on the DEM and rays are sent out from this point to all points in the model. Points (cells) that are found not to be hidden by other cells are coded accordingly to give a simple map (Figure 11.8). Because DEMs are often encoded directly from aerial photographs, the heights recorded may not take into account features such as woods or buildings in the true landform, so the results may need to be interpreted with care. In some cases the heights of landscape elements may be built into a DEM in order to model their effect on intervisibility in the landscape. Viewsheds can also be calculated from TINs (Li et al. 2004).

Estimating the intervisibility of sites is an important GIS application in simulators for pilot training, the location of microwave transmission stations, the appreciation of scenery, or the location of forest fire warning stations. The effects of errors in the DEM on the computation of viewsheds have been studied by Fisher (1995).

Shaded relief maps

Cartographers have developed many techniques for improving the visual qualities of maps, particularly in portraying relief differences in hilly and mountainous areas. One of the most successful of these is the method of relief shading that was developed largely by the Austrian and Swiss schools of cartography and which has its roots in chiaroscuro, the technique developed by Renaissance artists for using light and shade to portray three-dimensional objects. These hand methods relied

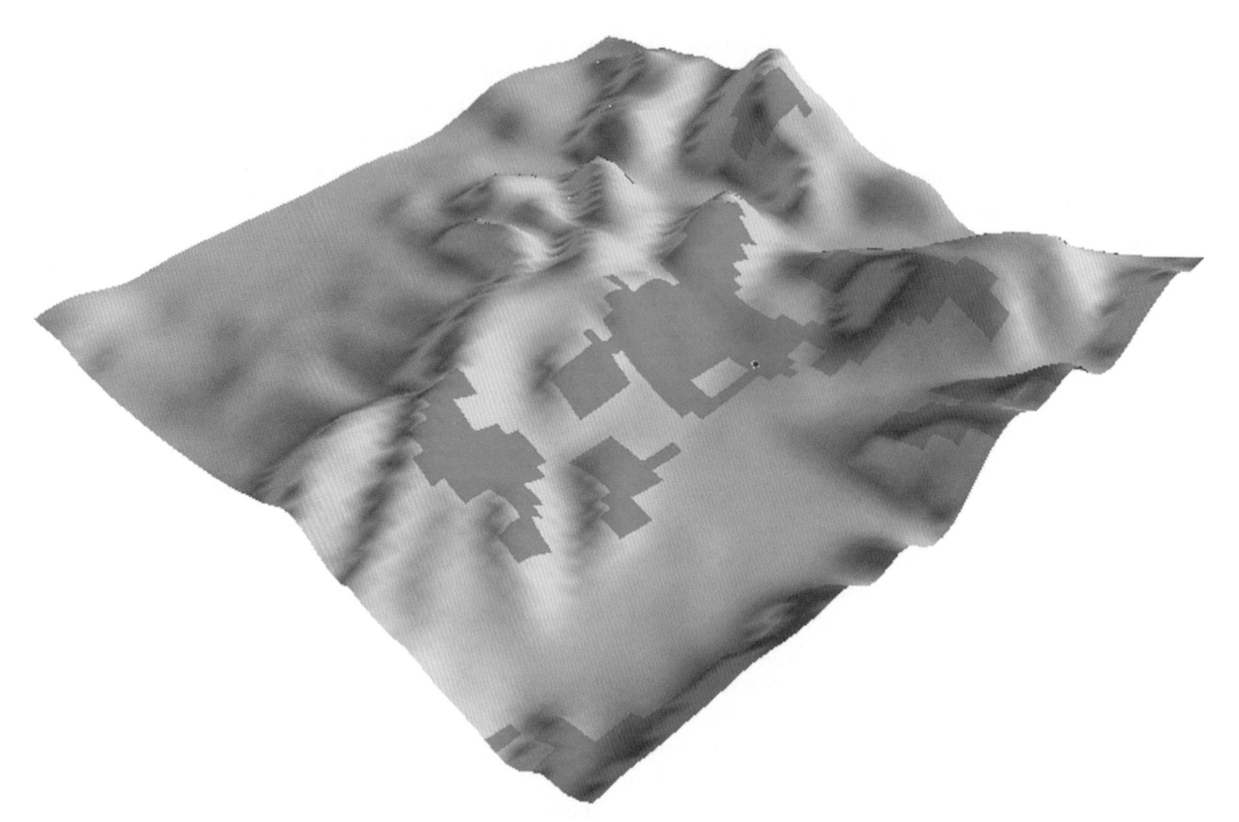

Figure 11.8 Viewshed (green area) from a lookout point (red dot) draped over a DEM

on hand shading and airbrush techniques to produce the effect desired; consequently the end product, though often visually very striking, was very expensive and was very dependent on the skills of the cartographer who, one suspects, was often also something of a mountaineer.

As digital maps became a possibility, many cartographers realized that it might be possible to produce shaded relief maps automatically, accurately, and reproducibly. The principle of automated shaded relief mapping is based on a model of what the terrain might look like if it were to be made of an ideal material, and illuminated from a given position. The final results resemble an aerial photograph because of the use of greyscales and continuous tone techniques for portrayal, but the shaded relief map computed from an altitude matrix differs from aerial photographs in many ways. First, the shaded relief map does not display terrain cover, only the digitized land surface. Second, the light source is usually chosen as being at an angle of 45° above the horizon in the north-west—a position that has much more to do with human faculties for perception than with astronomical reality. Third, the terrain model is usually smoothed and generalized because of the data-gathering process and will not show the fine details present in the aerial photograph.

The shaded relief map can be produced very simply. All that is required are the estimates of the orientation of a given surface element (i.e. the components of slope) and a model of how the surface element will reflect light when illuminated by a light source placed 45° above the horizon in the north-west. The apparent brightness of a surface element depends largely on its orientation with respect to the light source, and also to the material. Glossy surfaces will reflect more light than porous or matt surfaces. Most discussion in the development of computed shaded relief maps seems to have been generated by the problem of how to estimate reflectance (Horn 1981).

According to Horn (1981), the following method is sufficient to generate shaded relief maps of reasonable quality. The first step is to compute the slopes p, q at each cell in the x (east–west) and y (south–north) directions, as given in equations 10.9 and 10.10. These values are then converted to a reflectance value using an appropriate 'reflectance map'. This is a graph relating reflectance to the slopes p, q for the given reflectance model used. Horn suggests that the following formulations for reflectance give good results:

$$(\text{i})\, R(p,q) = 1/2 + 1/2(p' + a)/b$$
$$\text{where } p' = (p_o p + q_o q)/\sqrt{(p_o^2 + q_o^2)} \qquad 11.3$$

is the slope in the direction away from the light source. For a light source in the 'standard cartographic position'

(45° above the horizon in the north-west) $p_0 = 1/\sqrt{2}$ and $q_0 = -1/\sqrt{2}$.

The parameters a and b allow the choice of grey values for horizontal surfaces and the rate of change of grey with surface inclination; $a = 0$ and $b = 1/\sqrt{2}$ are recommended.

$$\text{(ii)}\, R(p,q) = 1/2 + 1/2(p' + a)/\sqrt{(b^2 + (p' + a)^2)} \quad 11.4$$

maps all possible slopes in the range 0–1.

Some formulations for reflectance are computationally complex and it may be more efficient to create a lookup table for converting slopes to reflectance. The reflectance value for each cell is then converted to a grey or colour scale for display (e.g. Figure 11.9). The computations for shaded relief are not complex and do not require large memories, except for the display, because all

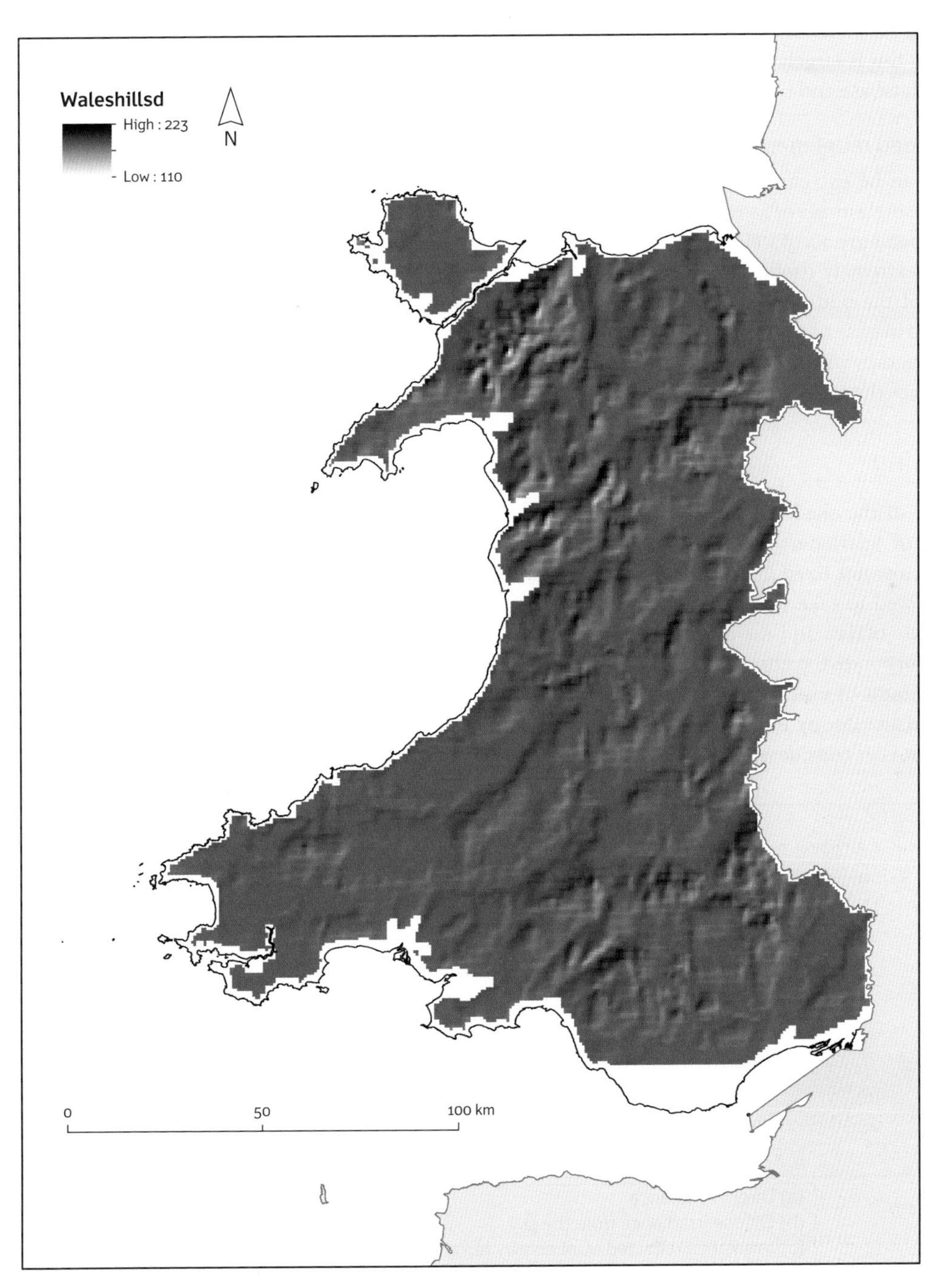

Figure 11.9 Hillshade map, derived from digital elevation model of Wales

calculations use no more data from the altitude matrix than is necessary to fill the kernel.

Shaded relief maps are produced from TINs in a similar way to that just described, with the exception that the reflectance is determined for each triangular facet instead of every cell. The facets are shaded by hatching the triangles with parallel lines oriented along the surface gradient whose separation varies with the intensity of the reflected light. The results strongly retain the structure of the triangular net, and in the authors' opinion give less realistic images than those produced from altitude matrices, though this is largely dependent on both the resolution of the database and the output device.

Applications of shaded relief maps

Shaded relief maps can be extremely useful by themselves for presenting a single image of terrain in which the three-dimensional aspects are accurately portrayed. Not only have they been extremely useful for giving three-dimensional images of the bodies of the solar system; they are widely applied in quantitative landform analysis. When used in combination with thematic information, they can greatly enhance the realism of the final map.

Irradiance mapping

This is the extension of the shaded relief principle to compute the amount of solar energy falling directly on a surface. The sun is now not fixed in any one position in the sky, but is allowed to take a position according to the latitude, the time of day, and the day of the year. There is a need to incorporate the effect of atmospheric absorption on the amount of energy actually received, and also to model the shadowing effect of terrain (the **sky view factor**), which is of considerable importance in hilly landscapes in winter or at the beginning or end of the day. Diffuse irradiance is more difficult to calculate, as are the exact effects of local reflection.

Information on topographic solar radiation models can be found in Wilson and Gallant (2000a) and the references cited in that work. Direct radiance received at any point is a function of solar zenith angle, solar flux at the top of the atmosphere (exoatmospheric flux), atmospheric transmittance, solar illumination angle, and sky obstruction (Figure 11.10). Zenith angle and exoatmospheric flux vary with time of day and year, and atmospheric transmittance varies as a complex function of atmospheric absorbers and scatterers (clouds, dust). Atmospheric transmittance also increases with altitude because of a decrease in the number of absorbers and scatterers.

For an atmospheric transmittance T_o the direct irradiance I on a slope is given by:

$$\begin{aligned} I &= \cos iS_0 \exp(-T_0 / \cos\theta_0) \\ &= \left[\cos\theta_0 \cos\beta + \sin\theta_0 \sin\beta \cos(\varphi_0 - A\right] \\ &\quad S_0 \times \exp(-T_0 / \cos\theta_0) \end{aligned} \tag{11.5}$$

where cos i is the cosine of the solar illumination on the slope, S_o is the exoatmospheric solar flux, θ_o is the solar zenith angle, φ_o is the solar azimuth, A is the azimuth of the slope, and β is the slope angle. Because both β and A are derived from digital elevation data, equation 11.3 describes the spatial variation of the dominant component of incident solar radiation (**irradiance**). As cos i varies with time of day and year, it is possible to compute both the spatial and temporal variation of irradiance (Figure 11.11). Note that the sky view factor limits the direct irradiation when the sun falls below the horizon and this needs to be included in the calculations

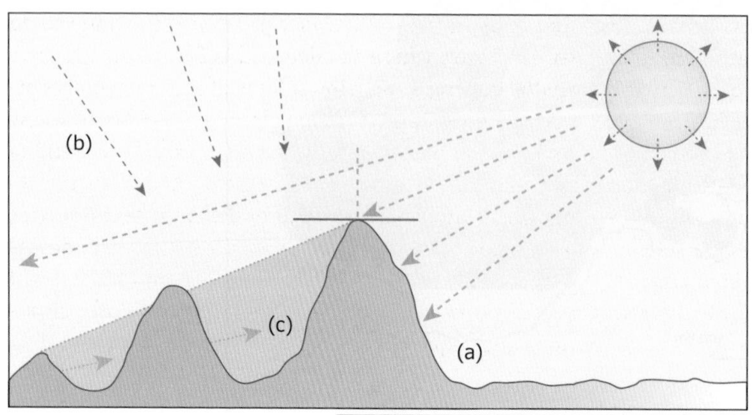

Figure 11.10 Computing solar irradiance for a slope

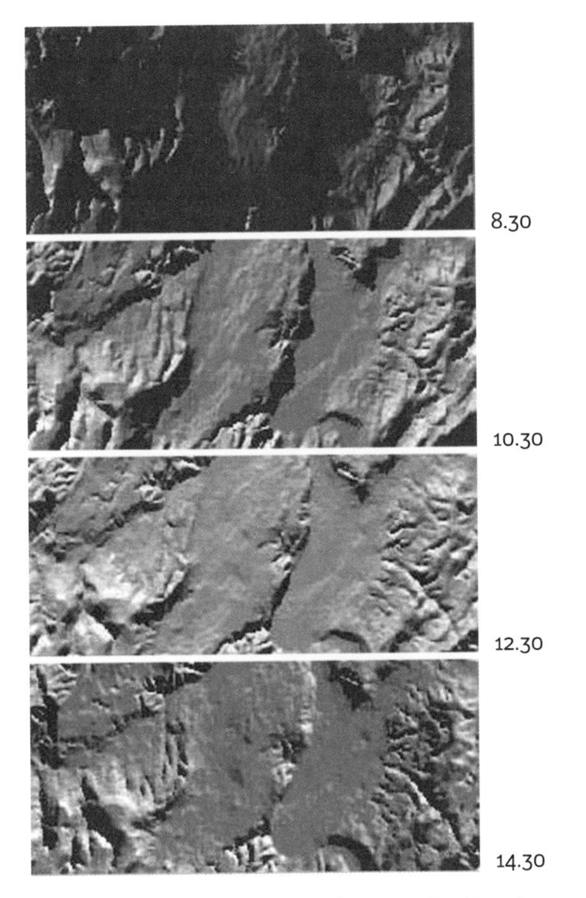

Figure 11.11 Variation in direct irradiance as a function of time of day (view from north)

(i.e. direct irradiation is only computable for those parts of the terrain directly illuminated or in the 'viewshed' of the sun).

It is a simple matter to integrate the daily or monthly estimates of irradiance for a whole season or year, and so to create a map that distinguishes sites in terms of the energy inputs for plant growth, home heating, or rock weathering. Combining a classified map of warm and cold sites with a map of site wetness derived from the upstream elements (see equation 10.14) enables maps of warm–wet, warm–dry, cold–wet, and cold–dry conditions to be made. Further combination with soil information and vegetation indices from classified satellite images can quickly provide a detailed reconnaissance model of the landscape which can be used for hypothesis generation and fieldwork planning.

11.5 Applications of DEMs

In this chapter, many applications in which DEMs are used are mentioned (see Table 11.1 for a summary of attributes which can be derived from DEMs). Recent published studies which make use of DEMs include Goswami et al. (2012), who use DEMs in an exploration of channel evolution. Kang et al. (2009) used DEMs to divide the Upper Mekong (Lancang River) into sub-basins as part of an analysis assessing impacts on the fish community in the river. In a study concerned with predicting the location of open-air rock art of the Coa River Valley, Aubrya et al. (2012) used DEMs to derive slope and aspect maps, and solar radiation and cost-weighted distances to watercourses. In addition, DEMs have been used to inform spatial interpolation of, for example, precipitation amounts, with the objective of increasing the accuracy of estimates (see, for example, Lloyd 2005).

11.6 Future developments

Remote sensing (encompassing photogrammetry) is likely to become the main source of terrain data at all scales of analysis. However, it is extremely unlikely that remote sensing will be able to provide data of sufficient accuracy for all applications in the near future. Ground survey, either independently or as ground control for remote sensing, is likely to retain a significant role for many applications. The advent of GPS, and recent developments in cost and accessibility, enable users to obtain their own positional data rapidly and at a fairly high level of accuracy. Therefore, the capacity to produce accurate DEMs over small areas has increased markedly in the last decade, and it will continue to do so. However, where the demand is for more extensive spatial coverage and a lower level of accuracy is acceptable, a wide range of free or low-cost DEMs of many parts of the world are now available, and this trend is continuing. At the present time, researchers are assessing the relative performance of different technologies in different circumstances. Further testing and development of technologies such as InSAR and LiDAR will provide tools that are better suited to specific situations and requirements.

Table 11.1 Summary of attributes that can be computed from DEMs and their application

Attribute	Definition	Applications
Elevation	Height above mean sea level or local reference	Potential energy determination; climatic variables—pressure, temperature, vegetation and soil trends, material volumes, cut-and-fill calculations
Slope	Rate of change of elevation	Steepness of terrain, overland and sub-surface flow, land capability classification, vegetation types, resistance to uphill transport, correction of remotely sensed images
Aspect	Compass direction of steepest downhill slope	Solar irradiance, evapotranspiration, vegetation attributes, correction of remotely sensed images
Profile curvature	Rate of change of slope	Flow acceleration, zones of enhanced erosion/deposition, vegetation, soil and land evaluation indices
Plan curvature	Rate of change of aspect	Converging/diverging flow, soil water properties
Local drain direction (ldd)	Direction of steepest downhill flow	Computing attributes of a catchment as a function of stream topology; assessing lateral transport of materials over locally defined network
Upstream elements/ area/ specific catchment area	Number of cells/area upstream of a given cell/upslope area per unit width of contour	Catchment areas upstream of a given location (if the outlet, area of whole catchment), volume of material draining our of catchment
Stream length	Length of longest path along ldd upstream of a given cell	Flow acceleration, erosion rates, sediment yield
Stream channel	Cells with flowing water/ cells with more than a given number of upstream elements	Flow intensity, location of flow, erosion/sedimentation
Ridge	Cells with no upstream contributing area	Drainage divides, vegetation studies, soil, erosion, geological analysis, and connectivity
Wetness index	ln(specific catchments area/ tan(slope))	Index of moisture retention
Stream Power Index	Specific catchment area * tan(slope)	Measure of the erosive power of overland flow
Sediment transport index (LS Factor)	$(n+1)\left(\frac{A_s}{22.13}\right)^n\left(\frac{\sin\beta}{0.0896}\right)^m$	Characterizes erosion and deposition processes (cf. USLE)
Catchment length	Distance from highest point to outlet	Overland flow attenuation
Viewshed	Zones of intervisibility	Stationing of microwave transmission towers, fire watch towers, hotels, military applications
Irradiance	Amount of solar energy received per unit area	Vegetation and soil studies, evapotranspiration, location of energy-saving buildings, shaded relief

Source: Adapted from Moore et al. 1993

11.7 Summary

Developments in remote sensing technology in the last decade make feasible the availability of high spatial resolution altitude matrices for most of the populated parts of the earth's surface. In addition, the reduction in cost of simple GPS systems enables individuals to make their own positional measurements and derive DEMs based on specific requirements. The situation is such that anyone who requires detailed models of the earth's surface can (or will soon be able to) acquire this data cheaply and quickly. This chapter has detailed methods for measuring landforms, and for visualizing various aspects of these landforms or assessing which parts of the landscape are visible from particular areas.

? Questions

1. Explain why digital elevation models are essential prerequisites for modelling environmental processes.
2. Outline appropriate methods for detailed mapping of landform in a forested area of 200 x 200 m^2.
3. Devise a suitable set of spatial analysis operations for deriving the best location for hiking trails in a national park, taking into account the wish to have good views but to avoid difficult or dangerous terrain.

→ Further reading

▼ Bannister, A., Raymond, S., and Baker, R. (1998). ***Surveying***. 7th edn. Pearson Education, Harlow.

▼ Li, Z., Zhu, Q., and Gold, C. (2004). ***Digital Terrain Modeling: Principles and Methodology***. CRC Press, Boca Raton, FL.

▼ Lillesand, T. M., Kiefer, R. W., and Chipman, J. W. (2008). ***Remote Sensing and Image Interpretation***. 6th edn. Wiley, Hoboken, NJ.

▼ Lloyd, C. D. (2004). Landform and earth surface. In P. M. Atkinson (ed.) Geoinformatics, in ***Encyclopedia of Life Support Systems***. Developed under the auspices of the UNESCO. EOLSS Publishers, Oxford. <http://www.eolss.net>.

▼ Wilson, J. P. (2012) Digital terrain modelling. ***Geomorphology***, 137: 107–21.

▼ Wilson, J. P. and Gallant, J. C. (eds.) (2000). ***Terrain Analysis: Principles and Applications***. Wiley, New York.

Space–Time Modelling and Error Propagation

The world we have considered so far is one of environments, landscapes, and interactions that change over space. Now in this chapter we will explore how dynamics over time can be included in our spatial modelling and analysis. Natural and human processes continually act on our environment at scales ranging from micro-scale rock weathering to global atmospheric interactions, and from individual humans through to international communities spanning the globe. Capturing those changes and the causes, interactions, and effects of natural and human processes through dynamic spatial modelling allows us to learn more about those environments. While every model is a limited abstraction of reality, simulations under present, past, and future conditions can help us understand subsequent states of environments that result from processes such as physical, chemical, biological, economic, and social reactions. Models thus have an important role in scientific explorations, management, and policy development.

Learning objectives

In this chapter we will focus on computational models representing space–time dynamics, which are developed in, or linked to, a GIS. By the end of the chapter, you will:

- **understand the various components and inputs to a computation model**
- **have learned how space–time processes may be represented in a model**
- **be able to describe how GIS may be used in space–time modelling**
- **understand how errors in data or process representations are propagated through modelling, and the impacts these have on resulting outputs.**

Modelling is behind so many of our day-to-day activities: from the shape of car we drive, or the weather forecast we listen to, through to the interest rates we pay on our loans. It impacts our understanding of the universe and even the fears we might hold for the future of planet Earth. Models are all simplified representations of a reality, yet with the advances in computational power that are now available, the processes and environments we can recreate and simulate in digital models seem to capture ever greater complexities of interactions, allowing us to explore how systems work and the reactions within. Models not only support our learning about current systems, they also allow us to explore new futures through investigating impacts on an area of any possible changes (a different policy, a change in demographics, or the effects of climate change on a country's water and food security).

We model for many reasons, but these fall roughly under two main headings:

1. To aid our understanding of a complex reality through simplifying, diagnosing, and examining interactions, causes, and effects.
2. To predict events, states, and outcomes under particular conditions.

In the following sections we will explore the basics of a computation model and then how these may be developed in a GIS environment.

12.1 Introducing computational modelling

Models come in various types and forms (conceptual, **physical**, and **analogous**) and use different devices to capture the reality that is of interest. For example, a scaled representation of a new housing development, made of cardboard (a physical model) such as the one shown in Figure 12.1 helps us understand and visualize the space, shape, and connections of its various elements and the surrounding environment. Each type of model has particular uses and brings insight and understanding to different areas of enquiry. In this chapter, the focus is on **computational models** as this is the environment of GIS.

Figure 12.1 Physical model of a new sustainable housing development (image copyright Centre for Sustainable Energy)

With this type of modelling, mathematical and logical operations are coded in computer binary (see Chapter 3) to represent the state of an environment and the processes involved. Simulations (experimentation with models) using varying scenarios and conditions can then be run to yield insights into dynamical changes and interactions, shown by numerical values for different variables.

Taking a very simple, spatially static model, we can represent mathematically the changes found across systems that come about as a reaction to variations in the conditions. For example, we can take an established statistical linear regression model relationship between the temperatures in the Swiss Alps and elevation: this is defined in equation 12.1. We can see from this equation that with increasing height it becomes colder:

$$T = 5.697 - 0.00443 * E \tag{12.1}$$

where T is in degrees Celsius and E is elevation in metres.

We can use this equation to determine the temperature across an area of the Swiss Alps. Taking a digital elevation model (see Chapter 11 for detailed examination of DEMs) as an input, the equation can be run for individual grid cells (each is treated as a separate entity). The resulting gridded temperature map show values across an area, and highlights the possibilities of applying a simple mathematical model to generating data for a new variable.

The art/science of formulating computational models requires us to consider a number of factors which will define how and what we represent. The goal is to produce a model that is realistic enough that it reflects the essential aspects of the phenomena being modelled, but simple enough that it can be translated into mathematical and computational formulae and code. This will be determined by answering various questions:

- Why do we want to model this?
- Where do we want to model and at what time and space scales?
- What is our understanding of the environment and processes acting in this area?
- What data exist that we can use in the modelling?
- What models exist already that could meet our needs?

Once these questions have been answered, we have a more bounded idea of the overall model environment, and the four basic stages of defining and developing a computation model can be followed:

a) Conceptual model—Describing, using words/diagrams, our understanding of how features and processes of a system work and interact.

b) Mathematical model—Translating the conceptual model into a series of mathematical equations that represent the various reactions and interactions between the variables that have been used to define the system. These may then be translated into numerical and analytical solutions where the complexity is great.

c) Computational model—Converting the mathematical equations into definitions and code that can be represented in a computer system, and establishing the input and output data set routines.

d) Linking models to reality—Comparing the model outputs to real observations is an important process, involving verification, calibration, and data assimilation to ensure the model represents as accurately as possible the world it is intended to represent.

The mathematical translation of a conceptual model involves defining the various elements of the representation—in particular the **forcing functions** and the processes acting—across defined time steps. It can be fundamentally represented as:

$$Z_{1...m} = f(I_{1...n}, P_{1...i}) \tag{12.2}$$

where $Z_{1...m}$ are the model output state variables, the model structure f is defined by a function or set of functions, forcing functions/external variables (inputs) $I_{1...n}$, with associated parameters $P_{1...i}$ being defined in two or three spatial dimensions.

In more detail, five main elements are used in defining a model mathematically and computationally (Jorgenson 2009):

1. ***State variables***—these describe the state of the system at a moment in time, resulting from processes acting on, and influenced by, the forcing functions. In a GIS, there are the derived points, lines, polygons, and raster values that are the outputs of the different model iterations, and variables might include the amount of precipitation, the level of groundwater in an aquifer, or the state of an ecosystem.
2. ***Forcing functions or external variables***—these are the data sets of independent variables that influence the system of study. In GIS terms, these may be stored as attribute values of points, lines, polygons, and rasters, and are the input variables, such as atmospheric physical and chemical conditions in climate change modelling, or economic and social settings that affect human activities.
3. ***Processes***—biological, chemical, physical, economic, and social—are represented by functions such as mathematical operations, behavioural rules, or logical equations (as given in Box 12.1). They describe the relationship between the forcing functions and the state variables, and between the state variables, and are based on theoretical or empirical relations (defined from real-world data).
4. ***Parameters***—coefficients in the mathematical representations of the processes described in 3, which represent the influence of a subprocess or other variable of influence that is not fully defined in 1, 2, or 3 and may be a constant. For example, in modelling run-off from a hillslope, parameters might be used for fixed or varying infiltration capacity and hydraulic conductivity of the soils.
5. ***Universal constants***—such as gravitational acceleration of 9.80665 m/s^2.

Taking water flowing across a hillslope as an example, there are many different processes at play, including the gravitational pull downhill; the friction and other stresses from the landscape surface that slow this movement down; the movement of water into the soil and then through to the groundwater systems or into the channel. The following data might be used:

- *State variables*: soil moisture levels; river flow; groundwater level.
- *Forcing functions*: precipitation; temperature; land cover; soil types; slope and elevation.
- *Processes:* flow over the surface; evapotranspiration; infiltration; flow through the soil system; flow in the groundwater.
- *Parameters*: these are coefficients that account for subprocesses such as flow resistance from vegetation cover, or hydraulic conductivity.
- *Universal constants*: gravitational constant as this influences the flow rate downslope and in the channel.

12.2 Capturing spatio-temporal dynamics in computation modelling

The ability to define and represent each of the five modelling elements depends on the state of our knowledge and the information we have available to describe them. This is particularly complex when variations over both space and time are included in this definition. Establishing the spatial and temporal data models that form the basis of the modelling will not only influence the form of the input data, but also how the processes are defined. This leads to different types of computational models: lumped, semi-distributed, or distributed; and black-box, physical, and probabilistic.

> **Box 12.1 Modelling operations**
>
> **a)** *Arithmetical operations*
>
> New attribute is the result of addition (+), subtraction (–), multiplication (*), division (/), raising to power (**), exponentiation (exp), logarithms (ln—natural, log—base 10), truncation, square root.
>
> **b)** *Trigonometric operations*
>
> New attribute is the sine (sin), cosine (cos), tangent (tan), or their inverse (arcsin, arccos, arctan), or is converted from degrees to radians or grad representation.
>
> **c)** *Statistical operations*
>
> New attribute is the ***mean***, ***mode***, ***median***, ***standard deviation***, ***variance***, ***minimum***, ***maximum***, ***range***, ***skewness***, ***kurtosis***, etc. of a given attribute represented by *n* entities.
>
> **d)** *Multivariate operations*
>
> New attribute is computed by a multivariate regression model, or ***principal component analysis***, ***factor analysis***, ***correspondence analysis*** transformation of multivariate data.
>
> **e)** *Rule-based operations*
>
> New attribute is computed based on various conditions being met/not met ***If*** ... ***Then*** ... ***Else*** statements.
>
> **f)** *Logical operations*
>
> New attribute is computed based on logical operations Truth or falsehood (0 or 1) resulting from union ($\vee$ logical OR), intersection ($\wedge$ logical AND), negation ($\neg$ logical NOT), and exclusion ($\underline{\vee}$ logical exclusive or XOR) of two or more sets.

Defining space and time representations

Defining the spatial dimension involves demarcating the extents of the study area as well as how the space within it is divided up and the resolution of the smallest unit in the model. The spatial data model adopted will influence how the various components are defined. Where the functions, variables, and processes are represented as spatially continuous (known as distributed modelling) then a raster approach is usually adopted (cf. Chapters 2 and 3). An alternative is where a series of distinct spatial units are defined and within each unit the variable values are the same for a given time step. This is known as **semi-distributed modelling**, and may be based on either aggregated grid cells or vector-based entities. Lumped modelling is a further simplification of the spatial representation in which fewer, more aggregated spatial units are defined. For example, in hydrological modelling a catchment may be divided into sub-basins and water flow responses are generated for each lumped unit, and these are then integrated to derive predictions for the dynamics of the whole system.

In defining the representation of time dimensions, the temporal resolution of the models is specified in terms of length of the model's time step and is usually based on equal time sequences across a period, or centred on an event. The time step and period of modelling can range from seconds and minutes through to decades and millennia. Where fine resolution time steps are used, or the modelling covers long periods, large numbers of data files may be used, requiring considerable processing and storage capabilities. This is typical of climate change analysis today.

Defining processes

Representing processes involves converting the conceptual understanding or theoretical knowledge into definitions based on mathematical or logical operations (given in Box 12.1). A number of modelling approaches may be used, with the representation reflecting factors such as the purpose of the model, the knowledge of the system, and available field data.

Many mathematical environmental models have used the relatively simple 'black-box' approach in which the real world is described through statistical regressions and correlations, as equation 12.1 exemplified. The definition of the process is based on a statistical or other simply defined relationship between one or more independent variables and their influence on dependent variables. This relationship is developed using measured data for both the forcing functions and the state variables. In the resulting, usually statistically derived, mathematical formulae, the processes are not explicitly defined but the influences of the forcing functions on the state variables are represented through parameter values which may be adjusted to account for the extent of this influence.

For example, **black-box modelling** of hillslope run-off processes is often centred on simple regression equations expressing the relationship between rainfall and run-off, with processes such as interception, infiltration, overland, and throughflow described through constant parameter values (see equation 12.3). Relationships defined by such regression equations are usually based on local field data (empirical) evidence, so the environment and its responses captured in the formulae apply only to this area. They should not be extrapolated or transferred to environments outside the ranges of the data originally used to define them.

$$\text{Run-off} = \text{Rainfall}^{\ a,\ b,\ c,\ d} \qquad 12.3$$

where a, b, c, d are numerical parameters that represent different processes affecting run-off.

In contrast, where there is access to greater knowledge or data, or more detailed insight is needed, a physically based (also called mechanistic) approach involving more complex sets of mathematical functions may be used. This has been used in a wide variety of applications, for instance to represent causes and effects in crop growth, air quality, groundwater movement, pesticide leaching, and epidemiological hazards. The dynamics of the system are based on theoretical knowledge of the dynamics, and are represented using differential or partial differential equations, which are resolved using analytical or numerical methods to obtain exact or approximate solutions.

Navier–Stokes equations, which describe the physics of the motion of fluids, are an example of differential equations used in some modelling systems today. The equations arise from applying Newton's second law and other influences on the movement of air and water. They mathematically represent fundamental processes and are used in modelling for weather forecasting or for exploring phenomena such as ocean currents. The differential equations are solved across an area using numerical and analytical approximations such as finite difference (regular grid) or finite element (irregular triangles, polygons, etc.)—see Section 2.6.

In this exploration so far, we have assumed that the state variables at any moment in time may be explained directly in terms of their previous state, the external forcing functions, and the process definitions. This approach may be expressed very simplistically as follows:

$$Z_{(t+1)} = aZ_{(t)} + by(t) \qquad 12.4$$

where:

$Z_{(t)}$ and $Z_{(t+1)}$ = state of system at times t and $t+1$

a and b = constants

$y(t)$ = forcing functions and process definitions.

The assumption in **deterministic modelling** is that there is only one solution for the state variable, given particular defined inputs. As such, it does not take on board the inherent uncertainty or variability in defining the processes and variables. The alternative approach, **probabilistic modelling**, allows many possible outputs to be generated for a given input and state. Models can then utilize the whole range of possible states of the variables, and the result is a probability distribution of the model outputs rather than one single value. The corresponding simplistic equation (cf. equation 12.4) in probabilistic modelling is:

$$Z_{(t+1)} = aZ_{(t)} + by_{(t)} + r \qquad 12.5$$

where the new term, r, is a **random variable**.

The randomness may be introduced in models through values for forcing functions, parameters, and process definitions. Methods such as Monte Carlo techniques have been used for exploring the complete range of outputs that a model can generate as a result of the probability distributions of both the input data and the model parameters. **Monte Carlo simulation** adds random values for r from a defined probability function, for any factor that has inherent uncertainty. It then calculates results for different iterations—each time using a different set of random values from the probability functions (such as normal, log-normal, Gaussian distributions)—so it may involve hundreds or even thousands of recalculations. The end result is a distribution of possible outcome values.

12.3 GIS-based computational modelling

GIS have been used as the basis for developing spatio-temporal dynamic modelling for many different applications. At the most basic level, GIS are loosely coupled to a model and used as the platform for generating spatial data for defining the external forcing functions and for visualizing the state variable outputs. In other applications, models are developed within the GIS using built-in spatial functionality, specialist modules, or dynamic modelling/scripting tools as well as the spatial data modelling capabilities to define the processes, forcing functions, and parameters. There are thus a variety of approaches that might be adopted—the following examples will highlight the different possibilities (Beven and Moore 1994; Goodchild et al. 1996; McDonnell 1996; Skidmore 2002; Maguire et al. 2005; Koomen et al. 2007; Pfieffer 2008; Brimicombe 2010; Khalema-Malebese and Ahmed 2012; O'Sullivan and Perry 2013; Wainwright and Mulligan 2013).

Loosely coupled models and GIS

Loosely coupled models may have different degrees of system integration with GIS. At the most basic, a GIS may be used to derive model parameters and variables' values from data layers already held in the database. This

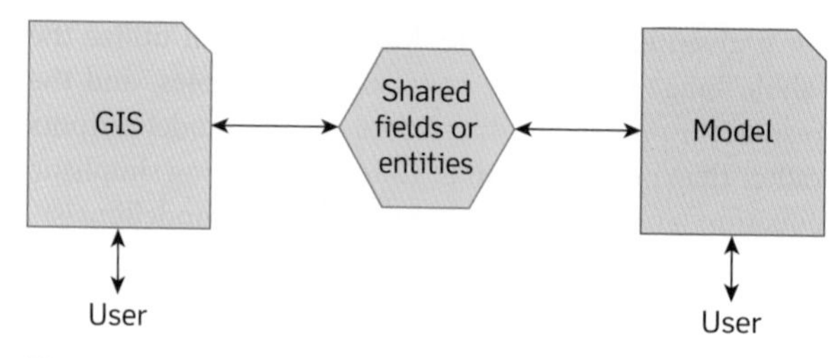

Figure 12.2 Loosely coupled GIS and model

approach harnesses the spatial data-handling capabilities of a GIS by loosely coupling this to more sophisticated time representation in dynamic computational software. The coupling of the GIS and dynamic models requires a cross-matching of the corresponding individual spatial units (entities or grid cells) used in each system.

Other aspects of the models, such as the domains, operators, and algebra, remain separate in loose coupling, so internally the two systems still retain individual identities. There is no sharing of the database or modelling functionality between the two systems, with interchange files passing data between the two systems (as shown in Figure 12.2). Varying degrees of subroutine development or scripting are required, depending on the nature of the GIS and modelling, to bolt the systems together. The link is often not seamless, with, for example, one system being exited before the other runs and the importing/exporting of files between the two.

The advantage of this approach is that the algorithms of the dynamic model are run separately, which allows the functionality and structure of the host programming language to be employed while the GIS permits a *spatially* distributed approach to be adopted. GIS thus serve an important role in preparing variable data for forcing functions and initial state conditions, and estimating parameters, using the various functions and capabilities highlighted in earlier chapters (McKinney and Cai 2002; Imam and Kushwaha 2013). For example, McDonnell (2000) linked a GIS to the dynamic modelling software Stella to explore the impacts of different dam locations and operational schedules on downstream water quantity and quality variables and biota. GIS may be used to derive areally averaged values for variables from various point data using the interpolation methods described in Chapters 8 and 9. Secondary attributes (see Chapters 10 and 11) may also be derived which characterize the spatial variability of specific forcing functions such as potential solar radiation ratio, slope, aspect, and topological data such as road or river network structure. GIS' spatial visualization capabilities are also often used to display the results of the modelling.

The linking of modelling to GIS can be used in many different contexts (Case study 12.1 highlights a case from the Middle East). Gret-Regamey and Straub (2006), for example, loosely linked a Bayesian network to a GIS to develop a spatially explicit risk assessment for avalanches in Switzerland. The Bayesian network linked a model-derived probability assessment of snow pressure at defined locations, with spatial data held in a GIS to estimate the probability of damage to buildings and transport infrastructure, with or without protection measures and strategies in place.

The loosely coupled approach is often used when model definitions—particularly of processes—are complex and the relatively limited functionality of many GIS does not support the mathematical or logical expressions required. Loose coupling is also popular when a model needs to be run many times (e.g. for multiple time steps or for many small spatial units), as these computations can often be processed much faster by stand-alone modelling software, without the overhead of the GIS. Loose coupling recognizes the advantages of the two systems—of GIS for storing large volumes of spatial data, and of models for rapidly computing the changing attribute values—and keeps the advantages of each. This, coupled with the limited capabilities of representing temporal variables in current GIS (discussed in Chapter 2) has ensured that loosely coupled applications are common.

Raster-based modelling with GIS

For some applications, model development is possible within a GIS, bringing with it the benefits of a single, integrated operations interface and data sharing (e.g. Pagelow and Olmedo 2005). The raster spatial data model is by far the most commonly adopted basis for this modelling. The approach builds directly on the ideas, formal logic, and primitive operatives set up under map algebra and the **Map Analysis Package** (MAP) for undertaking spatial modelling across two or more raster data layers (Tomlin 1983, 2012)—described in detail in Chapter 10. Using these concepts, spatial operations may be defined, based on one or more raster cells in the corresponding positions in the different data layers, to derive a new model output value. As the following list highlights, the first two are local operations with model outputs defined by the state, and the external functions and processes affecting a cell. The operators used to define the processes can be linked through scripting to derive complex process representations:

a) Point operations: operations calculated locally on individual cells or point locations.

b) Neighbourhood operations: operations in which a value is computed for a particular cell, based on the values of surrounding cells.

Case study 12.1
Example of loosely coupled modelling: the MAWRED programme

Given the natural aridity in the Middle East North Africa (MENA) region, managing water effectively and efficiently is a vital concern for decision-makers. With agriculture continuing to use the lion's share of the water, the challenge is to balance water and food security demands, which are often directly at odds with each other. With groundwater resources steadily declining, but management hampered by limited data availability in most countries, there is an urgent need to develop new insight through means other than field observations.

The ***MAWRED*** programme (Modelling and monitoring Agricultural and Water Resources Development) uses hydrological modelling supported by observations derived from remote sensing to give an overview of water resource availability. By linking to crop maps and irrigation water use models, an understanding of the current extent of the use of water in agriculture (currently largely unknown), and so the impacts of current policies, were derived. NASA's MENA-Land Data Assimilation System, a catchment land surface model developed for the MENA region, was run to generate groundwater, surface water, soil moisture, and evapotranspiration values at the sub-basin levels. The catchments were defined using digital elevation data in a GIS and transferred to the model. Irrigation area data were derived from the image processing of stacked remotely sensed data and then further analysed in a GIS for use in the hydrological modelling.

The initial model outputs were compared to groundwater values derived from the GRACE (Gravity Recovery and Climate Experiment) satellite system (see Figure 4.4d) for the region, using data assimilation methods within the model simulation exercises, to maximize the model's

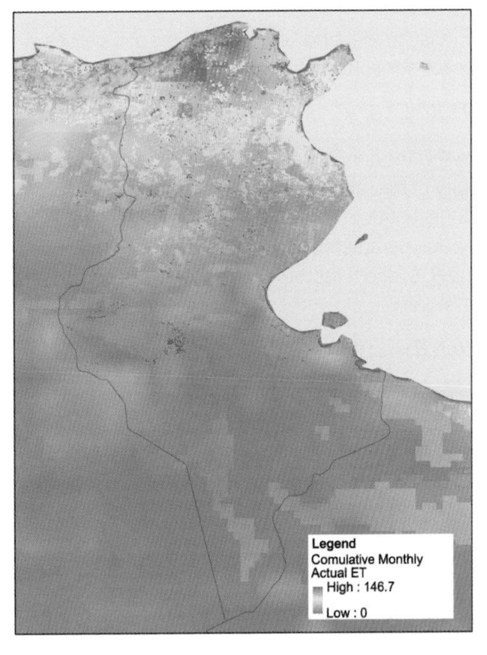

Figure 12.3 Modelling and satellite data were used to derive values for actual evapotranspiration over Tunisia, North Africa

accuracy. The final model outputs were then verified with field data where available. Once the most accurate results were obtained, monthly maps were generated by the model and the high-quality maps (see figure) and exchange files were created in the GIS so that decision-making organizations across the region could use this information in their own spatial analytical systems. (*Mawred* means 'the source' in Arabic; see http://www.mawredh2o.org for more details.)

c) Zonal operation: operates on cells with the same value, such as all the cells in a given land cover class, or urban zone.

d) Global operations: where an operation generates a value for the whole area.

e) Network operations: operations that generate values for nodes and arcs based on their topological connections in a network.

Building from these static (single time step) modelling concepts (see Chapter 10 for more detail and possible variable calculations), dynamic raster modelling incorporates temporal changes by allowing values to be computed for each time step on a cell-by-cell basis, or across neighbourhoods or regions. Values for the previous time step are used to deterministically calculate updated values for the spatial units for the next time step, following equation 12.4. The modelling thus extends non-temporal spatial models to those based on multiple time steps.

As an example of point-based raster modelling, potential soil erosion values from maize cultivation were generated for the Kisii District of Kenya introduced in Chapter 7. Agricultural intensification may increase soil losses here, and modelling can be used to identify which areas are most prone to this. Two different empirical erosion models were used to simulate the soil losses: the Universal Soil Loss Equation (USLE) (Wischmeier and Smith 1978) and the Soil Loss Estimation Model for Southern Africa (SLEMSA) (Stocking 1981; Elwell and Stocking 1982) (see Box 12.2). These are both well known and easy to use, and the data for both are readily available in many countries. The disadvantage of these empirical models, though, is that they are oversimplifications

Box 12.2 Empirical soil loss models

The Universal Soil Loss Equation (USLE)

USLE Wischmeier and Smith 1978) predicts erosion losses for agricultural land by the empirical relation:

$$A = R * K * L * S * C * P$$

where A is the annual soil loss in tonnes h^{-1}, R is the erosivity of the rainfall, K is the erodibility of the soil, L is the slope length in metres, S the slope in per cent, C is the cultivation parameter, and P the protection parameter.

The R, L, and S factors are derived from empirical regressions:

R factor: $R = 0.11\, abc + 66$

where a is the average annual precipitation in cm, b is the maximum day precipitation occurring once in two years, in cm, and c is the maximum total precipitation of a shower of one year occurring once in two years, also in cm.

L factor: $L = (l/22.1)^{1/2}$ where l is the slope length in metres.

S factor: $S = 0.0065s^2 + 0.0454s + 0.065$ where s is the slope as per cent.

The Soil Loss Estimation Model for Southern Africa (SLEMSA—Stocking 1981; Elwell and Stocking 1982):

Control variables:

E	Seasonal rainfall energy (J/m^2)
F	Soil erodibility (index)
i	Rainfall energy intercepted by crop (per cent)
S	Slope steepness (per cent)
L	Slope length (m)

Submodels:

Bare soil condition $K = \exp[(0.4681 + 0.7663F)$.
$\ln E + 2.884 - 8.1209F]$

Crop canopy $C = \exp[-0.06/]$

Topography $X = L^{0.5}(0.76 + 0.53S + 0.076S^2)/25.65$

Output: predicted mean annual soil loss (tonne ha^{-1})
$Z = KCX$

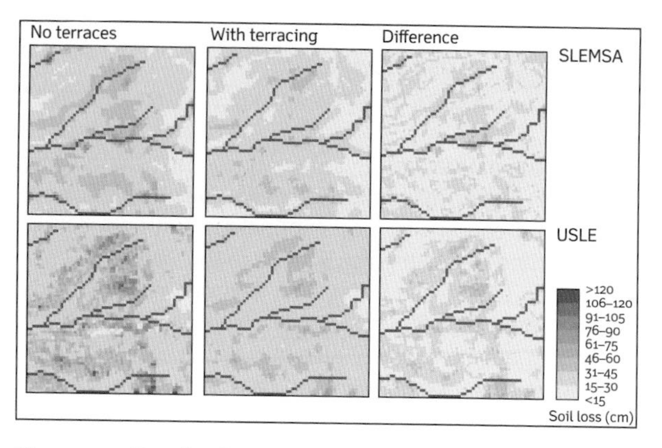

Figure 12.4 Results of grid-based soil erosion modelling using the SLEMSA and USLE models and with different land protection practices

of erosion processes, but specialist models are often location-specific or require data that are not readily available.

The models were run for the Kisii area, computing soil losses per grid cell. As the models are point-based, there is no accounting for the soil loss or gains in neighbouring grid cells. Simulations were run both with and without the inclusion of the 'trash line' terraces—a practical measure to reduce losses, used by the farmers and constructed with maize stalks along the contours. The resulting estimated soil losses, with and without the effects of the trash lines, suggest that erosion will be concentrated in certain areas (see Figure 12.4). Given that empirical equations were used in the analysis, it is important to verify the results using local field data as we have no way of knowing their accuracy otherwise.

The Kisii example highlights environmental modelling in a GIS involving relatively simple mathematical functions. Among the many commercial and open-source GIS there is a great variety in the operational functionality that is available to represent the definitions of processes across the raster environment. Many support a relatively limited range, with mathematical and logical operations most commonly offered (cf. Box 12.1). More purpose-developed GIS provide a developed modelling language that supports more sophisticated representations of processes and the influencing variables. One of the first raster GIS to incorporate spatio-temporal dynamic modelling was the open-source PCRaster, which incorporated a dynamic generic modelling language including libraries of model building blocks and analytical functions (van Deursen 1995; Wesseling et al. 1996; Burrough 1998; Karssenberg et al. 2001; Burrough et al. 2005).

The core modelling component of this system is the PCRcalc module, a raster map calculation engine. It provides a large number of functions and operators which take maps as their input and return maps as their results. These can be combined and nested to perform complex operations. Users script the data generation and process representations using its internal language PCRcalc, or through the more generic Python (Karssenberg et al. 2007). An example of PCRaster use is given in Box 12.3.

Box 12.3 Example of raster-based dynamic modelling using PCRaster

In the Netherlands, many heathlands formerly dominated by heather (Calluna) have become dominated by grasses (Deschampsia). These changes are thought to be induced by factors such as local nitrogen enrichment—eutrophication—and the influence of the heather beetle. During heather beetle outbreaks it is observed that Calluna is more severely affected. Further, it is hypothesized that outbreaks of the beetle are stimulated by nitrogen enrichment, and that they become more severe and occur more frequently under eutrophic conditions. No outbreaks are observed in young Calluna stands (< 5 years). During an outbreak of this beetle plague, Calluna plants die off almost completely over large areas, opening up canopy for enhanced growth of grasses.

The CalGIS (Calluna GIS model) was developed in PCRaster to simulate the heather–grasses competition in the Netherlands, using the soil map as the basic input map (Van Deursen and Heil 1994). It optimistically assumes sandy soils in the Netherlands to be potential heather habitat. The spatial dynamic behaviours of the grass, heather, and beetle are calculated over a 100-year period with each year represented as a file in a dynamic map stack.

For example, the heather growth is simulated using the following PCRcalc statements within a model script:

```
h_growth = (0.005 * nutrient * heather + 0.22 * heather)
           * (100-grass-heather)/100

h_death = 0.12 * heather

heather = min(max(heather + h_growth - h_
          death, 0), 100)
```

The various dynamics can be visualized and analysed using the 'Animate map stack' button or through time plots showing the spatially averaged development of the heather over the 100 time steps. The simulation can be run for different eutrophication scenarios by selecting another level for the nutrient variable.

Cellular automata modelling

A form of spatio-temporal modelling used increasingly in a variety of applications in recent years is **cellular automata** (CA). This form of modelling, often based on a regular grid of cells, is by definition spatial, but in recent CA models it has been developed either within a GIS or loosely coupled to one (Takeyama and Couclelis 1997; see Baetens and Baets 2012 for irregular tessellations). CA has been shown to simulate well a variety of real-world systems, including in studies of people and crowd movement, urban landscapes and growth, forest fire diffusion, and disease spreads (Clarke and Gaydos 1998; Batty et al. 1999; Geertman et al. 2007; Piyathamrongchai and Batty 2007; Stevens et al. 2007; Yassemi et al. 2008; Yeh and Li 2009; Collin et al. 2011; Liu et al. 2013). The unique character of this relatively simple form of modelling is that processes (often social, physical, or biological activities) are defined using sets of rules that affect the state of each grid cell, suiting applications where behaviours cannot be easily represented by strict mathematical equations. The end result is that complex macro-level phenomena (patterns) can emerge from relatively simple interactions between values of local grid cells. Geographical constraints or boundary conditions can also be represented implicitly through the arrangement of the grid cells.

The modelling is based on four major elements: cells, states, and neighbourhood and transition rules. Each grid cell has one of a finite number of states, with the initial state (and boundary conditions) specified through constructed or generated data, and these can have a large influence on the subsequent evolution of the system. With each time step, the new state of each cell is determined according to some fixed deterministic or probabilistic transition rule, based on its current state and that of the cells in its neighbourhood. The transition rule is usually the same for each cell, does not change over time, and is applied to the whole grid simultaneously. It is developed from typical operators—usually +, –, *, /, div, mod, and or not, or and, =, !=, <, >, ≤, ≥—and defined as a deterministic or probabilistic mathematical function, or:

If expression then Else statement

If expression then statement 1 Else statement 2

The definition of the neighbourhood itself—the zone of influence of a grid cell—can vary with the two typical configurations known as von Neumann (1966) and Moore (shown in Figures 12.5 and 12.6 respectively). It may also be in either one or two dimensions: in two dimensions it might be like Figure 12.5b or Figure 12.6b.

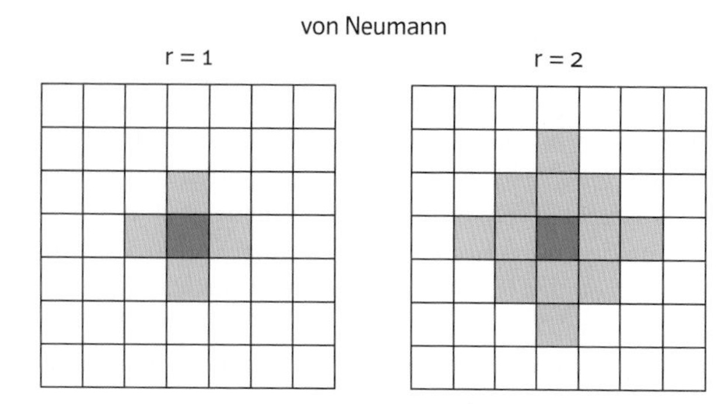

Figure 12.5 The von Neumann cellular automata neighbourhood configuration for (a) one and (b) two dimensions

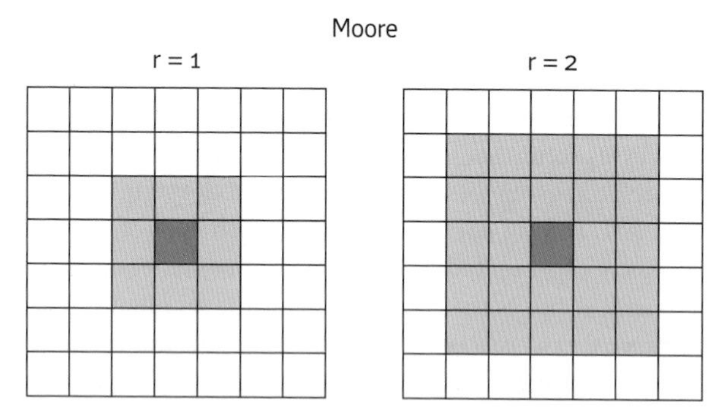

Figure 12.6 The Moore cellular automata neighbourhood configuration for (a) one and (b) two dimensions

A good example is the research of Yassemi et al. (2008) in which fire behaviour was explored using a raster-based model in a GIS, with CA modelling incorporated using its application programming language. The study is based on a deterministic model of fire spread with a CA Moore configuration of eight cells. The transition rules are based on the understanding that fires spread from one neighbouring cell to the central one only when the neighbour is completely burning. Thus the burn state of a cell is important in determining its impact on neighbouring cells, and this is defined as a ratio of burned to total area, with values ranging from 1 (completely burned) to 0 (unburned) along a continuous scale. The state of each cell is calculated at each time step, using trigonometry to derive the direction of travel and speed of the fire and so the area burned at the end of the time interval.

In the model, once a cell is ignited fire travels through it at a speed determined by variables such as fuel type, moisture content of the foliage, topographic and weather factors, and the type and duration of prediction. The GIS is used to derive this 'rate of speed vector' for individual cells, combining these different layers. The outputs are raster arrays of fire values at the end of each time interval, which are visualized using the GIS or using dynamic animation which effectively highlights the output fire dynamics.

Agent-based modelling

The modelling described so far has represented the reactions and interactions between forcing functions, processes, and states that are variable across an area. However, the effects on these environments of individual 'agents', such as humans or animals moving across and *interacting* with a landscape, are not included. Important developments over the last decade in **agent-based modelling** (ABM) within or loosely coupled to a GIS now allow these to be included in the simulations, bringing new insight and dimensions to the understanding of certain processes (see Heppenstall et al. 2012 for detailed coverage). Building on the body of knowledge developed from non-spatial ABM applications in areas such as financial markets or immune systems, successful applications using a GIS include studies of interactions of people in an urban environment, environmental degradation by humans, disease spread, invasive species transmission, and shifting cultures (Gimblett 2002; Batty 2005; Brown et al. 2005; Manson and Evans 2007; Crooks et al. 2008; Heppenstall et al. 2012; Torrens et al. 2011; Torrens 2012; Johnston 2013; Crooks and Wise 2013).

ABM is, like CA, structured on a gridded cell base with rules governing the reactions of interacting agents. Underlying environmental conditions only change in response to the agents' actions. Thus the most important part of this modelling centres on representing in detail the ABM agents and their behaviours, using individual rule-based profiles which define their interactions not only with the environment but also with each other. These can include conditional decision-making and other non-linear rules that distinguish them from mathematically based raster modelling systems. Their dynamics and relationships to the wider environment tend to be based on individual and neighbouring cells, and the overall scale of the modelling tends to be more local than the large landscapes of the raster modelling described earlier. It is through these rules and interactions that patterns of reactions and changes emerge, giving a complexity that is difficult to predict.

In recent work on urban growth, Kim and Batty (2011) used agent-based modelling to explore possible changes in the southern fringe of Seoul city, Korea. Taking an area of 25 × 25 km with cell sizes of 50 (i.e. 250 000 cells), potential scenarios were explored for the release of greenbelt land and the introduction of new transit stations for a high-speed train. The agent-based modelling links microeconomic residential location choice theory with urban growth, taking on board interactions between buyers and sellers and varying preferences amongst significant sets of agents. By representing various household location decisions, the modelling showed the emergence and evolution of urban growth structures arising from the spatial heterogeneity, and evolving over time.

12.4 Accounting for errors in modelling

In spatial analysis in general, and modelling in particular, it is important to understand the accuracy and uncertainty around the results as it is these that determine the extent to which they can be used and relied on in the application of interest. There are three major determinants of this accuracy:

a) the quality of the data (see Section 4.7)

b) the quality of the model, and

c) the way the data and the model interact through processing.

Most GIS or modelling systems do not provide analytical means to determine the effects of errors or uncertainty on the results of the model, so it is important to undertake further analysis, usually away from the modelling environment. It is particularly important to understand the uncertainties in the input spatial data and the propagation of any associated errors from this through the models during processing.

The effects of uncertainty from input variable data can be compounded by inaccuracy in elements of the model definition itself, such as the parameter coefficients that are included. It is useful to be able to quantify all the errors—δu—in the output maps of u. This knowledge can be used to improve either the model or the data collection.

The options for improvement include:

a) using 'better' methods for spatial interpolation or using numerical models instead of simple logic

b) collecting more data and optimizing sampling

c) collecting different data

d) using 'better' models—either models which are more complete or which give lower prediction errors, or are better matched to the spatial and temporal data resolution available

e) improving the model calibration

f) improving the spatial and/or temporal resolution by matching correlation structures.

Determining error propagation using Monte Carlo simulation

Given that (a) we are aware of the possibility of statistical errors in the data, and (b) we have the means to quantify these errors (using ordinary statistics or geostatistics, stochastic simulation (see Chapter 9) or retrospective validation with independent data), the question arises of how to use this information to quantify and then reduce the inaccuracies that may accrue in the results of computational models. Put simply, if a new attribute U is defined as a function of inputs A_1, A_2, ... A_n, we want to know what the error is that is associated with U, and what the contributions are from each A_n to that error. If we can solve this problem, we can attach a pedigree to the results of particular modelling and can compare the results of different scenarios with confidence.

The simplest, but very computer-intensive, approach to error propagation is to treat each variable or attribute as having a Gaussian (normal) **probability distribution function** (PDF) with known mean μ and variance o^2 for each entity or cell. In the simplest case we would use a single PDF for all cells in a layer and assume **stationarity**. If more information about spatial contiguity is available we can use conditional simulation to estimate cell-specific PDFs that reflect the location of known data points and the spatial correlation structure of the attributes (Chapter 9).

The arithmetical operation to derive new data is then carried out, not with the single mean values, but using a value that is drawn from the PDFs for each cell. To take care of the variation within the PDFs, the calculations are repeated many times (at least 100 times) to compute the mean result per entity or pixel and its standard deviation. The technique is popularly known as the Monte Carlo method, because of the random or chance element in each estimation.

Although the Monte Carlo method is computer-intensive (known as a 'brute force' technique), it provides interesting information about how possible errors in the data can affect the results of numerical operations in different parts of a geographic area. For example, in Chapter 10 the operations for computing the derivatives of a raster DEM such as slope were described, as well as the impact of RMS errors (root mean squared error) over the whole domain.

In estimating the uncertainty, users are then able to make more informed decisions on how to use the data sets and the results from the modelling exercises. For example, small relative errors in a DEM can strongly influence the locations of computed stream channels; so calibrating a model for these channels alone may optimize it for only one out of a large range of possible catchment characteristics. And the same holds true for validating a model. If validation measurements are made in locations where the probability is large that a stream flow of magnitude F corresponds to a given upstream contributing area, then the validation will be robust. If validation measurements are made in areas where the probabilities are lower, then the chance that the predictions and the validation measurements will converge is naturally smaller. This explains why in hydrological modelling a simple validation based on stream flow at the outlet of the catchment will be robust, while validation measurements made at randomly chosen locations in the catchment may be subject to considerable error.

Analytical approaches to error propagation

Although Monte Carlo methods of error analysis are straightforward and can be adapted to many kinds of numerical modelling, even today they require considerable computing resources. Fortunately, for many numerical models used in GIS to compute new attributes from existing properties of entities or cells this can be achieved using the standard statistical theory of error propagation (Heuvelink 1998). The problem is then that, for each entity or cell we should estimate, the error in the output value U as a function of the errors in the input values A_i:

$$U = f(A_i) \qquad 12.6$$

includes only arithmetical relationships (+, –, *, –s–, raising to powers, exponentiation, etc.).

Consider the situation in which the value of an attribute on map A is not exact but has an associated error term δ so that the value of the attribute cannot be better known than $A \pm \delta a_i$. A_i could be the value of readily available water in a soil mapping unit that is assumed to be statistically homogeneous. We wish to combine the readily available soil water with an estimate of irrigation effectiveness A, with an error $A_j \pm 8a_j$. If the attributes A_i and A_j are statistically independent, and if δa_i and δa_j are each of the order of 20 per cent, it can be shown that the error of total available water $8u$ in the computation of $U = (A_i + A_j)$ is of the order of 28 per cent. For cartographic overlay operations involving more than two steps, the increase in error can be explosive.

Box 12.4 presents the partial differential equations for the simple theory of error propagation. Using these equations we can examine how errors propagate through simple **bivariate** models with $a_1 = 10 \pm 1$ and $a_2 = 8 \pm 1$.

Box 12.4 Simple theory of error propagation

Considering only random, independent errors, for a relationship

$$u = f(a_1, a_2, a_3, \ldots a_j) \qquad B12.4.1$$

in which the a_js are all independent, Su, the standard deviation of u, is given by

$$Su = \left[\sum_{i=1}^{j} (u/a_i)^2 \cdot Sa_i^2\right]^{½} \qquad B12.4.2$$

and the standard error of u, SEu, is given by

$$SEu = \left[\sum_{i=1}^{j} (u/a_i)^2 \cdot SEa_i^2\right]^{½} \qquad B12.4.3$$

where SEa_i is the standard error of a_i.

These formulae hold when there is no correlation between the x_is. When they are correlated, an extra term must be added to express the increase in error in u due to correlation. This term is:

$$\left[\sum_{i=1}^{j}\sum_{j=2}^{j} \{\delta u/\delta a_i \cdot \delta u/\delta a_j \cdot Sa_i \cdot Sa_j \cdot r_{ij}\right] \qquad B12.4.4$$

Addition or difference operations—no correlation variables

Let $u = a_i \pm a_j; \pm \ldots$, then $\delta u/\delta a_1 = 1$, $\delta u/\delta a_2 = \pm 1$.

$$Su = \sqrt{(Sa_1^2 + Sa_2^2)} \qquad 12.7$$

so

$$u = 10 + 8 = 18$$

and

$$Su = \sqrt{(1 + 1)} = 1.414$$

The absolute error of u is greater than either a_1 or a_2, but in the case of addition, the relative error (1.414/18 = 8 per cent) is lower than for the original variates (10 and 12.5 per cent). For subtraction, the absolute error Su is the same, but the relative error is now much greater at (1.414/2 = 70 per cent). Whereas addition of two random numbers, and hence of two maps, can be thought of as a benign operation with respect to error propagation, subtraction can lead to explosive increases in relative errors, particularly when a_1 and a_2 are similar in value.

When a_2 is a constant (i.e. $u = a_1$ + constant), there is no difference in the variance of u and a_1. Adding or subtracting constants has no deleterious effect on errors.

Addition operations of correlated variables

When the variables $a_1, a_2, \ldots$ are correlated, the term given in equation B12.4.4 (in Box 12.4) must be included in the computation of the error of u. Let $u = a_1 + a_2$, in which ra_1a_2 expresses the correlation ($-1 \le r \le 1$) between a_1 and a_2.

$$Su = \sqrt{\{Sa_1^2 + Sa_2^2 + 2Sa_1 \, . \, Sa_2 \, . \, ra_1a_2\}} \qquad 12.8$$

and

$$Su = \sqrt{\{1 + 1 + 2 \bullet 1 \bullet 1 \bullet ra_1a_2\}}$$

If a_i and a_j are 100 per cent positively correlated, the error in u can be as much as, but not more than, the sum of the errors of a_i and a_j. If a_i and a_j are negatively correlated, then the error in u, Su, could be less than if a_i and a_j were independent.

Multiplication or division operations—no correlation variables

Let

$$u = a_1^c . a_2^d \qquad 12.9$$

where c and d are assumed exact constants. Then,

$$\delta u/\delta a_1 = ca_1^{(c-1)} . a_2^d \text{ and}$$
$$\delta u/\delta a_2 = da_1^c . a_2^{(d-1)}$$

so by equation B12.4.4

$$Su = \sqrt{\{c^2 . a_1^{2(c-1)} . a_2^{2d} . Sa_1^2 + d^2 . a_1^{2c} . a_2^{2(d-1)} . Sa_2^2\}} \qquad 12.10$$

Therefore if

$$u = a_1 . a_2$$

then

$$u = 8 * 10 = 80$$

and

$$Su = \sqrt{\{a_2^2 . Sa_1^2 + a_1^2 . Sa_2^2)}$$
$$= \sqrt{\{64 * 1 + 100 * 1)}$$
$$= \sqrt{164}$$
$$= 12.8$$

Multiplication not only raises the absolute error, but also the absolute error, in this case to 12.8/80 = 16%.

When a_j is a constant, c, i.e. $u = a_1 \bullet c$, the error propagation reduces to

$$Su = \sqrt{\{c^2 . Sa_1^2\}} \qquad 12.11$$

Raising to powers

For

$$u = Ca_1^c \qquad 12.12$$

where C and c are constants, note that a_1 is perfectly correlated with itself so that the error of u, Su, is given by

$$Su = \sqrt{\{C^2 . c^2 . a_1^{2(c-1)} . Sa_1^2\}} \qquad 12.13$$

For $a = 10 \pm 1$ in the expression $u = a_1^2$

$$u = 10^2 = 100$$

and

$$Su = \sqrt{\{(2a_l)^2 . Sa_1^2\}}$$
$$= \sqrt{\{20^2 . 1\}} = \sqrt{400}$$
$$= 20$$

Not only has the absolute error increased, but the relative error (= 20/100 = 20%) has also doubled.

Logarithmic operations and other relations

Let

$$u = C \ln a_i \qquad 12.14$$

then

$$\delta u/\delta a_i = C/a_i \qquad 12.15$$

so

$$Su = \sqrt{\{(C^2/a_i^2) . Sa_i^2\}} = C . Sa_i/a_i \qquad 12.16$$

Equation 12.15 shows that increase or decrease in error depends solely on the ratio of

$$C : a_i.$$

If

$$u = C \sin a_i$$

then

$$Su = C . Sa_i . \cos a_i \qquad 12.17$$

where Sa_i and a_i are in radians.

Simple example of estimating error propagation

Taking a simple example to highlight this analysis in action, a farmer may wish to estimate the uncertainty associated with the net returns from his wheat fields, knowing that the yields and the costs of management and harvesting vary spatially over the farm, and that there is also an uncertainty in the price he will receive. For each field, on a tonne per hectare basis, he wishes to evaluate the errors in his predictions of

$$\text{Net value } (N) = \text{yield } (Y) \times \text{price } (P) - \text{costs } (C) \quad 12.18$$

For each field, let

Y be 6 ± 2 t ha^{-1}

P 100 ± 10 currency units per tonne, and

C 40 ± 20 currency units ha^{-1}.

The gross value per hectare is:

$$\begin{aligned} G &= Y * P \\ &= 6 * 100 \\ &= 600 \text{ currency units ha}^{-1} \end{aligned} \quad 12.19$$

The uncertainty in the gross value is:

$$\begin{aligned} S_G &= \sqrt{\{P^2 \,.\, S_Y^2 + Y^2 \,.\, S_P^2\}} \\ &= \sqrt{\{10\,000 * 4 + 36 * 100\}} \\ &= 116.62 \text{ currency units ha}^{-1} \end{aligned} \quad 12.20$$

The net value is

$$\begin{aligned} N &= G - C \\ &= 600 - 40 \\ &= 560 \text{ currency units ha}^{-1} \end{aligned} \quad 12.21$$

and the uncertainty is

$$\begin{aligned} S_N &= \sqrt{\{S_G^2 + S_C^2\}} \\ &= \sqrt{\{13\,600 + 400\}} \\ &= 118.32 \text{ currency units} \end{aligned} \quad 12.22$$

The methods given above can easily be worked out on a pocket calculator for single equations with only two or three variables, but the computations become tedious when they must be repeated for a large number of cells or entities in a database, or when numerical models involve larger numbers of attributes, including correlation. Various software packages are now available to understand and trace errors through complex numerical models in 'point mode' that operate on the attributes of entities or on multiple raster overlays—the ADAM system is one useful example (e.g. Heuvelink et al. 1989; Heuvelink and Burrough 1993).

Consider a multiple regression model in which the a_i values are raster maps of input attributes and the b_i values are the coefficients:

$$u = b_0 + b_1 a_1 + b_2 a_2 + \ldots + b_n a_n \quad 12.23$$

Contributions to the error in output map u come from the errors associated with the model coefficient b_i and from the errors associated with the spatial variation and measurement errors (nugget; see Chapter 9) of the a_i input attributes.

ADAM uses this technique to apply the error propagation to each entity, which in the case of a raster map is the grid of pixels, but in other implementations could be polygons or other entities in vector mode. The regression model coefficients and their errors can be computed from sample data, and the same point samples can be used to create raster input maps of cell values and their errors. ADAM computes the model output and the errors as maps, and also as maps of the spatial variation of the error contribution from all coefficients and input variables.

12.5 Summary

This chapter has reviewed the fundamental components of computational modelling and the development of different approaches possible in a GIS environment. The growing use of the spatial analytical and display capabilities of GIS to support spatio-temporal modelling has been greatly facilitated by the transfer of data between systems using the widely adopted data exchange standards. Despite the benefits of GIS for data storage, retrieval, and display, modelling entirely within a GIS is still relatively limited by processing speed and functionality of GIS compared to specialist modelling software.

In the most widely used GIS, specialist raster modules are provided for the modelling but their functionality is still relatively limited for defining the complex processes of our natural and human environments.

The chapter also highlighted the importance of understanding error analysis and the impacts of uncertainty on modelling and analysis. These methods can also be applied to more general spatial data analysis. While inaccuracies can come through many different parts of the data-collection process, analysis, and in the model development, it is important that this is acknowledged and analysed so that the quality of the results is understood. Simple rules can be followed to reduce error propagation (Alonso 1968):

1. Optimize the data collection by maximizing the number of samples used and considering the impact of sampling methods.
2. Choose appropriate spatial interpolation methods.
3. Avoid inter-correlated variables.
4. Use addition functions in modelling where possible.
5. If you cannot add, then multiply or divide.
6. Avoid as far as possible taking differences or raising variables to powers.

The challenge now is to use this information in developing the best set of procedures and modelling tools to reduce the impact of uncertainty and inaccuracy. A really intelligent GIS would be able to carry out error propagation studies ***before*** a modelling operation is started, to estimate whether the methods and data chosen were likely to yield the results intended. It would report to the user where the major sources of error come from and would present him or her with a set of options which would achieve better results. This intelligent GIS has yet to be developed but may be an area of future effort, especially given the growing use of data collected by others in analysis and modelling exercises. Some possible areas where an intelligent GIS could help to reduce errors would be summarized as:

a) using optimal interpolation techniques
b) using the appropriate sampling density to collect data about a phenomenon
c) checking for outliers, subgroups, systematic bias, etc.
d) adopting appropriate classifications and class breakpoints
e) improving model calibration and reducing errors in input model parameters.

? Questions

1. If you were to design a modelling system to understand the impacts of climate change on your home region, (a) what input data would you use, (b) what process modelling would you include, and (c) how would you present the results to local politicians?
2. How you design a GIS-based modelling system for managing disasters: (a) earthquake, (b) hurricane, (c) train collision?
3. How do you think the errors in the output of any given model will depend on (a) the uncertainties in the regression equation and (b) the spatial variation in the data?

Further reading

▼ Beven, K. and Moore, I. D. (eds.) (1994). ***Terrain Analysis and Distributed Modelling in Hydrology**: Advances in Hydrological Processes*. Wiley, Chichester.

▼ Brimicombe, A. (2010). ***GIS, Environmental Modeling and Engineering***. 2nd edn. CRC Press, Boca Raton, FL.

▼ Harris, R. and Jarvis, C. (2011). ***Statistics for Geography and Environmental Science***. Prentice Hall, Harlow.

▼ Heppenstall, A. J., Crooks, A. T., See, L. M., and Batty, M. (eds.) (2012). ***Agent-based Models of Geographical Systems***. Springer Sciences, Dordrecht.

▼ Heuvelink, G. B. M. (1998). ***Error Propagation in Environmental Modelling with GIS***. Taylor & Francis, London.

▼ Koomen, E., Stillwell, J., Bakema, A., and Scholten, H. J. (eds.) (2007). ***Modelling Land-use Change: Progress and Applications***. Springer, Dordrecht.

▼ Lin, H. and Batty, M. (eds.) (2009). ***Virtual Geographic Environments***. Science Press, Beijing.

▼ Liu, Y. (2008). ***Modelling Urban Development with Geographical Information Systems and Cellular Automata***. CRC Press, Boca Raton, FL.

▼ Malczewski, J. (1999). ***GIS and Multicriteria Decision Analysis***. Wiley, New York.

▼ Skidmore, A. K. (2002). ***Environmental Modelling with GIS and Remote Sensing*** (Geographic Information Systems Workshop). CRC Press, Boca Raton, FL.

▼ Tomlin, C. D. (2012). ***GISystems and Cartographic Modeling***. ESRI Press, Redlands, CA.

▼ Vieux, B. E. (2004). ***Distributed Hydrologic Modelling using GIS***. 2nd edn. Kluwer, Dordrecht.

▼ Wainwright, J. and Mulligan, M. (eds.) (2013). ***Environmental Modelling: Finding Simplicity in Complexity***. Wiley-Blackwell, Chichester.

Fuzzy Sets and Fuzzy Geographical Objects

Up to this point we have assumed that real-world phenomena can be modelled either by exactly defined delineated entities like polygons or by smooth continuous fields. Uncertainty has been treated probabilistically, using conventional statistical methods and spatial statistics. In this chapter we present ways of dealing with uncertainty, complexity, and vagueness in terms of fuzzy, overlapping sets. Instead of probability, **fuzzy set theory** uses concepts of admitted possibility, which is described in terms of the fuzzy membership function. **Fuzzy membership functions** permit individuals to be partial members of different, overlapping sets. The sets can be defined exogenously, using the semantic import model, or can be computed from multivariate data using the methods of **fuzzy *k*-means**. If the original data are measured at point locations, membership functions can be mapped by interpolation: zones where different fuzzy set surfaces intersect are locations of 'confusion' and can be extracted to provide crisp boundaries. The methods are illustrated with applications from soil survey, land classification, pollution mapping, and vegetation science.

Learning objectives

In this chapter we will focus on how imprecise fuzzy ideas can be used in spatial data analysis. By the end of the chapter, you will:

- **understand how to deal with imprecision in overlapping attribute classes**
- **have learned how this understanding can be extended to geographical phenomena that are expressed either as continuous fields or as entities like polygons.**

13.1 Imprecision as a way of thought

In the process of classification and retrieval of spatial data we unconsciously use the basic laws of thought, first developed by Aristotle:

- The law of *identity* (everything is what it is—a house is a house).
- The law of *non-contradiction* (something and its negation cannot both be true—a house cannot be both house and not a house).
- The *principle of the excluded middle* (every statement is true or false—this house is lived in or it is not).

The principle of the excluded middle ensures that all statements in conventional logic can have only two values—true or false—which can be coded as 0 or 1. This assumption lies at the heart of most of mathematics and computer science (Barrow 1992; Kosko 1994), so naturally the paradigm is very deeply embedded in our tools for computation and data retrieval.

Many geographical phenomena, however, are not simple, clear-cut 'entities'. The patterns produced by natural and human processes vary over many spatial and temporal scales and the ensemble entities are defined by not one, but many interacting attributes. Consequently, it is often a very difficult practical problem to partition the real world into unique, non-overlapping sets. When our perception of reality yields something that is neither clearly an entity nor a continuous field, we use simple pragmatism to force the observation into one mode or the other, because we have had no means in GIS for dealing with entities otherwise. Even though we think we have defined classes exactly, we may not be able to assign individuals to the correct groups, either because the rules are ambiguous or because measurements cannot be made with sufficient accuracy.

Formal thought processes in Western logic have traditionally emphasized the paradigm of truth versus falsehood, so we have very little formal training on how to deal with overlapping concepts. The law of the excluded middle and its role in mathematical proof—*reductio ad absurdum*—has been of paramount importance in scientific and philosophical development. The rules of logic used in computer query languages are all based on exact ideas of truth or falsehood, which implies that all objects in a database fit the same paradigm.

In environmental data this is not necessarily so, however, and by admitting partial truth, or class overlap, the emerging paradoxes may be addressed (see e.g. Burrough and Frank 1996). We often use this approach in natural language when we are not forced by scientific or legal considerations to work with exact concepts. For example, consider the discrimination between tall and short people. At first, we might consider that a person who is 190 cm (6 ft 2.5 inches) in length is *tall*, and that one who is 150 cm (4 ft 11 inches) is *short* (Figure 13.1). But

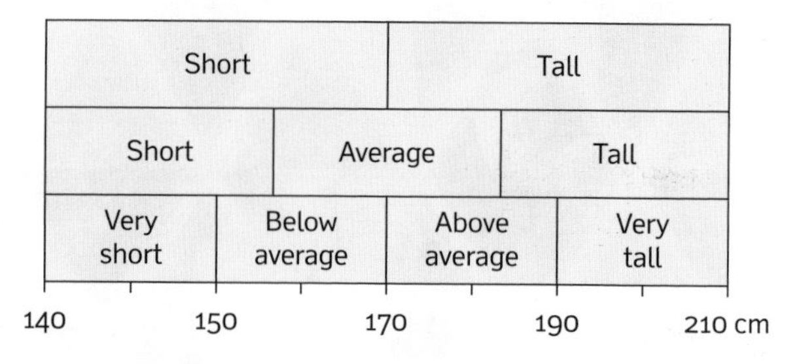

Figure 13.1 Natural language deals with fuzzy classes by redefinition and class splitting. In fuzzy sets, the grade of membership is expressed in terms of a scale that can vary continuously between 0 and 1.

is a person who is 170 cm long, tall, or short? If the tall/short class boundary is fixed halfway between these two extremes, we cannot decide, so in practice we take the middle ground and create a new set of classes—*short—average—tall*. This creates new problems for classifying people who are 154 cm or 186 cm in length, but we can easily create new classes in the areas of overlap to deal with the *below average* and *above average* cases.

Many users of geographical information have a clear notion (or central concept) of what they need. For example, river managers might need to answer questions such as 'Which areas are under threat of flooding?' or 'Which parts of the wetlands suffer from polluted discharges?' Such imprecisely formulated requests must then be translated into terms of the basic units of information available. Furthermore, not all information stored in the simple data model is exact. Many of the data collected during field survey are often described in seemingly vague terms—for example, vegetation can be described as vital, or partially vital—and so retain a strong flavour of qualitative ambiguity. Clearly, we need methods for specifying how we must deal with imprecise information and borderline cases.

Geographical phenomena are more complicated than many other multivariate defined entities because we must consider grouping both in *attribute space* and in *geographical space*. Grouping in attribute space determines whether all entities are of the same kind, and the problem of determining a group identity is one of choosing a classification based on attributes that leads to unambiguous, non-overlapping classes with clear rules for allocation. Grouping in geographical space determines whether entities (or multivariate fields) of a similar kind occupy contiguous regions.

13.2 Fuzzy sets and fuzzy objects

Fuzziness is a type of definable imprecision characterizing classes that for various reasons cannot have or do not have sharply defined boundaries—the development of these ideas emanated from Zadeh's ideas of fuzzy sets (Zadeh 1965). If one accepts that spatial processes interact over a wide range of spatial scales in ways that cannot be completely predicted, then one can appreciate the need for the 'fuzzy' concept in geographical information. Fuzzy set theory is a generalization and not a replacement for the better-known abstract set theory which is often referred to as Boolean logic (see Klir et al. 1998; Cox et al. 1999; Ragin 2000; Zimmerman 2001; Ross 2010 for further details).

Fuzziness is *not* a probabilistic attribute, in which the degree of membership of a set is linked to a given statistically defined probability function. Rather, it is an admission of *possibility* that an individual is a member of a set, or that a given statement is true. The assessment of the possibility can be based on subjective, intuitive ('expert') knowledge or preferences, but it could also be related to clearly defined uncertainties that have a basis in probability theory. For example, uncertainty in class allocation could be linked to the possibility of measurement errors of a certain magnitude.

In the traditional *crisp set*, all members match the class concept and the class boundaries are sharp. In many cases the boundary values of crisp sets are chosen either (a) on the basis of expert knowledge (e.g. boundary values of discriminating criteria chosen by custom, law, or an external taxonomy), or (b) using methods of numerical taxonomy. The degree to which an individual observation z is a member of the set is expressed by the membership function MF^{B}, which for crisp (Boolean) sets can only take the values 0 or 1. Note that z is used here as a general attribute value, as in regionalized variable theory (Chapter 9). Formally, we write:

$$MF^{B}(z)=1 \quad \text{if } b_1 \leq z \leq b_2$$
$$MF^{B}(z)=0 \quad \text{if } z < b_1 \text{ or } z > b_2 \qquad 13.1$$

where b_1 and b_2 define the exact boundaries of set A. For example, if the boundaries between 'unpolluted', 'moderately polluted', and polluted soil were to be set at $b_1 = 50$ units and $b_2 = 100$ units, then the membership function given in equation 13.1 defines all 'moderately polluted soils'.

In contrast, fuzzy sets admit the possibility of partial membership, so they are generalizations of crisp sets to situations where the class boundaries are not, or cannot be, sharply defined, as in the case of tall and short people given in Section 13.1. The same is true for an environmental property such as 'internal soil drainage', which embraces all conditions from total impermeability to excessively free draining. To state just what is, and what is not, 'a moderately well-drained soil' requires not strict allocation to an exactly defined class, but a qualitative judgement that by implication allows the possibility of partial membership.

A fuzzy set is defined mathematically as follows: if Z denotes a space of objects, then the fuzzy set A in Z is the set of ordered pairs

$$A=(z, MF_A^F(Z)) \text{ for all } z \varepsilon Z \qquad 13.2$$

where the membership function $MF_A^F(Z)$ is known as the 'grade of membership of z in A' and $z \varepsilon Z$ means that z belongs to Z. Usually $MF_A^F(Z)$ is a number in the range 0, 1, with 1 representing full membership of the set (e.g. the 'representative profile' or 'type') and 0 representing non-membership. The grades of membership of z in A reflect a kind of ordering that is not based on probability but on admitted possibility. The value of $MF_A^F(Z)$ of object z in A can be interpreted as the degree of compatibility of the predicate associated with set A and object z; in other words $MF_A^F(Z)$ of z in A specifies the extent to which z can be regarded as belonging to A. So, the value of $MF_A^F(Z)$ gives us a way of giving a graded answer to the question 'to what degree is observation z a member of class A?' Figure13.2 uses the method of Venn diagrams to illustrate the difference between crisp and fuzzy sets.

Put simply, in fuzzy sets, the grade of membership is expressed in terms of a scale that can vary continuously between 0 and 1. Individuals close to the core concept have values of the membership function close to or equal to 1; those further away have smaller values. Note that this immediately gets around the problems of the principle of the excluded middle—truth is *not* absolute—and individuals can, to different degrees, be members of more than one set. The problem is to determine the membership function unambiguously.

Setting up classifications in fuzzy sets can follow two different approaches. In the first (and simpler) approach, an a priori membership function is defined with which

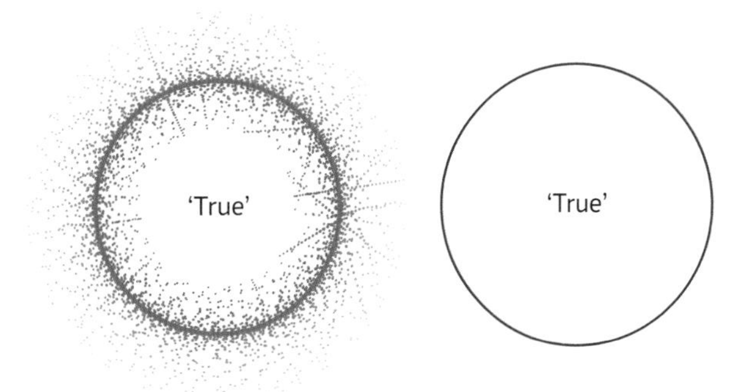

Figure 13.2 Images of fuzzy sets (left) and Boolean (crisp) sets (right)

individuals can be assigned a membership grade. This is known as the **semantic import approach** or model (SI). The second approach is analogous to cluster analysis and numerical taxonomy in that the value of the membership function is a function of the classifier used. One frequently used version of this model is known as the method of fuzzy *k*-means. Both methods can be usefully applied to spatial data.

13.3 Choosing the membership function 1: the semantic import approach

The semantic import approach is useful in situations where users have a very good, qualitative idea of how to group the data, but for various reasons have difficulties with the exactness associated with the standard Boolean model. The choice of fuzzy sets does not mean that one is opting out. The selection of boundaries for crisp sets and of class intervals can be an objective or a subjective process, depending on the way in which scientists agree to define classes. The same is just as true for assigning the membership function of a fuzzy set. The membership function should ensure that the grade of membership is 1 at the centre of the set and that it falls off in an appropriate way through the fuzzy boundaries to the region outside the set, where it takes the value 0. The point where the grade of membership is 0.5 is called the **'crossover point'**. The membership function must be defined in such a way that these conditions hold, so not all functions are possible.

Suitable membership functions for use with the SI approach

Just as there are various types of probability distribution (normal, log-normal, rectangular, hyperbolic, Poisson, etc.), so there can be different kinds of fuzzy membership functions (see Figure 13.3). These are used to determine the membership value at the edges of the set. Most common are the 'linear' MF^F and the 'sinusoidal' MF^F.

The linear MF^F is given by a pair of sloping lines that peak at $MF = 1$ for the central concept of the set, *c*, and have *MF* values = 0.5 at the boundaries (Figure 13.3b). The slope of the line gives the width of the fuzzy transition zone. Note that areas inside the sloping lines, but outside the Boolean rectangle are zones of *partial truth*.

The sinusoidal MF^F is given by:

$$MF_A^F(z) = \frac{1}{\left(1 + a(z-c)^2\right)} \quad \text{for } 0 \le z \le P \qquad 13.3$$

where *A* is the set in question, *a* is a parameter governing the shape of the function, and *c* defines the value of the property *z* at the central concept. By varying the value of *a*, the form of the membership function and the position of the crossover point can be fitted to Boolean sets of any width (Figure 13.3c). We call this membership function 'Model 1'.

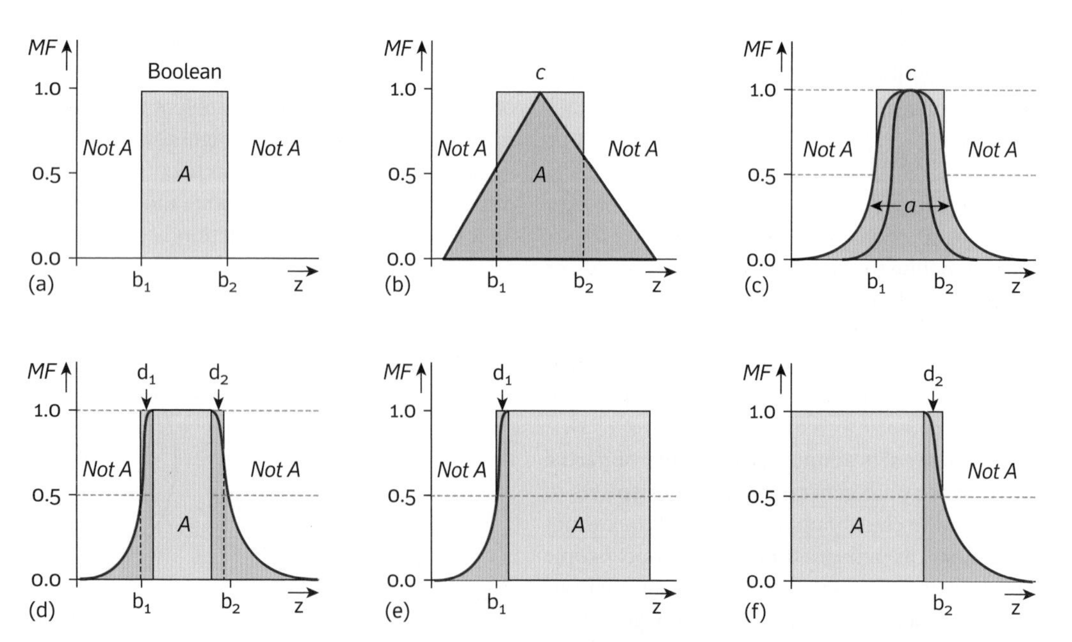

Figure 13.3 Boolean and fuzzy membership functions for the SI method with different membership functions: (a) Boolean, (b) linear, (c) sinusoidal, (d) fuzzy set Model 2, (e) asymmetric Model 3, and (f) asymmetric Model 4

For example, consider the universe of soil depths from 0 to 200 cm, in which we wish to distinguish 'deep' soils from 'shallow' soils and from 'very deep' soils. If we chose c = 100 cm as the ideal centre (or 'standard index') of the fuzzy set of deep soils, then a value of a = 0.0004 gives lower and upper crossover points of 50 cm and 150 cm, respectively, where $MF_A^F(z) = 0.5$. The values of z flanking the central concept from the crossover points (in this case from 50–100 cm and 100–50 cm) can be thought of as the *transition zones* surrounding the central concept of the fuzzy set.

Extending the definition of membership functions to sets with a range of values that meet the central concept

Often, it may be sensible to extend the central concept of a fuzzy set to include a range of possible values rather than a single value, and this idea is implicit in the definition of crisp sets (equation 13.1; Figure 13.3a). Equation 13.3 can easily be modified to handle central concepts that cover a range of values; instead of using the parameter a to define the form of the fuzzy membership function it is easier in practice to use a transition zone of width d_1 and d_2.

Figure 13.3d shows a fuzzy set with a central core region and upper and lower transition zones of width d_1 and d_2. Though the upper and lower transition zones to the fuzzy set could be equally broad, this is not necessary and the figure shows lower and upper transition zones of unequal width. We call this 'Model 2'. It seems intuitive to locate the transition zones so that the crossover points of the fuzzy set lie at the boundaries of the equivalent Boolean set, thereby ensuring that observations lying just inside a Boolean boundary are still not full members of the core of the set. This is sensible if observations have an associated error band which means that though the recorded value may lie within the Boolean set, the true value may be outside. Therefore, values of d_1 and d_2 are essentially *half-widths* of the transition zones because they give the width *inside* the Boolean boundary to the value of z corresponding with the central concept. This also ensures that selections made by Boolean and fuzzy sets can be strictly compared and that at the boundaries, b_1, b_2, $MF^F = 0.5$.

Model 2 is defined by three equations:

$$MF^F(z) = \frac{1}{1+\left(\frac{z-b_1-d_1}{d_1}\right)^2} \text{ if } z < b_1 + d_1 \qquad 13.4a$$

$$MF^F(z) = 1 \text{ if } b_1 + d_1 \leq z \leq b_2 - d_2 \qquad 13.4b$$

$$MF^F(z) = \frac{1}{1+\left(\frac{z-b_2+d_2}{d_2}\right)^2} \text{ if } z > b_2 - d_2 \qquad 13.4c$$

where $MF^F(z)$ is the value of the continuous membership function corresponding to the attribute value z. Note that if parameters d_1 and d_2 are zero, equation 13.4 yields the Boolean membership function (equation 13.1).

Asymmetric membership functions

In many situations only the lower or upper boundary of a class may have practical importance. This could be true, for example, in situations where we only wish to know if the soil is deep enough for a given purpose—if it exceeds this depth by 5 cm or 200 cm is immaterial. In these situations it is easy to use the various parts of equation 13.4 to describe the membership function on the lower or the upper side (Model 3, Figure 13.3e or Model 4, Figure 13.3f).

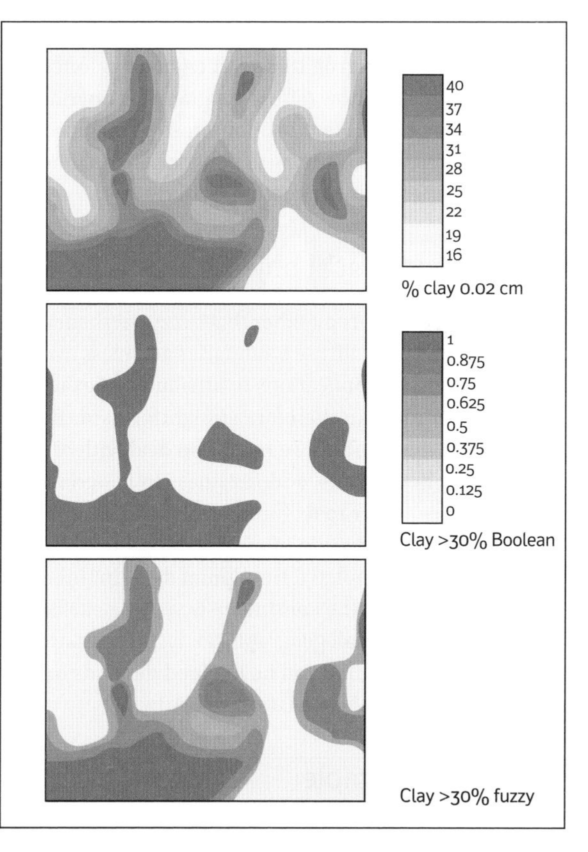

Figure 13.4 Boolean and fuzzy retrieval of soil texture

Box 13.1 Procedures for computing Boolean and fuzzy maps

Computing maps of soil with clay content > 30 per cent

Boolean:

CLAY_30.Boo = IF(CLAY.EST GE 30 THEN 1 ELSE 0)

Fuzzy:

CLAY_30.FUZ = IF(CLAY.EST GE 35 THEN 1 ELSE (1/(1 + ((CLAY.EST-30-s)/5)**2)))

where CLAY.EST is the interpolated map of soil clay content.

Choosing values for the width of the transition zones

There seem to be no hard and fast rules for choosing the values of the d_1 and d_2, but it seems sensible to relate the width of the transition zones to what is known about the precision of measuring the attribute of the phenomenon or object in question. For data measured at points, the width of the transition zones could reflect the known accuracy of the measurement technique: for grid data interpolated by kriging, the width of the transition zone could be given by the kriging standard error. If fuzzy membership functions are used to represent how diffuse geographical boundaries may be, the widths of the transition zones of membership functions related to geographical boundaries could be defined using expert knowledge from the terrain.

Fuzzy geographical objects

Before going further, remember that fuzzy set theory initially only deals with diffuse boundaries and class overlap in attribute space. Applications to attributes of exact geographical entities are therefore straightforward. But in geographical information we must also deal with diffuse geographical boundaries, and fuzziness in geographical objects can also apply to features such as the boundaries of polygons, or to variation in membership function values that can be interpolated from point data. In this case, it is useful to refer to the continuously varying surface as a fuzzy field. As with all geographical data, the ease with which fuzzy information can be mapped depends on the strength of the spatial correlation structures.

A practical example

To demonstrate the differences between Boolean and fuzzy sets, Figure 13.4a shows a map of the clay content of the C horizon of the soil in part of the Lacombe Experimental Farm, Alberta, interpolated by ordinary block kriging. The area mapped measures 755×545 m^2 with a 5 m resolution. Figure 13.4b shows the areas matching the Boolean class limits of >30 per cent clay, and Figure 13.4c shows the fuzzy classification using Model 3 with a lower transition zone width of 5 per cent clay (equivalent to a kriging standard deviation of 5 per cent clay). Box 13.1 gives the pseudo code for computing these maps and shows how easy it is to compute the fuzzy membership function values.

The semantic import approach: recapitulation

Unlike the simple Boolean set where membership values are discretely 0 or 1, the semantic import approach transforms the data to a continuous membership function ranging from 0 to 1. The value of the membership function gives the *degree* to which the entity belongs to the set in question. Compared with Boolean classification, where only the boundary values have to be chosen, the extra problems the semantic import approach brings are (a) choosing between a linear, sinusoidal, or other function for assessing class membership and (b) selecting the values of the dispersion indices d_1 and d_2. Therefore the semantic import approach needs more information than the conventional Boolean method, but this is amply repaid by the extra sensitivity in data analysis.

13.4 Operations on several fuzzy sets

Just as with Boolean sets, data in fuzzy sets can be manipulated using logical query methods to select and combine data from several sets, and standard query languages in relational database management systems have been modified to accept continuous logical operations (e.g. Kollias and Voliotis 1991). The basic operations on fuzzy subsets are similar to, and are a generalization of, the AND/OR/NOT/XOR and other operations used for Boolean sets (see Cox et al. 1999; Mordeson and Nair 2001; George 2007; Ross 2010 for more details).

Box 13.2 gives the main logical operations for fuzzy sets. Note that the integral sign does *not* mean 'summation' here, but 'for the universe of objects Z'. For our purposes, the most used operations will be union (maximize), intersection (minimize), negation (complement), and the convex combination (weighted sum). The concept of the convex combination is useful when linguistic

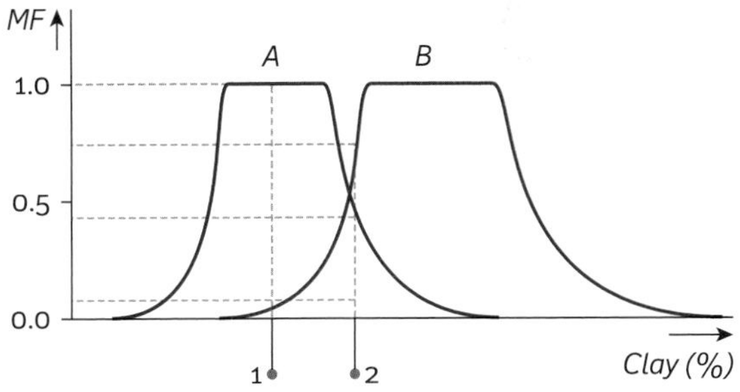

Figure 13.5 Two overlapping fuzzy membership functions for soil clay content

modifiers such as 'essentially' and 'typically' are to be used, or when different attributes can compensate for each other. This idea is particularly appropriate in land evaluation, as we shall see. Union, intersection, or convex combinations lead to the computation of a new MF value, which we call the **joint membership function** value or JMF.

It is now easy to see how an individual entity can be a member of one or more sets. Figure 13.5 shows the membership functions for two adjacent classes of clay texture of soil. Observation 1 is clearly in class *A*, and has a membership value of 1.0 for class *A* and 0.01 for class *B*. Observation 2 is near the class boundaries: it has a membership value of 0.4 in class *A* and a membership value of 0.7 in class *B*. The *JMF* for union (OR) for observation 1 is 1.0, and for observation 2 is 0.7; for intersection (AND) the *JMF* is 0.01 for observation 1 and 0.4 for observation 2. Note that if the two classes have different transition widths, then the membership values for a given observation do not have to sum to 1 for the semantic import approach. The same is true for classes defined using different attributes.

Joint membership values for interpolated attributes

Box 13.3 demonstrates how union and intersection work with classes of different attributes. Suppose we have data from soil profiles that give the clay content for three layers—0–20 cm, 30–40 cm, and 70–80 cm to an accuracy of ±5 per cent. Let us define a 'heavy-textured soil' as one with a clay content of ≥ 30 per cent in the first layer, ≥35 per cent in the second layer, and ≥40 per cent in the third layer. We can use a sinusoidal function with a dispersion value of 5 per cent (Model 3). In practice it

Box 13.2 Operations on fuzzy sets

1. Two fuzzy sets, *A* and *B*, are said to be equal ($A = B$) iff (where iff means if and only if)

$$\int_Z MF_A(z)/z = \int_Z MF_B(z)/z \qquad MF_A = MF_B$$

2. *A* is contained in *B* (*A B*) iff

$$\int_Z MF_A(z)/z \le \int_Z MF_B(z)/z$$

3. The union of fuzzy sets *A* and *B* is the smallest fuzzy subset containing both *A* and *B*

$$A \cup B = \int_Z (MF_A(z) \vee MF_B(z))/z$$

v is the symbol for max. Union corresponds to the connective 'OR'.

4. The intersection of fuzzy sets *A* and *B* is denoted by $A \cap B$ and is defined by:

$$A \cap B = \int_Z (MF_A(z) \wedge MF_B(z))/z$$

^ is the symbol for min. Intersection corresponds to the connective 'AND'.

5. The complement of *A* corresponding to NOT is denoted by A^{-1} and is defined by:

$$A^{-1} = \int_Z (1 - MF_A(z))/z$$

6. The product of *A* and *B* (a 'soft' AND) is defined by:

$$AB = \int_Z (MF_A(z).MF_B(z))/z$$

7. The bounded sum of *A* and *B* (a 'soft' OR) is defined by:

$$A \oplus B = \int_Z^1 \wedge (MF_A(z) + MF_B(z))/z$$

where + is the arithmetic sum.

8. The bounded difference is defined by:

$$A \ominus B = \int_Z^0 \vee (MF_A(z) - MF_B(z))/z$$

9. If $A_1, \ldots A_k$ are fuzzy subsets of *Z*, and $w_1, \ldots w_k$ are non-negative weights summing to unity, then the *convex combination* of $A_1, \ldots A_k$ is a fuzzy set *A* whose membership function is the weighted sum:

$$MF_A = w_1 MF_{A1} + \cdots w_k MF_{Ak}$$
$$= \sum_{i=1}^{k} w_i MF_{Ai} \text{ where } \sum_{i=1}^{k} w_i = 1,\ w_i > 0.$$

Box 13.3 Examples of minima and maxima in *MF* values for clay

Site	Original data			Fuzzy memberships			Boolean MF		Fuzzy	
#	C1	C2	C3	C1	C2	C3	AND	OR	AND	OR
1	26.7	26.9	32.8	**0.27**	0.13	<u>0.14</u>	0	0	0.14	0.07
2	30.8	16.8	45.3	0.59	**0.04**	<u>1.00</u>	0	1	0.04	1.00
3	39.8	42.6	45.6	**1.00**	1.00	<u>1.00</u>	1	1	1.00	1.00
4	32.6	46.8	46.9	**0.81**	<u>1.00</u>	<u>1.00</u>	1	1	0.81	1.00
5	16.9	46.7	52.2	**0.07**	<u>1.00</u>	<u>1.00</u>	0	1	0.07	1.00
6	48.8	34.9	54.8	<u>1.00</u>	**0.49**	<u>1.00</u>	1	1	0.49	1.00
7	20.0	24.5	30.7	**0.10**	0.09	<u>0.11</u>	0	0	0.09	0.11

Comparative minima are in bold; comparative maxima are underlined.

Limits: layer 1 ≥ 31%, layer 2 ≥ 35%, layer 3 ≥ 40%; widths 5%.

may be necessary to distinguish profiles in which only one of the three horizons is 'heavy' (clay anywhere in the profile) from profiles where all three layers are heavy (clay throughout). The first situation is a union (OR), and requires selection of the *maximum* MF value; the second is a union (AND) and requires selection of the *minimum*. The right-hand columns of the table show which profiles would be selected, and which rejected by both categories. Note that if the measurement error is indeed ±5 per cent, measured clay values of 5 per cent less than the Boolean boundary (corresponding to a $MF^F = 0.2$) would be possible members of the set of 'heavy soils', though they would be completely rejected by the Boolean selection.

Figure 13.6 shows the same analysis, but now applied to interpolated surfaces of the clay content for the Lacombe data to distinguish the situation of 'clay in some part of the area' (Figure 13.6a and b) from 'clay throughout the profile' (Figure 13.6c and d). Clearly, fuzzy classification is a trivial operation in geographical information systems that provide 'cartographic algebra'. Note that the operation would be essentially the same if the clay content of the three layers was expressed as soil polygons, but the appearance of the result would be dictated by the form of the mapping units.

If data are originally on a point support, one must decide whether to interpolate all data to a common grid first before carrying out fuzzy classification, or to classify the profile data first and then interpolate the membership functions. The second option is difficult for Boolean classes unless they are interpolated as indicator functions, but not for fuzzy MF values, although there may be theoretical problems because the fuzzy MF is not normally distributed or is constrained to the range 0–1. As in other areas of modelling, computational factors may decide which route is the best. For many users of geographical information systems it will be easier to apply the classification models to data that have already been interpolated.

Fuzzy classification with ordinal data

In Chapter 7 we showed how conventional Boolean logic is used in the 'top-down' classification of areas of land in terms of suitability to grow a crop. In this case the crop was maize and the location a small part of Kisii District in Kenya (Figure 7.5). Readers will recall that once the land characteristics (soil depth, soil type, slope classes) had been mapped, land qualities (water availability, oxygen availability, nutrient availability, and erosion hazard) were derived on a simple ternary ranking of 'poor' (or insufficient), 'moderate' (or just about sufficient), and 'good' (easily sufficient). The final land suitability classification was achieved by the simple rule of the most limiting factor. In other words, for every site or grid cell the worst ranking of any of the four land qualities determines the final ranking. The final result was that 84 per cent of the area was declared 'poor', 5 per cent 'moderate', and 11 per cent 'good'.

Inspection of the four maps of land qualities (Figure 7.5) shows that water availability and nutrient availability are the two land qualities most limiting for suitability to grow maize, yet these are the two attributes that can most easily be modified (by irrigation or by the addition of fertilizers or organic manures). As in the previous example with the clay content of the Lacombe soil, we can define a fuzzy membership function for each land quality. In this case the functions have been defined

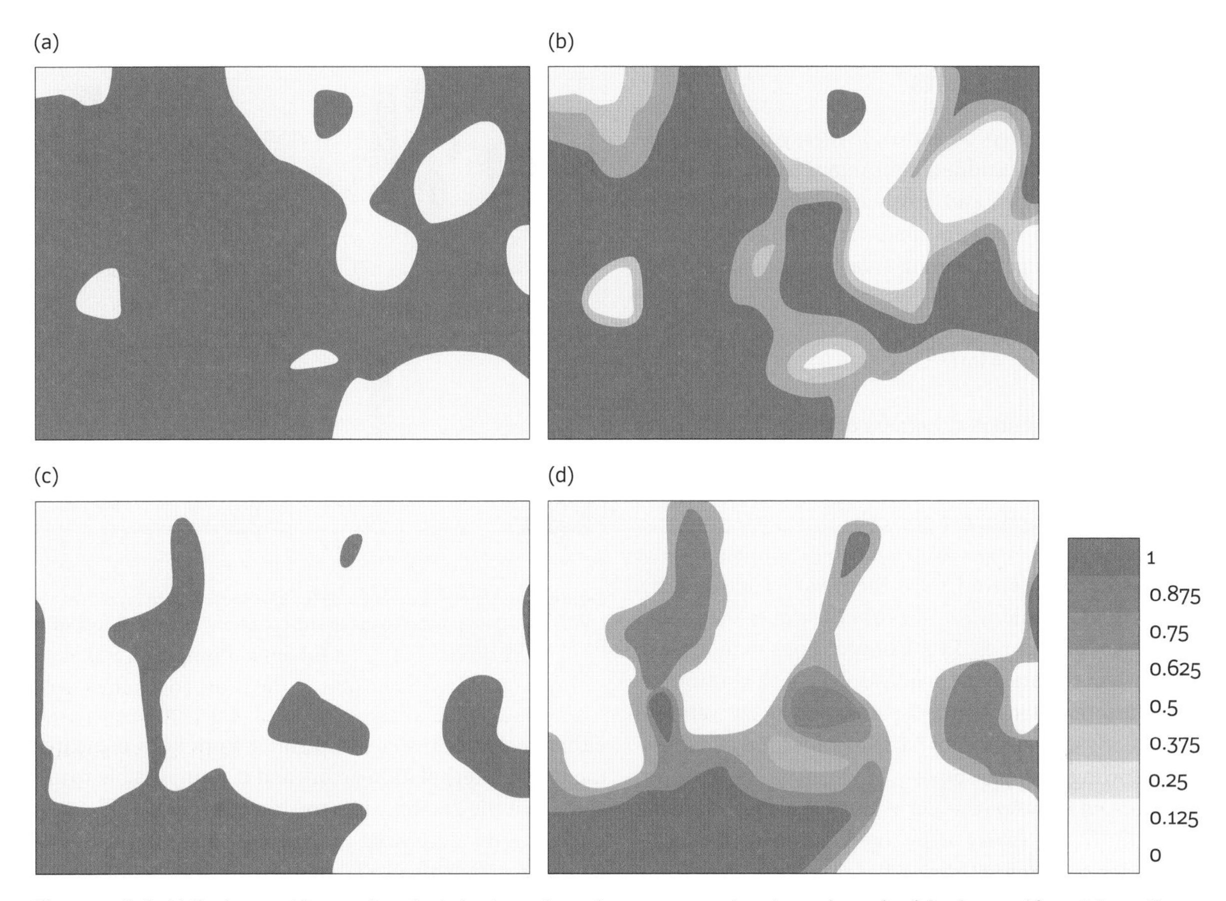

Figure 13.6 (a, b) Boolean and fuzzy union of sets to show where clay occurs anywhere in any layer; (c, d) Boolean and fuzzy intersection to show where clay occurs throughout the soil profile

so that the transition zones are equal to half a ranking. This distinguishes land qualities that are 'poor to moderate' from those that are 'moderate to good', for example.

Applying this idea to the Kisii land evaluation example yields the result presented in Figure 13.7. This reveals that, far from 84 per cent of the area being 'poor', actually some 71 per cent scores better than being 0.7 of the ideal concept of 'good sites'. The implications for land use are that with improved husbandry the landscape can be protected and be productive, not just written off.

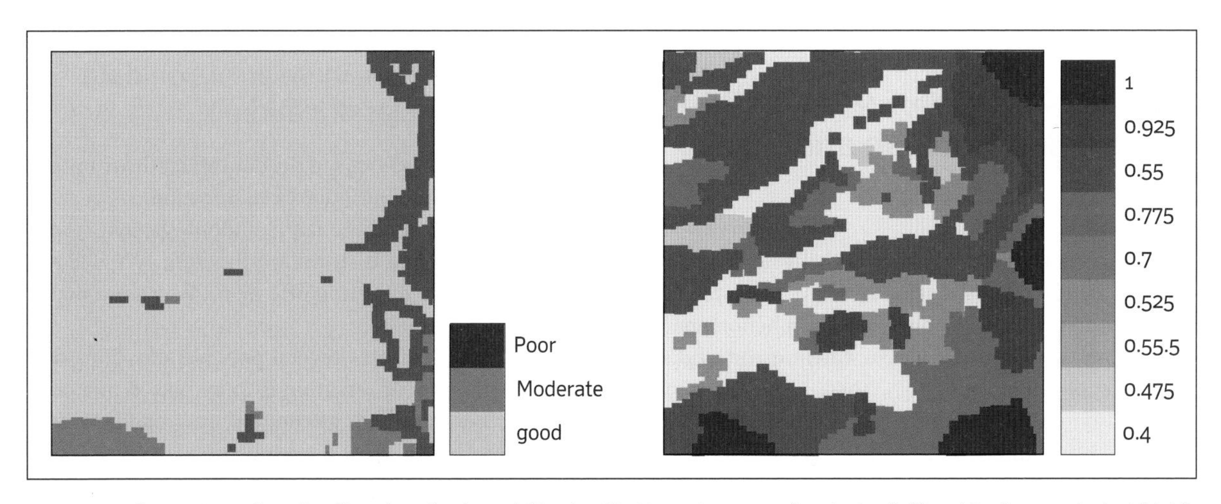

Figure 13.7 Comparison of results of top-down land capability classification using conventional rules (left) and the fuzzy equivalent (right)

Using fuzzy logic with continuous data such as drainage nets

As shown in Chapters 10 and 11, it is easy to derive secondary data from gridded continuous surfaces, slope, potential flow paths, and upstream contributing areas. As in Chapter 10, a wetness map can be computed from a map of upslope contributing areas and a map of slopes at each grid cell.

$$W = \ln(UPL * g / \tan \beta) \qquad 13.5$$

where W is the estimated wetness of the grid cell, UPL is the number of upslope elements discharging through the grid cell, g is the area covered by each grid cell, and β is the slope of the land at the grid cell.

Because maps of upstream contributing areas and the wetness indices derived from them vary continuously, there are no simple ways to extract features that could be termed 'boundaries'. Even contour lines do not make sense when the attribute in question is both continuous and clustered along limited pathways. Fuzzy sets are particularly appropriate for handling this kind of data, as the following example shows.

Figure 13.8 shows the derivation of a map showing where the higher values of the topologically derived wetness index combine with subsoil clay content. The limits of the clay content are taken as 40 per cent with a threshold of 5 per cent, and the wetness index limit is 14 with a threshold of 6. Figure 13.8c shows the Boolean retrieval of all wetness zones with index ≥14 lying on subsoil clay ≥40 per cent, and Figure 13.8d shows the fuzzy equivalent. Note that the fuzzy retrieval retains the whole of the drainage net in the areas of heavy clay, thereby preserving the topological continuity of the drainage, which is lost in the Boolean solution.

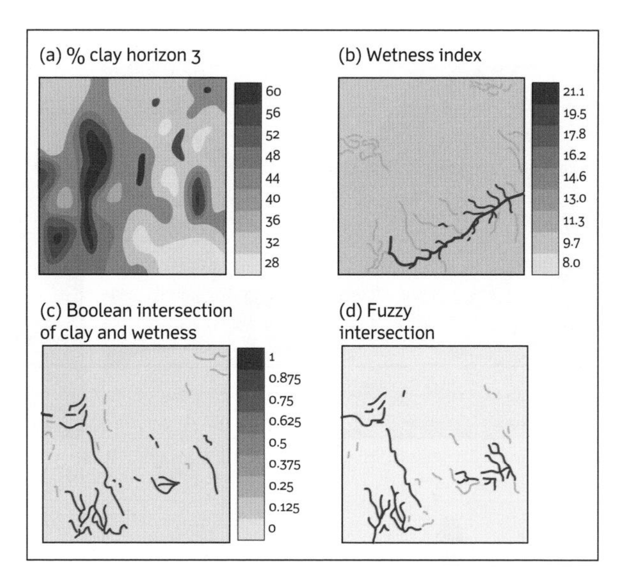

Figure 13.8 Using fuzzy methods for intersecting data layers with continuous data preserves detail and connectivity which is lost with Boolean selection

13.5 Error analysis of selections made using Boolean and fuzzy logic

As we have seen, a major problem with the Boolean model is that it assumes that the attributes have been described and measured exactly. In many natural phenomena this is not possible because of measurement errors and spatial variation (Burrough 1986, 1989; Goodchild 1989). Errors also increase because the spatial variation cannot be seen in its entirety, but must be reconstructed from data gathered at sample locations. The resulting maps (in either vector or raster form) suffer from both measurement and interpolation errors and it is impossible to assume that the database is error-free. Consequently, when uncertain data are used in logical or quantitative models, the results will also contain errors (Heuvelink 1998), especially when these have values near the boundaries of the Boolean classes. Using the Monte Carlo methods explained in Chapter 12, Heuvelink and Burrough (1993) analysed the propagation of errors through Boolean and fuzzy reclassification and retrieval, and demonstrated that fuzzy selection and reclassification was far less sensitive to small errors in the data, particularly when data values are near critical class boundaries.

The effects of attribute errors on data retrieval and classification

To understand how errors in the data affect the results of both Boolean and continuous classification, we replace the deterministic attribute z by a random variable Z, with a certain probability distribution function. We will denote the standard deviation of Z by σ_Z, which is a measure of the 'error' in Z.

Figure 13.9 illustrates several possible situations that can arise when an observation is to be placed in a class whose boundaries on the attribute Z are b_1 and b_2. Figure 13.9a shows the standard Boolean case when there is no error, so $\sigma_z = 0$. The attribute Z is in effect deterministic, so the individual observation either falls entirely within the class boundaries (right-hand bar) or it falls outside (left-hand bar). The corresponding values of the membership function are 1 or 0 respectively. Because σ_z is zero, the membership value *MF(Z)* is also error-free. Figure 13.9b shows the same situation, but the individual observation is now classified by a continuous membership function. The right-hand observation is within the class kernel so *MF(Z)* = 1. The left-hand observation is outside the kernel and the Boolean boundaries so it returns an *MF(Z)* < 0.5.

Now consider what happens when σ_z is non-zero. In Figure 13.9c, the *probability density* P_z of *Z* takes the characteristic bell-shaped form and lies well within the limits of the Boolean class. The probability that *Z* really falls outside the class limits is vanishingly small, so for all practical purposes *MF(Z)* = 1. The same situation occurs with the continuous membership function when P_z falls within the class kernel (Figure 13.9d). Equally clear-cut results follow when P_z lies well outside the class limits. So in these cases the error in the observation does *not* cause an error in the classification result. This is in itself a result that is worth noting.

The results are less than clear-cut when P_z straddles the Boolean class boundaries or the continuous transition zones. Figures 13.9e and 13.9f show the situation when the mean of *Z* equals b_1. In the Boolean case (Figure 13.9e) the distribution of *MF(Z)* becomes a discrete distribution with two possible values, 0 and 1. In this case the chances are equal that *Z* falls within or outside the class boundary, so the probability of obtaining each value is 0.5. In the continuous case shown in Figure 13.9f, the mean of *Z* equals b_1 and the distribution band is narrower than the transition zone. In this case *MF(Z)* is continuously distributed just as *Z* is. The mean of *MF(Z)* is about 0.5, and because the membership function varies steeply at b_1, the standard deviation of *MF(Z)* is much larger than for *Z*, though it is still substantially smaller than the Boolean standard deviation in Figure 13.9e.

Figures 13.9g and 13.9h show the situation when P_z mainly lies inside the Boolean limit b_1 but still with considerable overlap. In the Boolean case (Figure 13.9g) the effect is to unbalance the probabilities of returning a membership value of 0 or 1; the probability of returning a value of 1 is increased. In the continuous case, P_z runs into the class kernel so the resulting distribution is a mixed distribution: it is the combination of a continuous and a discrete distribution. There is a definite chance that the membership value will be exactly 1 in some cases, though values less than 1 can also occur.

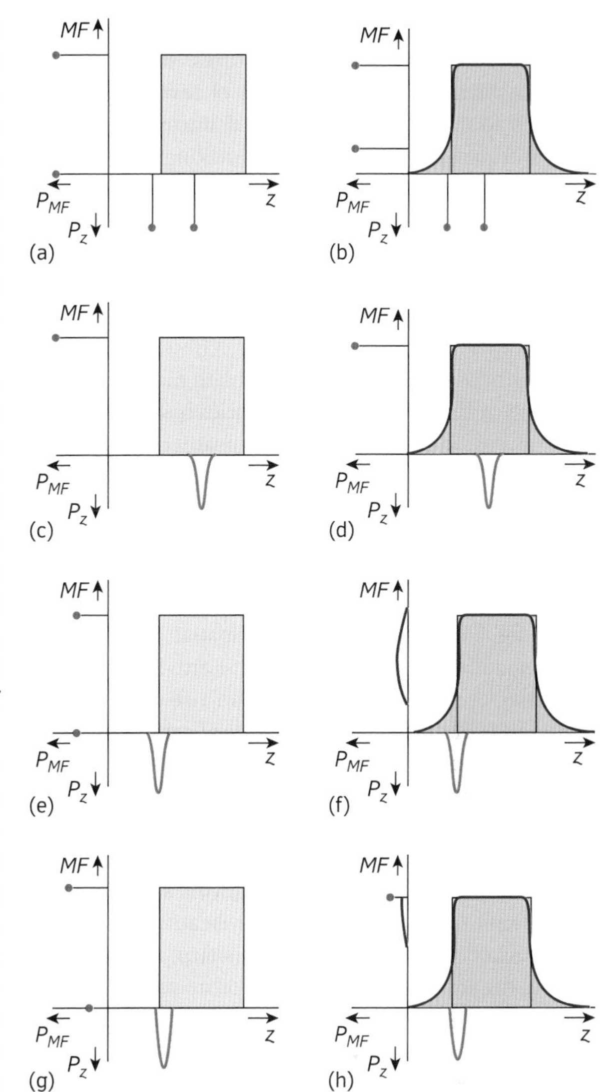

Figure 13.9 The effects of measurement errors on the results of Boolean classification (left column) and fuzzy classification (right column)

Although not shown, the situation where P_z is broader than the transition zone will generally yield mixed distributions of *MF(Z)* with peaks at 0 and/or 1.

13.6 Applying the SI approach to polygon boundaries

So far we have concentrated on classifying attributes, but many kinds of mapping involve the delineation of similar areas of land, often by examining its external appearance, either in the field or from aerial photographs or satellite images. Observable features, such as colour changes, breaks of slope, or edges of texture patterns, are used to infer 'boundaries'. The delineations are first drawn in pencil, then checked by making point observations within similarly classified areas, and then finalized in ink. Conventionally, the field sheets are redrawn as printing masters by cartographers who draw all soil, ecotope, and geological boundaries with a line of standard width (usually 0.2 mm) irrespective of the original nature of the boundary in the field. These artefacts are later digitized to yield digital choropleth maps on which, in theory at least, the drawn boundaries have zero width. This imposition of cartographic neatness suppresses useful information about the nature of spatial change and has misled generations of users of choropleth maps into thinking that soil, vegetation, or geological boundaries are always and everywhere sharp, and that the units are always homogeneous.

All field scientists know that spatial variations in soil, vegetation, or geology can occur either abruptly or gradually. A dyke intrusion can give rise to geological boundaries that are sharp at the scale of centimetres, but variations in texture of alluvial or aeolian deposits can occur over hundreds of metres or kilometres. It is not difficult to indicate whether a boundary identified during fieldwork or aerial photo interpretation is sharp or diffuse, because that can often readily be seen. It is also not difficult to attach an attribute indicating the relative or estimated width of a boundary (sharp, medium, diffuse) to the arcs of a polygon when soil maps are digitized in a vector system, nor is it impossible to draw boundary lines in different styles to indicate their width—in fact all these have been possible for years. The fact is that, apart from some concerns about the accuracy of boundary location, until recently few had thought it necessary and useful to include information about the nature of the boundary in a choropleth map (Lagacherie et al. 1996). Indeed, one objection to doing so was that it was impossible to compute the area of a polygon with diffuse boundaries.

By using the methods of fuzzy logic on polygon boundaries it is very simple to incorporate information about the nature of those boundaries and also to calculate sensible area measures. There are at least two separate approaches: the map unit approach and the individual boundary approach.

The map unit approach

The simplest approach is to assume that a type of boundary (diffuse or abrupt) can be uniquely attributed to each kind of map unit or polygon. For example, soil units in narrow floodplains might have sharp, well-defined boundaries whereas boundaries between loess and coarser aeolian deposits might occur over several hundred metres. Information about the type of boundary can be converted to parameters for a fuzzy membership function of type Model 2, 3, or 4 (equation 13.4a, b, and c), which is then applied to the distance from the drawn boundary. The widths of the transition zones must be chosen with the scale of the map in mind. For example, a sharp boundary on a 1:25 000 map drawn 0.2 mm thick covers 50 m, so the width of the spatial transition zone centred over the drawn boundary location would be 25 m. A diffuse boundary at the same scale might extend over 500 m and have a transition zone width of 250 m.

Fuzzy transition zones can be computed from polygon boundaries by first spreading isotropically and outwards from the original delineation (cf. Chapter 7) and then applying an SI model to indicate the external gradation of MF value from well inside the polygon ($MF = 1$) to the outside. As an example, consider the soil map fragment in Figure 13.10 (taken from the Soil Map of the Mabura Hill Forest reserve in Guyana, Jetten 1994). Figure 13.10a shows the plateau phase of the plinthisols. This unit has fairly diffuse boundaries, and as a first approximation let us assume that the true boundary has a locational uncertainty extending over 500 m—i.e. a transition zone of width 250 m.

The procedure to compute a fuzzy boundary for a single polygon in a raster representation is simple. One extracts the boundary with an edge filter and then applies a spread command to compute the zones around the boundary (see Chapter 10). The parameters of the membership function are selected so that the locations corresponding to the original drawn boundary are at the crossover value (i.e. $MF = 0.5$). The membership function is then applied so that those sites well within the original boundary receive a membership value of 1, those sites inside, but near the boundary receive a membership value between 0.5 and 1, and those sites outside the boundary receive a membership value below 0.5, concomitant with their distance from the boundary. Figure 13.10a shows the result, with the original boundary given in white. Clearly, it is now trivial to compute the area of the polygon that corresponds to any level of the membership function, so previously voiced doubts about estimating the area of polygons with fuzzy boundaries are no longer valid.

The procedure can be repeated for all map units, varying the width of the boundary to match the unit (Table 13.1). When all polygons are displayed with

(a) Plinthisols—plateau with diffuse fuzzy boundary (d_i = 250 m)

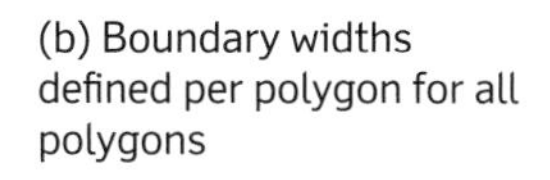
(b) Boundary widths defined per polygon for all polygons

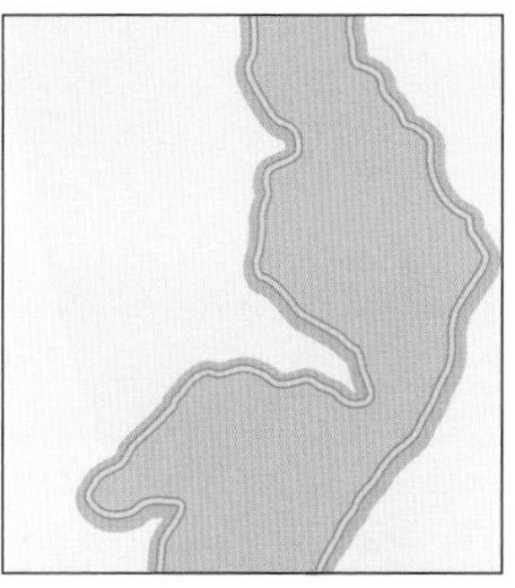

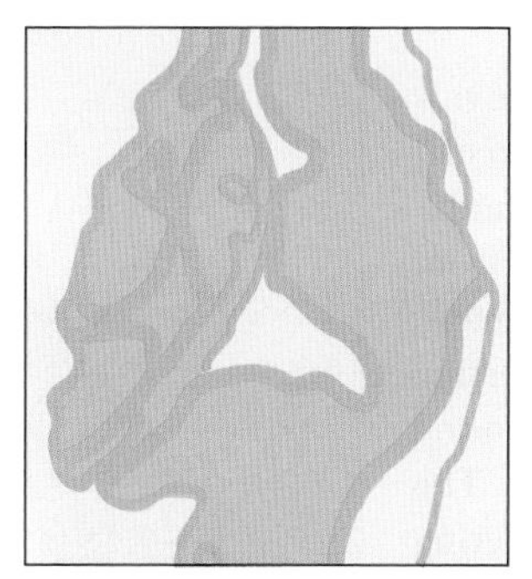

(c) Plinthisols—plateau with local internal boundary widths of sharp (50 m), medium (100 m), and diffuse (250 m)

(d) Boundary widths defined per boundary section for units 1, 7, and 8

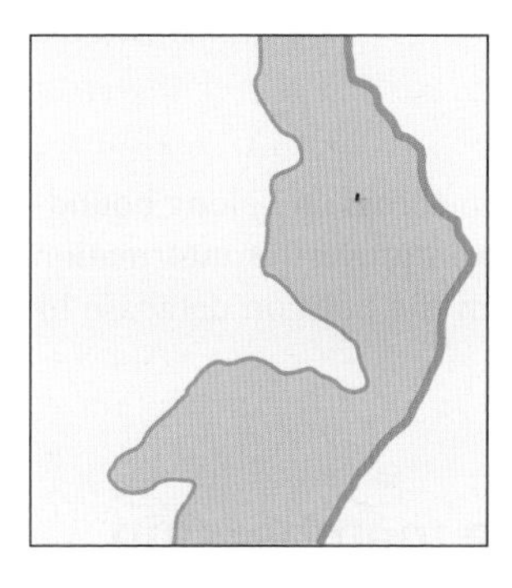

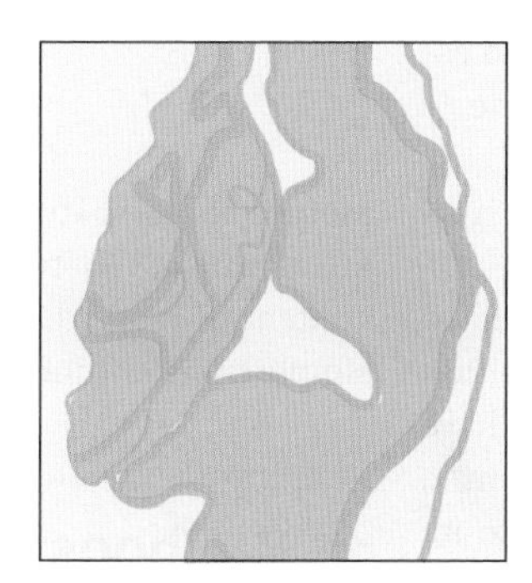

Figure 13.10 Using fuzzy membership functions to describe boundary widths

Table 13.1 Boundary widths per map unit

Soil unit	Boundary width d (m)
Albic arenosol	50
Gleyic arenosol	50
Histosol	25
Haplic ferralsol	50
Dystric fluvisol	25
Plinthisol—valley	50
Plinthisol—plateau	100
Plinthisol—hillside	100

fuzzy boundaries it is not sensible to display both the inner and the outer zones of transition for them all because the resulting map is difficult to read. Therefore it is better to confine the display of the boundary *JMF* to the internal transition zone (*JMF* $\geq$ 0.5). Figure 13.10b shows the result for the different map units when the boundary widths have been assigned according to Table 13.1.

The individual boundary approach

Commonly, areas of soil or vegetation do not have the same kind of boundary all the way round the occurrence. For example, a raised river terrace may have a diffuse boundary on the upper side and a sharp boundary on the lower side where the river has cut it away. Other boundaries might be sharp for part of their length and diffuse elsewhere. It is not difficult to add attributes to boundary sections and to spread these individually. The results are then pooled and fuzzy membership functions can be computed. Figure 13.10c shows the result of recognizing three kinds of boundary for the plinthisols—sharp (50 m), medium (100 m), and diffuse (250 m): Figure 13.10d shows this procedure applied to more units.

13.7 Combining fuzzy boundaries and fuzzy attributes

Besides making the map image more realistic and providing the means to compute which parts of a polygon can really be regarded as part of the core area, fuzzy boundaries can also be useful when derived maps are made from original geological, soil, vegetation, or land use maps. Conventionally, derived maps are made by converting the map unit class into a number or code representing the representative value of the attribute in question. The result is that if adjacent map units have widely differing values of an attribute, the transformed map shows a discontinuity at the boundary between them.

Information about the gradual variation of an attribute over the boundary between two dissimilar map units can be provided if the boundary membership functions of the two polygons are used to compute a weighted estimate of the attribute values over the boundary zone.

$$z_x = \frac{\sum(z_i * MF_i)}{\sum MF_i} \qquad 13.6$$

Figure 13.11 shows the weighting procedure used to compute variation in soil moisture across the transition from Plinthisol plateau to hillside soil units, which have a soil moisture supply of 480 mm/a and 400 mm/a, respectively, and boundaries of transition width 250 m. Figure 13.12 shows a close-up of the results compared with the usual Boolean lookup table equivalence.

Recapitulation: fuzzy polygon boundaries

The SI approach can be used to add information about the abruptness of boundaries to a polygon database. Information about how sharp or diffuse the boundaries are can be incorporated on a polygon-by-polygon or individual line segment basis. The results give a sensible picture about spatial variation across and along boundaries, which can be used to improve the information content of maps derived from the polygon database by simple transfer functions.

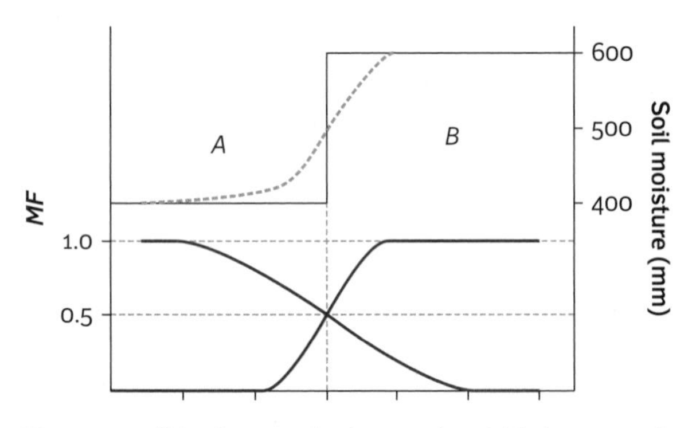

Figure 13.11 Using fuzzy overlap to compute weighted averages of attributes in the overlap zone

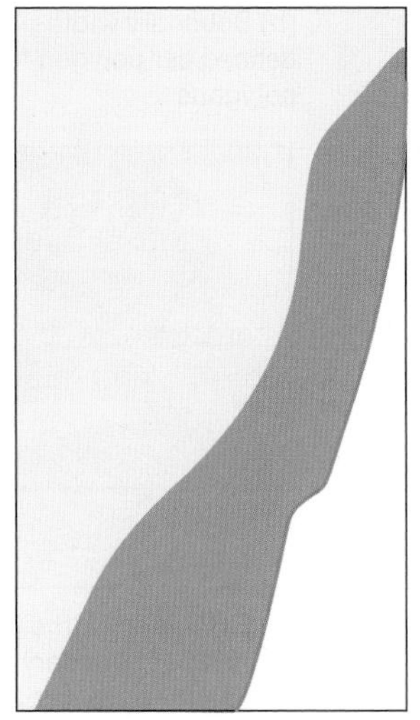
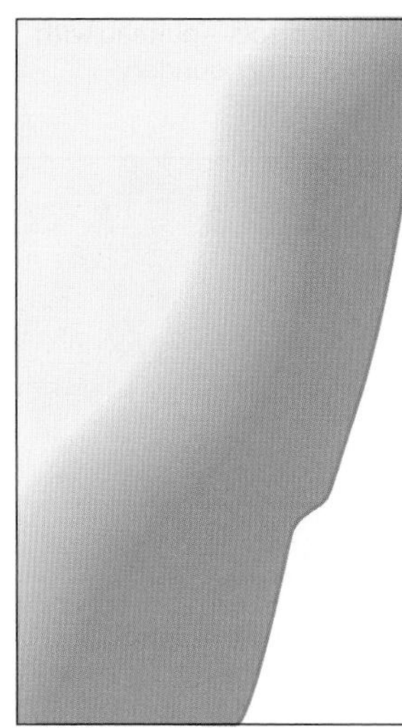

Figure 13.12 Conventional (left) and fuzzy (right) derived soil moisture maps for two adjacent polygons with a gradual shared boundary

13.8 Choosing the membership function 2: fuzzy *k*-means

Although the SI approach to creating fuzzy classes is extremely flexible, it can never be optimal. Very often users may not know which classification is useful or appropriate, and exploratory techniques that can suggest suitable polythetic, overlapping classes can be useful.

The fuzzy *k*-means approach is primarily aimed at data reduction and convenient information transfer and was developed in the field of pattern recognition (Bezdek et al. 1999; Bezdek 2012). Data reduction is realized by translating a multiple attribute description of an object into *k* (or *c*) membership values to *k* classes or clusters. The clusters are optimal in the sense that the multivariate within-cluster variance is minimal. Near-zero variance means that all objects have nearly equal attributes, which means a high density and small distances between them in attribute space. Conversely, large variance is equivalent to low density and large distances in attribute space. An optimal clustering procedure should identify the dense spots in attribute space as class centres, while the boundaries between classes in attribute space should be located in the lowest-density regions. Note that the fuzzy *k*-means approach initially says nothing about geographical contiguity of these optimal, overlapping classes.

Fuzzy *k*-means works by an iterative procedure that usually starts with an initial random allocation of the objects to be classified to *k* clusters (hence the name). Given the cluster allocation, the centre of each cluster (in terms of attribute values) is calculated as the average of the attributes of the objects. In the next step, the objects are reallocated among the classes according to the relative similarity between objects and clusters. The similarity index is usually a well-known distance measure: the Euclidian, Diagonal (attributes are scaled to have equal variance), or Mahalanobis (both variance and covariance are used for distance scaling) metrics are frequently used. Reallocation proceeds by iteration until a stable solution is reached where similar objects are grouped in one cluster (Figure 13.13).

Allocation of objects in conventional crisp *k*-means is always to the nearest cluster, with $MF = 1$ to this cluster and $MF = 0$ to all others. With fuzzy *k*-means, membership values may range between 0 and 1. Box 13.4 lists the algorithms used for calculating *MF* and the class centres. In ordinary *k*-means the *k* memberships of each object sum to 1. This causes a loss of degrees of freedom that may be compensated for by introducing an extragrade class. The parameter *q* is the so-called fuzzy exponent, which determines the amount of overlap in the cluster model. For $q \rightarrow 1$, allocation is crisp and no overlap is allowed. For large *q* there is complete overlap and all clusters are identical. Ideally *q* should be chosen to match the actual amount of overlap, which is generally unknown. Because distance is an overall measure of similarity, departure from the class centre of one attribute may be compensated by close correspondence of another. The effective relative weights of individual attributes in this valuation are determined by the type of distance measure used.

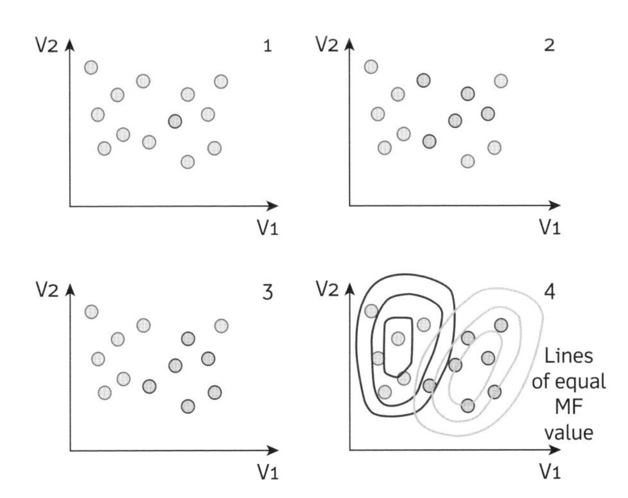

Figure 13.13 The formation of clusters using fuzzy *k*-means iteration

The net result of fuzzy *k*-means clustering is that individual multivariate objects (points, lines, or polygons) are assigned an MF value with respect to each of the *k* overlapping classes. The centroid of each class is chosen optimally with respect to the data. In the variant of the technique used here, the MF values *sum to 1* rather than individually as is the case with the SI approach. This means that rather than all sets having equal value, as with SI, the sets are ranked according to importance. This has obvious implications for the procedures for manipulating sets given in Box 13.2, and means that procedures such as intersection and convex combination need to be carried out with care on the sets derived by fuzzy *k*-means.

Because the values of the MF_k are in effect new attributes, the geographical distribution of each set can be displayed by conventional methods of mapping, including interpolation. However, because of class overlap it is not always easy to read a map when more than one interpolated surface is displayed, particularly when displays are limited to black and white. New methods for visualizing the results of fuzzy *k*-means classification on maps are clearly needed.

Examples of using fuzzy *k*-means and kriging to map overlapping, multivariate classes

The case study on heavy metal pollution on the Maas floodplains used in Chapters 8 and 9 demonstrated interpolation methods with a data set of soil samples taken from a regularly flooded area of the River Meuse floodplain in the south of the Netherlands, and data from a slightly larger area were used to illustrate costs and benefits in Chapter 10. Here we also use the larger subset of the data, but this time include all attributes (elevation, distance from the river, heavy metal concentrations (Cd, Zn, Pb, Hg), and organic material) to illustrate the multivariate fuzzy *k*-means approach.

Following the original flooding frequency map, the attribute data were classified into three classes with a fuzzy overlap $q = 1.5$. The fuzzy *k*-means classification yielded a new data set of memberships to each of the three classes for each site sampled. Table 13.2 presents summary statistics for the classes.

Figures 13.14a, b, and c show the interpolated surfaces for the three classes. There is a strong suggestion of a relation between the three fuzzy classes and the flooding

Box 13.4 Algorithms used in fuzzy *k*-means clustering

1. Membership μ of the *i*th object to the *c*th cluster in ordinary fuzzy *k*-means, with *d* the distance measure used for similarity, and the fuzzy exponent *q* determining the amount of fuzziness:

$$\mu_{ic} = \frac{\left[(d_{ic})^2\right]^{-1/(q-1)}}{\sum_{c'=1}^{k}\left[(d_{ic'})^2\right]^{-1/(q-1)}} \quad \text{B13.4.1}$$

2. Membership μ of the *i*th object to the *c*th cluster in fuzzy *k*-means with extragrades, *d* and *q* as above, and with α as a weighting factor between extragrade and ordinary classes:

$$\mu_{ic} = \frac{[(d_{ic})^2]^{-1/(q-1)}}{\sum_{c'=1}^{k}[(d_{ic'})^2]^{-1/(q-1)} + \left[\frac{1-\alpha}{\alpha}\cdot\sum_{c'=1}^{k}[(d_{ic'})^2\right]^{-1/q-1)}} \quad \text{B13.4.2}$$

3. Membership μ of the *i*th object the extragrade cluster in fuzzy *k*-means with extragrades:

$$\mu_{i\cdot} = \frac{\left[\frac{1-\alpha}{\alpha}\cdot\sum_{c'=1}^{k}(d_{ic\cdot})^2\right]^{-1/(q-1)}}{\sum_{c'=1}^{k}[(d_{ic\cdot})^2]^{-1/(q-1)}] + \left[\frac{1-\alpha}{\alpha}\cdot\sum_{c'=1}^{k}(d_{ic\cdot})^2\right]^{-1/(q-1)}} \quad \text{B13.4.3}$$

4. *j*th attribute value *z* of the *j*th cluster in ordinary fuzzy *k*-means:

$$z_{ci} = \frac{\sum_{i=1}^{n}(\mu_{ic})^{q*zij}}{\sum_{i=1}^{n}(\mu_{ic})^{q}} \quad \text{B13.4.4}$$

5. *j*th attribute value *z* of the *c*th cluster in fuzzy *k*-means with extragrades:

$$z_{cj} = \frac{\sum_{i=1}^{n}\left[(\mu_{ic})^{a} - \frac{1-\alpha}{\alpha}\cdot(d_{ic})^{-4}\cdot(\mu_{i*})^{q}\right]*z_{ij}}{\sum_{i=1}^{n}\left[(\mu_{ic})^{a} - \frac{1-\alpha}{\alpha}\cdot(d_{ic})^{-4}\cdot(\mu_{i*})^{q}\right]} \quad \text{B13.4.5}$$

frequency classes: that flood frequency class 1 covers similar areas to fuzzy class 3, flooding frequency class 3 covers a similar area to fuzzy class 1, and fuzzy class 3 occupies areas to the south and north. Note the distribution of these classes over the part of the area used in Chapters 8 and 9, which is outlined in Figure 13.14e. In that area only two fuzzy classes are really important, and this was one of the conclusions reached in the anovar analysis of the flood frequency mapping of the smaller area.

13.9 Class overlap, confusion, and geographical boundaries

The concept of 'confusion' may help us to combine interpolated maps of fuzzy memberships into easy to understand crisp zones. If a site has a membership value near 1 in one of the fuzzy classes, it is clear to which class it belongs. But if the *MF* values for two or more classes are similar, it is by no means clear as to which class the site should be allocated. The situation is therefore confusing. Let us define the degree of class overlap in attribute space in terms of a *confusion index* for each observation, grid cell, or object; two possible forms are:

$$CI_1 = 1-(MF_{\max} - MF_{mx2}) \quad 13.7$$

or

$$CI_2 = MF_{\max} / MF_{mx2} \quad 13.8$$

Using equation 13.7 to compute the *CI* for both sets of data provides interesting information on the spatial contiguity of the interpolated classes. The floodplain pollution classes yield well-defined zones of large *CI* where dominance of one class is replaced by another (Figure 13.14d). In this case the zones of large *CI* are so thin that they can easily be refined to polygon boundaries. An effective procedure is to compute those cells

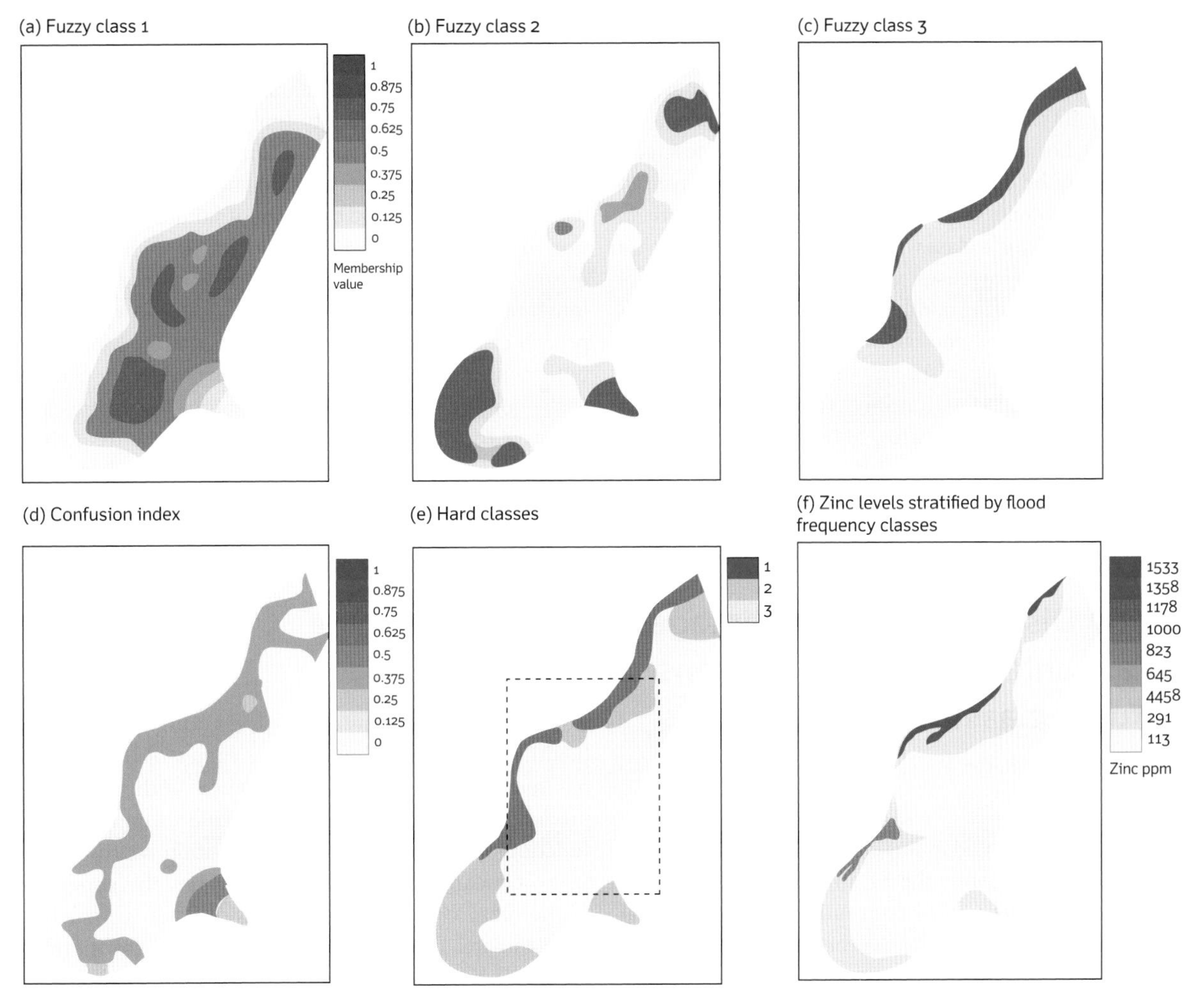

Figure 13.14 (a–c) Maps of interpolated fuzzy *k*-means membership values for the three soil pollution classes; (d) of the confusion index; (e) crisp map showing where each class is dominant; (f) map of zinc levels obtained by stratified kriging for comparison with (e)

having the local maximum profile convexity and extract them as a Boolean data type. Reclassifying all cells within these boundaries yields the choropleth map Figure 13.14e. The striking similarity of the three mapping units obtained in this way to the original flooding frequency classes is provided by comparing Figure 13.14e with the flooding frequency-stratified kriging map of zinc (Figure 13.14f). Clearly, when surfaces of continuous membership functions are strongly contiguous, computing the confusion index is a useful way to delineate the zones that are dominated by any given class.

Confusion indices and the SI approach

Although the geographical expression of the confusion index has been presented above using the results of fuzzy *k*-means analyses, the same methods can be used to explore the spatial expression of SI classes. Indeed, by

Table 13.2 Statistics of fuzzy classes, floodplain pollution

Class	Minimum membership	Maximum membership	Mean membership	Variogram type	Nugget C_0	Sill C_1	Range (m)	Ratio C_1/C_0
1	.0002	.999	.468	Sph	.0405	.267	1170	6.59
2	.001	.9983	.314	Sph	.0424	.0904	420	2.13
3	.0001	.9972	.217	Sph	.0590	.103	1170	175

repeating the SI classification for a range of transition zone widths d_i one can examine the relations between geographical location and the core of a class in attribute space. If transition zone widths d_i are zero there is no confusion, because there is no overlap in attribute space. Classes may be spatially disjoint, however. As overlap increases, so confusion will become more widespread. Areas in the core of a class will not experience an increase in confusion as d increases: areas near the edge of a class will. So, by computing confusion indices for a range of d values (class overlap) one can obtain maps of geographical locations that are in the core of the attribute classes (and therefore stable with respect to class overlap), and areas that are boundary cases, both geographically and in attribute space. If the result is a map in which the CI is low everywhere, we may conclude that delineation is not useful.

13.10 Discussion: the advantages, disadvantages, and applications of fuzzy classification

Advantages

The material presented in this chapter has demonstrated the advantages of both the SI approach and fuzzy k-means classification over conventional data retrieval and classification methods. It has been shown how exact classification loses information and increases the chance of classification errors when data are corrupted by inexactness, which is usually the case with environmental data. It has also shown that applying the SI approach to exactly delineated polygons can improve their information content, as long as information about the nature of the sharpness or diffuseness of the identified boundaries is available.

As with the SI approach, the overlapping classes and gradual boundaries resulting from the fuzzy k-means approach appear to be more congruent with reality than conventional methods. The fuzzy membership values carry through more of the initial information than crisp classification. Membership values can easily be interpolated over space, and the resulting variogram analysis and patterns demonstrate clearly if a given attribute class also has a coherent geographical distribution.

In contrast to the SI approach with predefined classes and class boundaries, the fuzzy k-means approach yields locally optimal classes which are not necessarily based on assumptions of linearity (unlike conventional data-reduction techniques, such as principal component and correspondence analysis). While linear correlations may sometimes be enhanced by means of suitable transformation functions (logarithmic, polynomial, etc.), attributes of soil, sediment, or water often co-vary in a complex, inherently non-linear manner. The fuzzy k-means approach captures this non-linear co-variation. Moreover, both continuous and categorical attribute data can be easily combined. In general, ordination techniques are sensitive to deviations from normality of the frequency distributions of the attributes, including bimodality and outlying values. Because of the fuzziness allowed, the fuzzy k-means approach appears to deal accurately with both ends of the spectrum from co-variation to discrete clustering.

Disadvantages

As with most parametric methods, the greatest difficulties come with choosing the values of the control parameters to obtain the best results. With the SI approach, the user must choose the kind of membership function, boundary values, and transition widths. With fuzzy k-means using multiple attributes, the goodness of the fuzzy clustering obtained is difficult to visualize and evaluate. This relates to the choice of k, the number of classes, and q, the amount of overlap or fuzziness allowed. The added difficulty is that the optimal fuzziness may depend on the number of classes and vice versa. A formal approach using diagnostic functionals is possible, but not always successful. Scientific insight or compliance with existing classification schemes may help. Other problems with fuzzy k-means are the choice of attributes and which distance measure to apply, which determine the degree and shape of fuzziness in the model and the way different attributes are compromised. It should be noted that the degree of fuzziness is always assumed equal for all classes.

While the fuzzy k-means procedure may seem to result in less arbitrary shaped fuzzy boundaries than with the SI approach, this is only a matter because the method is less direct. Adequate cartographic techniques to map k classes (or $k + 1$ extragrade) concurrently still have to be developed for both SI and fuzzy k-means.

13.11 Summary

The decision to use SI or fuzzy *k*-means approaches in continuous classification depends on the context of the problem and also on the level of prior information (Figure 13.15). There is a growing literature that illustrates the practical advantages of both SI and fuzzy *k*-means approaches (Burrough 1989; Burrough et al. 2001; Oberthür 2000; Triantafilis et al. 2003; Gorsevski et al. 2003; Braimoh et al. 2004; Schmidt and Hewitt 2004; Hagen-Zanker et al. 2005).

In situations where a well-defined and functional classification scheme exists, the straightforward SI approach offers great advantages, but the precise definition of the fuzzy boundaries requires some special attention. Studies reported here have shown that SI continuous classes are more robust and less prone to errors and extremes than simple Boolean classes that use the same attribute boundaries. The results can be mapped showing the relations between gradual changes in attribute values/membership functions in both data space and geographical space.

In the fuzzy *k*-means approach, the criteria that distinguish between the ultimate classes are a result of the analysis rather than input to the model. These criteria may be complex non-linear functions of the original attributes, so there is no clear-cut relation between membership values and attribute values. While objects having identical attribute values will end up with identical memberships, the reverse is not generally true because of the mutual compensation possible among attributes. Since data reduction

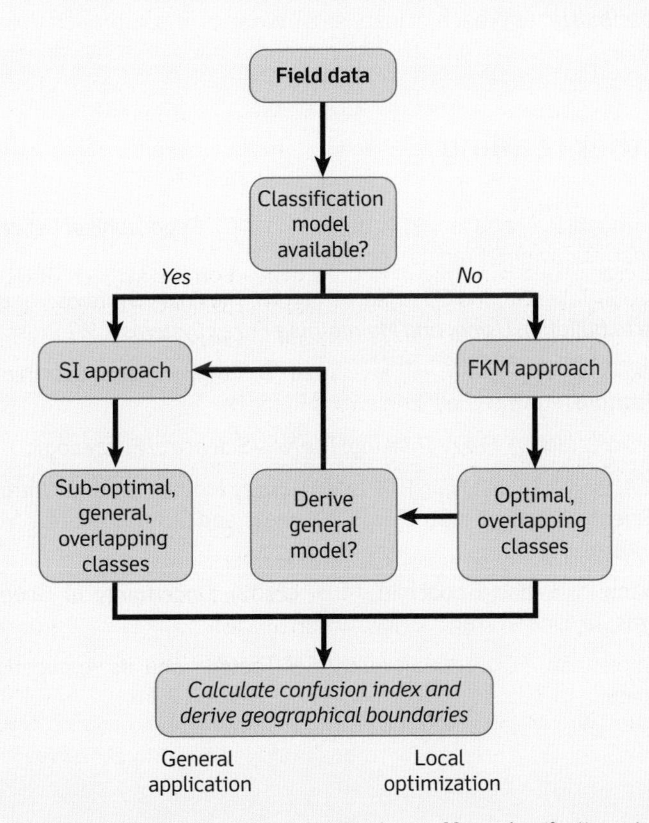

Figure 13.15 Flow chart governing the choice of fuzzy classification using either the SI or the fuzzy *k*-means approach

is one of the main reasons for using class memberships instead of the original attribute values, some loss of information will always be faced.

The fuzzy ***k***-means approach is appropriate when information about the number and definition of classes is lacking. Fuzzy ***k***-means methods yield sets of optimal, overlapping classes that can also be mapped in data space and in geographical space. The results can be used as (a) a once-only analysis of a limited area to identify the lineage of (spatial) variability, (b) as a means to set up classes so that membership functions can be interpolated, which in turn can predict attribute values at unsampled locations, or (c) as a means to define better the simple membership functions that are to be used in straightforward SI analyses.

? Questions

1. Which geographical phenomena cannot be modelled satisfactorily by crisp entities, as explained in Chapter 2? What alternative data models might be used to represent these phenomena?
2. How you would go about measuring the width of geographical boundaries in practice: (a) in a landscape, (b) in a city?
3. What value is there in using fuzzy membership functions in sequential data-retrieval operations and sieve mapping?
4. What are the differences between a ***probabilistic*** treatment of uncertainty and the ***possibilistic*** approach of fuzzy sets? When is one approach to be preferred above the other?

Further reading

▼ Burrough, P. A. and Frank, A. U. (eds.) (1996). ***Geographical Objects with Indeterminate Boundaries***. Taylor & Francis, London.

▼ Cox, E., Taber, R., and O'Hagan, M. (1999). ***The Fuzzy Systems Handbook: A Practitioner's Guide to Building, Using, and Maintaining Fuzzy Systems***. CRC Press, Boca Raton, FL.

▼ Klir, G. J., St Clair, Y. H., and Yuan, B. (1998). ***Fuzzy Set Theory: Foundations and Applications***. Prentice Hall, Englewood Cliffs, NJ.

▼ Kosko, B. (1994). ***Fuzzy Thinking: The New Science of Fuzzy Logic***. Harper Collins, London.

▼ Mordeson, J. N. and Nair, P. S. (2001). ***Fuzzy Mathematics: An Introduction for Engineers and Scientists***. 2nd edn. Studies in Fuzziness and Soft Computing, Vol. 20. Physica Verlag, Heidelberg.

▼ Zhang, J. X. and Goodchild, M. F. (2002). ***Uncertainty in Geographical Information Systems***. Taylor and Francis, New York.

▼ Zimmerman, H. J. (2001). ***Fuzzy Set Theory—and its Applications***. 4th edn. Kluwer, Dordrecht.

GIS, Transformations, and Future Developments

This book has explored the scientific fundamentals behind GIS and the data that are used in such systems. It has laid out the conceptual, mathematical, and computational nuts and bolts that are behind the screen of the GIS we are using. The early chapters considered geographical phenomena and the way people perceive them and build conceptual models of space. These ideas are the starting point for understanding the structures and systems for handling spatial data in a GIS. We then proceeded to demonstrate how the different data models of exact entities and continuous fields are modelled in a computer and how databases containing digital spatial data can be created. These data models, algorithms, and graphic capabilities have transformed the way we think about the earth and how we analyse and use spatial information, whether at the local, national, or global scale.

We then showed how these data can be retrieved and analysed, and how many different methods of spatial analysis provide the means to derive new information from raw data. The possibilities were explored using logical and numerical models that take these data and then provide new understanding and insights as well as support the development of new hypotheses for testing. The increasingly complex data handling and analyses supported by GIS has meant that they have been used in policy development, decision-making, and research in a range of political, economic, and academic settings. The result has been the adoption of these systems by a broader base of users than ever before across the globe. There is a need to consider GIS as part of a bigger picture, and this chapter will consider the fundamental axioms of their use as well as the policy and legal frameworks around spatial data before looking forward to how these systems and their use will develop in the coming years.

Learning objectives

By the end of this chapter, you will:

- **understand the importance and nature of spatial data policies**
- **be able to describe some of the major legal and regulatory challenges affecting spatial data.**

14.1 The fundamental axioms and procedures of GIS use

GIS have transformed societies in various ways, and in turn society has transformed the development of the systems, the data needs, and how they are used. Throughout this book we have shown how advances both in the GIS industry and in information technology more generally have transformed beyond recognition the way we interact with and use spatial data. People with no experience of GIS as an academic subject or from technical training are using these systems in their day-to-day lives, and terms like 'citizen cartographer' and 'crowdsourced data' are becoming part of the field (Sui and Goodchild 2012). This is adding many new dimensions to how we think about and use spatial data to enrich our lives and enhance decision-making processes.

It is possible to capture on an overarching, formal basis the fundamental axioms and procedures underlying GIS use: these are given in Box 14.1. They encapsulate the fundamentals of how we generate and use

Box 14.1 Axioms and procedures for handling data in GIS

The following axioms and procedures summarize the ideas presented in this book.

1. It is necessary to identify some kind of discretization such as ***entities*** (individuals) that carry the data. In GIS the primitive entities are ***points, lines, polygons***, and ***pixels*** (grid elements). Complex entities having a defined internal structure can be built from sets of points, lines, and polygons.
2. All fundamental entities are defined in terms of their ***geographical location*** (spatial coordinates or geometry), their ***attributes*** (properties), and ***relationships*** (topology). These relationships may be purely geometrical (with respect to spatial relations or neighbours), or hierarchical (with respect to attributes), or both.
3. Individuals (***entities***) are distinguishable from one another by their attributes, by their location, or by their internal or external relationships. In the simple, static view, individuals or 'objects' are usually assumed to be internally homogeneous unless they are a representation of a mathematical surface or unless they are complex objects built out of the primitives. In most cases GIS usually only distinguish objects that are internally homogeneous and that are delineated by crisp boundaries. A more complex GIS allows intelligence about inexactness in objects in which the attributes, the relationships, or the location and delineation are subject to uncertainty or error.
4. Both entities and attributes can be classified into useful categories.
5. The propositional calculus (Boolean algebra) can be used to perform logical operations on an entity, its attributes, its relations, and the groups to which it belongs.
6. In GIS the propositional calculus is extended to take account of: distance; direction; connectivity (topology); adjacency; proximity; superposition; group membership; ownership of other entities.

   Intelligent GIS extend the propositional calculus to take account of non-exact data.
7. New entities (or sets of entities) can be created by geometrical union or intersection of existing entities (line intersection, polygon overlay).
8. New complex entities or objects can be created from the basic point, line, area, or pixel entities.
9. New attributes can be derived from existing attributes by means of logical and/or mathematical procedures or models.

   New attribute $U = f(A, B, C, \ldots)$

   The mathematical operations include all kinds of arithmetic (addition, subtraction, multiplication, division, trigonometry, differentiation and integration, etc.) depending on data type—see Box 12.1.
10. New attributes can also be derived from existing topological relations and from geometric properties (e.g. linked to, or size, shape, area, and perimeter), or by interpolation.
11. Entities having certain defined sets of attributes may be kept in separate subdata sets called data planes or overlays.
12. Data at the same XYZt coordinate can be linked to all data planes (the principle of the common basis of location).
13. Data linked to any single XYZt coordinate may refer only to an individual at that coordinate, or to the whole of an individual in or on which that point is located.
14. New attribute values at any XYZt location can be derived from a function of the surroundings (e.g. computation of slope, aspect, connectivity).

geographical data in GIS, and in essence synthesize the main elements of the preceding chapters. These were first defined in the first edition of this book (Burrough 1998) and still hold true today, even with the revolutions in data development and advances in GIS technology.

14.2 Policies and legal frameworks of geographical data

While the axioms and procedures remain the same, the changes in collection of geographical data and use of GIS across multitudes of organizations and individuals has meant the roles of government and governance have evolved in the last few decades. The changes bring with them many issues centred on policy and legal frameworks that were not so pressing when governments were the main generators of data. The growing role of the private sector and associated markets for supplying data and services add to the data-generation activities of governments, and there is a need to develop these within overall policy and legal frameworks that support efficiency and efficacy as well as protection of property and people. In addition, developments in social media and in various business activities ensure vast amounts of location-linked personal and experiential data are now collected that reveal information about us. While some have hailed the changes as the 'democratization' of geographical data, a number of areas of concern have resulted that need to be addressed by government policies or legal and regulatory frameworks.

Government policies

Governmental organizations at many levels—national, subnational, and regional—have controlled and managed geographical data collection, dissemination, and use over many centuries. Today, the role of governments has expanded, requiring strategic policy and legal and regulatory frameworks in many related areas to support maximizing the use of this information wealth, while at the same time protecting the privacy of individuals. Policies need to provide a strategic framework and courses of action to cover not only geographical data provided by governments themselves but by other organizations, so that enterprise and initiatives leading to innovations and further development are supported.

One of the important responsibilities of governments is to maintain and deliver the national geographical information base, requiring strategies and standards to support the coherent collection, maintenance, and dissemination of all government data, across the various levels of administration. This involves not only establishing the national georeferencing system outlined in Chapter 3, but also national plans for earth observations and data-collection exercises that reflect the need of many groups of users and developers, including the private sector and community groups. Guidelines to bring this about might include the following:

- Data are collected once and used many times.
- Data are collected to open standards.
- Metadata of the data are stored in a consistent form so that data can be found and used easily.
- Data are maintained to defined standards.
- Data are accessible to the market for that data.
- Data are easily licensed for public and commercial reuse.

To bring coherence and maximize sharing of data collected in administrative areas, spatial data infrastructures (SDIs) have been developed in hundreds of jurisdictions across the world (see Section 4.4 for more detail) (Harvey and Tulloch 2006; Rajabifard et al. 2006; Elwood 2008; Nedovic-Budic et al. 2011). An SDI is centred on connected but decentralized systems of data repositories that conform within a strategic framework to various agreements on data/metadata and technology standards, as well as governance and institutional arrangements. The first initiatives by the US federal government to establish national SDIs began in the 1990s and this has now been replicated in many different countries and at different levels of governance. This enables geographical information to be used by many users who would not otherwise have access to this information. International organizations, as well as regional bodies such as the European Union, are also developing their own SDIs to bring about seamless data sets that can be queried and accessed wherever a user is located. Although these ideas are often readily embraced in countries, the development and implementation of SDIs bas proved challenging for many in terms of model development, funding, and importantly the coordinated and strong leadership required (Nedovic-Budic et al. 2011; Cooper et al. 2013; Hendriks et al. 2012).

While SDI concepts have been embraced by many countries to provide strategic solutions to wide geographical data usage, other areas of policy and legal coverage, with the requisite governance arrangements needed to manage geographical data effectively, are arguably lagging behind. As has been stated many times in this book, developments in technology and the way GIS are used have changed radically in the last decade, and various emerging legal and policy issues have not yet been adequately addressed in many countries.

Ownership, copyrights, and commodification of geographical data

Related areas of increasing importance are centred on legal aspects of geographical data such as who owns them, the nature of intellectual property rights surrounding them, and what liability issues arise from using them.

Chapters 1 and 4 described how the early geographical data collectors tended to be official representatives of government, whose tasks were based on protecting the interests of a government or individual. With developments in map-making, many of the data-collecting government agencies took on additional roles in establishing common geographical referencing frameworks which have formed the basis for all involved in these activities. Today, geographical data collectors embrace broad sections of society ranging from individuals through to community organizations, the private sector, and various levels of government.

The ownership of geographical data in those earlier times was much more easily understood as organizations were involved with both collection and dissemination. Ownership and copyright vested in that organization. Today the primary data collectors are not the only organizations producing and supplying data. New data may be developed based on primary data collected by one organization with enhanced content provided by the subsequent supplier. The ownership and copyright of these data will vary between legal jurisdictions, but with most ownership will be awarded to the enhancing organization as long as it can be proven that originality and creativity has been added to the new product. Difficult questions arise around the ownership rights of an organization whose main operations are not geographical data generation, but which wishes to sell information it collects as part of doing business. For example, can a utility company sell details of our mobile phone, water, or energy use to other organizations? This will vary depending on the legal agreements in place, with provisions often included as part of the terms and conditions customers accept.

Clarity in ownership and copyright is usually agreed between supplier and user through the licensing terms of data supply. The licences will often not only state details of ownership of the data, but also define what operations may be undertaken with it, including any involving payment of fees. For example, a licence may prohibit you from developing from these data products and then selling them commercially.

A major development in the last decade has been the growth of a commercial geographical data industry that today is estimated to be worth hundreds of millions of dollars annually. This has grown rapidly in response to the needs of the user communities, and today offers many forms of data sets and products that have been collected and developed. Government agencies in some countries are also now charging for data as an important part of cost-recovery initiatives to offset the enormous expenses involved in collecting the data.

The data available commercially have many different sources including commercial satellite and airborne systems, field sensor and survey systems, information collected through the delivery of utility and commercial services, and many other forms of information. Some of the data are generated from existing data. These new data sets are charged for and range in price, often reflecting the status of the individual or organizational user. Commercial organizations will usually pay higher fees for the data than researchers, not-for-profit organizations, and similar organizations. In all cases the extent of the fees charged will influence who will be able to afford to use this data, highlighting the potential issues of access.

This 'commodification' of data by both commercial and government agencies has been the subject of debate (Longley et al. 2010: 466 summarizes the various arguments). The role of governments in particular, the use of public funds, innovation stimulation, influences of the user and access, are some of the key issues central to the discussions. Other arguments focus on the data products being offered and how these meet the needs of the various user communities. The types of data products needed today by ever more wide-ranging GIS-aware user communities might not be met by traditional data-collection agencies in either subject area or the form available. The commercial sector can bridge that data needs gap as readily as a forward thinking government agency. At present the involvement of the private sector in influencing government data-collection strategies is limited in many countries, and the links along the information supply chain tend to be ad hoc and reactive.

Privacy and geographical data

Linked to ownership are the important discussions on data privacy rights, which have received increasing attention in the past few years. With geographical data collection and use more pervasive, easier, and more commercially important, examination of who has access to information on individual citizens has become an increasing subject of debate. Vast data collections that have location/individual person identifiers, including from social media, fine resolution satellite imagery, and many forms of commercial data, are now available either for free or on a cost basis. Whistle-blowers have highlighted—to growing public alarm—some of the extent of who is watching whom, and how data that are assumed to be private and shared only with selected people may be accessed by certain organizations.

The controversy and responses to any violations of privacy are as varied as individuals themselves, with arguments centred on the balance between the public's right to access information versus the individual person's right to privacy (Elwood and Leszczynski 2011). The issue of privacy tends to apply to a subset of geographical data which centre on details of individuals and their immediate environment. Data that are made publicly available highlighting details (through images or explicit information) related to a person's social or economic status are particularly sensitive. Similarly, any disclosure of personal data on their day-to-day actions and interactions may be felt by some as an intrusion and violation of their freedom, and should be not part of any organization's data-collection activities.

The strength of people's feelings has been shown on various occasions, such as the reaction to the publication of a map in 2012 showing the names and addresses of all handgun permit-holders in New York's Westchester and Rockland counties, that was posted on a newspaper's website (Maas and Levs 2012). The map, published by the local *The Journal News*, allowed readers to zoom in on red dots that indicated which residents were licensed to own pistols or revolvers. It prompted thousands of responses arguing many sides of the debate.

The ongoing reaction to Google's Street View project also highlights people's responses to any perceived infringement of their privacy. The global exercise of collecting data using cameras mounted on vehicles or people from public property has resulted in millions of street-level panoramic images which are used in Google Maps and various smartphone apps. In some countries reactions have centred on the basic legality of the exercise, where laws exist prohibiting filming an individual without consent on public property for the purpose of public display. At an individual level there have been angry responses relating to not only the detailed display of property but also the fact that people are captured in these images—sometimes engaging in activities they do not wish to be seen publicly doing. It raises the question of when the scale of data collection becomes intrusive.

With geographical data there are few policy or legal frameworks to protect personal anonymity, unlike for other sensitive data such as medical records or financial data. The rapid development in geographical data-collection and dissemination devices has taken place with little legal response to protect any definitions of privacy. Previously, socio-economic, demographic, and other data sets such as those derived from censuses were released in spatially aggregated forms that gave local information but protected individuals' confidentiality and privacy. With increasingly sophisticated data collection and data mining/analysis now possible across and between data sets in this era of Big Data, information may be disaggregated, and individuals re-identified, undermining any efforts for protection of privacy. Furthermore, while regulations and policies are in place to control how socially aggregated data collected by government agencies are disseminated, these often do not exist for data available from other suppliers. Geomasking techniques, which provide privacy protection for individual address information while maintaining geographical resolution for mapping purposes, may be used to certain degrees. However, a balance is needed to support geographical data being made available for users to bring new insight, yet at the same time to defend various privacy rights of people. The legal framework tends to be complex, largely underdeveloped, and ongoing.

Access to data

The proliferation of data sets available through the Internet from a plethora of different data providers has created unprecedented opportunities for using geographical data. However, not all data are freely available, and for any organization or individual the ability to access data from a provider, whether government, commercial, civilian, or military, will be subject to different types of controls:

- legal/regulatory
- security
- economic
- physical.

These are imposed for many reasons, including classified levels of security or sensitivity for data of national or commercial significance and where payments for data sets are required. The various debates on charging for data were discussed earlier in this chapter. Access controls can also be put in place to bring a balance between protecting the privacy of individuals with respect to personal information held about themselves by institutions and giving individuals a right of access to that information. Where geographical data restrictions do exist, these may be relaxed during times of disasters or other crises so that all data, including military and commercially held information, can be used to help in such a situation.

Access controls are put in place by governments and businesses, and within organizations. These controls might restrict access to data for particular areas, for particular data resolutions, or given attribute layers or geographical object types. At a national level, access to government geographical data along with other forms of information may be controlled by laws and regulations, with the nature of restrictions and censorship varying widely across countries and jurisdictions. In some countries, freedom of information acts give the public the right to access various types of information, while in others there are greater controls although not explicitly defined in policies or laws. Remotely sensed data can often be subject to many controls as national centres are part of a country's military system.

With rapid changes in data-collection technology, these regulatory controls and laws can quickly become out of date. For example, India's Remote Sensing Data Policy (RSDP) was issued in 2001 and restricted the dissemination of data of less than 5.8 m resolution; after that permission is required from the authority concerned. With the ready availability of high-resolution geographical data today and the desire for the vast volumes of data collected by national and state government agencies, a National Data Sharing and Accessibility Policy is being brought into place (http://www.dst.gov.in/nsdi.html).

Different access controls also exist within the same organization, with users restricted to various parts of a database depending on the classified level of sensitivity or privacy of the information they are authorized to use. Access control models, such as role-based models, have been used in GIS to restrict different users' access based on various kinds of spatial constraints including layer constraints, region constraints, and spatial object constraints.

Away from regulatory and economic access controls, users' abilities to tap into the readily available data may also be hindered by physical restrictions. In many parts of the world, Internet speed does not support the downloading of data sets. Even within a country some users may not be able to access data as they are held at government agencies in national or regional capital cities. This often affects the most vulnerable communities, meaning that their ability to inform and influence decision-making using evidence-based reasoning can be restricted relative to those who can access the available data. Ideas of information asymmetry in GIS user communities, where one party has better access to information than another, have been little explored in GIS.

Liability

In Chapters 4 and 12 the occurrence of error in geographical data, resulting possibly from a number of sources, was detailed. The result can be that errors in location or attribute values may occur. Users need to be aware of any limitations to the data, so that subsequent analysis and decision-making can take this uncertainty on board. Where some form of loss, harm, or adverse outcome arises from poor decisions based on errors in the data, the organization supplying it may be liable as a result of their perceived negligence or recklessness.

At the centre of possible liability suits is that the duty of care is expected of a professional individual and that the extent of that responsibility may be defined in an agreement, by law or by a form of code of professional practice. The ability to sue for liability will be determined by the nature of the jurisdiction—with highly variable legal and regulatory systems in place, if at all. There are limited regulations in place setting standards for geographical data, so it is difficult to prove that data sets do not conform in terms or accuracy etc. There are also limitations if the data are being used for a purpose for which they were not originally collected or distributed. There is a problem of matching expectations of the data with the realities of collection and generation challenges.

In any legal case there is a need to determine whether the provision of GIS data is a service or a product, and the case will vary between law suits depending on the nature of the engagement between the parties. Tort law, which is used in cases where it is thought a civil wrong might have occurred which unfairly causes someone else to suffer loss or harm resulting, is often the basis for attaching legal liability to GIS producers (see Phillips 1999 for an overview). These include theories of strict liability and negligence applied in the areas of charting, computerized technology, and publishing (see Box 14.2 for an example).

Box 14.2 Data distribution liability statements

Many organizations now issue a data distribution liability statement to clarify their positions. The following statement from the US National Park Service is a clear example of these and highlights typical clauses that are found today (http://www.nps.gov/gis/liability.htm).

The National Park Service shall not be held liable for improper or incorrect use of the data described and/or contained herein. These data and related graphics are not legal documents and are not intended to be used as such.

The information contained in these data is dynamic and may change over time. The data are not better than the original sources from which they were derived. It is the responsibility of the data user to use the data appropriately and consistent within the limitations of geospatial data in general and these data in particular. The related graphics are intended to aid the data user in acquiring relevant data; it is not appropriate to use the related graphics as data.

The National Park Service gives no warranty, expressed or implied, as to the accuracy, reliability, or completeness of these data. It is strongly recommended that these data are directly acquired from an NPS server and not indirectly through other sources which may have changed the data in some way. Although these data have been processed successfully on a computer system at the National Park Service, no warranty expressed or implied is made regarding the utility of the data on another system or for general or scientific purposes, nor shall the act of distribution constitute any such warranty. This disclaimer applies both to individual use of the data and aggregate use with other data.

14.3 Future GIS transformations

The uses of GIS today would have been difficult to predict when the last edition of this book was printed. When advances and breakthroughs have been made there was often a coming together of transformations in technology and societies that would have been hard to imagine. Who would have thought that millions of the world's citizens would openly share their innermost thoughts and feelings with complete strangers? Social media sites have challenged our ideas of privacy and data sharing in ways that no sociologist would have been able to predict. Another challenging idea is today's developments in driverless cars which will rely on detailed road information in tandem with many forms of sensors to keep us safe and take us to our destinations. Geographical data sets produced using, for example, GPS-enabled smartphones and initiatives such as OpenStreetMap (http://www.openstreetmap.org) are indicative of the ways in which geographical data production and supply is being democratized. Given the developments in so many technological areas that might produce or use geographical data, it is probably not wise to gaze metaphorically into a crystal ball to predict future GIS transformations. There are, however, opportunities and challenges today that will require developments in GIS technology, analytical methods, and practices.

Taming the fire hose of big-data

The possibilities and challenges that arise from the Big Data era mean that, while never before have we had so much data on the world and its people, we do not necessarily have the means to handle, interrogate, and analyse this tsunami of information. The data now readily available are at global through to individual citizen level and in many different forms, not just alphanumeric values linked to locations. Data today may capture information on a location using a photographic image, video, a scanned hand-drawn map, and sound as well as through textual and numerical values. Sensors and imaging devices are ever more widely used, capturing streams of data at variable space and time resolutions, including information in three dimensions and on underground environments. Spatio-temporal models, as described in Chapter 12, are producing vast amounts of new data under past, current, and future conditions. The underlying question in this era of Big Data is therefore: how can we structure, handle, and derive knowledge from these zettabytes of data (Janowicz 2012)?

Given the extent of the data sets, it is likely that data-mining methods are going to be a vital component in extracting information. For geographical data, deriving information that links geographical properties to attribute values will require further developments in using statistical and modelling methods (Connors et al. 2013). Techniques and methods that can not only find the patterns, associations, causal linkages, and other features, but are also able to take on board the uncertainty and inaccuracy of the different data sets and their conceptual frames of reference, will be important. Not all data collectors or data sets are equal, and determining the quality, accuracy, and bias is difficult and complex when the data comes from multiple combined sources.

Computer infrastructure developments will be also vital and this will involve thinking about storage and processing power alongside development of algorithms that can help put structure into data sets to speed up querying and analysis. Storing and accessing these almost inconceivable volumes of data will require us to think once again about database structures, optimization routines, visualization methods and technologies, and where and how we physically locate the repositories.

Helping those that do not have access

Many people in the world have access to a wealth of geographical data that makes their lives easier—from the simple, such as knowing in real time where the nearest coffee shop is, to understanding the flood risk in where we want to live. But there are many millions of people who do not have these capabilities, and this might be for many different reasons that range from technology and economics through to culture, language, and perceptions. For example, concepts embodied in GIS reflect the thinking of technically oriented people, so isolating some users as they are not part of this epistemic community. Accessibility, quality, and usefulness of data resources might also limit their use (Elwood 2008). Users may be also isolated from using the spatial information as so many data sets are collected and disseminated using the English language. Although great strides have been made in many countries by both governments and commercial companies to supply data in local languages, this can mean limits to the information sets available.

For many reasons the result is an increasing difference between the enablement of the spatially data rich and poor. The importance of this influences many areas of individuals' lives and engagement in decision-making processes that affect them. That data are used to influence public sentiment through framing agendas and discourses, and so bringing acceptance for ideas that a government or other agent wants to bring into play, is increasingly acknowledged as important. Where citizens are not able to access the base data and even other information sources, their ability to engage is more limited (McDonnell 2008). For some citizen communities, access and the ability to contribute to map/knowledge development may be emancipatory, but there is also evidence that it can upset psychological and societal balance and reify existing power relations (Elwood 2006a, 2006b, 2011; Wright et al. 2009; Newman et al. 2010; Elwood and Leszczynski 2012; Perkins 2013). It is important to continue to increase our understanding of the impacts on many areas within and between societies.

The possibilities for addressing this challenge are manifold. One technological fix that has been used to great effect in other areas of life such as banking and utility bill payments, is the mobile phone. The rate of penetration of the mobile into the lives of those in even the remotest communities is the swiftest ever. The development of more advanced GIS for mobile technologies would offer significantly greater access to data through satellite communications than is currently possible. Apps for mobiles are already providing global data sets that can be used for simple querying and viewing, but maybe in the future more of the functionality of GIS can be made available—and in the local language.

Building user capacity to exploit our GIS futures

The GIS user communities today show how many different forms of capacity are enabling increased use. For many users, familiarity with forms of interface in their mobile devices is more important than following a formal course, with self-learning through interaction central to capacity development. For professional users and researchers there is a continual need to develop, understand, and use new developments in analytical and computation methods, and this is only likely to increase as new GIS technologies and practices are made available to embrace the challenges of Big Data and other developments. This capacity building is as likely to be done through learning from web-based materials (formal training and blogs/users' websites) as from traditional training courses. There is a continual need to question and explore possibilities for effective knowledge development across different communities in the world. The web has allowed people who are isolated from GIS centres of excellence to learn and grow their skills with the increasing material available. Trends in this will continue, with new possibilities through sound- and vision-enabled global connectivity to close gaps between technology and human capacity.

Overcoming the space–time representation conundrum

In this brave new world of handheld devices and access to terabytes through a tablet, some GIS challenges remain even after decades of research. Spatial data models that support true space–time representation are still not available even though many of the current and potential uses of GIS are predicated on this. There is a need for forward-thinking on possible database structures that could take on board the temporal dimension for many data areas. As increasing supplies of data are generated from a wide range of fields that have both space and time identifiers, databases even outside of GIS will need to take these requirements into account to reduce redundancy and clumsiness in the systems.

How will we use our GIS in the future?

How we will interact with our GIS in the future is open to much speculation (Goodchild 2010 offers some thoughts on this topic). The ideas of augmented reality have been with us for a while and are already used in many areas of geographical analysis, such as experiencing a proposed housing development or exploring the deep geologies of our earth. Developments in gaming technologies may potentially further support these advances, bringing more human-based operability and interfaces, and allowing us to explore and experience areas in a different way than through maps and animations. This will support many areas of intelligent design.

With developments in different related technology areas the design of the form of the hardware devices is likely to evolve. Tablets are now widely used as GIS hardware devices and are very different from the traditional personal computer, but in the future new materials are likely to alter the type of IT we can build and its shape and design. Recent developments in battery technology could revolutionize the way we power our GIS and the freedom of movements possible. In a recent article, a team of researchers (Pikul et al. 2013) have shown how 3D electrodes can be used to build 'microbatteries' that are many times smaller than commercially available options, or the same size but many times more powerful. These can also be recharged 1000 times faster than competing technology today, freeing us for longer from the demands of access to fixed power outlets.

* 14.4 Summary

GIS is an important tool for use in solving problems which require the manipulation and analysis of geographical data. Many of today's ambitions of improving economic and social well-being and environmental conditions require a spatial awareness and understanding of processes and form. In an increasingly interdependent and information-rich world, governments, policymakers, and citizens face the common problem of bringing expert knowledge to bear on decision-making. Smart cities, and we hope smartly managed sustainable natural environments, will depend on the developments in GIS to support initiatives in our ever more populous world (e.g. Batty et al. 2014; Seto et al. 2012; Spiegel et al. 2012; Tao 2013). Policymakers need basic information about the societies they govern—about how current policies are working, possible alternatives, and their likely costs and consequences. Citizens increasingly demand the same, and non-governmental organizations (NGOs) have grown to be an integral part of the response to this increased demand for information. Some of the vital challenges today such as food and water security, climate change, land degradation, marine pollution, biological diversity, poverty, urbanization, or demographic change require geographical data to be integrated, modelled, and analysed to derive insight and understanding. Overcoming these obstacles will have major implications for communities and decision-makers. GIS will continue to offer answers, but its successful use will be dependent on the questions posed, the data available, and the institutional support which is needed to bring about any changes.

Today decision-makers and the general public are increasingly besieged by more information than they can possibly use. The problem is that this information can be unsystematic, unreliable, and/or tainted by the interests of those who are disseminating it. Developments in GIS research and academic thinking, as well as in practical applications, will help to harness the possibilities that this data-rich world offers.

These will need to be further supported by enhancing the policy, governance, and legal frameworks to ensure that these data are developed and managed appropriately and available to all who need them.

Further reading

- Longley, P. A., Goodchild, M. F., Maguire, D. J., and Rhind, D. W. (2010). ***Geographic Information Systems and Science***. 3rd edn. Wiley, Hoboken, NJ.
- Peng, Z.-H. and Tsou, M.-H. (2003). ***Internet GIS: Distributed Geographical Information Services for the Internet and Wireless Networks***. Wiley, Hoboken, NJ.
- Pickle, J. (1993). ***Ground Truth: The Social Implications of Geographic Information Systems***. Guildford Press, New York.
- Wise, S. (2014). ***GIS Fundamentals***. 2nd edn. CRC Press, Boca Raton, FL.

APPENDIX

Glossary of Commonly Used GIS Terms

ABM, *see* agent-based modelling.

Absolute georeference, the referencing in space of the location of a point using a predefined coordinate system such as latitude and longitude or a national grid.

Absolute space, a theoretical framework in which space is defined as a fixed and immovable reference point, which is not affected by what occupies it or occurs within it.

Abstraction, individual, distinct items are derived from the division of observable real-world phenomena.

Access time, a measure of the time interval between the instant that data are called from storage and the instant that delivery is complete.

Accumulation operator, a function used to derive a sum of the water flow that has built up upstream of a point on a digital elevation model.

Accuracy, conformance to a recognizable standard or true value. The statistical meaning of accuracy is the degree by which an estimated mean differs from the true mean.

Aerial photography, digital or optical images/photographs taken of the earth's surface from a position a distance above it, usually from an aeroplane.

Agent, an active, autonomous, and reactive representation of a human or other entity that is used in modelling and analysis.

Agent-based modelling, a form of computer modelling that represents one or more agents and their actions, reactions, and interactions with the defined environment, which are controlled by various defined rules.

Algorithm, a set of rules for solving a problem. An algorithm must be specified before the rules can be written in a computer language.

Alphanumeric characters, the combined set of letters, numbers, and special characters such as those found on a keyboard.

Altitude matrix, a grid of elevation values.

Analogue 1. Representation of information by a continuously varying signal (contrast with discretized signals such as digital data). 2. A model representation of a physical variable or phenomenon by another variable which displays proportional relationships over a specified range—e.g. using a map to describe an area on the earth's surface.

Analysis of variance (ANOVA), a series of statistical tests that are used to find out if there is a significant difference in values of a variable, for example air temperature, between different sample data sets.

Anisotropic, describes a spatial phenomenon that has different physical properties or actions in different directions.

ANOVA, *see* analysis of variance.

API, *see* application programming interface.

App, abbreviation for application program.

Application program, a program or set of computer programs designed for a specific task.

Application programming interface (API), a set of computing tools, routines, and protocols for building software applications.

Arc, a complex line connecting a sequence of coordinate points. Also known as a chain or string.

Archival storage, magnetic and optical media (tapes, removable disks) used to store programs and data outside the normal addressable memory units of the computer.

Area, a fundamental unit of geographic information (a geographic primitive) which is a measure of a particular extent of the earth's surface (*see also* point, line, and polygon).

Area cartograms, maps in which the size shown for a particular geographical area, for example a country, is determined by the value of the variable being represented, such as annual rainfall.

Array, a series of addressable data elements in the form of a grid or matrix.

Aspect, the direction to which a hillside faces, for example north-west.

Attribute, non-graphic information associated with a point, line, or area element in a GIS.

Autocorrelation, autocovariance, statistical concepts expressing the degree to which the value of an attribute at spatially adjacent points varies with the distance or time separating the observations.

Autocovariogram, a visual graphic of autocorrelation statistics, showing the degree to which an attribute of spatially adjacent points varies with distance between the various observations.

Automated cartography, the process of drawing maps with the aid of computer-driven display devices such as plotters and graphics screens. The term does not imply any information processing.

Average, *see* mean.

Azimuth 1. Used in the definition of the position of an object in the sky, such as a star, and is the angle between a reference point (such as due north) on a horizontal plane (such as the earth's surface) and the position the object would be at if it was projected onto that same plane. 2. Azimuthal map projection projects points from the earth's surface onto a flat plane surface that is positioned at a tangent to it.

B-spline, or basis-spline, a form of mathematical function for fitting a generated surface through known data points, involving a linear combination of control points.

B-trees, a generalization of binary trees.

Best linear unbiased estimate (BLUE), a term associated with kriging interpolation. Kriging is 'best' in that it aims to minimize the error variance. It is linear since estimates made using kriging are weighted linear combinations of the sample data. It is unbiased as the expected error is zero.

Bias, systematic errors that are consistent and may be corrected through the use of a simple transformation.

Big Data, a popular term used to describe the exponential growth and availability of data today, both structured and unstructured, and characterized by large volumes, high velocity in generation, wide variety of subject areas, high variability and complexity in the information, and which challenges existing database systems, and analytical and presentational methods and capabilities.

Binary arithmetic, the mathematics of calculating in powers of 2.

Binary coded decimal, the expression of each digit of a decimal number in terms of a set of bits.

Binary trees, a data compression and indexing technique which subdivides the spatial or attribute data into a series of levels.

Bit, the smallest unit of information that can be stored and processed in a computer. A bit may have two values: 0 or 1—i.e. YES/NO, TRUE/FALSE, ON/OFF.

Bit map, a pattern of bits (i.e. ON/OFF) on a grid, stored in memory and used to generate an image on a raster scan display.

Bivariate, using two different variables. The term is often used to describe modelling and analysis using two variables, such as using both elevation and precipitation to explain the amount of stream flow.

Black-box modelling, a computational model representation of a system in which input, output, and processes are characterized mathematically without explicitly defining the actual workings.

Block codes, a compact method of storing data in raster databases which simplifies the region into a series of two-dimensional blocks of various dimensions, e.g. 2 × 2, 3 × 3. The record in the database records the origin, the bottom left, and the side of each square.

Block diagram, a quasi-three-dimensional display of a surface.

Block kriging, the prediction of attribute values for square blocks of land using the kriging interpolation.

Boolean algebra, a form of algebra in which statements involving variables and operators AND, OR, NOT, are established that produce results as truth values—*true* or *false*—usually denoted 1 and 0, respectively.

Boolean operators, operators based on logic which are used to query two or more sets of data. The operators allow inclusion, exclusion, intersection, and differences in the data to be determined.

Break points, points on the graph of a continuous variable (e.g. splines) where there is an abrupt change in direction.

Buffering, the creation of a zone of specified width around a point, line, or area. The buffer is a new polygon which is used in queries to determine which entities occur within or outside the defined area.

Bug, an error in a computer program or in a piece of electronics that causes it to function improperly.

Byte, a group of contiguous bits, usually eight, that represents a basic unit of information which is operated on as a unit. The number of bytes is used to measure the capacity of memory and storage units, e.g. 256 kbytes, 300 Mbytes.

C++, a high-level programming language often used to write graphics programs.

CA, *see* cellular automata.

Cadastral map, a map showing the precise boundaries and size of land parcels.

Cartesian coordinates, locations established with reference to two axes at right angles to each other (x axis is horizontal and y vertical), that may be negative or positive values, and are relative to the origin point (0, 0).

Cartograms, maps, often of countries, in which data such as population are used instead of the areal size to define the shape and extent of the individual map units, giving geographically distorted but often challenging and powerful representations.

Cartographic communication, the use of visual properties such as colour, shape, and pattern in maps to communicate information about geographical data and spatial relationships.

Cartographic modelling, the procedure of using algebraic techniques to build models for spatial analysis.

Cartographic principles, a set of norms, procedures, and rules used in the design, compilation, and construction of maps that ensure clarity, accessibility to the user, organization, and balance in the final product.

Cartographic visualization, the use of graphical representations across different technologies, based on mapping techniques and practices.

Cartography, the art and science of drawing charts and maps.

Category, a specifically defined division within a classification schema.

Cell, the basic element of spatial information in the raster/grid description of spatial entities (*see* pixel).

Cellular automata (CA), a form of modelling based on grid cells in which the processes are defined using sets of rules that affect the state of each grid cell. This form of modelling suits applications where behaviours cannot be easily represented by strict mathematical equations.

Census area, district, tract, ward, a uniquely defined area unit in population censuses, for which data are aggregated from the individual responses to a community-based output.

Centroid, the geometric centre of a polygon in a database, against which attribute data is often linked to.

Chain, *see* arc.

Chain codes, a compact method of storing data in raster databases which simplifies the boundary of a region in terms of a sequential series of north, south, east, or west directional vectors grid, and the number of cells in each direction.

Character, an alphabetical, numerical, or special graphic symbol that is treated as a single unit of data.

Chorochromatic map, a map in which the area is divided into a series of zones which are each displayed using a single colour or shading.

Choropleth map, a map consisting of a series of single-valued, uniform areas separated by abrupt boundaries.

Citizen cartographers, ordinary people who collect geographical data and create maps, usually using online tools and GNSS/GPS-enabled technologies, and often making these available online for other users.

Classification, the process of assigning items to a group or set according to their attributes.

Clip, an operation in a GIS in which a polygon is overlaid onto one or more layers and targeted features are isolated and extracted from within the area outlined by the polygon. The areas outside this polygon are discarded.

Cloud computing, a group of computing server hardware connected through network infrastructure, usually the Internet, allowing users to run programs and store data away from their personal computers.

Clump, the fusion of neighbouring cells which are of the same class into larger units.

Code, a set of specific symbols and rules for representing data and programs so that they can be understood by the computer.

Coefficient of determination, a statistical measure of how well data points fit a statistical model such as that resulting from linear regression. It is denoted as R^2.

Cokriging, estimation of a regionalized variable using observations of that variable supplemented by observations of one or more additional variables from within the same geographical area, thereby reducing the estimation variance if the original variable has been undersampled.

Command, an instruction sent from the keyboard or other control device to execute a computer program.

Command language, an English-like language for sending commands for complicated program sequences to the computer.

Complete spatial randomness (CSR), a term used in statistics to refer to points occurring in a fashion that suggests no order or structure to the events.

Composite map, a single map created by joining together several maps that have been generated separately.

Computational model, a representation of part of reality in which the environment and processes acting on it are defined digitally and through computer commands.

Computer assisted/aided cartography (CAC), the use of computer hardware and specific software for making maps and charts.

Computer graphics, a general term embracing any computing activity that results in graphic images.

Computer word, a set of bits (typically 16 or 32) that occupies a single storage location and is treated by the computer as a unit of information.

Computing environment, the total range of hardware and software facilities provided by a computer and its operating system.

Concavity, a surface that curves inwards; often used to denote a hillside where there is hollowing in.

Conceptual model, the abstraction, representation, and ordering of phenomena using the mind.

Conditional simulation, the simulation of a single random function that honours the data values at the sampling points.

Configuration, a particular combination of computer hardware and software for a certain class of application tasks.

Conformal projection, a map projection that preserves local angles, while other properties such as distance may be distorted.

Confusion index, a measure of the relative dominance of the membership values assigned to an individual for two or more fuzzy classes.

Conical projection, a map project based on a cone shape of flat paper in which the lines of longitude are mapped as straight lines from an apex point, while the lines of latitude are circular arcs at increasing distances from this point.

Connectivity, the linking of different spatial, mostly linear, units into complex chains.

Contiguity, adjacent spatial units touching to form an unbroken chain or surface.

Contour, a line connecting points of equal elevation.

Convexity, a surface that curves outwards; used to describe hillslopes that bulge out.

Convolution, the conversion of values from one grid to another which is different in terms of size or orientation.

Cookie cutter, *see* clip.

Coordinate system, a geometric system which defines a point numerically according to an established referencing structure for the two or three dimensions of space.

Correlation, a statistical measure of how much one data value for a variable depends on that of another from two or more data sets.

Covering, an overlay operation where the cover feature is placed on top of features in a second data layer (overlapping features in the second layer are replaced).

Cross-hatching, the technique of shading areas on a map with a given pattern of lines or symbols.

Cross-validation, a validation method in which observations are dropped one at a time from a sample size n, and n estimates are computed from the remaining $(n - 1)$ observations. The statistics of the estimates are used to evaluate the goodness of fit. In geostatistics this is done using the kriging equations to check the variogram with respect to the sample data.

Crossover point, the point at which there is a 50 per cent possibility that something belongs to a particular fuzzy class.

Crowdsourced geographical data, data collected by a large group of people, often not specialists, that may be solicited or unsolicited, usually involving online sharing.

Cylindrical projection, a map project based on a cylinder shape of flat paper in which the lines of latitude and longitude are both mapped as straight lines, producing a rectangular grid.

D8 (deterministic) algorithm, used in DEM analysis to determine the direction of flow down a slope from a grid cell taking into account the elevation, and so slope, of the surrounding eight pixels. The flow from this one grid cell is thus coded as in an east, north-east, north, north-west, west, south-west, south, or south-east direction.

Data analysis models, a series of commands that when combined perform a particular kind of data analysis.

Data exchange standards, widely accepted schema for structuring data so that they can be shared easily and used in a variety of different software systems.

Data model, the abstraction and representation of real-world phenomena according to a formalized, conceptual schema, which is usually implemented using the geographical primitives of points, lines, and polygons, or discretized continuous fields.

Data record, *see* tuple.

Data structure, the organization of data in ways suitable for computer storage and manipulation.

Data types, the classification of different data according to their characteristics—for example Boolean (0/1), nominal, ordinal, integer, scalar (real), directional, or topological data types—their function, and degree of precision.

Database, a collection of interrelated information, usually stored on some form of mass-storage system such as magnetic tape or disk. A GIS database includes data about the position and the attributes of geographical features that have been coded as points, lines, areas, pixels, or grid cells.

Database management system (DBMS), a set of computer programs for organizing the information in a database. Typically, a DBMS contains routines for data input, verification, storage, retrieval, and combination.

Delaunay triangulation, the graph obtained by joining pairs of points whose polyhedra are Thiessen (Voronoi/Dirichlet) divisions of the plane.

DEM, *see* digital elevation model.

Demographics, the study of human populations, usually based on quantitative data and statistical analysis.

Deterministic modelling, modelling where, given the same state of initial conditions and variable values, the model will perform the same way and give the same output values.

Device, a piece of equipment external to the computer designed for a specific function such as data input, data storage, or data output.

Differentiable continuous surface, the representation of a continuously varying phenomenon using scalar or integer data so that the rate of change across and within the area may be derived.

Differential GNSS/GPS, where accuracy in defining location is enhanced through using fixed, ground-based reference stations.

Digital, the representation of data in discrete, quantized units or digits.

Digital elevation model (DEM), a quantitative model of a part of the earth's surface in digital form. Also known as digital terrain model (DTM).

Digital orthophoto, a geometrically corrected, or orthorectified, digital aerial image.

Digitize, (noun) a pair of XY coordinates; (verb) to encode map coordinates in digital form.

Digitizer, a device for entering the spatial coordinates of mapped features from a map or document to the computer. A pointer device—a cursor, puck, or mouse—is used to locate key points.

Dirichlet tessellation, *see* Thiessen polygons.

Discretization/discretized, the process of dividing an area into a series of self-contained units.

Disjunctive kriging, a non-linear distribution-dependent estimator for regionalized variables that do not have simple (Gaussian) distributions. It is the most demanding kriging method in terms of computer resources, mathematical understanding, and stationarity conditions.

Distributed processing, the placement of hardware processors where needed, instead of concentrating all computing power in a large central CPU.

Distribution (frequency), summarizes the number of times a value or range of values is found in a sample.

Diversity, the number of different values in a moving window.

Double precision, typically refers to the use in 32-bit word computers of a double word of 64 bits to represent real numbers to a precision of approximately 16 significant digits.

Drainage net, used in DEM analysis, and is a derived, generated network structure of linked flow directions down a hillside.

Draping, a graphic visualization where data from one variable is superimposed onto a three-dimensional surface of the same area to show how it varies over the terrain.

Drift, a trend in the data.

Edge matching, allows the connection of polygons and lines for the same features in adjacent maps to be joined to create a seamless new map.

Edit, to remove errors from or to modify a computer file of a program, a digitized map, or a file containing attribute data.

Electromagnetic radiation, radiant energy emitted by all bodies above 0 degrees kelvin (−273°C) that travels as waves from its source and occurs across wide ranges in frequency. It includes x-rays, ultraviolet, visible light, infrared including thermal infrared, microwaves, and radio waves.

Element, a fundamental geographical unit of information, such as a point, line, area, or pixel. May also be known as an 'entity'.

Elevation, the height on the earth's surface above a reference point; most commonly taken from average local sea level or a measured or generated surface such as an ellipsoid or geoid.

Ellipsoid, mathematical model for the shape of the earth, taking account of flattening at the poles.

Entities, distinct units of a real-world phenomenon.

Entity conceptual model, a high-level schema based on the representation of the spatial objects and features of the world and their relationships, using the building blocks of points, lines, and polygons.

Ephemeral output, a visualization such as a map that lasts a very short time.

Equal area projection, a map projection that preserves the size of areas, while other properties such as distance or angles may be distorted.

Equidistant projection, a map projection that preserves the values for distances across the map, while other properties such as angles and areas may be distorted.

Equipotential surface, made of up lines on which every point on a given line is at the same potential. In geography it is used to depict the earth's surface which best coincides with mean sea level, which is defined as sea level if the oceans where at rest and only subject to terrestrial gravity.

Erase, an operator used to cut a feature (the erase feature) from another data layer.

Exact interpolator, an interpolation method that predicts a value of an attribute at a sample point that is identical to the observed value.

Experimental variogram, an estimate of a (semi-)variogram based on sampling.

Extrapolation, the estimation of the values of an attribute at unsampled points outside an area that is covered by existing measurements.

Feature planes, a series of separate different classes of phenomena which are often used as the basis of the different overlays.

Field 1. A type or class of data. 2. In a database, a set of records containing information (*compare with* tuple).

Field conceptual model, a high-level schema based on the representation of the spatial objects and features of the world and their relationships, using the building blocks of tessellated polygons such as squares or triangles.

Field observations, where using data-collection methods, you observe and record information on environments or people in 'real' locations.

File, a collection of related information in a computer that can be accessed by a unique name. Files may be stored on tapes or disks.

Filter, in raster graphics, a mathematically defined operation for removing long-range (high-pass) or short-range (low-pass) variation. Used for removing unwanted components from a signal or spatial pattern.

Finite difference modelling, a numerical modelling technique used with data held in regular grid form, in which algebraic equations are used to solve changes in a variable at each location.

Finite element modelling, a numerical modelling technique used with data held in irregular grid (usually triangular) form, in which algebraic equations are used to solve changes in a variable at each location.

Flattening, a mathematical value (usually represented as f) by which a spheroid is compressed to give an ellipsoid shape. It is generated through a ratio formula based on values for the semi-major and semi-minor axis; for the commonly used WGS84 ellipsoid representation of the earth's surface, the $1/f$ value is 298.26.

Font, symbolism used for drawing a line, or the name of a typeface used for displaying text.

Forcing functions, used to represent variables that are determined only as a function of time, and by the values of other variables in a model. An example is precipitation as a forcing function in a hydrological model. The precipitation changes over time but is not influenced by the flows of water taking place.

Format, the way in which data are systematically arranged for transmission between computers, or between a computer and a device. Standard format systems are used for many purposes.

Fourier analysis, a method of dissociating time series or spatial data into sets of sine and cosine waves.

Fractal, an object having a fractional dimension; one which has variation that is self-similar at all scales, in which the final level of detail is never reached and never can be reached by increasing the scale at which observations are made.

Fuzzy *k*-means, a clustering technique which partitions data points into clusters/sets, with the degree of 'belongingness' or membership (valued between 0 and 1) of that point determined by comparing it to the representative value for that cluster.

Fuzzy logic, a computational approach centred on the idea of 'degrees of truth' rather than the black/white true or false approach of most systems.

Fuzzy membership functions, membership functions which permit individuals to be partial members of different, overlapping sets.

Fuzzy set theory, a set of objects in which the membership of the set is expressed in terms of a continuous membership function having values between 0 and 1. Unlike Boolean sets, fuzzy sets can overlap and an individual can be a member of more than one overlapping sets to different degrees.

Gap, the distance between two graphic entities (usually lines) on a digitized map. Gaps may arise through errors made while digitizing or scanning the lines on a map.

Generalization, the process of reducing detail on a map as a consequence of reducing the map scale. The process can be semi-automated for certain kinds of data, such as topographical features, but requires more insight for thematic maps.

Geocoding, the activity of defining the position of geographical objects relative to a standard reference grid.

Geodemographic segmentation, classification of populations into areas which have common characteristics; examples are areas where there are many retired people with professional backgrounds and who have considerable disposable income, or areas with many young families who are purchasing their home using a mortgage.

Geodesy, the science of measuring and defining the earth's shape, its gravity fields, and its orientation in space.

Geodetical survey, the determination of the position of points on the earth's surface accounting for its curvature, rotation, and gravitational field.

Geographical attributes, information about the spatial location of an object.

Geographical coordinates, a reference system based on coordinates defined by latitude and longitude values, enabling every position on the earth's surface to be uniquely indentified.

Geographical data, data that record the location and a value characterizing the phenomenon on the earth's surface.

Geographical data collectors, people, who may be trained or not, who record information on some phenomena about the world.

Geographical data model, formalized schema for representing data which has both location and characteristic.

Geographical information system, a set of computer tools for collecting, storing, retrieving at will, transforming, and displaying spatial data from the real world for a particular set of purposes.

Geographical primitives, the smallest units of spatial information: in vector form these are points, lines, and areas (polygons); in raster form they are pixels (2D) and voxels (3D).

Geographically weighted regression, develops the statistical method of linear regression, to take into account changes in phenomena over an area that will affect the ability to predict. It constructs a model for each observation (or any other location) in a data set, weighting data points within a search area that is defined by the analyst. The result is a map of different regression coefficients across the area.

Geoid, a representation of the earth based on a model of global mean sea level. It is used in particular to measure surface elevation levels.

Geometry, an ambiguous term in GIS contexts. It is often used to describe the way in which a real-world entity is represented geometrically in a spatial database and the shape of an entity can be described in terms of the coordinates of the nodes or vertices and the lines which connect them.

Geomorphology, the study of landforms and processes acting on the earth's surface.

Geophysics, the study of the behavior and materials of the earth and planets using quantitative physics methods.

Geoservices, activities centred on the collection, generation, analysis, presentation, and distribution of geographical data undertaken by professionals, often based in commercial companies.

Geostatistical methods, statistical techniques and models that use ideas centred around probability and spatial structure to understand variability and estimate new values for points in unknown areas based on existing data sets.

Geotagging, the addition of location data to data collected using, for instance, cameras, videos, SMS, and other media ouputs.

Geovisualization, the creation through various tools and techniques of different images, maps, diagrams, and animations of geographical data, supporting enhanced analysis and insight development.

Geoweb, Internet-based systems for viewing, querying, displaying, and interacting with diverse geographical data sets using standard tools found in GIS such as for selecting, overlaying, measuring, and locating. Public and private sector organizations are developing their own geowebs to allow people to query and use the data they hold.

Global interpolators, geostatistical interpolation techniques that use all the points in a data set to provide estimates for unknown locations.

Global navigation satellite systems (GNSS), a set of satellites in geostationary earth orbits used to help determine geographic location anywhere on the earth by means of portable electronic receivers. The systems in operation today are USA's GPS and Russia's GLONASS, with other countries and regions aiming to bring their own into operation in the coming decade.

Global positioning system (GPS), the USA's NAVSTAR system made up of a constellation of 32 satellites, providing geographic location data for anywhere on the earth by means of portable electronic receivers.

Global referencing system, a structured frame of reference for defining locations on the earth's surface.

Gnomonic projection, a map projection which displays all lines of longitude as straight lines, and is produced as if the earth's surface is projected on a flat piece of paper placed at a tangent to it.

GNSS, *see* global navigation satellite systems.

Goodness of fit, a technique for quantifying how well a statistical model fits the real-world observations.

GPS, *see* global positioning systems.

Gradient, how steep a line or hillslope is; derived by dividing the change in height between the top and bottom points by the horizontal distance between them.

Granularity, the scale or level of detail in a data set.

Grid 1. A set of regularly spaced sample points. 2. A tessellation by squares. 3. In cartography, an exact set of reference lines over the earth's surface. 4. In utility mapping, the distribution network of the utility resources, e.g. electricity or telephone lines.

Grid map, a map in which the information is carried in the form of regular squares (also called a raster).

Hard copy, a copy of a graphics or map image on paper or other stable material.

Hard data, data obtained by direct measurement.

Hardware, the physical components of a GIS—the computer, tablet, printers, etc.

Hexadecimal system, the representation of numbers and letters using base 16 alphanumeric values.

Hierarchical structuring, a method of arranging objects or information so that the units are connected by a hierarchically defined pathway. From above to below, relations are one-to-many.

Histogram, a diagram showing the number of samples that fall in each contiguously defined size class of the attribute studied.

Hole effect, a condition in which the variogram does not increase monotonically beyond the range. The cause may be real or pseudo periodicities in the sample data.

Horizontal datum, the defined reference shape for the earth's surface from which all subsequent specific measurements of positions on the earth's surface are made. These can vary between countries.

Hypsometry, the measurement of the elevation of the earth's surface with respect to sea level.

Identity overlay, an overlay operator which retains all input layer features, but only those parts of the identity layer which fall within the input layer.

Inclusion problem, the problem of determining whether an object (e.g. a point) is located within a particular area.

Indexed files, files of data records in which pointers, based on a particular ordering such as alphabetic, are used to speed up accessing and searching instructions.

Indicator kriging, a kriging interpolation method which is non-linear and in which the original data are transformed from a continuous to a binary scale.

Inexact interpolator, an interpolation method that provides estimates at data locations that are not necessarily the same as the original measurements.

Input, (noun) the data entered into a computer system; (verb) the process of entering data.

Input device, a hardware component for data entry; *see* digitizer, keyboard, scanner.

Integer data, a number without a decimal component; a means of handling such numbers in the computer which requires less space and proceeds more quickly than with numbers having information after the decimal point (real numbers).

Interferometric radar, a technique using radar data to generate maps of the earth's surface using differences in the phase of the waves returning to the satellite or aircraft.

Interpolation, estimation of the values of an attribute at unsampled points from measurements made at surrounding sites.

Intersection 1. Geometric: the crossing of lines or polygons to form new units. 2. Logic: the combination of data from two Boolean sets using the AND operator.

Intervisibility, maps or images showing only the features and landscapes that can be seen from a particular vantage point.

Intrinsic hypothesis, a form of spatial stationarity which is less restrictive than second order stationarity, in which the stationarity requirements are confined to the first differences and not the underlying regionalized variable. The intrinsic hypothesis is useful for modelling regionalized variables in which the form of the variogram is a function of domain size.

Inverse distance weighting (IDW), a statistical technique that uses combinations of values from known scattered points to estimate data for unknown points. The relative distance from each known point used to the unknown point influences the extent by which it determines the new value generated.

Irradiance map, a map which shows how much light (solar or other source) falls across a particular feature, with annual, seasonal or daily maps used in various applications.

Island, a polygon that is located completely within another polygon with no contact between the various arcs or points making up each entity.

Isoline, a line which joins points of equal value.

Isopleth map, a map displaying the distribution of an attribute in terms of lines connecting points of equal value; *see* contour, *contrast with* choropleth map.

Isotropic, an adjective to describe something that has the same physical properties or actions in all directions.

Join 1. (verb) to connect two or more separately digitized maps; 2. (noun) the junction between two such maps, sometimes visible as a result of imperfections in the data.

Joint membership function, a fuzzy membership set that may be defined across different map layers using various other fuzzy membership sets, e.g. 'good neighbourhood', which may be defined by combining other fuzzy sets such as 'good schools', 'close-to-shops', 'affordable healthcare', 'low crime rates'.

Kriging, the name (after D. G. Krige) for a suite of interpolation techniques that use regionalized variable theory to incorporate information about the stochastic aspects of spatial variation when estimating interpolation weights.

Kriging standard deviation, the square root of the kriging variance.

Kriging variance, a measure of the uncertainty of estimation for values predicted by kriging.

Kriging with an external drift, a kriging method where the mean of the variable being mapped is represented using some exhaustively sampled secondary variable.

Land registry system, the structure in place to record who owns land and other rights associated with it, providing a formal system of protection under a particular jurisdiction. It is usually managed by a government agency that formally records the title of the land and any subsequent changes to it, such as through sales or other means of transfer.

Landsat, a series of earth resource scanning satellites launched by the USA.

Laplacian filter, a spatial filter which can be used in edge detection.

Layer, a logical separation of mapped information according to theme. Many GIS allow the user to choose and work on a single layer or any combination of layers at a time.

Legend, the part of a map explaining the meaning of the symbols used to code the depicted geographical elements.

Library, a collection of standard, frequently used computer subroutines or symbols in digital form.

LiDAR, a technology used in remote sensing systems to measure distances using laser light transmission and detection onto a surface. It has been widely used to generate detailed maps of the earth's surface, particularly at the local scale.

Line, one of the basic geographical primitives, defined by at least two pairs of XY coordinates.

Line of best fit, a line that, if drawn through a scatter plot of data points, would provide the best fit to all of them.

Linear interpolator, a method whereby the weights assigned to different data points are computed using a linear function of distance between sets of data points and the point to be predicated.

Linear proximity determination, computation of straight line distances from cells to other cells in a raster grid.

Local drain direction (ldd), the direction of steepest downhill slope as determined from a gridded DEM.

Local interpolators, interpolations which make use of some spatial subset (e.g. the nearest *n* observations) to make estimates.

Local referencing system, a structured framework used in defining locations within a local area.

Location, a specific position or point in physical space.

Loosely coupled models, the linking of a GIS to a dynamic model where the two systems are run separately but the data from each is exchanged between them to bring added value.

Macro language, a simple high-level programming language with which the user can manipulate the commands in a GIS.

Major axis of variogram, for a directional variogram, the direction in which there is greatest spatial continuity.

Majority, *see* mode.

Map 1. A hand-drawn or printed document describing the spatial distribution of geographical features in terms of a recognizable and agreed symbolism. 2. A collection of digital information about a part of the earth's surface.

MAP (Map Analysis Package), a computer program written by C. D. Tomlin for analysing spatial data coded in the form of grid cells.

Map algebra, a set-based algebra for manipulating geographic data (see Tomlin 1983).

Map projection, the basic system of coordinates used to describe the spatial distribution of elements in a GIS.

Map symbology, the symbols used to represent real-world features on a map.

Mapping unit, a set of areas drawn on a map to represent a well-defined feature or set of features. Mapping units are described by the map legend.

Marked point pattern, a point pattern with information attached to each point—an example is the heights of trees attached to tree locations.

Maximum likelihood, a method embodying probability theory for fitting a mathematical model to a set of data.

MBRs, *see* minimal bounding rectangles.

Mean, also referred to as average, is the central value of a data set, calculated by adding together all the individual values and then dividing by the number of instances in the list.

Median, the middle value of a data set if all the records are ordered according to their values.

Menu, a list of available options displayed on the computer screen that the user can choose from by using the keyboard or a device such as a mouse.

Mercator, a widely used map projection, based on a cylinder, that was developed by the Dutch cartographer Gerardus Mercator in the sixteenth century.

Metadata, information about the provenance, resolution, availability, age, ownership, price, copyright, and other matters concerning digital spatial data that is easily available to potential users.

Minimal bounding rectangles (MBRs), a method used in GIS databases to help speed up querying, with the various data points enclosed within a series of boxes to structure the space more efficiently.

Mode, the most commonly occurring value in a data set.

Model 1. A representation of attributes or features of the earth's surface in a digital database. 2. A set of algorithms written in computer code that describe a given physical process or natural phenomenon of the earth's surface.

3. A function fitted to an experimental variogram derived from sample data. 4. A. statistical distribution or a conceptualization of spatial variation.

Module, a separate and distinct piece of hardware or software that can be connected with other modules to form a system.

Monte Carlo simulation, a computer modelling technique that takes on board various uncertainties to generate, through the introduction of randomness, a series of possible outcomes—a probability distribution of outcomes, rather than just one set of values.

Moran's *I* coefficient, a statistical measure of the degree to which data values tend to be similar to neighbouring data values.

Morton ordering, a technique for reducing the geographical referencing of grid data to one dimension by following a set 'Z' shape directional pattern through the cells.

Mosaicing, the merging of two or more raster data sets to produce a new, larger, raster layer.

Multicriteria evaluation, selection of areas which fulfil multiple conditions represented by a set of data layers; such approaches are likely to be based on overlays.

Naïve knowledge, *see* tacit knowledge.

National grid, the map reference system used to identify all locations for a particular country, often developed and managed by the national mapping agency. It forms the reference basis from which subsequent maps are generated.

National mapping authority (NMA), the organization, usually public sector, responsible for generating the key maps for a country, particularly topographic and cadastral maps. In some countries the NMA is also responsible for establishing the horizontal and vertical data and the map referencing framework used in mapping for the country.

Natural breaks scheme, a classification scheme used to divide values into a set of classes; the approach is applied widely to select classes in mapping a continuously varying property such as populations of areas.

Negatively skewed, a frequency distribution which is characterized by a tail of (relatively) small values.

Neogeography, a term that captures the informal use of geographical methods and tools by non-experts, often individuals, facilitated through the Internet, web mapping, and mobile devices.

Nested sampling, the measurement of data at a series of points whose locations are hierarchically structured.

Network 1. Two or more interconnected computer systems for the implementation of specific functions. 2. A set of interconnected lines or arcs.

Network database structure, a method of arranging data in a database so that explicit connections and relations are defined by links or pointers of a many-to-many type.

Neural networks, or artificial neural networks, are computational models which are inspired by the central nervous systems of animals. Such approaches have been used, for example, in the classification of remotely sensed images.

NMA, *see* national mapping authority.

Node, the point at which arcs (lines, chains, strings in a polygon network) are joined. Nodes carry information about the topology of the polygons.

Noise, irregular variations or error, usually short-range, that cannot be easily explained or associated with major mapped features or processes.

Nominal data, data that can be assigned a class and counted, but cannot be ordered, such as gender, or class of ecological habitat.

Non-differentiable continuous surface, representation of a continuously varying phenomenon using binary, nominal, or ordinal data types that do not support the calculation of the rate of change across and within an area.

Non-linear kriging, *see* indicator kriging, disjunctive kriging.

Non-linear proximity determination, computation of distances from cells to other cells in a raster grid using friction surfaces.

Non-removable storage, computer data storage which cannot be moved easily; this includes hard disks.

Non-transitive variogram, a variogram in which the sill is not reached within the domain of interest (*see* intrinsic hypothesis).

Normalization, methods that are used to reduce redundancy and improve efficiency in a database.

Nugget, in kriging and variogram modelling, that part of the variance of a regionalized variable that has no spatial component (variation due to measurement errors and short-range spatial variation at distances within the smallest inter-sample spacing).

Numerical taxonomy, quantitative methods for classifying data using computed estimates of similarity.

Object code, a computer program that has been translated into machine readable code by a complier.

Object-oriented database structure, the organization of data within a database defined by a series of predefined objects and their properties and behavioural characteristics.

OGC, *see* Open GIS Consortium.

Ontology, a branch of philosophy which considers what exists and the nature of reality.

Open GIS Consortium (OGC), a voluntary international organization that sets standards for geographical data in terms of data sharing, content, and data processing (http://www.opengeospatial.org/ogc).

Operating system (OS), the control program that coordinates all the activities of a computer system.

Optimal estimator, an estimator for minimizing the value of a given criterion function; in kriging this is the estimation variance.

Ordered sequential files, a file of data records which are in sequence according to some structuring method such as the alphabet.

Ordinal data, data that are structured according to various classes that may be ranked, with values recorded for the number of occurrences within each class. An example would be age groups such as 0–5, 6–10,11–15 years, and data recorded as to how many are in each class in a sample set.

Ordinary kriging, a method for interpolating data values from sample data using regionalized variable theory in which the prediction weights are derived from a fitted variogram model.

Ordinary least squares, a method for estimating the unknown parameters in a linear regression model.

Orthophotos, a scale-correct photomap created by geometrically correcting aerial photographs or satellite images.

Output, the results of processing data in a GIS: maps, tables, screen images, tape files.

Overlay 1. (verb) the process of stacking digital representations of various spatial data on top of each other so that each position in the area covered can be analysed in terms of these data; 2. (noun) a data plane containing a related set of geographic data in digital form.

Paint, to fill in an area with a given symbolism on a raster display device (*see* cross-hatching).

Parallax, the apparent change in the relative position of objects which is caused by a change in viewing position; these displacements can be used as a basis for constructing 3D images from overlapping areal images.

Parameters, in mathematical modelling, a constant or variable term used in the definition of a function but not its general nature. It is often used in the calibration of models to represent subprocesses that are not defined explicitly.

Participatory GIS (PGIS), associated with the opening up of GIS technologies and data (both access and creation of new data) to a wide user group; PGIS is linked to notions of the democratization of GIS.

Peano–Hilbert ordering, a technique for reducing the geographical referencing of grid data to one dimension by following a recursive route through the cells.

Peripheral, a hardware device that is not part of the central computer.

Permanent outputs, paper maps or images designed to be finalized outputs.

PGIS, *see* participatory GIS.

Photogrammetry, a series of techniques for measuring position and altitude from aerial photographs or images using a stereoscope or stereo plotter.

Photomosaic, a collection of aerial photographs which are joined to form a contiguous view of an area.

Physical model, a scaled copy of the real world made from physical material. The geometry of the representation is accurate but the scale varies, often intentionally, to highlight a particular property such as the height of mountains.

Pit, a depression on the surface of a digital elevation model, which may represent a real feature or may be an artefact of the gridding.

Pixel, contraction of picture element; smallest unit of information in a grid cell map or scanner image.

Plan curvature (convexity and concavity), measures topographic convergence and divergence—the propensity of water to converge as it flows across a surface.

Planar projection, *see* azimuthal projection.

Plotter, any device for drawing maps and figures.

Point, one of the basic geographical primitives, defined by a pair of XY coordinates.

Point interpolation, spatial interpolation from point (as opposed to area) data.

Point operations, computation of new raster images on a cell-by-cell basis (e.g. by adding together three co-located images).

Point pattern, a set of locations which correspond to events (e.g. trees or people with a particular disease).

Polygon, a multisided figure representing an area on a map; a geographic primitive involving a series of lines joined to create an areal unit.

Polygon intersection, *see* intersection.

Polygon overlay, *see* overlay.

Polygon overlay and intersection, the creation of new polygons (entities) by the process of overlaying and intersecting the boundaries from two or more vector representations of area entities.

Polynomial, an expression having a finite number of terms of the form $ax + bx^2 + \ldots n$.

Positively skewed, a frequency distribution which is characterized by a tail of (relatively) large values.

Precision 1. Degree of accuracy of numerical representation; generally refers to the number of significant digits of information to the right of the decimal point. 2. Statistical: the degree of variation around the mean.

Prime Meridian, a north–south line of longitude, defined as Longitude Zero (0° 0′ 0″), which passes through Greenwich in London. All geographical coordinates are measured and defined in terms of their angles east or west from this line.

Principal components analysis (PCA), a method of analysing multivariate data in order to express their variation in terms of a minimum number of principal components or linear combinations of the original, partially correlated variables.

Prism 1. A polygonal solid; a polyhedron having parallel, polygonal, and congruent bases and sides that are parallelograms. 2. Sometimes used in GIS to indicate a 3D solid body delineated by irregular polygonal faces.

Probabilistic modelling, computer modelling where randomness is introduced to give a range of possible outcomes reflecting the uncertainty of an environment and its representation.

Probability, the chance of an event or occurrence.

Probability distribution function, a real-valued function (in the range 0, 1) whose integral over a set gives the probability of a random variable having a value within the set.

Profile curvature (concavity and convexity), the sideways view, in two-dimensional form, through a slope when measured down the steepest gradient.

Program, a set of instructions directing the computer to perform a task.

Projection, *see* map projection.

Proximity, the closeness of one item to another.

Pseudo cross-variogram : the cross-variogram allows for the analysis of spatial dependency between two variables. It requires that both properties are observed at the same locations. With the pseudo cross-variogram this requirement is relaxed and data on both variables are not required at the same locations.

Public Participation GIS (PPGIS), *see* participatory GIS (PGIS).

Puck, a handheld device for entering data from a digitizer, which usually has a window with accurately engraved crosshairs, and several buttons for entering associated data.

Pycnophylactic interpolation, the transfer of data values from irregular areas to grids whereby the cells which overlap the areas have the same total population as the areas (this is termed 'mass preservation').

Quadrant, a quarter of a circle measured in units of 90 degrees.

Quadratic polynomial, one in which the highest degree of terms is 2.

Quadtree, a data structure for thematic information in a raster database that seeks to minimize storage.

R-tree, a spatial indexing technique which groups entities according to their proximity by using minimal bounding rectangles. Hierarchies of rectangles may be established. When querying the database, any search is directed to the rectangle and any subsequent lower-level ones which contain the item of interest.

Radargrammetry, a series of techniques for measuring position and altitude from aerial radar images using a stereoscope or stereo plotter.

Random variable, a variable whose value is subject to chance, so will be different at different times. This variation can be captured through probability or probability density functions.

Range 1. In arithmetic, the difference between the largest and smallest values in a set. 2. In geostatistics, the distance at which a transitive variogram ceases to increase monotonically.

Raster, a regular grid of cells covering an area.

Raster data model, a formal structure for recording geographical data in the form of tessellating units such as grid squares.

Raster data structure, a database containing all mapped, spatial information in the form of regular grid cells.

Raster map, a map encoded in the form of a regular array of cells.

Raster to vector conversion, *see* vectorization.

Rasterization, the process of converting an image of lines and polygons from vector representation to a gridded representation.

Real data, numbers that have both an integer and a decimal component (scalars).

Real time, tasks or functions executed so rapidly that the user gets an impression of continuous visual feedback.

Realization, an equi-probable result of stochastic simulation based on a known probability distribution function.

Record, a set of attributes relating to a geographical entity; a set of related, contiguous data in a computer file.

Rectangular coordinates, *see* Cartesian coordinates.

Redundancy, the inclusion of data in a database that contribute little to the information content.

Region, a set of loci or points that have a certain value of an attribute in common.

Regionalized variable, a single-valued function defined over a metric space (a set of coordinates) that represents the variation of natural phenomena that are too irregular at the scale of interest to be modelled analytically.

Regression, a method for exploring the relationship between a single (dependent) variable and one or more independent variables.

Relational database model, a method of structuring data in the form of sets of records or tuples so that relations between different entities and attributes can be used for data access and transformation.

Relative georeferencing, the referencing in space of the location of a point to a local base station rather than to a global grid.

Relative space, a theoretical system in which space is defined by what occupies it or occurs within it.

Remote sensing systems, the collection of data without coming into physical contact with it. Satellites and aeroplanes are platforms most often used for this.

Representative fraction, a ratio which relates the size of features in the real world to their size on the map.

Resampling, a technique for transforming a raster image from one particular scale and projection to another.

Residuals, the difference between a model (e.g. regression) estimate and the true value.

Resolution, the smallest spacing between two displayed or processed elements; the smallest size of feature that can be mapped or sampled.

Response time, the time that elapses between sending a command to the computer and receiving the result at the workstation.

Rho8 (random) algorithm, a statistical version of the D8 algorithm which was introduced to represent better the stochastic aspects of terrain.

Run-length codes, a compact method of storing data in raster databases which simplifies the grid on a row-by-row basis by coding the start and end values of contiguous cells for each class.

Sampling, the technique of taking a series of measurements to obtain a satisfactory representation of the real-world phenomenon being studied.

Satellite sensor, technology for collecting values of electromagnetic radiation received within various defined wavebands, that is mounted on a satellite system.

Scalar data, data collected using positive and negative real numbers with decimals, with a range to infinity.

Scale, the relation between the size of an object on a map and its size in the real world.

Scanner, a device for converting images from maps, photographs, or from part of the real world into digital form. The scanning head is made up of a light or other energy source and a sensing device which records digital values of light reflected back from the surface. They are usually drum or flatbed-based devices.

Scenario, a result of a numerical simulation model in which certain input data may be given values to represent conditions not yet observed. Scenarios are often used to compare forecasts of how landscape changes may turn out.

Scripts, in computing, programs written for a particular environment that can automate and execute a task that would normally be undertaken by a human operator. An example would be scripts to automatically download on a daily basis a data set from a particular website.

SDIs, *see* spatial data infrastructures.

Semantic import approach, a method used in fuzzy set analysis to define a function using information already known of an area.

Semi-distributed modelling, an approach to modelling where there is a partial representation of the heterogeneity of an environment through the definition of a series of distinct spatial units across an area, that are linked but within each unit the variable values are the same.

Semi-major axis, used in the definition of an ellipsoid: the longest radius in an ellipse.

Semi-minor axis, used in the definition of an ellipsoid: the shortest radius in an ellipse.

Semivariogram 1. Given two locations $\mathbf{x}$ and $(\mathbf{x} + \mathbf{h})$, a measure of one-half of the mean square differences (the semivariance) produced by assigning the value $z(\mathbf{x} + \mathbf{h})$ to the value $z(\mathbf{x})$, where $\mathbf{h}$ (known as the lag) is the inter-sample distance. 2. A graph of semivariance versus lag $\mathbf{h}$.

Semivariogram model, one of a series of mathematical functions that are permitted for fitting the points on an experimental variogram (e.g. linear, spherical, exponential, Gaussian).

Shaded relief map, a means of portraying relief differences on a surface with a given illumination position (angle and height).

Shortest path algorithm, commonly refers to the algorithm developed by Dijkstra in 1959. In this approach, links, connected to a starting node, are selected that have the shortest path back to that starting node. This approach is much more efficient than searching all possible routes.

Sill, the maximum level of semivariance reached by a transitive semivariogram.

Simple kriging, an interpolation technique in which the prediction of values is based on a generalized linear regression under the assumption of second order stationarity and a known mean.

Simulation, using the digital model of the landscape in a GIS for studying the possible outcomes of various processes expressed in the form of mathematical models.

Sink, *see* pit.

Sky view factor, the amount of sky visible when viewed from the ground upwards.

Sliver, a narrow gap between two lines created erroneously by digitizing or by the vectorization software of a scanner.

Slope, gradient and aspect together.

Slopelength operator, similar to the accumulation operator but it computes a new attribute of a cell as the sum of the original cell value and the upstream cells multiplied by the distance travelled over the network.

Smoothing, a set of procedures for removing short-range, erratic variation from lines, surfaces, or data series.

Smoothing spline, a method of fitting a smooth polynomial function through erratic data to capture the long-range variation and to suppress local components.

Soft data, data obtained by inspection, intuition, or from other parties. Not measured directly and therefore often judged to be less reliable than hard data.

Software, general name for computer programs and programming languages.

Source code, a computer program that has been written in an English-language-like computer language. It must be compiled to yield the object code before it can be run on the computer.

Spatial autocorrelation, concerns the relationship between data values and neighbouring data values—if they tend to be similar then this indicates positive spatial autocorrelation (spatial dependence), while if they tend to be dissimilar this indicates negative spatial autocorrelation.

Spatial browsers, a web browser designed for accessing and querying spatial information.

Spatial contiguity, the condition where objects are adjacent to one another in space.

Spatial data, data that record the location and a value characterizing the phenomenon.

Spatial data infrastructure (SDI), an integrating framework of standards for geographical data which defines many aspects of database development, use, and institutional arrangements to ensure the data collected is widely accessible and used. It is often defined at a national level.

Spatial dependence, reflects the tendency for neighbouring observations to be more alike than those farther apart.

Spatial distribution, the spatial arrangement of observations.

Spatial filtering, the process of altering data values in a grid using the neighbours of each cell (e.g. taking the mean of values around each cell to produce a new smoothed grid).

Spatial function, measures which are some function of the area or neighbourhood surrounding a given cell (e.g. spatial filters).

Spatial geometry, the representation of features in an area through an ordered set of primitive elements of points, and lines and arcs.

Spatial interaction data, data which relate to flows or moves between places (e.g. the numbers of people who migrated between one place and another).

Spatial neighbourhood, the area around an observation—a neighbourhood could be based on simple proximity or some form of distance decay.

Spatial orientation, the compass direction in which an image or other data set is orientated.

Spheroid, a geometric representation of the shape of the earth.

Spike 1. An overshoot line created erroneously by a scanner and its raster–vector software. 2. An anomalous data point that protrudes above or below an interpolated surface representing the distribution of the value of an attribute over an area.

Spline, a polynomial curve of surface used to represent spatial variation smoothly.

Spurious polygons, polygons which are created when two or more vector layers are overlaid and there is a mismatch in common boundaries as they are represented differently in the layers.

SQL, *see* Structured Query Language.

State variable, a dynamically changing characteristic in an environment, such as stream flow, that represents its changing state over time in response to changes in environmental conditions or processes acting on it.

Stationarity, a statistical name for expressing degrees of invariance in the properties of random functions; it refers to the statistical model, and not to the data. Most commonly used to indicate invariance of first differences (*see* intrinsic hypothesis).

Statistical moments, first order is the mean; second order are the variance, the covariance, and the semivariance.

Stereo plotter, a device for extracting information about the elevation of landform from stereoscopic aerial photographs. The results are sets of X, Y, and Z coordinates.

Stochastic imaging, *see* conditional simulation.

Stochastic simulation, simulation using a probabilistic model to generate a range of allowable data values.

Storage, the parts of the computer system used for storing data and programs.

Stratified kriging, interpolation by any kriging method within a set of strata or divisions of the land into different classes.

Stream Power Index, a measure of the erosive power of flowing water.

Stream tube methods, used to determine the amount of flow as a fraction of the area of the source pixel entering each pixel downstream.

Structured Query Language (SQL), a standard language for interrogating and managing relational databases.

Support, this is the term used in geoststistics for the area or volume of the sample on which measurements are made (e.g. a volume of soil or water, or a pixel in a remotely sensed image).

Surface derivatives, products derived from topographic surfaces; the most common are gradient and aspect (together making up the slope).

Surface topology, relates to the connections between surface features and can be derived through computing drainage networks and catchment delineation.

SYMAP (SYnagraphic MAPping program), the original grid-cell mapping program developed by Howard T. Fisher at Harvard.

Syntax, a set of rules governing the way statements can be used in a computer language.

Tacit knowledge, informal knowledge which is developed by individuals and communities through experience and transferred wisdom, and is not easily codified and transferred verbally, digitally, or through writing.

Taxonomic classes, used to group soils, plants, or animals which share common characteristics.

Tessellation, the process of dividing an area into smaller, contiguous tiles with no gaps in between them.

Thematic map, a map displaying selected kinds of information relating to specific themes, such as soil, land use, population density, suitability for arable crops, and so on. Many thematic maps are also choropleth maps, but when the attribute is modelled by a continuous field, representation by isolines or colour scales is more appropriate.

Theodolite, an instrument for measuring angles in surveying.

Thick client systems, computer or software systems in which most of the computer processing takes place locally with little use of remote server systems across a network.

Thiessen polygons, a tessellation of the plane such that any given location is assigned to a title according to the minimum distance between it and a single, previously sampled point. Also known as Dirichlet tessellation or Voronoi polygons.

Thin client systems, computer or software systems that depend on most of the computer processing taking place on remote server systems across a network.

Thin-plate splines, can be conceptualized as surfaces that are fitted to some local data subset; thin-plate spline interpolation is widely used in GIS contexts.

Tile, a part of the database in a GIS representing a contiguous part of the earth's surface. By splitting a study area into tiles, considerable savings in access times and improvements in system performance can be achieved.

Tiling, the creation of a seamless spatial coverage by joining contiguous areas (tiles) together.

Time series analysis, the analysis of measurements of some property across time; an example is the analysis of precipitation amounts at one location for multiple time points.

TIN, *see* triangular irregular network.

Topographical map, a map showing the surface features of the earth's surface (contours, roads, rivers, houses, etc.) in great accuracy and detail relative to the map scale used.

Topology, a term used to refer to the continuity of space and spatial properties, such as connectivity, that are unaffected by continuous distortion. In the representation of vector entities, connectivity is defined *explicitly* by a directed pointer between records describing things that are somehow linked in space (for example a junction between two roads). In regular and irregular tessellations of continuous surfaces (e.g. grids) the topological property of connectivity between different locations may only be *implicitly* defined by the spatial rate of change of attribute values over the grid. The topology (connectivity) of gridded surfaces can be revealed by computing first, second, or higher order derivatives of the surface (see Chapter 8).

Total station, an instrument used for high-precision survey of spatial positions.

Transect, a set of sampling points arranged along a straight line.

Transfer function 1. A numerical method of transferring spatial data from one projection to another. 2. A numerical model for computing new attribute values from existing data using regression models or other algorithms.

Transformation/transform (image), the process of changing the scale (spatial resolution), map projection, or orientation of a mapped image.

Transitive variogram, a semivariogram having a range and a sill.

Trend surface analysis, methods for exploring the functional relationship between attributes and the geographical coordinates of the sample points.

Triangular irregular network (TIN), a vector data structure for representing geographical information that is modelled as a continuous field (usually elevation) which uses tessellated triangles of irregular shape (*see* Delaunay triangulation).

Tuple, a set of values of attributes pertaining to a given item in a database. Also known as a data record.

Union 1. Database: the joining of two or more data sets together. 2. Boolean logic: the joining of two sets using the 'OR' operator.

Universal constant, a value for a fundamental physical constant that is thought to be universal, such as the speed of light, or Newton's gravitational constant.

Universal kriging, simple kriging of the residuals of a regionalized variable after systematic variation has been modelled by a drift or trend surface.

Universal Soil Loss Equation, an empirical equation that links the key variables in the environment and the human activities that bring about soil erosion, such as the slope of the land, the nature of the rainfall and soils, and the crop production methods.

Universal Transverse Mercator (UTM), a global mapping system in which the globe is divided into 60 narrow zones of 6 degrees of longitude, which are each projected onto a cylindrical, conformal map projection. The resulting coordinate system is based on a rectangular (Cartesian) grid that supports decimal definitions.

Upscaling, the aggregation of spatial data from small areas to larger areas.

Upstream element map, a map showing the cumulative catchment areas for each cell according to the topology of the local drain direction map.

Utilities, a term encompassing public or private infrastructure such as water pipes, sewerage, telephone, electricity, and gas networks.

Utility, a term for system capabilities and features for processing data.

UTM, *see* Universal Transverse Mercator.

Validation points, locations at which estimates are made where true values are known but withheld—this enables computation of errors of estimation (*see also* cross-validation).

Variance/mean ratio (VMR), in point pattern analysis the VMR provides a means of assessing how far the point pattern is clustered or dispersed. A value of less than 1 suggests a dispersed point pattern, while a value of greater than 1 suggests a clustered point pattern.

Variogram, a common term for semivariogram.

Vector 1. Physics: a quantity having both magnitude and direction. 2. GIS: the representation of spatial data by points, lines, and polygons.

Vector data model, a formal structure for recording geographical data in the form of points, lines, and polygon units.

Vector data structure, a means of coding and storing point, line, and areal information in the form of units of data expressing magnitude, direction, and connectivity.

Vector to raster conversion, *see* rasterization.

Vectorization, the conversion of point, line, and area data from a grid to a vector representation.

Vertical datum, the defined point from which all elevation points are subsequently measured—often average sea level or a geoid surface. These are predominantly defined globally, regionally, and nationally.

Viewshed, those parts of the landscape that can be seen from a particular point.

Volunteered geographic information (VGI), *see* crowdsourced geographical data.

Voronoi polygons, *see* Thiessen polygons.

Voxels, three-dimensional, cubic units of space.

Weighted moving average, value of an attribute computed for a given point as an average of the values at surrounding data points, taking account of their distance or importance.

Wetness index map, a map indicating the degree of wetness of cells; flat areas are more likely to be wet, while steep convex areas are more likely to be dry.

WGS84, *see* World Geodetic System 84.

Window, an area (usually square) that is used to capture the data needed to compute derived attributes from an original map.

Word, *see* computer word.

Workstation, a minicomputer or high-level personal computer used for local computations; it is often connected to other computers by a network. The operating system used for workstations is often Unix or Windows NT.

World Geodetic System 84 (WGS84), a global standard reference system used to define the shape of the earth (ellipsoid), its gravitational surface of equipotential (geoid), and the coordinate system. It is the basis for technologies such as GPS/GNSS to position a location. Most GIS have this as a standard reference system that can be adopted by the user.

Zero, the origin of all coordinates defined in an absolute system—where the X, Y, and Z axes intersect.

Zonal operators, overlay operators for summarizing sets of values within zones—for example, the mean population within a set of large administrative areas.

Zoom, a capability for proportionately enlarging or reducing the scale of a figure or maps displayed on a computer screen or VDU.

References

Abel, D. J. and Mark, D. M. (1990). A comparative analysis of some two-dimensional orderings. *International Journal of Geographical Information Systems*, 4: 21–31.

Agarwal, P. (2005). Ontological consideration in GIScience. *International Journal of Geographical Information Science*, 19: 501–36.

Aguilar, F. J., Agüera, F., Aguilar, M. A., and Carvajal, F. (2005). Effects of terrain morphology, sampling density, and interpolation methods on grid DEM accuracy. *Photogrammetric Engineering & Remote Sensing*, 71: 805–16.

Allaby, M. (2013). *A Dictionary of Geology and Earth Sciences*. Oxford University Press, Oxford.

Alonso, W. A. (1968). *The Quality of Data and the Choice and Design of Predictive Models*. Highway Research Board Special Report 97, pp. 178–92.

Armstrong, M. (1998). *Basic Linear Geostatistics*. Springer-Verlag, Berlin.

Atkinson, P. M. and Lloyd, C. D. (2001). Assessing uncertainty in the United Kingdom Nitrogen Dioxide Monitoring Network with ordinary and indicator kriging. In P. Monestiez, D. Allard, and R. Froidevaux (eds.) *GeoENV III: Geostatistics for Environmental Applications*. Kluwer Academic Publishers, Dordrecht, pp. 33–44.

Atkinson, P. M. and Lloyd, C. D. (2007). Non-stationary variogram models for geostatistical sampling optimisation: an empirical investigation using elevation data. *Computers and Geosciences*, 33 (10): 1285–1300.

Atkinson, P. M. and Lloyd, C. D. (2009). Geostatistics and spatial interpolation. In A. S. Fotheringham and P. A. Rogerson (eds.) *The SAGE Handbook of Spatial Analysis*. SAGE Publications, London, pp. 159–81.

Atkinson, P. M. and Tatnall, A. R. (1997). Introduction: neural networks in remote sensing. *International Journal of Remote Sensing*, 18: 699–709.

Aubrya, T., Luis, L., and Dimuccio, L. A. (2012). Nature vs. culture: present-day spatial distribution and preservation of open-air rock art in the Coa and Douro River Valleys (Portugal). *Journal of Archaeological Science*, 39: 848–66.

Baetens, J. M. and de Baets, B. (2012). Cellular automata on irregular tessellations. *Dynamical Systems*, 27: 411–30.

Bailey, T. C. and Gatrell, A. C. (1995). *Interactive Spatial Data Analysis*. Longman Scientific and Technical, Harlow.

Ball, M. (2010). *What's the distinction between crowdsourcing, volunteered geographic information and authoritative data?*, Spatial Sustain website <http://www.sensysmag.com/spatialsustain/whats-the-distinction-between-crowdsourcing-volunteered-geographic-information-and-authoritative-data.html> (accessed 3/5/2013).

Bannister, A., Raymond, S., and Baker, R. (1998). *Surveying*. 7th edn. Pearson Education, Harlow.

Barreto, C., Fastovsky, D., and Sheehan, P. (2003). A model for integrating the public into scientific research. *Journal of Geoscience Education*, 50: 71–5.

Barrow, J. D. (1992). *Pi in the Sky*. Oxford University Press, Oxford.

Batty, M. (2005) *Cities and Complexity: Understanding Cities with Cellular Automata, Agent-based Models, and Fractals*. MIT Press, Cambridge, MA.

Batty, M. (2008). Virtual reality in GIS. In J. P. Wilson and A. S. Fotheringham (eds.) *Handbook of Geographic Information Science*. Blackwell, Malden, MA, pp. 317–34.

Batty, M., Axhausen, K. W., Giannotti, F., Pozdnoukhov, A., Bazzani, A., Wachowicz, M., Ouzounis, G., and Portugali, Y. (2014). Smart cities of the future. *The European Physical Journal Special Topics*, 214: 481–518.

Batty, M. and Xie, Y. (1994a). Modelling inside GIS. Part 1: model structures, exploratory spatial data analysis, and aggregation. *International Journal of Geographical Information Systems*, 8: 291–307.

Batty, M. and Xie, Y. (1994b). Modelling inside GIS. Part 2: Selecting and calibrating urban models using ARC-INFO. *International Journal of Geographical Information Systems*, 8: 451–70.

Batty, M., Xie, Y., and Sun, Z. (1999). Modeling urban dynamics through GIS-based cellular automata. *Computers, Environment and Urban Systems*, 99: 205–33.

Bedard, Y., Merrett, T., and Han, J. (2009). Fundamentals of spatial data warehousing of geographic knowledge discovery. In H. J. Miller and J. Han (eds.) *Geographic Data Mining and Knowledge Discovery*. CRC Press, Boca Raton, FL, pp. 45–68.

Beek, K. J. (1978). *Land Evaluation for Agricultural Development*. International Institute for Land Reclamation and Improvement, Pub. 23, Wageningen.

Bell, N., Schuurman, N., and Hayes, M. V. (2007). Using GIS-based methods of multicriteria analysis to construct socio-economic deprivation indices. *International Journal of Health Geographics*, 6: 17 <http://www.ij-healthgeographics.com/content/6/1/17> (accessed 11/11/2014).

Beven, K. and Moore, I. D. (eds.) (1994). *Terrain Analysis and Distributed Modelling in Hydrology*. Advances in Hydrological Processes. Wiley, Chichester.

Bezdek, J. C. (2012). *Pattern Recognition with Fuzzy Objective Function Algorithms* (reprint). Springer, London.

Bezdek, J. C., Keller, J., Krishnapuram, R., and Pal, N. R. (1999). *Fuzzy Models and Algorithms for Pattern Recognition and Image Processing*. Springer, New York.

Bivand, R. S., Pebesma, E. J., and Gómez-Rubio, V. (2008). *Applied Spatial Data Analysis with R*. Springer, New York.

Blackie, W. G. (2010). *The Imperial Gazetteer: A General Dictionary of Geography, Physical, Political, Statistical and Descriptive*. Blackie and Sons, Glasgow.

Boerma, P. N., Henneman, G. R., Kauffman, J. H., and Verwey, H. E. (1974). *Detailed Soil Survey of the Marongo area*. Preliminary Report No. 3. Training project in Pedology. Agricultural University, Wageningen.

Bothwell, J. and Yuan, M. (2012). A spatiotemporal GIS framework applied to the analysis of changes in temperature patterns. *Transactions in GIS*, 16 (9): 901–19.

Braimoh, A. K., Vlek, P. L., and Stein, A. (2004). Land evaluation for maize based on fuzzy set and interpolation. *Environmental Management*, 33: 226–38.

Brimicombe, A. (2010). *GIS, Environmental Modeling and Engineering*. 2nd edn. CRC Press, Boca Raton, FL.

Brown, D. G., Riolo, R., Robinson, D. T., North, M., and Rand, W. (2005). Spatial process and data models. Toward integration of agent-based models and GIS. *Journal of Geographical Systems*, 7: 25–47.

Burrough, P. A. (1986). *Principles of Geographical Information Systems for Land Resources Assessment*. Oxford University Press, Oxford.

Burrough, P. A. (1989). Fuzzy mathematical methods for soil survey and land evaluation. *Journal of Soil Science*, 40: 477–92.

Burrough, P. A. (1998). Dynamic modelling and geocomputation. In P. Longley, S. M. Brooks, R. A. McDonnell, and W. D. Macmillan (eds.) *Geocomputation: A Primer*. John Wiley, Chichester, pp. 165–91.

Burrough, P. A. (2001). GIS and geostatistics: essential partners for spatial analysis. *Environmental and Ecological Statistics*, 8: 361–77.

Burrough, P. A. and Frank, A. U. (eds.) (1996). *Geographical Objects with Indeterminate Boundaries*. Taylor & Francis, London.

Burrough, P. A., Karssenberg, D., and van Deursen, W. (2005). Environmental modeling with PC RASTER. In D. J. Maguire, M. Batty, and M. F. Goodchild (eds.) *GIS, Spatial Analysis and Modeling*. ESRI Press, Redlands, CA, pp. 333–56.

Burrough, P. A. and McDonnell, R. A. (1998). *Principles of Geographical Information Systems*. 2nd edn. Oxford University Press, Oxford.

Burrough, P. A., Wilson, J. P., van Gaans, P. M., and Hansen, A. (2001). Fuzzy k-mean classification of topo-climatic data as an aid to forestry mapping in the Greater Yellowstone Area, USA. *Landscape Ecology*, 16: 523–46.

Camara, G., Fonseca, F., Monteiro, A. M., and Onsrud, H. (2006). Networks of innovation and the establishment of a spatial data infrastructure in Brazil. *Information and Technology Development*, 12: 255–72.

Camelli, F., Lien, J.-M., Shen, D., Wong, D. W., Rice, M., Löhner, R., and Yang, C. (2012). Generating seamless surfaces for transport and dispersion modeling in GIS. *GeoInformatica*, 16: 307–27.

Campbell, J. B. and Wynne, R. H. (2011). *Introduction to Remote Sensing*. 5th edn. Guilford Press, New York.

Carver, S. J. (1991). Integrating multi-criteria evaluation with geographical information systems. *International Journal of Geographical Information Systems*, 5: 321–40.

Castella, J.-C., Kam, S. P., Quang, D. D., Verburg, P. H., and Hoanh, C. T. (2007). Combining top-down and bottom-up modelling approaches of land use/cover change to support public policies: Application to sustainable management of natural resources in northern Vietnam. *Land Use Policy*, 24: 531–45.

Castree, N., Kitchin, R., and Rogers, A. (2012). *A Dictionary of Human Geography*. Oxford University Press, Oxford.

Chang, F., Dean, J., Ghemawat, S., Hsieh, W. C., Wallach, D. C., Burrows, M., Chandra, T., Fikes, A., and Gruber, R. A. (2006). Bigtable: A Distributed Storage System for Structured Data <http://static.googleusercontent.com/external_content/untrusted_dlcp/research.google.com/en//archive/bigtable-osdi06.pdf> (accessed 12/4/2013).

Chernoff, H. (1973). The use of faces to represent points in *k*-dimensional space graphically. *Journal of the American Statistical Association*, 68: 361–8.

Chilès, J. P. and Delfiner, P. (2012). *Geostatistics: Modelling Spatial Uncertainty*. 2nd edn. Wiley, New York.

Chrisman, N. R. (1999). What does GIS mean? *Transactions in GIS*, 3: 172–86.

Cinderby, S., Snell, C., and Forrester, J. (2013). Participatory GIS and its application in governance: the example of air quality and the implications for noise pollution. *Local Environment: The International Journal of Justice and Sustainability*, 13: 309–20.

Clarke, K. C. and Gaydos, L. J. (1998). Loose-coupling cellular automaton model and GIS: long-term urban growth prediction for San Francisco and Washington/Baltimore. *International Journal of Geographical Information Science*, 12 (7): 699–714.

Cockings, S., Martin, D., and Leung, S. (2010). Population 24/7: building space–time specific population surface models. In M. Hakley, J. Morley, and H. Rahemtulla (eds.) *Proceedings of the GIS Research UK 18th Annual Conference GISRUK 2010. University College* London, London, pp. 41–8.

Codd, E. F. (1968). *Cellular Automata*. Academic Press, New York.

Coleman, A. M. and Webber, W. D. (2008). GeoSpatial infrastructure at the US Department of Energy's Hanford site: a review of existing conditions and a proposed action for development of a spatial data infrastructure. *Journal of Map and Geography Libraries*, 4: 83–95.

Collin, A., Bernardin, D., and Sero-Guillaume, O. (2011). A physical-based cellular automaton model for forest fire propagation. *Combustion Science and Technology*, 183: 347–69.

Connors, J. P., Lei, A., and Kelly, M. (2013). Citizen science in the age of neogeography: utilizing volunteered geographic information for environmental monitoring. *Annals of the Association of American Geographers*, 102: 1267–89.

Cooper, A. K., Moellering, H., Hjelmager, J., Rapant, P., Delgado, T., Laurent, D., Coetzee, S., Danko, D. M., Düren, U., Iwaniak, A., Brodeur, J., Abad, P., Huet, M., and Rajabifard, A. (2013). A spatial data infrastructure model from the computational viewpoint. *International Journal of Geographical Information Science*, 27 (6): 1133–51, DOI: 10.1080/13658816.2012.741239

Costa-Cabral, M. and Burges, S. J. (1993). Digital Elevation Model Networks (DEMON): a model of flow over hillslopes for computation of contributing and dispersal areas. *Water Resources Research*, 30 (6): 1681–92.

Couclelis, H. (1992). People manipulate objects (but cultivate fields): beyond the raster-vector debate in GIS. In A. U. Frank, I. Campari, and U. Formentini (eds.) *Theories and Methods of Spatio-Temporal Reasoning in Geographic Space*. Lecture Notes in Computer Science 639. Springer Verlag, Berlin, pp. 65–77.

Cox, E., Taber, R., and O'Hagan, M. (1999). *The Fuzzy Systems Handbook: A Practitioner's Guide to Building, Using, and Maintaining Fuzzy Systems*. CRC Press, Boca Raton, FL.

Crampton, J. W. (2009). Cartography: performative, participatory, political. *Progress in Human Geography*, 33: 840–8.

Cromley, E. K. and McLafferty, S. L. (2011). *GIS and Public Health*. 2nd edn. Guilford Press, New York.

Crooks, A., Castle, C., and Batty, M. (2008). Key challenges in agent-based modelling for geo-spatial simulation. *Computers, Environment and Urban Systems*, 32: 417–30.

Crooks, A. T. and Wise, S. (2013). GIS and agent-based models for humanitarian assistance. *Computers, Environment and Urban Systems*, 41: 100–11.

Crosetto, M. and Tarantola, S. (2001). Uncertainty and sensitivity analysis: tools for GIS-based model implementation. *International Journal of Geographical Information Science*, 15: 415–37.

Dahl, O.-J. and Nygaard, K. (1966). SIMULA—an algol-based simulation language. *Communications of the ACM*, 9: 671–8.

Date, C. J. (1995). *An Introduction to Database Systems*. 6th edn. Addison-Wesley, Reading, MA.

Davis, J. C. (2002). *Statistics and Data Analysis in Geology*. 3rd edn. John Wiley & Sons, New York.

de Smith, M. J., Goodchild, M. F., and Longley, P. A. (2007). *Geospatial Analysis: A Comprehensive Guide to Principles, Techniques and Software Tools*. 2nd edn. Matador, Leicester.

Densham, P. J. and Armstrong, M. P. (2003). Integrative data structures for collaborative modeling and visualisation in spatial decision support systems. In C. M. B. Medeiros (ed.) Advanced Geographic Information Systems, in *Encyclopedia of Life Support Systems (EOLSS)*. Developed under the Auspices of the UNESCO. EOLSS Publishers, Oxford <http://www.eolss.net> (accessed 11/11/2014).

Deutsch, C. and Journel, A. G. (1998). *GSLIB: Geostatistical Software Library and User's Guide*. 2nd edition. Oxford University Press, New York.

Devillers, R., Stein, A., Bédard, Y., Chrisman, N., Fisher, P., and Wenzhong, S. (2010). Thirty years of research on spatial data quality: achievements, failures, and opportunities. *Transactions in GIS*, 14: 387–400.

Di Baldassarre, G. (2012). *Floods in a Changing Climate: Inundation Modelling*. Cambridge University Press, Cambridge.

Diggle, P. (2003). *Statistical Analysis of Spatial Point Patterns*. 2nd edition. Arnold, London.

Dodge, M. and Kitchin, R. (2013). Crowd-sourced cartography: mapping experience and knowledge. *Environment and Planning A*, 45: 19–36.

Dodge, M., Kitchin, R., and Perkins, C. (eds.) (2011). *The Map Reader: Theories of Mapping Practice and Cartographic Representation*. John Wiley, Winchester.

Doherty, P., Guo, Q., Liu, Y., Wieczorek, J., and Doke, J. (2011). Georeferencing incidents from locality descriptions and its applications: a case study from yosemite national park search and rescue. *Transactions in GIS*, 15: 775–93.

Dorling, D. (1994). Cartograms for visualizing human geography. In H. M. Hearnshaw and D. J. Unwin (eds.) *Visualization in Geographical Information Systems*. Wiley, Chichester, pp. 85–102.

Douglas, D. H. and Peucker, T. K. (1973). Algorithms for the reduction of the number of points required to represent a digitized line or its caricature. *Canadian Cartographer,* 10 (2): 112–22.

Duffield, B. S. and Coppock, J. T. (1975). The delineation of recreational landscapes: the role of a computer-based information system. *Transactions of the Institute of British Geographers,* 66: 141–8.

Dungan, J. (1998). Spatial prediction of vegetation quantities using ground and image data. *International Journal of Remote Sensing,* 19: 267–85.

Dunn, C. E. (2007). Participatory GIS—a people's GIS? *Progress in Human Geography,* 31: 616–37

Dykes, J., MacEachren, A. M., and Kraak, M.-J. (eds.) (2005). *Exploring Geovisualization.* Elsevier, Amsterdam.

Ebdon, D. (1985). *Statistics in Geography.* 2nd edition. Blackwell, Oxford.

Elwell, H. A. and Stocking, M. A. (1982). Developing a simple yet practical method of soil-loss estimation. *Tropical Agriculture,* 59: 43–8.

Elwood, S. (2006a). Critical issues in participatory GIS: deconstructions, reconstructions, and new research directions. *Transactions in GIS,* 10: 693–708.

Elwood, S. (2006b). Negotiating knowledge production: the everyday inclusions, exclusions, and contradictions of participatory GIS research. *The Professional Geographer,* 58: 197–208.

Elwood, S. (2008). Grassroots groups as stakeholders in spatial data infrastructures: challenges and opportunities for local data development and sharing. *International Journal of Geographic Information Science,* 22: 71–90.

Elwood, S. (2011). Participatory approaches in GIS and society research: foundations, practices, and future directions. In R. McMaster, H. Couclelis, and T. Nyerges (eds.) *The SAGE Handbook of GIS and Society.* Sage Publications, London, pp. 381–99.

Elwood, S. and Leszczynski, A. (2011). Privacy reconsidered: new representations, data practices, and the geoweb. *Geoforum,* 42: 6–15.

Elwood, S. and Leszczynski, A. (2012). New spatial media, new knowledge politics. *Transactions of the Institute of British Geographers,* 38: 544–59.

Evans, I. S. (1980). An integrated system of terrain analysis and slope mapping. *Zeitschrift fur Geomorphologie Suppl.* 36: 274–95.

Evans, S., Hudson-Smith, A., and Batty, M. (2012). 3-D GIS: Virtual London and beyond. *Cybergeo: European Journal of Geography,* Selection of items from SAGEO 2005, article 359 <http://cybergeo.revues.org/2871> (accessed 11/11/2014); DOI: 10.4000/cybergeo.2871

FAO (1976). A framework for land evaluation. *Soils Bull.* 32. FAO Rome and International Institute for Land Reclamation and Improvement, Pub. 22.

Farr, T. G., Rosen, P. A., Caro, E., Crippen, R., Duren, R., Hensley, S., Kobrick, M., Paller, M., Rodriguez, E., Roth, L., Seal, D., Shaffer, S., Shimada, J., Umland, J., Werner, M., Oskin, M., Burbank, D., and Alsdorf, D. (2007). The shuttle radar topography mission. *Reviews of Geophysics,* 45: RG2004.

Fisher, P. F. (1995). An exploration of probable viewsheds in landscape planning. *Environment and Planning B: Planning and Design,* 22: 527–46.

Fisher, W. D. (1958). On grouping for maximum homogeneity. *Journal of the American Statistical Society,* 53: 789–98.

Fotheringham, A. S., Brunsdon, C., and Charlton, M. (2002). *Geographically Weighted Regression: The Analysis of Spatially Varying Relationships.* John Wiley & Sons, Chichester.

Fowler, A., Whyatt, J. D., Davies, G., and Ellis, R. (2013). How reliable are citizen-derived scientific data? Assessing the quality of contrail observations made by the general public. *Transactions in GIS,* 17: 488–506.

Fried, J. S., Brown, D. G., Zweifler, M. O., and Gold, M. A. (2000). Mapping contributing areas for stormwater discharge to streams using terrain analysis. In J. P. Wilson and J. C. Gallant (eds.) *Terrain Analysis: Principles and Applications.* John Wiley, New York, pp. 183–203.

Frihida, A., Marceau, D. J., and Theriault, M. (2002). Spatio-temporal object-oriented data model for disaggregate travel behavior. *Transactions in GIS,* 6: 277–94.

Gahegan, M. (1989). An efficient use of quadtrees in geographical information systems. *International Journal of Geographical Information Systems,* 3: 201–14.

Gahegan, M. (2008). Multivariate geovisualization. In J. P. Wilson and A. S. Fotheringham (eds.) *Handbook of Geographic Information Science.* Blackwell, Malden, MA, pp. 292–316.

Gallay, M., Lloyd, C. D., McKinley, J., and Barry, L. (2013). Assessing modern ground survey methods and airborne laser scanning for digital terrain modelling: a case study from the Lake District, England. *Computers and Geosciences,* 51: 216–27.

Geertman, S., Hagoort, M., and Ottens, H. (2007). Spatial-temporal specific neighbourhood rules for cellular automata land-use modelling. *International Journal of Geographical Information Systems,* 21: 547–68.

Gelernter, J. and Mushegian, N. (2011). Geo-parsing messages from microtext. *Transactions in GIS,* 15: 753–73.

George, M. (2007). *Fuzzy Mathematics: Applications in Economics.* Campus Books International.

Ghermandi, A. and Nunes, P. A. L. D. (2013). A global map of coastal recreation values: results from a spatially explicit meta-analysis. *Ecological Economics,* 86: 1–15.

Gimblett, R. (ed.) (2002). *Integrating Geographic Information Systems and Agent-based modeling techniques for simulating social and ecological process*. Sante Fe Institute Studies in the Sciences of Complexity. Oxford University Press, New York.

Goldberg, A. and Robson, D. (1983). *Smalltalk-80*. Addison-Wesley, Reading, MA.

Goodchild, M. F. (1989). Modeling error in objects and fields. In M. F. Goodchild and S. Gopal (eds.) *Accuracy of Spatial Databases*. Taylor & Francis, London, pp. 107–13.

Goodchild, M. F. (1992). Geographical Information Science. *International Journal of Geographical Information Systems*, 6: 31–45.

Goodchild, M. F. (2001). Geographic information systems. In N. J. Smelser and P. B. Baltes (eds.) *International Encyclopedia of the Social and Behavioral Sciences*. Pergamon, Oxford, pp. 6175–82 [368].

Goodchild, M. F. (2007a). Citizens as sensors: the world of volunteered geography. *Geojournal*, 69: 211–22.

Goodchild, M. F. (2007b). Towards a general theory of geographic representation in GIS. *International Journal of Geographical Information Science*, 21: 239–60.

Goodchild, M. F. (2009). NeoGeography and the nature of geographic expertise. *Journal of Location Based Services*, 3: 82–96.

Goodchild, M. F. (2010). Twenty years of progress: GIScience in 2010. *Journal of Spatial Information Science*, 1: 3–20.

Goodchild, M. F., Steyaert, L. T., Parks, B. O., Johnston, C., Maidment, D., Crane, M., and Glendinning, S. (eds.) (1996). *GIS and Environmental Modeling: Progress and Research Issues*. GIS World Books, Fort Collins, CO.

Goovaerts, P. (1997). *Geostatistics for Natural Resources Evaluation*. Oxford University Press, New York.

Gorsevski, P. V., Gessler, P. E., and Jankowski, P. (2003). Integrating a fuzzy *k*-means classification and Bayesian approach for spatial prediction of landslide hazard. *Journal of Geographical Systems*, 5: 223–51.

Goswami, R., Brocklehurst, S. H., and Mitchell, N. C. (2012). Erosion of a tectonically uplifting coastal landscape, NE Sicily, Italy. *Geomorphology*, 171: 114–26.

Graham, M. (2009). Neography and the palimpsests of place: Web 2.0 and the construction of a virtual earth. *Tijdscrift voor Economische en Social geographie*, 101: 422–36.

Graham, M., Stephens, M., and Hale, S. (2013). Mapping the geoweb: a geography of Twitter. *Environment and Planning A*, 45 (1): 100–2.

Gregory, I. N. and Hardie, A. (2011). Visual GISting: bringing together corpus linguistics and geographical information systems. *Literary and Linguistic Computing*, 26: 297–314.

Gret-Regamey, A. and Straub, D. (2006). Spatially explicit avalanche risk assessment linking Bayesian networks to a GIS. *Natural Hazards and Earth System Science*, 6: 911–26.

Guo, Q., Liu, Y., and Wieczorek, J. (2008). Georeferencing locality descriptions and computing associated uncertainty using a probabilistic approach. *International Journal of Geographical Information Science*, 22: 1067–90.

Hagen-Zanker, A., Straatmen, B., and Uljee, I. (2005). Further developments of a fuzzy set map comparison approach. *International Journal of Geographical Information Science*, 19: 769–85.

Haining, R. (2003). *Spatial Data Analysis: Theory and Practice*. Cambridge University Press, Cambridge.

Haklay, M. (2010). How good is volunteered geographical information? A comparative study of OpenStreetMap and Ordnance Survey datasets. *Environment and Planning B: Planning and Design*, 37: 682–703.

Haklay, M. (2013). Neogeography and the delusion of democratisation. *Environment and Planning A*, 45 (1): 55–69.

Haklay, M., Basiouka, S., Antoniou, V., and Ather, A. (2010). How many volunteers does it take to map an area well? The validity of Linus' Law to volunteered geographic information. *The Cartographic Journal*, 47: 315–22.

Harbaugh, A. W. (2005). *MODFLOW-2005, the U.S. Geological Survey Modular Ground-Water Model—the Ground-Water Flow Process*. Techniques and Methods 6-A16. U.S. Geological Survey.

Harley, J. B. (1988). Maps, knowledge, and power. In D. Cosgrove and S. Daniels (eds.) *The Iconography of Landscape: Essays on the Symbolic Representation, Design and Use of Past Environments*. Cambridge University Press, Cambridge, pp. 277–312.

Harley, J. B. (1989). Deconstructing maps. *Cartographica*, 26: 1–20.

Harris, R. and Jarvis, C. (2011). *Statistics for Geography and Environmental Science*. Prentice Hall, Harlow.

Harris, R., Sleight, P., and Webber, R. (2005). *Geodemographics: GIS and Neighbourhood Targeting*. Wiley, Chichester.

Harrower, M. and Brewer, C. (2003). ColorBrewer.org: an online tool for selecting colour schemes for maps. *The Cartographic Journal*, 40: 27–37.

Harvey, F. (2008). *A Primer of GIS: Fundamental Geographic and Cartographic Concepts*. The Guildford Press, New York.

Harvey, F. and Tulloch, D. (2006). Local-government data sharing: evaluating the foundations of spatial data infrastructures. *International Journal of Geographical Information Science*, 20: 743–68.

Hendriks, P. H. J., Dessers, E., and van Hootegem, G. (2012). Reconsidering the definition of a spatial data infrastructure. *International Journal of Geographical Information Science*, 26 (8): 1479–94.

Hengl, T., Heuvelink, G. B. M., and Rossiter, D. G. (2007) About regression-kriging: from equations to case studies. *Computers and Geosciences*, 33: 1301–15.

Heppenstall, A. J., Crooks, A. T., See, L. M., and Batty, M. (eds.) (2012). *Agent-based Models of Geographical Systems*. Springer Sciences, Dordrecht.

Heuvelink, G. B. M. (1998). *Error Propagation in Environmental Modelling with GIS*. Taylor & Francis, London.

Heuvelink, G. B. M. and Burrough, P. A. (1993). Error propagation in cartographic modeling using Boolean logic and continuous classification. *International Journal of Geographical Information Systems*, 7: 231–46.

Heuvelink, G. B. M., Burrough, P. A., and Stein, A. (1989). Propagation of errors in spatial modelling with GIS. *International Journal of Geographical Information Systems*, 3 (4): 303–22.

Heywood, I., Cornelius, S., and Carver, S. (2011). *An Introduction to Geographical Information Systems*. 4th edn. Pearson, Harlow.

Hills, G. A. (1961). *The Ecological Basis for Land Use Planning*. Research Report 46. Ontario Department of Lands and Forests, Canada.

Hodgkiss, A. G. (1981). *Understanding Maps*. Dawson, Folkstone.

Hollenstein, L. and Purves, R. S. (2010). Exploring place through user-generated content: using Flickr tags to describe city cores. *Journal of Spatial Information Science*, 1: 21–48.

Hopkins, L. D. (1977). Methods for generating land suitability maps: a comparative evaluation. *Journal of the American Institute of Planners*, 43 (4): 386–400.

Horn, B. K. P. (1981). Hill shading and the reflectance map. *Proceedings of the IEEE*, 69 (1): 14–47.

Howe, J. (2009). *Crowdsourcing: How the Power of the Crowd is Driving the Future of Business*. Random House Business Books, London.

Hudson-Smith, A. and **Crooks, A. T.** (2009). The renaissance of geographic information: neogeography, gaming and second life. In H. Lin and M. Batty (eds.) *Virtual Geographic Environments*. Science Press, Beijing, pp. 25–36. (Reprinted by ESRI Press 2011).

Hutchinson, M. F. (1989). A new procedure for gridding elevation and stream line data with automatic removal of spurious pits. *Journal of Hydrology*, 106: 211–32.

Hutchinson, M. F. (1995). Interpolating mean rainfall using thin plate smoothing splines. *International Journal of Geographical Information Systems*, 9: 385–404.

Imam, E. and Kushwaha, S. P. S. (2013). Habitat suitability modeling for Gaur (Bos gaurus) using multiple logistic regression, remote sensing and GIS. *Journal of Applied Animal Research*, 41: 189–99.

Isaaks, E. H. and Srivastava, R. M. (1989). *An Introduction to Applied Geostatistics*. Oxford University Press, New York.

Jaber, S. M., Ibrahim, K. M., and Al-Muhtaseb, M. (2013). Comparative evaluation of the most common kriging techniques for measuring mineral resources using geographic information systems. *GIScience & Remote Sensing*, 50: 93–111.

Janelle, D. G. (2012). Space-adjusting technologies and the social ecologies of place: review and research agenda. *International Journal of Geographical Information Science*, DOI:10.1080/13658816.2012.713958.

Janowicz, K. (2012). Observation-driven geo-ontology engineering. *Transactions in GIS*, 16: 351–74.

Jenks, G. F. and Caspall, F. C. (1971). Error on choroplethic maps: definition, measurement, reductions. *Annals of the Association of American Geographers*, 61: 217–44.

Jetten, V. (1994). *Modelling the Effects of Logging on the Water Balance of a Tropical Rain Forest: A Study in Guyana*. Tropenbos Series 6. University of Utrecht, Utrecht.

Jiang, B. and Liu, X. (2012). Scaling of geographic space from the perspective of city and field blocks and using volunteered geographic information. *International Journal of Geographical Information Science*, 26: DOI: 10.1080/13658816.2011.575074

Johnston, K. M. (ed.) (2013). *Agent Analyst: Agent-based Modeling in ArcGIS*. ESRI Press, Redlands, CA.

Jones, C. (1997). *Geographical Information Systems and Computer Cartography*. Longman, Harlow.

Jones, K. H. (1998). A comparison of algorithms used to compute hill slopes and aspects as a property of the DEM. *Computers and Geosciences*, 24: 315–23.

Jones, L. D. (2006). Monitoring landslides in hazardous terrain using terrestrial LiDAR; an example from Montserrat. *Quarterly Journal of Engineering Geology and Hydrogeology*, 39: 371–3.

Jongman, R. H. G., ter Braak, C. J. F., and van Tongeren, O. F. R. (1995). *Data Analysis in Community and Landscape Ecology*. Cambridge University Press, Cambridge.

Jorgenson, S. E. (2009). Environmental models and simulation, in Environmental Systems II. InBT *Encyclopedia of Life Support Systems* (*EOLSS*). Developed under the auspices of the UNESCO. EOLSS Publishers, Oxford <http://www.eolss.net> (accessed 11/11/2014).

Kang, B., He, D. M., Perrett, L., Wang, H. Y., Hu, W. X., Deng, W. D., and Wu, Y. F. (2009). Fish and fisheries in the Upper Mekong: current assessment of the fish community, threats and conservation. *Reviews in Fish Biology and Fisheries*, 19: 465–80.

Karssenberg, D., Burrough, P. A., Sluiter, R., and de Jong, K. (2001). The PCRaster software and course materials for teaching numerical modelling in the environmental sciences. *Transactions in GIS*, 5: 99–110.

Karssenberg, D. and De Jong, K. (2005). Dynamic environmental model in GIS: 2. Error propagation. *International Journal of Geographical Information Science*, 19: 623–37.

Karssenberg, D., De Jong, K., and Van der Kwast, J. (2007). Modelling landscape dynamics with Python. *International Journal of Geographical Information Science*, 21: 483–95.

Kerry, R., Oliver, M. A., and Frogbrook, Z. L. (2010). Sampling in precision agriculture. In M. A. Oliver (ed.) *Geostatistical Applications for Precision Agriculture*. Springer, Dordrecht, pp. 35–63.

Khalema-Malebese, L. and Ahmed, F. (2012). *Hydrological Flow Modelling Using Geographic Information Systems (GIS): Application of GIS to Model the Flow and Accumulation of Water in order to Determine Suitable Dam Sites*. Lambert Academic Press, Saarbrücken, Germany.

Kia, M. B., Pirasteh, S., Pradhan, B., Mahmud, A. R., Sulaiman, W. N. A., and Moradi, A. (2012) An artificial neural network model for flood simulation using GIS: Johor River Basin, Malaysia. *Environmental Earth Sciences*, 67: 251–64.

Kim, D. and Batty, M. (2011). *Modeling Urban Growth: An Agent-based Microeconomic Approach to Urban Dynamics and Spatial Policy Simulation*. UCL Working Paper Series, Paper 165. University College London, London.

Klir, G. J., St Clair, Y. H., and Yuan, B. (1998). *Fuzzy Set Theory: Foundations and Applications*. Prentice Hall, Englewood Cliffs, NJ.

Kollias, V. J. and Voliotis, A. (1991). Fuzzy reasoning in the development of geographical information systems. Frsis: a prototype soil information system with fuzzy retrieval capabilities. *International Journal of Geographical Information Systems*, 5: 209–24.

Koomen, E., Stillwell, J., Bakema, A., and Scholten, H. J. (eds.) (2007). *Modelling Land-Use Change: Progress and Applications*. Springer, Dordrecht.

Kosko, B. (1994). *Fuzzy Thinking: The New Science of Fuzzy Logic*. Harper Collins, London.

Kothuri, R. K. V., Ravada, S., and Abugov, D. (2002). Quadtree and r-tree indexes in oracle spatial: a comparison using GIS data. In *Proceedings of ACM SIGMOD Conference 4–6 June 2002, Madison, WI*, pp. 546–57.

Kraak, M.-J. (2008). Visualising spatial distributions. In P. A. Longley, M. F. Goodchild, D. J. Maguire, and D. W. Rhind (eds.) *Geographical Information Systems: Principles, Techniques, Management, and Applications*. 2nd edn, abridged. Wiley, Hoboken, NJ, pp. 49–65.

Krygier, J. B. (2002). A praxis of public participation GIS and visualization. In T. M. Harris, W. J. Craig, and D. Weiner (eds.) *Community Participation and Geographical Information Systems*. CRC Press, Boca Raton, FL, pp. 330–45.

Kubiak, M. and Dzieszko, P. (2012). Using thermal remote sensing in environmental studies. *Transactions in GIS*, 16: 715–32.

Lagacherie, P., Andrieux, P., and Bouzigues, R. (1996). *Fuzziness and uncertainty of soil boundaries: from reality to coding in GIS*. In P. A. Burrough and A. U. Frank (eds.) *Geographical Objects with Indeterminate Boundaries*. Taylor & Francis, London, pp. 275–86.

Langran, G. (1992). *Time in Geographic Information Systems*. Taylor & Francis, London.

Lees, B. G. (1996). Sampling strategies for machine learning using GIS. In M. F. Goodchild, L. T. Steyaert, B. O. Parks, C. Johnston, D. Maidment, M. Crane, and S. Glendinning (eds.) *GIS and Environmental Modeling: Progress and Research Issues*. GIS World Books, Fort Collins, CO, pp. 39–42.

Li, Z., Zhu, Q., and Gold, C. (2004). *Digital Terrain Modeling: Principles and Methodology*. CRC Press, Boca Raton, FL.

Lillesand, T. M., Kiefer, R. W., and Chipman, J. W. (2008). *Remote Sensing and Image Interpretation*. 6th edn. Wiley, Hoboken, NJ.

Lin, H. and Batty, M. (eds.) (2009a). *Virtual Geographic Environments*. Science Press, Beijing.

Lin, H. and Batty, M. (2009b). *Virtual geographic environments: a primer*. In H. Lin and M. Batty (eds.) *Virtual Geographic Environments*. Science Press, Beijing, pp. 1–10.

Liu, X., Ma, L., Li, X., Ai, B., Li, S., and He, Z. (2013). Simulating urban growth by integrating landscape expansion index (LEI) and cellular automata. *International Journal of Geographical Information Science*, DOI: 10.1080/13658816.2013.831097

Liu, Y. (2008). *Modelling Urban Development with Geographical Information Systems and Cellular Automata*. CRC Press, Boca Raton, FL.

Lloyd, C. D. (2004). Landform and earth surface. In P. M. Atkinson (ed.) Geoinformatics, in *Encyclopedia of Life Support Systems (EOLSS)*. Developed under the auspices of the UNESCO. EOLSS Publishers, Oxford. <http://www.eolss.net> (accessed 11/11/2014).

Lloyd, C. D. (2005). Assessing the effect of integrating elevation data into the estimation of monthly precipitation in Great Britain. *Journal of Hydrology*, 308: 128–50.

Lloyd, C. D. (2010a). Nonstationary models for exploring and mapping monthly precipitation in the United Kingdom. *International Journal of Climatology*, 30: 390–405.

Lloyd, C. D. (2010b). *Spatial Data Analysis: An Introduction for GIS Users*. Oxford University Press, Oxford.

Lloyd, C. D. (2011). *Local Models for Spatial Analysis*. 2nd edn. CRC Press, Boca Raton, FL.

Lloyd, C. D. (2014). *Exploring Spatial Scale in Geography.* John Wiley and Sons, Chichester.

Lloyd, C. D. and Atkinson, P. M. (2004). Increased accuracy of geostatistical prediction of nitrogen dioxide in the United Kingdom with secondary data. *International Journal of Applied Earth Observation and Geoinformation,* 5: 293–305.

Lloyd, C. D. and Lilley, K. D. (2009). Cartographic veracity in medieval mapping: analyzing geographical variation in the Gough Map of Great Britain. *Annals of the Association of American Geographers,* 99: 27–48.

Lohr, U. (1998) Digital elevation models by laser scanning. *Photogrammetric Record,* 16: 105–9.

Long, J. A. and Nelson, T. A. (2012). A review of quantitative methods for movement data. *International Journal of Geographical Science,* DOI: 10.1080/13658816.2012.682578

Longley, P. A., Goodchild, M. F., Maguire, D. J., and Rhind, D. W. (2010). *Geographic Information Systems and Science.* 3rd edn. Wiley, Hoboken, NJ.

Maas, K. C. and Levs, J. (2012). *Newspaper sparks outrage for publishing names, addresses of gun permit holders.* CNN website <http://edition.cnn.com/2012/12/25/us/new-york-gun-permit-map/index.html> (accessed 3/9/2013).

McDonnell, R. A. (1996). Including the spatial dimension: a review of the use of GIS in hydrology. *Progress in Physical Geography,* 21: 159–77.

McDonnell, R. A. (2000). Hierarchical modelling of the environmental impacts of river impoundment based on a GIS. *Hydrological Processes,* 14: 2123–42.

McDonnell, R. A. (2008). Challenges for integrated water resources management: how do we provide the knowledge to support truly integrated thinking? *International Journal of Water Resources Development,* 24: 131–43.

MacEachren, A. M., Robinson, A., Hopper, S., Gardner, S., Murray, R., Gahegan, M., and Hetzler, E. (2005). Visualizing geospatial information uncertainty: what we know and what we need to know. *Cartography and Geographic Information Science,* 32: 139–60.

McHarg, I. L. (1969). *Design with Nature.* Doubleday/Natural History Press, New York.

McKinney, D. C. and Cai, X. (2002). Linking GIS and water resources management models: an object-oriented method. *Environmental Modelling and Software,* 17: 413–25.

Maguire, D., Batty, M., and Goodchild, M. F. (eds.) (2005). *GIS, Spatial Analysis and Modeling.* Environmental Systems Research Institute Inc., Redlands, CA.

Malczewski, J. (1999). *GIS and Multicriteria Decision Analysis.* Wiley, New York.

Malczewski, J. (2006). GIS-based multicriteria decision analysis: a survey of the literature. *International Journal of Geographical Information Science,* 20: 703–26.

Maling, D. H. (1992). *Coordinate Systems and Map Projections.* Pergamon, Oxford.

Manolopoulos, Y., Nanopoulos, A., Papadopoulos, A. N., and Theodoridis, Y. (2010). *R-Trees: Theory and Applications.* Springer Verlag, London.

Manson, S. M. and Evans, T. (2007). Agent-based modeling of deforestation in southern Yucatan, Mexico, and reforestation in the Midwest United States. *Proceedings of National Academy of Science,* 104: 20678–83.

Mark, D. M., Lauzon, J. P., and Cebrian, J. A. (1989). A review of quadtree-based strategies for interfacing coverage data with digital elevation models in grid form. *International Journal of Geographical Information Systems,* 3: 3–14.

Martin, D. (1996). An assessment of surface and zonal models of population. *International Journal of Geographical Information Systems,* 10: 973–89.

Martin, D., Dorling, D., and Mitchell, R. (2002). Linking censuses through time: problems and solutions. *Area,* 34: 82–91.

Martin, D. J., Lloyd, C. D., and Shuttleworth, I. G. (2011) Evaluation of gridded population models using 2001 Northern Ireland census data. *Environment and Planning A,* 43 (8): 1965–80.

Mather, P. and Koch, M. (2011). *Computer Processing of Remotely-Sensed Images: An Introduction.* 4th edn. Wiley-Blackwell, Oxford.

Meiri, S., Dayan, T., Simberloff, D., and Grenyer, R. (2009). Life on the edge: carnivore body size variation is all over the place. *Proceedings of the Royal Society B,* 276 (1661): 1469–76.

Meng, Q., Liu, Z., and Borders, B. E. (2013). Assessment of regression kriging for spatial interpolation—comparisons of seven GIS interpolation methods. *Cartography and Geographic Information Science,* 40: 28–39.

Mennis, J. (2010). Multidimensional map algebra: design and implementation of a spatio-temporal GIS processing language. *Transactions in GIS,* 14: 1–21.

Middelkoop, H. (2000). Heavy-metal pollution of the river Rhine and Meuse floodplains in the Netherlands. *Geologie en Mijnbouw / Netherlands Journal of Geosciences,* 79 (4): 411–28.

Miesch, A. T. (1975). Variograms and variance components in geochemistry and ore evaluation. *Geological Society of America Memoir,* 142: 333–40.

Mitasova, H., Mitas, L., Brown, W. M., Gerdes, D. P., Kosinovsky, I., and Baker, T. (1995). Modelling spatially and temporally distributed phenomena: new methods and tools for GRASS GIS. *International Journal of Geographical Information Systems,* 9: 433–46.

Monmonier, M. (1990). Strategies for the visualization of geographic time-series data. *Cartographica,* 37: 30–45.

Monmonier, M. (1993). *How to Lie with Maps.* University of Chicago Press, Chicago.

Moore, I. D. (1996). Hydrological modelling and GIS. In M. F. Goodchild, L. T. Steyaert, B. O. Parks, C. Johnston, D. Maidment, M. Crane, and S. Glendinning (eds.) *GIS and Environmental Modeling: Progress and Research Issues.* GIS World Books, Fort Collins, CO, pp. 143–8.

Moore, I. D., Turner, A. K., Wilson, J. P., Jenson, S., and Band, L. (1993). GIS and land-surface-subsurface process modelling. In M. F. Goodchild, B. O. Parks, and L. T. Steyaert (eds.) *Environmental Modeling with GIS.* Oxford University Press, New York, pp. 196–230.

Moran, P. A .P. (1950). Notes on continuous stochastic phenomena. *Biometrika*, 37: 17–23.

Mordeson, J. N. and Nair, P. S. (2001). *Fuzzy Mathematics: An Introduction for Engineers and Scientists.* 2nd edn. Studies in Fuzziness and Soft Computing, Vol. 20. Physica Verlag, Heiderlberg.

Muralikrishnan, S., Muralikrishna, I. V., Manjunath, A. S., and Rao, K. M. M. (2007). Spatially integrated approach for terrain modelling and analysis for mobile communication applications. *Geocarto International*, 22: 297–307.

National Geodetic Survey, *Frequently Asked Questions* <http://www.ngs.noaa.gov/faq.shtml> (accessed 23/8/2013).

Nedovic-Budic, Z., Crompvoets, J., and Georgiadou, Y. (eds.) (2011). *Spatial Data Infrastructures in Context: North and South.* CRC Press, London.

Newman, G., Zimmerman, D., Crall, A., Laituri, M., Graham, J., and Stapel, L. (2010). User-friendly web mapping: Lessons from a Citizen Science Website. *International Journal of Geographical Information Science*, 24 (12): 1851–69.

Nonejuie, G. (2010). *An analysis of the role of dynamic participatory Geographical information systems in supporting different water stakeholders in Rayong Basin Thailand.* Unpublished DPhil thesis, University of Oxford.

Oberthür, T., Dobermann, A., and Aylward, M. (2000). Using auxiliary information to adjust fuzzy membership functions for improved mapping of soil qualities. *International Journal of Geographical Information Science*, 14: 431–54.

OCHA (2011). *Libya Crisis Map* <http://libyacrisismap.net> (accessed 18/10/2011).

Oliver, A., Montero, G., Montenegro, R., Rodríguez, E., Escobar, J. M., and Perez-Foguet, A. (2012). Finite element simulation of a local scale air quality model over complex terrain. *Advances in Science and Research*, 8: 105–13.

O'Sullivan, D. and Perry, G. L. W. (2013). *Spatial Simulation: Exploring Pattern and Process.* Wiley-Blackwell, Chichester.

O'Sullivan, D. and Unwin, D. J. (2010). *Geographic Information Analysis.* 2nd edn. Wiley, Hoboken, NJ.

Oxera (2013). *What is the economic impact of geoservices?* Oxera Consulting <http://www.oxera.com/Publications/Reports/2013/What-is-the-economic-impact-of-Geo-services-.aspx> (accessed 12/9/2013).

Pagelow, M. and Olmedo, M. T. C. (2005). Possibilities and limits of prospective GIS land cover modeling—a compared case study: Garrotxes (France) and Alta Alpujarra Granadina (Spain). *International Journal of Geographical Science*, 19: 697–722.

Pan, M., Li, Z., Gao, Z., Yang, Y., and Wu, G. (2012). 3-D geological modeling—concept, methods and key techniques. *Acta Geologica Sinica* (English edition), 86: 1031–6.

Pavlidis, T. (1982). *Algorithms for Graphics and Image Processing.* Springer Verlag, Berlin.

Pebesma, E. and Wesseling, C. G. (1998). GSTAT, a program for geostatistical modelling, prediction and simulation. *Computers and Geosciences*, 24: 17–31.

Perkins, C. (2013). Plotting practices and politics: (im)mutable narratives in OpenStreetMap. *Annals of the American Association of Geographers*, DOI: 10.1111/tran.12022

Peucker, T. K., Fowler, R. J., Little, J. J., and Mark, D. M. (1978). The triangulated irregular network. In *Proc. of the DTM Symposium, American Society of Photogrammetry—American Congress on Survey and Mapping, St Louis, MO*, pp. 24–31.

Peuquet, D. J. (1984). A conceptual framework and comparison of spatial data models. *Cartographica*, 21: 66.

Pfeiffer, D. U., Robinson, T. P., Stevenson, M., and Stevens, K. B. (2008). *Spatial Analysis in Epidemiology.* Oxford University Press, Oxford.

Phillips, J. L. (1999). Information liability: the possible chilling effect of tort claims against producers of geographic information systems data. *Florida State Law Review*, 26: 743–81.

Pikul, J. H., Zhang, H. G., Cho, J., Braun, P. V., and King, W. P. (2013). High-power lithium ion microbatteries from interdigitated three-dimensional bicontinuous nanoporous electrodes. *Nature Communications*, 4: Article 1732, DOI:10.1038/ncomms2747.

Piyathamrongchai, K. and Batty, M. (2007). Integrating cellular automata and regional dynamics using GIS. In E. Koomen, J. Stillwell, A. Bakema, H. J. Scholten (eds.) *Modelling Land-Use Change: Progress and Applications.* The GeoJournal Library, Vol. 90. Springer, Dordrecht, pp. 259–79.

Ragin, C. (2000). *Fuzzy-Set Social Science.* 2nd edn. University of Chicago Press, Chicago.

Rajabifard, A., Binns, A., Masser, I., and Williamson, I. (2006). The role of sub-national government and the private sector in future spatial data infrastructures. *International Journal of Geographical Information Science*, 20: 727–41.

Rayfield, E. J., Barrett, P. M., McDonnell, R. A., and Willis, K. J. (2005). A geographical information system (GIS) study of

Triassic vertebrate biochronology. *Geological Magazine*, 142: 1–28.

Reitsma, F. (2012). Revisiting the 'Is GIScience a science?' debate (or quite possibly scientific gerrymandering). *International Journal of Geographical Information Science*, 27 (2): 211–21, DOI:10.1080/13658816.2012.674529.

Reumann, K. and Witkam, A. P. M. (1974). Optimizing curve segmentation in computer graphics. In *International Computing Symposium 1973*. North-Holland, Amsterdam, pp. 467–72.

Rhind, D. (1977). Computer-aided cartography. *Transaction Institute of British Geographers*, 2: 71–96.

Ritsema Van Eck, J. R. (1993). *Analyse van transport netwerken in GIS vor sociaal-geografisch onderzoek*. Netherlands Geographical Studies 164, Koninklijk Nederlands Aardrijkskundige Genootschap, University of Utrecht.

Rivoirard, J. (1994). *Introduction to Disjunctive Kriging and Non-linear Geostatistics*. Clarendon Press, Oxford.

Robinson, A. H., Morrison, J. L., Muehrcke, P. C., Kimerling, A. J., and Guptill, S. C. (1995). *Elements of Cartography*. 6th edn. Wiley, New York.

Rodell, M., Famiglietti, J. S., Chambers, D. P., and Wahr, J. (2011). Contributions of GRACE to climate monitoring. *Bulletin of the American Meteorological Society*, 92: S49–S52.

Rogerson, P. A. (2006). *Statistical Methods for Geography. A Student's Guide*. 2nd edn. SAGE Publications, London.

Rosen, P. A., Hensley, S., Joughin, I. R., Fuk, K. L., Madsen, S. N., Rodriguez, E., and Goldstein, R. M. (2000). Synthetic aperture radar interferometry. *Proceedings of the IEEE*, 88: 333–82.

Rosenfeld, A. (1980). *Tree structures for region representation*. In H. Freeman and G. G. Pieroni (eds.) *Map Data Processing*. Academic Press, New York, pp. 137–50.

Ross, T. J. (2010). *Fuzzy Logic with Engineering Applications*, 3rd edn. John Wiley, Chichester.

Rossiter, D. G. (1996). A theoretical framework for land evaluation. *Geoderma*, 72: 165–90.

Ruzickova, K. (2012). Guest editorial: digital surface models for the geosciences. *Transactions in GIS*, 16: 599–601.

Samet, H. (1990). *The Design and Analysis of Spatial Data Structures*. Addison-Wesley, Reading, MA.

Schmidt, J. and Hewitt, A. (2004). Fuzzy land element classification from DTMs based on geometry and terrain position. *Geoderma*, 121: 243–56.

Seto, K. C., Reenberg, A., Boone, C. G., Fragkias, M., Haase, D., Langanke, T., Marcotuillio, P., Munroe, D. K., Olah, B., and Simon, D. (2012). Urban land teleconnections and sustainability. *Proceedings of the National Academy of Sciences*, 109: 7687–92.

Sheehan, D. E. (1979). A discussion of the SYMAP program. Harvard Library of Computer Graphics. In *Mapping Collection, Vol. ii: Mapping Software and Cartographic Databases*. Harvard, Cambridge, MA, pp. 167–79.

Sheppard, E. (1995). GIS and society: towards a research agenda. *Cartography and Geographic Information Systems*, 22: 5–16.

Shi, W., Cheung, C.-K., and Tong, X. (2004). Modeling error propagation in vector-based overlay analysis. *ISPRS Journal of Photogrammetry and Remote Sensing*, 59: 47–59.

Shih, F. Y. (2009). *Image Processing and Mathematical Morphology: Fundamentals and Applications*. CRC Press, Boca Raton, FL.

Skidmore, A. K. (1989). A comparison of techniques for calculating gradient and aspect from a gridded digital elevation model. *International Journal of Geographical Information Systems*, 3: 323–34.

Skidmore, A. K. (2002). *Environmental Modelling with GIS and Remote Sensing* (Geographic Information Systems Workshop). CRC Press, Boca Raton, FL.

Skidmore, A. K., Franklin, J., Dawson, T. P., and Pilesjo, P. (2012). Geospatial tools address emerging issues in spatial ecology: a review and commentary on the special issue. *International Journal of Geographical Information Science*, 25: 337–65.

Smith, B. and Mark, D. M. (2001). Geographical categories: an ontological investigation. *International Journal of Geographical Science*, 15: 591–612.

Smith, D. D. and Wischmeier, W. H. (1957). Factors affecting sheet and rill erosion. *Transactions*, American Geophysical Union, 38: 889–96.

Snyder, J. P. (1987). *Map Projections—A Working Manual*. United States Geological Survey Professional Paper 1395. United States Government Printing Office, Washington, DC.

Sonwalkar, M., Fang, L., and Sun, D. (2010) Use of NDVI dataset for a GIS based analysis: a sample study of TAR Creek superfund site. *Ecological Informatics*, 5: 484–91.

Spiegel, S. J., Ribeiro, C. A. A. S., Sousa, R., and Veiga, M. M. (2012). Mapping spaces of environmental dispute: GIS, mining and surveillance in the Amazon. *Annals of the American Association of Geographers*, 102: 320–49.

Steiner, D. and Matt, O. F. (1972). *Computer Program for the Production of Shaded Choropleth and Isarithmic Maps on a Line Printer: User's manual*. University of Waterloo, Ontario.

Steiniger, S. and Hunter, A. J. S. (2013). The 2012 free and open source GIS software map—a guide to facilitate research, development, and adoption. *Computers, Environment and Urban Systems*, 39: 136–50.

Steinitz, C. and Brown, H. J. (1981). A computer modelling approach to managing urban expansion. *Geo-processing*, 1: 341–75.

Stevens, D., Dragicevic, S., and Rothley, K. (2007). iCity: a GIS-CA modelling tool for urban planning and decision making. *Environmental Modelling & Software*, 22: 761–73.

Stocking, M. (1981). *A Working Model for the Estimation of Soil Loss Suitable for Underdeveloped Areas*. Development Studies Occasional Paper No. 15. University of East Anglia, Norwich.

Sui, D. and Goodchild, M. F. (2012). The convergence of GIS and social media: challenges for GIScience. *International Journal of Geographical Information Science*, 25 (11): 1737–48.

Takeyama, T. and Couclelis, H. (1997). Map dynamics: integrating cellular automata and GIS through GEO-Algebra. *International Journal of Geographical Information Science*, 11: 73–92.

Tao, W. (2013). Interdisciplinary urban GIS for smart cities: advancements and opportunities. *Geo-spatial Information Science*, 16: 25–34.

Tapia, A. H., Bajpai, K., Jansen, J., Yen, J., and Giles, L. (2011). Seeking the trustworthy tweet: can microblogged data fit the information needs of disaster response and humanitarian relief organizations. In *The 8th International Conference on Information Systems for Crisis Response and Management (ISCRAM)*. Lisbon, Portugal.

Teicholz, E. and Berry, B. J. L. (1983). *Computer Graphics and Environmental Planning*. Prentice Hall, Englewood Cliffs, NJ.

Thomas, D. S. F. and Goudie, A. G. (2000). *The Dictionary of Physical Geography*. Oxford University Press, Oxford.

Tobler, W. R. (1970). A computer movie simulating urban growth in the Detroit region. *Economic Geography*, 46: 234–40.

Tobler, W. (1979). Smooth pycnophylactic interpolation for geographic regions. *Journal of the American Statistical Association*, 74 (367): 519–36.

Tobler, W. (1995). The resel-based GIS. *International Journal of Geographical Information Systems*, 9: 95–100.

Tomlin, C. D. (1983). A map algebra. In *Proceedings of the Harvard Computer Conference 31 July-4 August 1983, Cambridge, MA*.

Tomlin, C. D. (2012). *GISystems and Cartographic Modeling*. ESRI Press, Redlands, CA.

Tomlinson, R. F., Calkins, H. W., and Marble, D. F. (1976). *Computer Handling of Geographic Data*. UNESCO, Geneva.

Torrens, P. M. (2012). Moving agent pedestrians through space and time. *Annals of the Association of American Geographers*, 102: DOI: 10.1080/00045608.2011.595658

Torrens, P. M., Li, X., and Griffin, W. A. (2011). Building agent-based walking models by machine-learning on diverse databases of space-time trajectory samples. *Transactions in GIS*, 15: 67–94.

Triantafilis, J., Odeh, I. O. A., Minasny, B., and McBratney, A. B. (2003). Elucidation of physiographic and hydrogeological features of the lower Namoi valley using fuzzy *k*-means classification of EM34 data. *Environmental Modelling & Software*, 18 (7): 667–80.

Turner, A. (2006). *Introduction to Neogeography*. O'Reilly Media, Inc., Sebastopol, CA.

United States Census (2013). <http://www.census.gov> (accessed 23/9/2013).

United States Geological Survey (2013). *Global Geographic Information Systems: What is a GIS* <http://webgis.wr.usgs.gov/globalgis/tutorials/what_is_gis.htm> (accessed 23/9/2013).

Ushahidi (2009). *Haiti Crisis Map* <http://haiti.ushahidi.com> (accessed 12/10/2011).

van Deursen, W. P. A. (1995). *Geographical Information Systems and Dynamic Models*. PhD thesis, NGS Publication 190, Utrecht University.

van Deursen, W. P. A. and Heil, G. W. (1994). Analysis of heathland dynamics using a spatial distributed GIS model. *Scripta Geobotanica*, 21: 17–27.

Vieux, B. E. (2004). *Distributed Hydrologic Modelling using GIS*. 2nd edn. Kluwer, Dordrecht.

Vieweg, S., Hughes, A. L., Starbird, K., and Palen, L. (2010). Microblogging during two natural hazards events: what Twitter may contribute to situational awareness. In *Proceedings of the 28th International Conference on Human Factors in Computing Systems, Atlanta, GA*, pp. 1079–88.

von Neumann, J. (1966). *The Theory of Self-Reproducing Automata*, ed author="A. W. Burks.".A. W. Burks. University of Illinois Press, Illinois.

Wainwright, J. and Mulligan, M. (eds.) 2013. *Environmental Modelling: Finding Simplicity in Complexity*. Wiley-Blackwell, Chichester.

Warf, B. and Sui, D. (2010). From GIS to neogeography: ontological implications and theories of truth. *Annals of GIS*, 16: 197–209.

Webster, R. and Oliver, M. A. (1990). *Statistical Methods in Soil & Land Resources Survey*. Oxford University Press, Oxford.

Webster, R. and Oliver, M. A. (2007). *Geostatistics for Environmental Scientists*. 2nd edn. John Wiley & Sons, Chichester.

Wesseling, C. G., Karssenberg, D., Burrough, P. A., and van Deursen, W. P. A. (1996). Integrating dynamic environmental models in GIS: the development of a dynamic modelling language. *Transactions in GIS*, 1: 40–8.

Wieczorek, J., Guo, Q., and Hijmans, R. (2004). The point-radius method for georeferencing locality descriptions and calculating associated uncertainty. *International Journal of Geographical Information Science*, 18: 745–67.

Wielemaker, W. G. and Boxem, H. W. (1982). *Soils of the Kisii area, Kenya.* Agricultural Research Report 922. PUDOC/ Agricultural University, Wageningen.

Wilson, J. P. (2012). Digital terrain modelling. *Geomorphology*, 137: 107–21.

Wilson, J. P. and Gallant, J. C. (2000a). *Digital terrain analysis.* In J. P. Wilson and J. C. Gallant (eds.) *Terrain Analysis: Principles and Applications.* John Wiley, New York, pp. 1–27.

Wilson, J. P. and Gallant, J. C. (2000b). Primary topographic attributes. In J. P. Wilson and J. C. Gallant (eds.) *Terrain Analysis: Principles and Applications.* John Wiley, New York, pp. 51–85.

Wilson, J. P. and Gallant, J. C. (2000c). Secondary topographic attributes. In J. P. Wilson and J. C. Gallant (eds.) *Terrain Analysis: Principles and Applications.* John Wiley, New York, pp. 87–131.

Wilson, M. W. and Graham, M. (2013). Situating neogeography. *Environment and Planning A*, 45 (1): 3–9.

Winchell, M. F., Jackson, S. H., Wadley, A. M., and Srinivasan, R. (2008). Extension and validation of a geographic information system-based method for calculating the Revised Universal Soil Loss Equation length-slope factor for erosion risk assessments in large watersheds. *Journal of Soil and Water Conservation*, 63: 105–11.

Wischmeier, W. H. and Smith, D. D. (1978). *Predicting Rainfall Erosion Losses.* Agricultural Handbook 537. USDA, Washington, DC.

Wise, S. (2002) *GIS Basics.* Taylor & Francis, London.

Wood, J. (2009). *The LandSerf Manual* <http://staff.city.ac.uk/~jwo/landserf/landserf230/doc/landserfManual.pdf> (accessed 11/11/2014).

Wood, J., Dykes, J., and Slingsby, A. (2010). Visualization of origins, destinations and flows with OD maps. *The Cartograpic Journal*, 47: 117–29.

Wood, M. (1994). *The traditional map as a visualisation technique.* In H. M. Hearnshaw and D. J. Unwin (eds.) *Visualization in Geographical Information Systems.* Wiley, Chichester, pp. 9–17.

Worboys, M. F. (2005). Relational databases and beyond. In P. A. Longley, M. F. Goodchild, D. J. Maguire, and D. W. Rhind (eds.) *Geographical Information Systems, Principles and Technical Issues.* Abridged edn. John Wiley, Hoboken, NJ, pp. 163–74.

Worboys, M. F. and Duckham, M. (2004). *GIS: A Computing Perspective.* CRC Press, Boca Raton, FL.

Wright, D. J., Duncan, S. L., and Lach, D. (2009). Social power and GIS technology: a review and assessment of approaches for natural resource management. *Annals of the Association of American Geographers*, 99: 254–72.

Wright, D. and Wang, S. (2011). The emergence of spatial cyberinfrastructure. *Proceedings of the National Academy of Sciences*, 108: 5488–91.

Yamazaki, D., Baugh, C. A., Bates, P. D., Kanae, S., Alsdorf, D. E., and Oki, T. (2012). Adjustment of a spaceborne DEM for use in floodplain hydrodynamic modeling. *Journal of Hydrology*, 436-7: 81–91.

Yassemi, S., Dragićević, S., and Schmidt, M. (2008). Design and implementation of an integrated GIS-based cellular automata model to characterize forest fire behavior. *Ecological Modelling*, 210: 71–84.

Yeh, A. G. and Li, X. (2009). Cellular automata and GIS for urban planning. In M. Madden (ed.) *Manual of Geographic Information Systems.* The American Society for Photogrammetry and Remote Sensing, Betheseda, MD, pp. 591–620.

Zadeh, L. A. (1965). Fuzzy sets. *Information and Control*, 8 (3): 338–53.

Zevenbergen, L. W. and Thorne, C. R. (1987). Quantitative analysis of land surface topography. *Earth Surface Processes and Landforms*, 12: 47–56.

Zhou, Q. and Liu, X. (2004). Error analysis on grid-based slope and aspect algorithms. *Photogrammetric Engineering and Remote Sensing*, 70: 957–62.

Zimmerman, H. J. (2001). *Fuzzy Set Theory—and its Applications.* 4th edn. Kluwer, Dordrecht.

Index

accessibility 46, 72, 86, 108, 138, 236, 247, 289, 292, 294
accumulation operator 226–7
accuracy/precision 9, 22, 25, 27, 37, 39–40, 47, 54, 57, 71–3, 78, 80, 82, 86–8, 94, 113, 151, 162, 191, 193, 196–8, 232–3, 235–40, 242, 244, 246–7, 253, 257–8, 261, 265, 267–8, 272–3, 278, 284, 292–3
aerial photography 11, 14, 28, 74–5, 82–3, 135, 148, 213, 223, 235, 238–9, 242–4, 278
agent-based modelling (ABM) 39, 260–1
agents 2, 39, 77, 260–1
algorithms 54, 58–9, 64, 134, 142, 154, 203, 220–1, 225–6, 240, 243, 256, 281–2, 287, 294
  D8 224
  Douglas–Peucker 84, 141
  FD8 224
  finite difference 219
  photogrammetric 83
  point-in-polygon 140
  Reumann–Witkam 84
  Rho8 224
  shortest path 138
  slope-weighted 224
  stream-tube 224
  thinning 84
  weeding 56, 84
  Zevenbergen–Thorne 221
altitude matrix 134, 201–2, 219, 221, 224, 227–8, 233–5, 238–40, 244, 246
analogue model 30
analysis 2, 11, 21, 39, 54, 101, 104, 112, 121, 127–8, 134, 144, 150, 175, 202–5, 217, 226, 233, 243, 257, 261, 264–5, 267, 287
application programming interface (API) 48, 50
application programs 3, 48, 260
arc 53–4, 82
arithmetic 23, 46–7, 95, 120, 129, 131–3, 204, 230, 261–3, 288
aspect 18, 26, 35, 201, 219–22, 224, 228–30, 232–4, 247–8, 256, 288
attribute operations 128, 144
attributes 2–3, 6–7, 9, 11, 15, 17, 22, 28–30, 32–41, 43, 56, 128, 132, 134, 139, 152, 158, 280
autocorrelation/correlation 87, 111, 117–19, 151, 171–4, 176, 182, 185, 191, 195, 261, 272
automation 11–12, 56, 71, 83–4, 86–7, 106, 116, 138, 177, 224–5, 227, 230, 233, 239–40, 244
axioms 30, 39, 43, 128, 287–9

B-tree 58, 60
best linear unbiased estimate (BLUE) 172
Big Data 1, 49, 88, 291, 293–4
binary arithmetic 46–7
binary trees 63
black-box modelling 254
block codes, *see* codes
block diagram 103, 230, 233, 235
Boolean
  algebra 129, 139, 288
  logic 48, 65, 127, 130, 135, 269, 274, 276
  operations 129, 131–2
  values 22
boundaries 11, 16–17, 22–3, 25, 28, 30, 32–3, 36, 38–40, 51, 52, 55–6, 58, 62–3, 75, 77, 80, 82–3, 85–7, 137, 282
break points 164–5
browsers 8
buffering/buffers 128, 137–8, 141–4, 228, 230

cartograms 101–2, 108
cartography 9, 12, 26, 28, 71, 165, 241, 243–4, 262, 278, 284, 288
  communication 92
  computer-assisted 11
  design 94, 99, 101, 108
  distances 101
  modelling 202
  principles 91–2, 96
categories 2, 80, 92, 94, 96–7, 108, 193, 288
cellular automata 218, 259–60
chain codes, *see* codes
citizen cartographers 8
cloud computing 4, 7, 49
clumping 203, 228
codes 7, 46–8, 67, 73, 132, 202, 224, 243, 252–3
  block 61–3, 65
  chain 61–2, 227
  run-length 52, 60–3, 65, 148
coefficient of determination 114, 116, 120
computational modelling 252, 255
conceptual models 22, 29–32, 46, 152, 253, 287
configuration 111, 120, 189
confusion index 282–3, 285
connectivity 23, 28, 47, 54–5, 128, 137–8, 226, 228, 242, 248, 276, 288, 294
contiguity 119, 152, 161, 190, 201, 228, 261, 282
continuous fields 17–18, 22, 29–30, 32–6, 39, 48, 52, 59, 65, 92, 112, 125, 128, 148, 201–3, 218, 229–30, 267–8, 287
contours 36, 83, 135, 147–50, 163, 165, 177, 191, 194–5, 202, 230, 232–3, 235–6, 239–40, 242–3, 248, 258, 276
convexity 201, 219, 221, 233, 283
convolution 148, 204, 240–1
cookie cutter 140–2, 144, 227
coordinates, 2, 22–3, 26–8, 30, 33, 47, 69, 72–3, 75, 83–6, 155, 236, 288
crisp sets 269–71

data 18, 25, 27, 91, 101, 106–7, 117, 121, 138, 149, 152, 168, 270
  access 2–4, 14, 16, 21, 45, 47–8, 52, 58, 67, 71, 77–81, 86, 92, 108, 112, 235, 255, 289–92, 294
  analysis 7, 12, 65, 72, 112, 120, 125, 128, 134, 144, 192, 202, 233, 265, 267, 272
  Boolean 23, 132, 283
  classification 77, 80, 113, 285
  collection/collectors 1, 4, 6–7, 12, 14, 21, 48, 71, 74, 76–81, 86, 88, 113, 148, 150, 198, 261, 265, 268–9, 290–3
  continuous 33, 53, 187, 276
  crowdsourced 77, 80, 288
  demographic 28, 291, 295
  digital 5–6, 11, 26, 28–9, 45, 67, 69, 71, 79, 82, 85, 242
  editing 82, 84, 87, 230
  exchange 26, 30, 69, 79–81, 264
  input 4–6, 16–17, 24, 39, 45, 52, 56, 69–70, 75, 82–4, 86–7, 134, 171, 189, 191, 194, 202–3, 240, 251, 253–5, 261–2
  metadata 69, 78–81, 289

data (*continued*)
modelling/models 6, 8, 12, 21–2, 29–30, 32–6, 38–43, 45, 47–8, 52–4, 58–9, 67, 81, 86, 88, 127–8, 148, 253–6, 268, 287, 294
nominal 22–3, 39–41, 112, 129–32, 194–5, 203–4, 206
ordinal 22, 23, 39, 112, 130, 132, 204, 206, 274
quality 41, 78, 80–1, 84, 86, 88, 144, 194, 242, 261
raster 33, 36, 45, 52–4, 60–3, 65, 67, 82–6, 134, 148, 207, 217, 233, 235, 240, 242, 256
records 9, 38, 46, 48, 50, 54, 87
retrieval 6, 8, 48, 50–1, 111–12, 128–9, 131, 143–4, 268, 276, 284
scalar 22–3, 40, 131–2
spatial data infrastructure (SDIs) 81, 289
spatio-temporal 38–9, 66, 106, 255, 258
storage 5–6, 12, 49, 53, 233, 264
structures/structuring 8, 30, 33, 35, 46–8, 52–64, 67, 82, 84, 134, 148, 194–5, 233
three-dimensional 35–6, 75–6, 230
transfer 4, 7, 12, 80, 264
transformations 5–6, 84, 86, 114, 153
types 15, 21–3, 39–41, 43, 49, 51–2, 58, 60–3, 76, 83, 96, 111, 129–32, 148, 203, 283, 288, 290
updating 6, 12, 46, 48, 52, 54, 87, 230
vector 22, 32, 45, 49, 52–5, 57–60, 67, 82–3, 106, 128–9, 242
verification 5, 69, 82, 84, 253, 257–8
database management systems (DBMS) 6, 45, 48–52, 67, 69
NoSQL 49–50
relational 48, 50, 52, 66, 129, 203, 272
databases 5–6, 12, 15–18, 30–1, 33, 39–40, 43, 45–6, 57, 65–7, 85–6, 106
digital 7–8, 69, 72, 75, 127, 287
hybrid 51, 56, 82, 143
indexing 58–61
land cover 41
object-oriented 50–1, 82, 84, 143
raster 60, 82
relational 48, 50–2, 60, 82, 84
structures 47–8, 51, 81
vector 50, 53, 83–4
Delaunay triangulation 57, 59, 160, 201, 233, 235
dependence 117, 119, 148, 152, 155, 172, 175, 177, 191
derivatives 18, 31, 35, 163–4, 166, 201, 203, 219, 221–2, 229, 231, 261
deterministic approaches/methods 87, 147, 151, 160, 192, 194–5, 198, 255, 259–60
differential GPS 238
digital elevation models (DEM) 18, 58, 61, 65, 83, 96, 98, 103–4, 112, 135, 148, 150, 166, 191, 195, 201–2, 207–10, 212, 219, 221–9, 231–3, 235–45, 247–8, 252, 257, 261–2
digitizers 5, 82–3
dimensions 76, 253–4
distribution 2, 9, 18, 120–1, 135
drainage 54, 131, 201, 203, 224–5, 229, 248, 269, 276
networks 34, 201, 203, 223–5, 227, 229, 231, 276

edge matching 85
electromagnetic radiation 11, 74
electronic distance measurement (EDM) 237
elevation 9–10, 18, 25, 30, 32–3, 35–6, 58, 61, 65, 71, 75–6, 83, 113, 115, 117, 120–1, 128, 132, 134, 148, 150–1, 158–9, 166–7, 172, 176, 192–3, 195–7, 207–10, 212, 219–20, 225, 229, 231–3, 235–6, 238–40, 242–3, 248, 252–3, 257, 281
ellipsoids 23–6, 73, 176
entities 7, 17–18, 22–3, 29–30, 32–9, 41, 47–8, 50, 52–6, 60, 65, 69, 77, 83–5, 87, 113, 120, 125, 127–9, 131–5, 137–41, 143, 148, 162, 202–3, 207, 228–30, 254, 256, 262, 264, 267–8, 272, 287–8
entity overlap 139
entity-based model 22, 38, 42
equipotential surface 25
error propagation 87, 261–5
errors 6, 15, 17–18, 36, 43, 47, 55, 57–8, 80, 84, 86–8, 111, 131–2, 141, 144, 149, 152, 154–5, 158, 165–6, 168, 172–4, 177, 179, 181, 185, 190, 193–5, 197, 219, 221, 223–5, 228, 239–40, 242–3, 251, 261–5, 269, 271–2, 274, 276–7, 284–5, 288, 292
extrapolation 148, 255

Famine Early Warning Systems Network (FEWS NET) 14
field conceptual models 22
field observations 5, 257
fields 29–30, 32, 48, 50–2, 74, 256, 268
filters/filtering 29, 54, 140, 201, 203–17, 219, 222, 227, 230, 243, 278
Laplacian 207
finite difference modelling 35, 219–20, 255
finite element modelling 35, 201–2, 255
functions 202, 255
fuzzy
boundaries 270, 278–9, 284–5
classification 268, 272, 274, 276–7, 280–5
k-means 187, 267, 270, 280–6
logic 18, 134, 276, 278
membership 130, 267, 270–4, 278–9, 282, 284
objects 267–8, 272
sets 267–3, 275–6
transition zone 270, 278

generalization 37, 54, 86, 99, 132–4, 269, 272
geodemographic segmentation 134
geographic information and society (GISoc) 3
geographical information science (GISc) 3, 9
geometry 21, 23, 51, 288
georeferencing 15, 23–4, 28, 71, 73, 77, 80, 85, 236, 289
geoservices 1–2
geostatistics 18, 150–1, 171–2, 175, 177, 179, 186–7, 191–3, 198, 243, 261
geotagging 108
geoweb 108
global navigation satellite system (GNSS) 5, 12, 25, 71–3, 78, 237
global positioning system (GPS) 1, 12, 25–6, 72–3, 77–8, 86, 92, 149, 235, 237–9, 247–8, 293
granularity 34, 37–8
graphics 54–5, 57, 99, 230, 287
graphs 6, 50, 114–15, 120, 244
grids 11, 23, 26–7, 33–7, 53–4, 58, 60–5, 166, 217
ground-based survey 236, 247

hardware 3–5, 7, 12, 54, 82, 295
hierarchies 22, 28, 33, 37–9, 45, 48, 50–1, 54, 58–60, 63–5, 288
histogram 85, 112, 114, 120, 154, 158, 243
hole effect 175
hypsometric surface 35, 148, 219, 235

image processing 75, 83, 257
imprecision 267–8

indicator functions 148, 151, 188, 274
individual boundary approach 278–9
intensity 75, 121–3, 198, 246, 248
interactions 29, 36, 54, 107, 127–8, 235
interferometric radar 235, 238
interpolators/interpolation 29, 41, 61, 65, 80, 82–3, 105, 135, 160, 174–6, 179–82, 185–9, 191–2, 194–6, 203–4, 225, 229–30, 232–3, 235, 238–43, 247, 256, 265, 267, 272–4, 276, 281–4, 286, 288
  areal 147, 150, 162
  bilinear 241
  choropleth 11
  deterministic 147, 167, 172
  exact 150, 162–3, 165, 177, 195
  geostatistical 171–2, 193
  global 151–2, 155, 166
  inexact 150, 158
  inverse distance 163–4
  isoline 11
  linear 163
  local 151, 166, 240–1
  point 148
  pycnophylactic 161–3
  spatial 18, 86, 112, 147–8, 168, 198, 247, 261
  spline 147, 163, 165–6, 168, 197, 225
  surface 165
inverse distance weighting (IDW) 119, 147, 151, 160, 163, 166–8, 172, 194, 196–8
irradiance 18, 201, 203, 229, 231, 233, 243, 246–8
islands 32, 55–7, 140, 233
isolines 11, 36, 96, 103, 198, 230
isopleths 148

K function 123–4
kernel 122, 204, 219, 241, 246, 277
kriging 151, 172, 196, 281
  block 171, 179, 181, 191, 197, 272
  cokriging 171, 182–6, 188, 193, 195, 198
  cross-validation 179
  disjunctive 188
  indicator 171, 181, 187–8, 191, 195, 289
  log-normal 181
  multivariate 187
  non-linear 181
  ordinary 177, 179–81, 183, 185, 187, 190–3
  point 179–83, 185, 191, 193
  probabilistic 187
  regression 187
  simple 181, 187, 190
  standard deviation 177, 180, 185, 193
  stratified 171, 182, 193, 283
kriging (*continued*)
  universal 171, 186–7, 193
  variance 177

land registry 23, 40
landscapes 2, 11, 24, 29, 39, 41, 43, 71, 75, 87, 99, 107, 135, 139, 149, 152, 155, 224–5, 229, 232, 235, 242–3, 246–7, 251, 259–60, 275
Light detection and ranging (LiDAR) 70, 74–5, 104, 238–9, 247
line of best fit 114–17
locations 107, 118, 120
loosely coupled models 255–7, 259–60

macro language 202
map algebra 201–2, 256
Map Analysis Package (MAP) 11, 256
map units 52, 134, 148, 152–4, 278–80
mapping points 91–3
maps 69, 71, 84, 91
  cadastral 2, 9, 12
  chorochromatic 36, 41, 96
  choropleth 11, 36, 41, 52, 96, 160, 179, 192, 278, 283
  line of sight 235, 243
  projections 6, 15, 26–7, 60–1, 75–6, 99, 230, 235, 241–2
  symbology 92
  topographical 9, 12, 223, 242
media
  digital 5–6
  physical 79
  social 1, 12, 21, 48, 71, 78, 86, 88, 289, 291, 293
minimal bounding rectangle (MBR) 59
modelling 6, 39, 65, 67, 87, 144, 148, 182, 195, 201, 229, 251–3, 255–7, 259, 264
modules 3, 7, 11, 35, 112, 172, 239, 255, 258, 265
Moran's *I* autocorrelation coefficient 117–19
mosaicing 75, 85, 242
movement 3, 35, 38, 65–6, 107, 117, 134, 201, 232, 253, 255, 259, 295
multicriteria evaluation 18, 134
multivariate clustering 134

national grid 23, 27
national mapping agencies (NMAs) 9, 12, 25, 71–2, 76–7, 81, 235, 240
natural breaks scheme 96
Navigation Satellite Timing and Ranging (NAVSTAR) 73
neighbourhood 203, 288
neogeography 12, 108
neural networks 129–30, 134
NoSQL 49
Open GIS Consortium 7, 81
OpenStreetMap project 8, 77, 108, 293
operating systems 7, 14
operations 61, 144, 201–3, 256
operators 83, 129, 133, 139–40, 142, 144, 201, 204, 217–18, 226–9, 243, 256, 258–9
optimizing sampling 174, 198, 261
orderings 64, 67
  Morton 65–6
  Peano–Hilbert 65–6
ordinary least squares 115
Ordnance Survey 27, 76, 81
orientation 92
orthophotos 75, 83, 232–3, 241–2
output 5–7, 86–7, 91, 101, 106–8

patterns 57, 87, 91, 111–12, 132, 148, 174, 176, 202, 206
phenomena 9–10, 21–2, 28–9, 31–2, 38, 43, 46, 53, 60, 71, 127, 267–8, 287
photogrammetry 9, 12, 75, 83, 235, 238–9, 242, 247
point operations 201–3, 256
point patterns 92–3, 112, 121–5
polygons 28, 30, 32–4, 36–7, 40–1
  boundaries 52, 58, 165, 202, 277–8, 280, 282
  networks 56–7, 84, 233
  Thiessen 147, 160–3, 167
  weird 55–7
prediction 152–3, 158, 160
probability 130, 148–9, 187, 189, 256, 262, 269, 277
  distribution 120, 190, 255
probability distribution function (PDF) 190, 255, 261, 269, 276
proximity 23, 47, 127–9, 137, 143, 163, 165, 230
  determination 201

quadrat analysis 122
quadtrees 34–5, 60–5, 67, 148

R-tree 58–9
radar-based systems 238–9
radargrammetry 238–9
random variables 172–3, 224–6
randomness 121
raster-based models 171, 219, 233, 256, 259–60
rasterization 82–3, 85–6, 202
referencing 23–4, 26–8, 64, 69, 71, 77, 80, 82, 88, 105, 117, 198, 289
regionalized variables 172–3, 185–6, 202, 269
regression 113–15, 120, 125, 130, 133–4, 147, 150, 152, 155–9, 166–7, 181, 186–7, 192–4, 220, 252, 254, 258, 264

relationships 23, 28, 111, 120, 128, 288
remote sensing 11–12, 36, 41, 65, 72, 74–5, 77, 83, 92, 104, 106, 128, 134, 148–50, 198, 202–3, 207, 224, 227, 229, 232–3, 236, 238–40, 242–3, 247–8, 257, 292
resolution 17, 33–7, 41, 53–4, 60, 63–5, 72, 77, 80, 83–5, 87, 148, 163, 198, 202, 207–8, 213, 221, 225, 227, 232–3, 235, 239–40, 242–3, 246, 248, 254, 261, 272, 291–3

sampling 147, 149
satellites 1, 4–5, 11–12, 14, 24–5, 41, 69–75, 83, 148, 202, 235, 237–8, 241–2, 247, 257, 278, 290–1, 294
scale/scaling 1, 6, 12, 15, 22–4, 26, 30, 37, 39, 48, 51, 54–5, 60, 71–2, 76–7, 79–80, 83, 85–8, 92, 94–9, 121–4, 150, 153, 174–6, 185, 187, 191, 226, 237, 239–42, 247, 251–2, 260, 268–9, 278, 281, 291
scanners 5, 41, 82–3, 235, 240
semantic import approach (SI) 267, 270, 272–3, 277–8, 280–1, 283–6
sensors 5–6, 11, 28, 63, 72, 74–5, 77, 83, 87, 150, 225, 290, 293
shaded relief 18, 201, 203, 230–3, 235, 240, 243–6, 248
simulation 54, 134, 188, 230, 232, 251–2, 257–60
  conditional 171, 190–3, 195, 261
  Monte Carlo 190–1, 195, 224, 255, 261–2, 276
  stochastic 189, 261
skewness 114, 120, 130, 153, 181, 254
slopelength operator 227
slopes 9, 18, 31, 35–6, 58, 103, 114–15, 120, 135–6, 152, 155, 174, 201, 208, 219–24, 226–8, 230, 232–3, 235, 237, 240, 244–8, 253–4, 256, 258, 261, 270, 274, 276, 278, 288
Soil Loss Estimation Model for Southern Africa (SLEMSA) 257–8
space–time 32, 38, 103, 251, 294
spatio-temporal dynamics 66, 253, 255, 258
spatio-temporal modelling 38–9, 66, 253, 255, 258–9, 264, 293
stationarity 149, 173, 181, 191, 195, 261
statistical analysis/methods 9, 101–2, 106, 111–12, 114, 116, 120, 129–30, 133, 151–2, 158, 179, 232, 267, 293
statistics 11–12, 28, 43, 77, 87, 96, 112–13, 117–18, 120–1, 125, 129, 134, 149–50, 152, 154–5, 218, 224, 230, 240, 252, 254, 261–2, 269, 281, 283
stereoplotters 82–3
structure 87, 152, 160, 172, 176, 182
Structured Query Language (SQL) 48–50, 129
surface derivatives 201, 231
surveys 5, 9, 11–12, 22–5, 27, 35, 40–1, 46, 69, 71–2, 76–7, 80–1, 86–7, 92, 97, 99, 133, 135, 138, 141, 148, 150, 232, 235–6, 239, 242, 247, 267–8, 290
SYMAP (SYnagraphic MAPping system) 11

Taarifa Platform 78
tessellation 22, 32–6, 41, 43, 60, 103, 148, 240, 259
thin-plate splines (TPS) 147, 151, 164–8, 195–7
three-dimensional models 14, 35–6, 75, 105, 165, 226, 232, 243, 246
topology 2–3, 6, 22, 28, 33–4, 36, 38–41, 46, 52–8, 60, 85–7, 127–30, 132, 134, 137–8, 143–4, 160, 166, 201–3, 223, 225–6, 229, 233, 235, 248, 256–7, 276, 288
total station 235, 237–8
transfer function 133, 135, 152, 158, 280
transformations 287, 293
trapezoidal rule 129, 165
trend surface 150, 152, 155–7, 163, 166–8, 172, 194
triangular irregular network (TIN) 34–5, 57–9, 83, 148, 160, 201, 222, 224, 229, 233, 235, 237, 243, 246
two-dimensional space 64, 202
units 52, 65, 131, 134–5, 152, 177, 228, 254, 256–7
Universal Soil Loss Equation (USLE) 226, 248, 257–8
Universal Transverse Mercator (UTM) 26–7, 241

variance 113, 118, 130, 133, 149, 158, 171–86, 188, 190–1, 196–7, 254, 261, 263, 280–1
  analysis of variance (ANOVA) 120, 152–7, 169, 182, 198
  variance/mean ratio (VMR) 122
variation 9, 29–30, 33, 36, 87, 121, 123, 148–50, 152, 155, 160–2, 168, 171–7, 182–3, 191–3, 242–3, 246, 256, 264–5, 276, 278, 280, 286
variograms 87, 114, 177–80, 187–90, 192, 195–6, 197–8, 204, 283–4
  anisotropic 176, 181, 185, 193
  autocovariograms 171
  cross- 182–6
  experimental 171, 174–6, 181, 191
  isotropic 176, 193
  local 176
  nested 181
  non-transitive 174
  transitive 174
vector data 49
  model 22, 32, 45
  structures 53
vectorization 83–5, 207, 227
viewsheds 201, 203, 232, 243–4, 247–8
virtual cities 103–5
visualization 17–18, 66, 78, 91–4, 101–8, 112, 147, 231, 256, 294
  dynamic 106
volunteered geographic information (VGI) 77

weighted moving average 174, 177